Mathematical Logic

Mathematical Logic

Stephen Cole Kleene

DOVER PUBLICATIONS, INC.
Mineola, New York

Bibliographical Note

This Dover edition, first published in 2002, is an unabridged republication of the work originally published in 1967 by John Wiley & Sons, Inc., New York.

Library of Congress Cataloging-in-Publication Data

Kleene, Stephen Cole, 1909–
Mathematical logic / Stephen Cole Kleene.
p. cm.
Originally published: New York : Wiley, 1967.
Includes bibliographical references and index.
ISBN-13: 978-0-486-42533-7 (pbk.)
ISBN-10: 0-486-42533-9 (pbk.)
1. Mathematics—Philosophy. 2. Logic, Symbolic and mathematical. I. Title.

QA9.A1 K54 2002
511.3—dc21

2002034823

Manufactured in the United States by LSC Communications
4500057578
www.doverpublications.com

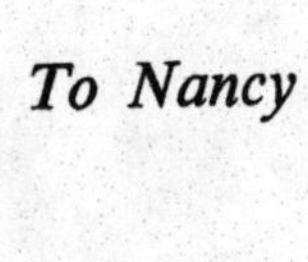

To Nancy

PREFACE

After the appearance in 1952 of my "Introduction to Metamathematics", written for students at the first-year graduate level, I had no expectation of writing another text. But various occasions arose which required me to think about how to present parts of the same material more briefly, to a more general audience, or to students at an earlier educational level.* These newer expositions were received well enough that I was persuaded to prepare the present book for undergraduate students in the Junior year.

In "Introduction to Metamathematics", the study of mathematical logic begins properly only in Chapter V (with some definitions in Chapter IV). Graduate students in mathematics can cover rapidly the introductory material which precedes it there. But for less advanced students or in shorter courses, too much time would be used by such a thorough introduction. I am now convinced that it is also sound pedagogically and scientifically to start doing logic (correctly) right from the beginning, even if not all the reasons for doing it nor all the criteria governing how it is done have been enumerated in advance. The rest of the "introduction" can come later.

On this basis, Part I (Chapters I–III) of the present book gives quite a thorough, yet elementary, treatment of mathematical logic of first order (substantially equivalent to Chapters V–VII and § 73 of IM). The treatment does not stop at formulating logic in one way and practicing with that, as one might do at an even more elementary level. Modern logicians characteristically work with their material in a flexible manner, using different formulations, and passing from one to another, as suits the purpose at hand. Thus in Part I the student will first meet model theory (truth tables) in a fuller treatment than in IM, then Hilbert-type proof theory (with postulates, including modus ponens), and thirdly proof theory handled through derived rules. The principal derived rules are essentially the ones, akin to those in Gentzen's natural deduction systems, which I have been using in teaching logic since 1936. (In Chapter VI, a Gentzen-type sequent system will be introduced, as a fourth formulation of logic.)

Part II of the book is intended to supplement Part I, by providing greater depth of understanding of Part I and an introduction to some of

* Cf. the last five items under my name in the bibliography (pp. 378–379).

the newer ideas and more profound results of logical research in the present century. In Part II the treatment is less elementary than in Part I. According to the time available and the level of the class, the material in Part II may be surveyed or studied more intensively. I have never believed that even the average student is benefitted by avoiding entirely material that only the better students can be counted on to appreciate fully. However, trial of the material in classes has shown that, omitting the starred sections, a class can in a semester get part way into Chapter VI.

Specifically, Chapter IV is the postponed "introduction" (abridged from IM Chapters I–III) plus an introduction and prospectus to the study of formal number theory (Chapters IV, VIII of IM). Chapter V surveys the famous incompleteness and undecidability results of Gödel, Church and others, using the Turing machine concept, without always giving detailed proofs. (Thus an overview is provided of some of the principal results of IM § 42 and Part III, without the detailed theory developed there.) These two chapters concern the foundations of mathematics more than pure logic.

In Chapter VI the emphasis reverts to logic. Gödel's completeness theorem and Gentzen's theorem (besides theorems of Löwenheim, Skolem, Herbrand, Henkin, Beth, Craig and A. Robinson) are obtained, using an approach which has been in the literature only since 1955. There are more compact treatments of Gödel's completeness theorem. The one used here is thought to have the merit, for an introduction to the subject, that almost from the very beginning it should both be clear in what direction one is going and plausible that with patience in managing details one can thus reach the goal. Besides, this approach gives Gentzen's theorem quickly, though nonconstructively. (This chapter corresponds to Part IV of IM, but there are considerable differences in the approach and selection of topics.)

Chapters IV and V can be omitted in order to facilitate completing Chapter VI in a semester course emphasizing logic without the foundations of mathematics. (A few points from Chapters IV and V can then be picked up as needed in Chapter VI, and only a few items in Chapter VI will have to be omitted.)

A fair number of exercises is supplied; but especially in Part II they do not illustrate all the topics. A course taught from this book is not intended to be primarily a problem course. The student should be disabused of the idea Freshman calculus students often have that the text is of no importance except as it helps him to do the exercises. Mastering definitions is especially important to a proper understanding of the enterprise.

I am grateful to H. William Oliver and Edward Pols for taking the notes on my lectures at NSF Summer Institutes at Williams in 1956 and Bowdoin in 1961, respectively. The 1956 lectures and notes were reworked in 1961, and the latter have been reworked and very much expanded in this book. Among the added subjects are ones suggested by H. Jerome Keisler, Georg Kreisel and Julius R. Weinberg. Keisler also suggested improvements, and supplied some eight exercises, after teaching from a draft of the book. Weinberg, William W. Boone, Burton Dreben and Jean van Heijenoort helped with some references. Particularly, the last two persons are my source for a more accurate assessment of the contributions of Löwenheim, Skolem and Herbrand than has hitherto prevailed in the literature. Finally, I thank William E. Ritter for reading the printer's proof independently of myself, and for suggesting improvements made in proof.

S. C. Kleene

Madison, Wisconsin
October, 1966

CONTENTS

PART I. ELEMENTARY MATHEMATICAL LOGIC

Mathematical Logic

PART I

ELEMENTARY MATHEMATICAL LOGIC

CHAPTER I

THE PROPOSITIONAL CALCULUS

§ 1. Linguistic considerations: formulas. *Mathematical logic* (also called *symbolic logic*) is logic treated by mathematical methods. But our title has a double meaning, since we shall be studying the logic that is used in mathematics.

Logic has the important function of saying what follows from what. Every development of mathematics makes use of logic. A familiar example is the presentation of geometry in Euclid's "Elements" (c. 330–320 B.C.), in which theorems are deduced by logic from axioms (or postulates). But any orderly arrangement of the content of mathematics would exhibit logical connections. Similarly, logic is used in organizing scientific knowledge, and as a tool of reasoning and argumentation in daily life.

Now we are proposing to study logic, and indeed by mathematical methods. Here we are confronted by a bit of a paradox. For, how can we *treat* logic mathematically (or in any systematic way) without *using* logic in the treatment?

The solution of this paradox is simple, though it will take some time before we can appreciate fully how it works. We simply put the logic that we are studying into one compartment, and the logic that we are using to study it in another. Instead of "compartments", we can speak of "languages". When we are studying logic, the logic we are studying will pertain to one language, which we call the *object language*, because this language (including its logic) is an object of our study. Our study of this language and its logic, including our use of logic in carrying out the study, we regard as taking place in another language, which we call the *observer's language*.[1] Or we may speak of the *object logic* and the *observer's logic*.

It will be very important as we proceed to keep in mind this distinction between the logic we are studying (the object logic) and our use of logic in studying it (the observer's logic). To any student who is not ready to do so,

[1] In the literature, this language is usually called the "metalanguage" or the "syntax language". However, both these names often carry a connotation about the scope of the study or the type of the methods used in it. Cf. pp. 62–65 (especially bottom p. 63) of our "Introduction to Metamathematics" 1952b (hereafter cited as "IM").[10] To avoid such a connotation when it is not intended, we are adopting "observer's language".

In a textbook on Russian written in English, Russian is the object language and English is the observer's language.

we suggest that he close the book now, and pick some other subject instead, such as acrostics or beekeeping.

All of logic, like all of physics or all of history, constitutes a very rich and varied discipline. We follow the usual strategy for approaching such disciplines, by picking a small and manageable portion to treat first, after which we can extend our treatment to include some more.

The portion of logic we study first deals with connections between propositions which depend only on how some propositions are constructed out of other propositions that are employed intact, as building blocks, in the construction. This part of logic is called *propositional logic* or the *propositional calculus*.

We deal with propositions through declarative sentences which express them is some language (the object language); the propositions are the meanings of the sentences.[2] Declarative sentences express propositions (while interrogatory sentences ask questions and imperative sentences express commands). The same proposition may be expressed by different (declarative) sentences. Thus "John loves Jane" and "Jane is loved by John" express the same proposition, but "John loves Mary" expresses a different proposition. Under the usual definition of $>$ from $<$, the two sentences "$5 < 3$" and "$3 > 5$" express the same proposition (which happens to be false), namely that increasing 5 by a suitable positive quantity will give 3; but "$5^2 - 4^2 = 10$" expresses a different proposition (also false). Each of "$5 < 3$", "$3 > 5$" and "$5^2 - 4^2 = 10$" asserts something about the outcome of a mathematical process, which is the same process in the first two cases, but a different one in the third. "$3 - 2 = 1$" and "$(481 - 581) + 101 = 1$" express two different propositions (both true).

We save time, and retain flexibility for the applications, by not now describing any particular object language. (Examples will be given later.)

Throughout this chapter, we shall simply assume that we are dealing with one or another object language in which there is a class of (declarative) sentences, consisting of certain sentences (the aforementioned building blocks) and all the further sentences that can be built from them by certain operations, as we describe next. These sentences we call *formulas*, in deference to the use of mathematical symbolism in them or at least in our names for them.

First, in this language there are to be some unambiguously constituted sentences, whose internal structure we shall ignore (for our study of the propositional calculus) except for the purpose of identifying the sentences.

[2] Hence some writers call this part of logic "sentential logic" or the "sentential calculus".

We call *these* sentences *prime formulas* or *atoms*; and we denote them by capital Roman letters from late in the alphabet, as "P", "Q", "R", . . . , "P_1", "P_2", "P_3", ... Distinct such letters shall represent distinct atoms, each of which is to retain its identity throughout any particular investigation in the propositional calculus.

Second, the language is to provide five particular constructions or operations for building new sentences from given sentences. Starting with the prime formulas or atoms, we can use these operations, over and over again, to build other sentences, called *composite formulas* or *molecules*, as follows. (The prime formulas and the composite formulas together constitute the *formulas*.)[3] If each of A and B is a given *formula* (i.e. either a prime formula, or a composite formula already constructed), then A $\sim$ B, A $\supset$ B, A & B and A $\vee$ B are (*composite*) *formulas*. If A is a given *formula*, then $\neg$A is a (*composite*) *formula*. (The first four operations are "binary" operations, the last is "unary".)

The symbols "$\sim$", "$\supset$", "&", "$\vee$" and "$\neg$" are called *propositional connectives*.[4] They can be read by using the words shown at the right in the following table; but the symbols are easier to write and manipulate.[5]

Equivalence	$\sim$	"(is) equivalent (to)", "if and only if"
Implication	$\supset$	"implies", "if . . . then . . .", "only if"
Conjunction	&	"and"
Disjunction	$\vee$	"or", ". . . or . . . or both", "and/or"
Negation	$\neg$	"not"

Here we must mention the fact that the natural word languages, such as English, suffer from ambiguities. (Of this, more will be said later.) Logicians are therefore prone to build special symbolic languages. Our

[3] The analogy with chemistry limps a little, since we use "molecule" only for a formula which is not an atom, whereas in chemistry an atom may sometimes be a molecule (e.g. helium He).

[4] Other symbols are often used in the literature, the most common being "$\equiv$", "$\leftrightarrow$" or "$\overset{\leftarrow}{\rightarrow}$" for our "$\sim$"; "$\rightarrow$" for "$\supset$"; "." (sometimes omitted) or "$\wedge$" for "&"; "$\sim$", "$-$" or "$\bar{\ }$" (thus: $\bar{A}$) for "$\neg$". The symbols "$\vee$" and "$\neg$" need not be made as large as in the type used in this book.

[5] Anyone who doubts the advantages of symbols (in their proper place) is invited to solve the equation $x^2 + 3x - 2 = 0$ by completing the square (as taught in high school), *but* doing all the work in words. We start him off by stating the equation in words: The square of the unknown, increased by three times the unknown, and diminished by two, is equal to zero.

Anyone who doubts that *apt* choices of mathematical symbolism have played a major role in the modern development of mathematics and science is invited to multiply 416 by 144, *but* doing all the manipulations in Roman numerals. His problem is thus to multiply CDXVI by CXLIV. Cf. Time magazine, vol. 67 no. 12 (March 19, 1956), p. 83.

unspecified object language may be such a symbolic language, having symbols "$\sim$", "$\supset$", "&", "V" and "$\neg$" which play roles described accurately below but suggested or approximately described by the words. It comes to nearly the same thing to think of the object language as a suitably restricted and regulated *part* of a natural language, such as English; then "$\sim$", "$\supset$", "&", "V" and "$\neg$" can be thought of as names in the observer's language for the verbal expressions at the right in the table.[6]

The names at the left in the table above apply to the propositional connectives or to the formulas constructed using them. Thus "&" is our symbol for conjunction; and A & B is a conjunction, namely the conjunction of A and B. Also A $\supset$ B is the implication by A of B; etc.

So that there will be no ambiguity as to which formulas are the atoms, we now stipulate that none of the atoms be of any of the five forms A $\sim$ B, A $\supset$ B, A & B, A V B and $\neg$A which the molecules have.[7] For example, "Socrates is a man", "John loves Jane", "John loves Mary", "$5 < 3$", "$3 > 5$", "$a+b = c$" and "$a > 0$" (where "a", "b", "c" stand for numbers) could be atoms; then "John loves Jane or John loves Mary", "$\neg\, 5 < 3$" and "$5 < 3 \sim 3 > 5$" would be molecules.

We are using capital Roman letters from the beginning of the alphabet, as "A", "B", "C", . . . , "A_1", "A_2", "A_3", . . . , to stand for any formulas, not necessarily prime. Distinct such letters "A", "B", "C", . . . , "A_1",

[6] If "$\sim$", "$\supset$", etc. are symbols in the object language, then, when we write "A $\sim$ B", "A $\supset$ B", etc., we have a mixture of the two languages, since "A", "B" are names in the observer's language for formulas in the object language, while "$\sim$", "$\supset$", etc. are symbols of the object language itself. However, it should be clear enough here what is meant: "A $\supset$ B" is a name in the observer's language for the formula in the object language which results by infixing the symbol "$\supset$" of the object language between the two formulas in the object language which are named in the observer's language by "A" and "B" respectively. The mixing of languages disappears if we agree that "$\supset$" can serve as a name for itself in these contexts, and generally whenever we need a name in the observer's language for the symbol "$\supset$" of the object language. We say then that "$\supset$" is being used *autonymously* (after Carnap 1934).[10] We confine this usage to "$\sim$", "$\supset$", etc. and other *symbols* of a symbolic or partially symbolic object language when it should be clear that we are talking *about* expressions in that language.

Ordinary English words will not be used autonymously. Hence, when they are used outside of quotation marks, they must be understood as in the observer's language.

If we should wish to name the sentence A $\supset$ B but using the words "if . . . then . . ." instead of the symbol $\supset$ (here "$\supset$" is used autonymously), we would write " "if A then B" ". (The name of the sentence is what is inside the outer quotes; the whole is the name of that name.) Here again there is a mixture of languages (inside the inner quotes), but the meaning should be clear: the sentence named is the one obtained by replacing the letters "A" and "B" by the sentences they name.

[7] If any of the sentences we originally proposed to take as prime were already of those forms, we could start over by dissecting them into components not of those forms and using those components instead in our list of atoms.

"A_2", "A_3", ... need *not* represent distinct formulas (in contrast to "P", "Q", "R", "P_1", "P_2", "P_3", ..., which represent *distinct prime* formulas).

Composite formulas would sometimes be ambiguous as to the manner in which the symbols are to be associated if we did not introduce parentheses. So we shall write "(A ⊃ B) ⊃ C" or "A ⊃ (B ⊃ C)" and not simply "A ⊃ B ⊃ C". However we can minimize the need for parentheses by assigning decreasing ranks to our propositional connectives, in the order listed:[8]

$$\sim, \supset, \&, \vee, \neg.$$

Where there would otherwise be two ways of construing a formula, the connective with the greater rank reaches further. Thus "A ⊃ B & C" shall mean A ⊃ (B & C), and "C ∼ A & B ⊃ C" shall mean C ∼ ((A & B) ⊃ C). The unary operator ¬ being ranked last, "¬A ∨ B" shall mean (¬A) ∨ B rather than ¬(A ∨ B), and "¬¬A ⊃ A" shall mean (¬(¬A)) ⊃ A. This practice is familiar in algebra, where "$a + bc^2 = d$" means $(a + (b(c^2))) = d$.

EXAMPLE 1. In "A ⊃ (B ⊃ C)" the letters "A", "B", "C" stand for formulas constructed from P, Q, R, ..., P_1, P_2, P_3, ... (i.e. from the atoms named by "P", "Q", "R", ..., "P_1", "P_2", "P_3") using (zero or more times) ∼, ⊃, &, ∨, ¬, and (as required) parentheses. For example, A ⊃ (B ⊃ C) might be the particular formula P ⊃ (Q ∨ R ⊃ (R ⊃ ¬P)) (so A is P, B is Q ∨ R and C is R ⊃ ¬P). Here a second pair of parentheses has been inserted, but our ranking of the symbols makes parentheses around Q ∨ R superfluous. The parentheses enable us to see how the formula was constructed, starting from the atoms P, Q, R, and using five steps of composition to introduce the five numbered *occurrences* of propositional connectives, thus:

$$\begin{array}{lllll}
 & & & & P \\
 & Q & R & R & \neg_2 P \\
 & Q \vee_1 R & & R \supset_3 \neg_2 P & \\
P & Q \vee_1 R \supset_4 (R \supset_3 \neg_2 P) & & & \\
P \supset_5 (Q \vee_1 R \supset_4 (R \supset_3 \neg_2 P)) & & & &
\end{array}$$

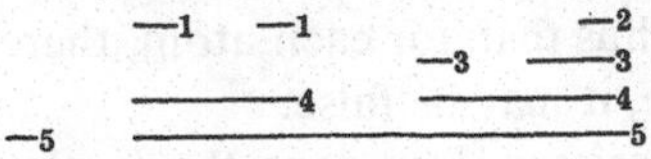

—1 —1 —2 —3 —3 —4 —4 —5 —5

[8] Some authors instead rank ∨ ahead of &; this is done in Algol and some other computer-programming languages. We shall rarely use our ranking between & and ∨ (which follows Hilbert and Bernays 1934 and IM).[1] Some authors (as Whitehead and Russell 1910–13) replace parentheses to a greater or lesser extent by dots ".", ":", "∴" used in the manner of punctuation marks in English.

When we say (next) that, via the ranking, "A ⊃ B & C" shall mean A & (B ⊃ C), we mean that "A ⊃ B & C" becomes a name for the formula A & (B ⊃ C) which "A & (B ⊃ C)" already named.

In a manner obvious from the parentheses or the construction, each occurrence of a connective "connects" or "applies to" or "operates on" one or two parts of the formula, called the *scope* of that (occurrence of a) connective. The scopes of the connectives are shown here by correspondingly numbered underlines; thus the scope of $\supset_4$ consists of the two parts $Q \vee R$ and $R \supset \neg P$.

EXERCISE 1.1. Identify the scope of each (occurrence of a) propositional connective: (a) $P \supset \neg P \sim \neg P$. (b) $\neg P \,\&\, Q \sim R \,\&\, \neg\neg(P \vee Q) \supset S$.

§ 2. Model theory: truth tables, validity. Not only are we restricting ourselves in this chapter to the study of the logic of propositions. But also in this and later chapters we shall concern ourselves primarily with a certain kind of logic, called *classical logic*.

Since the discovery of non-Euclidean geometries by Lobatchevsky (1829) and Bolyai (1833), it has been clear that different systems of geometry are conceptually equally possible. (We shall say a little more about this in § 36.) Similarly, there are different systems of logic. Different theories can be deduced from the same mathematical postulates, the differences depending on the system of logic used to make the deductions. The classical logic, like the Euclidean geometry, is the simplest and the most commonly used in mathematics, science and daily life. In this book we shall find the space for only brief indications of other kinds of logic.[9]

Thus far we have assumed about each prime formula or atom only that it can be identified; i.e. that each time it occurs it can be recognized as the same, and as different from other atoms.

Now we make one further assumption about the atoms, which is characteristic of classical logic. We assume that each atom (or the proposition it expresses) is either *true* or *false* but not both.

We are not assuming that we *know* of each atom *whether* it is true or false. That knowledge would require us to look into the constitution of the atoms, or to consider facts to which they allude under an agreed interpretation of the words or symbols, none of which is within our purview *in the propositional calculus*.

Our assumption is thus that, for each atom, there are exactly two possibilities: it may be true, it may be false.

The question now arises: How does the truth or falsity (*truth value*) of a composite formula or molecule depend upon the truth value(s) of its component prime formula(s) or atom(s)? This will be determined by repeated use of five definitions, given by the following tables. These tables relate the truth value of each molecule to the truth value(s) of its *immediate*

[9] *Some* other kinds of logic would require other propositional connectives than the five we introduced in § 1, e.g. □ and ◊ end § 12.

component(s). In the left-hand columns we list all the possible *assignments* of truth t and falsity f to the immediate component(s). Then in the line (or row) for a given assignment, we show the resulting truth value of each molecule in the column headed by that molecule.

A	B	$A \sim B$	$A \supset B$	$A \& B$	$A \vee B$
t	t	t	t	t	t
t	f	f	f	f	t
f	t	f	t	f	t
f	f	t	t	f	f

A	$\neg A$
t	f
f	t

Thus $A \sim B$ is true exactly when A and B have the same truth value (hence the reading "equivalent", i.e. "equal valued", for $\sim$); $A \supset B$ is false exactly when A is true and B is false; $A \& B$ is true exactly when A and B are both true; $A \vee B$ is false exactly when both are false; and $\neg A$ is true exactly when A is false.

Some controversy has arisen about the name "implication", and the reading "implies", for our $\supset$. Say A is "the moon is made of green cheese" and B is "2+2=5". Then $A \supset B$ is true under our table (because A is false), even though there is no connection of ideas between A and B. Similarly, if B is "2+2=4", $A \supset B$ is true (because B is true), quite apart from whether A bears any relationship to "2+2=4". This is considered paradoxical by some writers (Lewis 1912, 1917, Lewis and Langford 1932).[10]

In modern mathematics the name "multiplication" is often used for various mathematical operations that behave more or less analogously to the arithmetical one called "multiplication". Similarly, we find it convenient to use the name "implication" to designate the operation defined by the second truth table above; and then in our logical discussions we usually read $A \supset B$ as "A implies B", even though "if A then B" or "A only if B" probably renders the meaning better in everyday English. This "implication", and the present "equivalence", are called more specifically "material implication" and "material equivalence".[11]

It is, of course, possible to be interested in other senses of "implication"; but then one must have recourse to ways of defining it other than by a

[10] A date appearing in conjunction with a person's name ordinarily constitutes a reference to the bibliography at the end of the book. The few exceptions are dates of old and well-known works not primarily in logic and dates not associated with publication.

[11] In everyday English, ". . . if . . . then" functions grammatically as a conjunction like "and" and "or" (connecting sentences), while "implies" is a transitive verb (connecting nouns). From this standpoint, when we use "if A then B" and "A implies B" interchangeably, we can regard the latter as short for " "A" implies "B" " or "that A implies that B".

"two-valued truth table". Our definition is the only reasonable one with such a table.[12]

A related question is why we should want to assert a material implication A ⊃ B, when if A is true we could more simply assert B (or if our hearers don't know that A is true, we could more informatively assert A & B), and if A is false we could more simply say nothing. But we ordinarily assert sentences of the form "If A, then B" when we don't know whether A is true or not. For example, before an election I might say [1] "If our candidate for President carries the state by 500,000, then our man for the Senate will also win". This form of statement enables me to predict what will happen in *one* eventuality without attempting to say more. If it turns out that our candidate for President doesn't carry the state by 500,000, my prediction will not have been proved false. Since we are committed here to a two-valued logic, my statement should then be regarded as true, though perhaps uninteresting. If when I speak, returns are in showing our candidate ahead by a safe 500,000, I would more likely say [1a] "Our man for the Senate will win" or [1b] "Since our candidate for President is carrying the state by 500,000, our man for the Senate will win". But [1] would not have become false, just partially redundant and thus unnatural for me to say (unless I haven't heard the latest election returns).

To take an analogous mathematical example, suppose a positive integer $n > 1$ is written on a piece of paper in *your* pocket, and I do not know what the integer is. I could truthfully say [2] "If n is odd, then $x^n + y^n$ can be factored". By saying this, I am claiming that, *when* you produce the paper showing the value of n, then I will be able to factor $x^n + y^n$ for the n you produce if it turns out to be odd (and I am making *no claim* about the factorability of $x^n + y^n$ in the contrary case). Thus, if you have bet that I am wrong, to settle the bet you produce the number n. If for example it is 3, I then show you the factorization $(x + y)(x^2 - xy + y^2)$, and you pay me. If for example it is 4 (or 6), I win automatically.

These examples should make it clear that material implication A ⊃ B ("If A, then B") is a useful and natural form of expression.[13] Similar remarks apply to material equivalence A ∼ B.

[12] Then ordinary usage certainly requires "If A, then B" to be true when A and B are both true, and to be false when A is true but B is false. So only our choice of t in the third and fourth lines can be questioned. But if we changed t to f in both these lines, we would simply get a synonym for &; in the third line only, for ∼. If we changed t to f in the fourth line only, we would lose the useful property of our implication that "If A, then B" and "If not B, then not A" are true under exactly the same circumstances (which will appear later as *12a in Theorem 2).

[13] We are talking about the use of "If . . . , then . . ." with verbs in the indicative mood. Grammar allows also contrary-to-fact conditionals with verbs in the subjunctive mood. These are "If A, then B" sentences where the falsity of A (which is indicated) does not make the whole true irrespective of what B is. Say the $n > 1$ in your pocket

Likewise, when we don't know whether A is true or not, and don't know whether B is true or not, it can be useful to assert "A or B" or in symbols A ∨ B. If we already knew that A is true, it would be simpler and more informative to say "A"; etc. Our disjunction A ∨ B, defined by the fourth truth table, is the "inclusive disjunction" or "nonexclusive disjunction", which is true when A is true or B is true or both A and B are true. This is more useful to us than the "exclusive disjunction", expressed by the words "A or B but not both", which instead has f also in the first row. While English is ambiguous, Latin is clear, using "vel" for the inclusive disjunction and "aut" for the exclusive disjunction. The symbol ∨ comes from the first letter of "vel".

We postpone further discussion of the relation between our symbols and ordinary language to the end of the chapter.

We now illustrate the repeated use of the above tables by computing the truth table for P ⊃ (Q ∨ R ⊃ (R ⊃ ¬P)). The result is shown first as (1) below, then the details of the computation of the third line (or row). For this line, we first substitute for the atoms P, Q, R the respective values t, f, t assigned to them for that line. Then we compute the values of innermost composite parts repeatedly. Thus by the table for ∨, f ∨ t is t; by the table for ¬, ¬t is f; and by the table for ⊃, t ⊃ f is f (which we use three times). The successive stages in this computation are shown in successive lines for clarity, and are then summarized in a single line.

Completed truth table:

(1)

	P	Q	R	P ⊃ (Q ∨ R ⊃ (R ⊃ ¬P))
1.	t	t	t	f
2.	t	t	f	t
3.	t	f	t	f
4.	t	f	f	t
5.	f	t	t	t
6.	f	t	f	t
7.	f	f	t	t
8.	f	f	f	t

---

is already known to me and is 4. I could say truthfully [2′] "If the n had been odd, I could have factored $x^n + y^n$"; but I could not say truthfully [2″] "If the n had been odd, I could have factored $x^{n+1} + y^{n+1}$". (With n odd, $x^{n+1} + y^{n+1}$ may or may not be factorable.) A contrary-to-fact conditional "If A, then B" makes an assertion about a "hypothetical situation" analogous to the actual situation but differing from it by A holding. Sunday morning quarterbacks and Wednesday morning politicians find them useful. We gave a mathematical example, because we can be positive in affirming [2′] and refraining from affirming [2″], whereas in football and politics matters are more controversial.

Computation of the third line:

$$
\begin{array}{l}
P \supset (Q \vee R \supset (R \supset \neg P)) \\
t \supset (f \vee t \supset (t \supset \neg t)) \\
t \supset (\ \ t \ \ \supset (t \supset f\ )) \\
t \supset (\ \ t \ \ \supset \ \ f \ \ ) \\
t \supset \qquad f \\
\quad f \\
\hline
t\ f\ \ f\ t\ t\ f\ \ t\ f\ f\ t
\end{array}
$$

The computation process illustrated just now constitutes a mechanical procedure by which we can compute the truth table for any formula E, or more specifically the truth table for E using (or "entered from") a given list $P_1, \ldots, P_n$ of the prime components of E. (In (1), we used the list P, Q, R; we could have used instead Q, P, R or Q, R, P, etc. to obtain different tabulations of the same collection of eight computation results.) In the trivial case that E is a prime formula P, the computation takes zero steps, and the value column is identical with the column of assignments to P.

In practice it is not always necessary to apply the procedure in full detail. Thus the observation that A ⊃ B is t whenever A is f (irrespective of the truth value of B) suffices to justify our entering t in the last four lines of the above table without further ado.

There are formulas for which the value column in the truth table will contain only t's, for example P & ¬P ⊃ (Q ∨ R ⊃ (R ⊃ ¬P)), P ⊃ ¬P ∼ ¬P and P ⊃ P, as the reader may verify (Exercise 2.2). The order of listing the prime components does not matter here. (Why?) Such formulas are therefore always true, regardless of the truth or falsity of their prime components. Without knowing the truth values of the prime components, we can nevertheless say that the composite formula is true. Such formulas are said to be *valid*, or to be *identically true*, or (after Wittgenstein 1921) to be *tautologies* (in, or of, the propositional calculus).

To give a verbal example, the proposition "If I am going too fast, then I am going too fast" is true on the basis of the propositional calculus; indeed, it has the form P ⊃ P. But the proposition "I am going too fast", if true, is true on other grounds.

It might seem that the valid formulas or tautologies are the least interesting, because from one point of view they give no information. My admitting that "If I am going too fast, then I am going too fast" can hardly give any of you much satisfaction. But it will appear as we proceed that the tautologies are important.

EXERCISES. 2.1. Find the truth tables: (a) $\neg P \vee Q$. (Compare this with the table for $P \supset Q$.) (b) $(\neg P \vee Q) \,\&\, (R \supset (P \sim Q))$. (c) $Q \supset P \vee Q$. Are any of these formulas valid?

2.2. Verify that $P \,\&\, \neg P \supset (Q \vee R \supset (R \supset \neg P))$, $P \supset \neg P \sim \neg P$ and $P \supset P$ are valid.

2.3. Show that the following formulas are valid. To reduce the work, observe that the whole implication $A \supset B$ fails to be valid only if you can pick truth values for P, Q (or for P, Q, R, S) which simultaneously make B take the value f and A the value t. Consider all the choices of values that make B f, and verify that none of them make A t.

(a) $((P \supset Q) \supset P) \supset P$. (Peirce's law, 1885.)

(b) $((P \supset R) \,\&\, (Q \supset S)) \,\&\, (\neg R \vee \neg S) \supset \neg P \vee \neg Q$.

(c) $(P \supset Q) \supset (\neg Q \supset \neg P)$.[14]

2.4. Show the following not valid by computing just one suitable line of the table: (a) $P \vee Q \supset P \,\&\, Q$. (b) $(P \supset Q) \supset (Q \supset P)$.[14]

2.5. Find formulas composed from P, Q, R whose truth tables have the following value columns: (a) f f f f t f f f. (Use a method applicable to any truth table with just one t.) (b) t f f f t f t f. (First use a method applicable to any table with more than one t. Can you find a shorter formula with the same table?) (c) f f f f f f f f.

§ 3. Model theory: the substitution rule, a collection of valid formulas. The definition of validity provides us with an automatic way of deciding as to the validity of any formula: simply compute its truth table, and see whether we get all t's. This is a very fortunate situation, and one should not hesitate to do this in any case of doubt.

However, computing truth tables of formulas at random would be a rather slow way of discovering valid formulas. Anyone not familiar with simple examples of valid formulas and with methods for proceeding to others (whether or not he has officially studied logic) would properly be described as sluggish in his mental processes.

One simple principle is this. In *defining* validity, we use a truth table entered from the prime components, so as to take into account all the structure of the formula available to the propositional calculus. However, to *establish* validity, we may not need to dissect a formula all the way down to its prime components or atoms. If we get all t's in a table entered from (values of) components not necessarily prime, we can be sure it is valid. For example, $P \,\&\, \neg P \supset P \,\&\, \neg P$ is of the form $A \supset A$; Table (a) (below) entered from A gives all t's; hence the formula is valid. For, in computing each line of Table (b) entered from P (as is called for under the definition of validity), the first part of the computation consists in

[14] $\neg Q \supset \neg P$ is the *contrapositive* of $P \supset Q$, and $Q \supset P$ is the *converse* of $P \supset Q$.

computing the value of P & ¬P, i.e. of the A. Then the rest of the computation consists in computing the value of the whole from that of A (shown underlined in Table (b)); but this we have already done with result t in computing Line 2 of Table (a). Table (a) is the same as Table (c)

(a)

A	A	⊃	A
t	t	t	t
f	$\underline{\text{f}}$	$\underline{\text{t}}$	$\underline{\text{f}}$

(b)

P	P	&	¬	P	⊃	P	&	¬	P
t	t	$\underline{\text{f}}$	f	t	$\underline{\text{t}}$	t	$\underline{\text{f}}$	f	t
f	f	$\underline{\text{f}}$	t	f	$\underline{\text{t}}$	f	$\underline{\text{f}}$	t	f

(c)

P	P	⊃	P
t	t	t	t
f	f	t	f

except for the notation; instead of saying we construct a table for A ⊃ A entered from A, it comes to the same thing to say that we verify the validity of P ⊃ P, and then substitute A (i.e. P & ¬P) for P in P ⊃ P. This reasoning gives the following theorem, in which we write "⊧ E" as a short way of saying "E is valid".[15]

THEOREM 1. (Substitution for atoms.) *Let* E *be a formula containing only the atoms* $P_1, \ldots, P_n$, *and let* E* *come from* E *by substituting formulas* $A_1, \ldots, A_n$ *simultaneously for* $P_1, \ldots, P_n$, *respectively. If* ⊧ E, *then* ⊧ E*.

On the other hand, to show by truth tables that a formula is *not* valid, the tables must in general be entered from the prime components. For example, P & ¬P ⊃ Q is of the form A ⊃ B. The table for A ⊃ B entered from A and B (§ 2) does not have all t's (in other words, P ⊃ Q is not valid). But P & ¬P ⊃ Q is valid. This example shows that the converse of Theorem 1, namely "If ⊧ E*, then ⊧ E", does not hold.

Returning to the example preceding Theorem 1, since Table (a) entered from A has all t's (or equivalently (c) has all t's), we shall have all t's for *every* formula of the form A ⊃ A, not just the particular one we took there with P & ¬P as the A. This is included in the theorem; for, with E fixed and "⊧ E" established, we can apply the theorem with any choice of $A_1, \ldots, A_n$.

[15] Expressions containing "⊧" ("⊧ E" here and "$A_1, \ldots, A_m$ ⊧ B" in § 7) are not formulas of the object language, but expressions of the observer's language, used in writing concisely certain statements about formulas. The definition of "formula" for the propositional calculus was concluded in § 1, and allows only ∼, ⊃, &, V, ¬ (as symbols of the object language) to be used in building up formulas from the atoms P, Q, R, . . . , P_1, P_2, P_3, ... Now "⊧" is a symbol of the observer's language, and hence stands outside every formula, and outranks ∼, ⊃, &, V, ¬; thus "⊧ A ∼ B" means "⊧ (A ∼ B)" rather than "(⊧ A) ∼ B".

We use this principle to establish in the next theorem a collection of forms of valid formulas.[16]

For example, as *1 we give the result just established.

As 5b, we claim that, for each choice of formulas (built up from P, Q, R, . . . , P_1, P_2, P_3, . . .) as the A and B, the resulting formula B ⊃ A ∨ B is valid. For, in Exercise 2.1 (c) we saw that Q ⊃ P ∨ Q is valid; and hence by Theorem 1 B ⊃ A ∨ B is valid.

In the same way, every one of the results in Theorem 2 can be proved automatically, by first verifying the validity of the particular formula which has P, Q, R in place of A, B, C, and then using Theorem 1 (or equivalently, by constructing a truth table entered from A, B, C).

The student may accordingly take the whole list on faith, as he does a table of square roots or trigonometric functions or integrals.

We intend that the student should acquire the ability to use these results, and indeed should learn enough of them so that he can operate without Theorem 2 before him. However, we do not ask him to learn the list outright *now*, but rather to use it for reference and in doing so to become familiar with the results most often used.[17]

THEOREM 2. *For any choice of formulas* A, B, C:

1a. ⊨ A ⊃ (B ⊃ A).

1b. ⊨ (A ⊃ B) ⊃ ((A ⊃ (B ⊃ C)) ⊃ (A ⊃ C)).

3. ⊨ A ⊃ (B ⊃ A & B).

4a. ⊨ A & B ⊃ A.

4b. ⊨ A & B ⊃ B.

5a. ⊨ A ⊃ A ∨ B.

5b. ⊨ B ⊃ A ∨ B.

6 . ⊨ (A ⊃ C) ⊃ ((B ⊃ C) ⊃ (A ∨ B ⊃ C)).

[16] Most of these results have equivalents in IM (= Kleene "Introduction to Metamathematics" 1952b). With a few exceptions, we employ the numbering of IM. This will facilitate using IM as a reference work supplementing the present book, or the present book as an introduction to IM. (This accounts for the gaps and other irregularities in the numbering in Theorem 2. The present 9a, 10a, 10b, *4a, *12a, *55c, *63a do not correspond to like-numbered results in IM; and *55a, *55b correspond to *63, *62 of IM, but are renumbered here to come earlier.)

The meaning of "°" on 8, *12a, etc. will be explained at the end of § 12.

[17] The developments below should assist the reader in becoming familiar with these and other results. We shall make various applications, and establish interconnections, which should help to fix them in mind. For some of them we shall give new proofs to make the results more meaningful.

Following Church 1956 p. 73, *49 may be called more specifically the "complete law of double negation"; 8 the "law of double negation" simply; and the converse of 8 the "converse law of double negation".[14] Similarly, *12a is the "complete law of contraposition"; with ∼ replaced by ⊃, the "law of contraposition"; by ⊂, the "converse law of contraposition". Also, 1a is the "law of affirmation of the consequent" (cf. *10a); and *7 is the "law of reductio ad absurum". With ∼ replaced by ⊃, *4a is "importation"; by ⊂, "exportation". (When replacing ∼ here, supply parentheses.)

7. $\models (A \supset B) \supset ((A \supset \neg B) \supset \neg A)$.
8°. $\models \neg\neg A \supset A$.

9a. $\models (A \supset B) \supset ((B \supset A) \supset (A \sim B))$.
10a. $\models (A \sim B) \supset (A \supset B)$.
10b. $\models (A \sim B) \supset (B \supset A)$.

(Introductions and eliminations of logical symbols.)

*1. $\models A \supset A$.
*2. $\models (A \supset B) \supset ((B \supset C) \supset (A \supset C))$.

*3. $\models A \supset (B \supset C) \sim B \supset (A \supset C)$.
*4a. $\models A \supset (B \supset C) \sim A \,\&\, B \supset C$.

(Principle of identity, chain inference, interchange of premises, importation and exportation.)

*10a. $\models \neg A \supset (A \supset B)$.
*12a°. $\models A \supset B \sim \neg B \supset \neg A$.

(Denial of the antecedent, contraposition.)

*19. $\models A \sim A$.
*20. $\models (A \sim B) \sim (B \sim A)$.

*21. $\models (A \sim B) \,\&\, (B \sim C) \supset (A \sim C)$.

(Reflexive, symmetric and transitive properties of equivalence.)

*31. $\models (A \,\&\, B) \,\&\, C \sim A \,\&\, (B \,\&\, C)$.
*32. $\models (A \vee B) \vee C \sim A \vee (B \vee C)$.

*33. $\models A \,\&\, B \sim B \,\&\, A$.
*34. $\models A \vee B \sim B \vee A$.

*35. $\models A \,\&\, (B \vee C) \sim (A \,\&\, B) \vee (A \,\&\, C)$.
*36. $\models A \vee (B \,\&\, C) \sim (A \vee B) \,\&\, (A \vee C)$.

*37. $\models A \,\&\, A \sim A$.
*38. $\models A \vee A \sim A$.

*39. $\models A \,\&\, (A \vee B) \sim A$.
*40. $\models A \vee (A \,\&\, B) \sim A$.

(Associative, commutative, distributive, idempotent and elimination laws.)

*49°. $\models \neg\neg A \sim A$.

*50. $\models \neg(A \,\&\, \neg A)$.
*51°. $\models A \vee \neg A$.

(Law of double negation, denial of contradiction, law of the excluded middle.)

*55a. $\models \neg(A \vee B) \sim \neg A \,\&\, \neg B$.
*55b°. $\models \neg(A \,\&\, B) \sim \neg A \vee \neg B$.

*55c°. $\models \neg(A \supset B) \sim A \,\&\, \neg B$.

(De Morgan's laws 1847,[18] negation of an implication.)

*56°. $\models A \vee B \sim \neg(\neg A \,\&\, \neg B)$.
*57°. $\models A \,\&\, B \sim \neg(\neg A \vee \neg B)$.

*58°. $\models A \supset B \sim \neg(A \,\&\, \neg B)$.
*59°. $\models A \supset B \sim \neg A \vee B$.

*60°. $\models A \,\&\, B \sim \neg(A \supset \neg B)$.
*61°. $\models A \vee B \sim \neg A \supset B$.

*63a. $\models (A \sim B) \sim (A \supset B) \,\&\, (B \supset A)$.

(Expressions for some connectives in terms of others.)

[18] In verbal form, these go back at least to Ockham ("Summa Logicae", 1323–9). Cf. Łukasiewicz 1934, Bocheński 1956.

EXERCISES. 3.1. Redo the illustration preceding Theorem 1 (with Tables (a), (b), (c)) to show that $P \vee \neg Q \supset P \vee \neg Q$ is valid (i.e. taking $P \vee \neg Q$ instead of $P \& \neg P$ as the A).

3.2. Establish 1a, 4a, 6, 7, *50, *51 by the automatic method (indicated above), except using shortcuts when you can in the truth table computations.

3.3. Show that, if a table for a formula entered from components not necessarily prime has all f's, then the formula is not valid. (Cf. the first remark following Theorem 1.)

§ 4. Model theory: implication and equivalence. Suppose the truth table for a formula E is constructed as in § 2 by using exactly its prime components $P_1, \ldots, P_n$, and suppose that a new table is constructed for E using additional atoms $P_{n+1}, \ldots, P_{n+m}$ not in E. Then the new table differs from the original table only in that the value column of the new table splits into 2^m parts, corresponding to the 2^m assignments of t's and f's to the atoms $P_{n+1}, \ldots, P_{n+m}$ which do not occur in E. Each of these 2^m parts is a duplicate of the value column of the original table, since the same computation (based only on the assignments to $P_1, \ldots, P_n$) is used in each part. For example with $n = 2$ and $m = 1$, Tables (e), (f), (g) below have been constructed by entering from three atoms, although the formula at the head of each of those tables contains just two atoms.

	P_1	P_2	P_3	(d) $(P_1 \vee P_2) \& (P_1 \supset P_3)$	(e) $P_2 \vee P_3$	(f) $P_1 \& P_3$	(g) $P_2 \supset P_2 \vee P_3$
1.	t	t	t	t	t	t	t
2.	t	t	f	f	t	f	t
3.	t	f	t	t	t	t	t
4.	t	f	f	f	f	f	t
5.	f	t	t	t	t	f	t
6.	f	t	f	t	t	f	t
7.	f	f	t	f	t	f	t
8.	f	f	f	f	f	f	t

In Tables (e) and (g), Lines 5–8 (P_1 is f) are duplicates respectively of Lines 1–4 (P_1 is t); and in Table (f), Lines 3, 4, 7, 8 (P_2 is f) are duplicates respectively of Lines 1, 2, 5, 6 (P_2 is t).

In particular, if the table for a formula E entered from only its prime components contains only t's, then so does the table for E entered using given additional atoms; and conversely. (This is illustrated by Table (g).) Thus, $\models$ E if and only if the table for E entered from any particular list

$P_1, \ldots, P_n$ (containing at least all the prime components of E) has only t's.

In Theorems 3 and 4, we shall compare truth tables for A and B (also for A ⊃ B or A ∼ B). To make this easy, we shall enter each table from one list of atoms $P_1, \ldots, P_n$, including all that occur in either of A and B. So if A and B do not contain the same atoms, the table for A or for B is entered from more atoms than occur in it. By the preceding discussion, it will make no difference if the list $P_1, \ldots, P_n$ contains still more atoms.

THEOREM 3. *If* ⊨ A *and* ⊨ A ⊃ B, *then* ⊨ B.

PROOF. Consider any assignment of t's and f's to a list $P_1, \ldots, P_n$ of atoms as described. The computation of the corresponding value of A ⊃ B consists in first computing the values of A and B, and thence computing the value of A ⊃ B by the table for ⊃ (beginning of § 2). By the hypotheses that ⊨ A and ⊨ A ⊃ B, both the value obtained for A and the final value for A ⊃ B are t. From the table for ⊃, this can only be the case when Line 1 of that table applies, and in Line 1 of that table B is also t. Since this is the case for each assignment to $P_1, \ldots, P_n$, the formula B receives the value t for all assignments, i.e. ⊨ B, as was to be shown.

THEOREM 4. (a) *For each assignment*, A ∼ B *is* t *if and only if* A *and* B *have the same truth value*. Hence: (b)° ⊨ A ∼ B *if and only if* A *and* B *have the same truth table*.

PROOF. Consider any formulas A and B. (a) In the computation of the value of A ∼ B for a given assignment of t's and f's to $P_1, \ldots, P_n$, the first part consists in computing values of A and B, after which the computation is concluded by entering the basic table for A ∼ B in § 2 with the resulting values of A and B. From that table we see that A ∼ B is t if and only if the values computed for A and B are the same. (b) So the table for our A ∼ B has all t's, exactly if, for *every* assignment, A and B have the same value.

EXAMPLE 2°. By (b) of the theorem with the result of Exercise 2.1 (a), ⊨ P ⊃ Q ∼ ¬P ∨ Q (and ⊨ ¬P ∨ Q ∼ P ⊃ Q). Thence by substitution (Theorem 1), ⊨ A ⊃ B ∼ ¬A ∨ B (and ⊨ ¬A ∨ B ∼ A ⊃ B). Thus we reprove *59 of Theorem 2. (This proof differs from the one suggested in § 3 only in that we now take into account the general principle stated as Theorem 4 (b), instead of (like robots) separately completing the computation of P ⊃ Q ∼ ¬P ∨ Q in each line.

THEOREM 5. (Replacement theorem.) *Let* C_A *be a formula containing a formula* A *as a specified (consecutive) part, and let* C_B *come from* C_A *by replacing that part by a formula* B. *If* ⊨ A ∼ B, *then* ⊨ $C_A \sim C_B$.

PROOF. Assume $\models A \sim B$. Then by Theorem 4 (b), A and B have the same table. Hence if, in the computation of a given line of the table for C_A, we replace the computation of the specified part A by a computation of B instead, the outcome will be unchanged. Thus C_B has the same table as C_A; so by Theorem 4 (b), $\models C_A \sim C_B$.

EXAMPLE 3°. From Example 2 by Theorem 5,

$$\models \neg P \vee Q \supset (P \supset \underline{\neg P \vee Q}) \sim \neg P \vee Q \supset (P \supset (P \supset Q)).$$

The part A of C_A is underlined. In writing C_B, a pair of parentheses is required which was unnecessary in C_A.

By a "consecutive" formula part A of C_A we are understanding a formula part A which is consecutive before parentheses are omitted, and whose value is thus computed in the course of computing the value of the whole C_A. Thus $P \vee Q$ does not occur as a consecutive part of $\neg P \vee Q \supset (P \supset \neg P \vee Q)$, as becomes clear upon restoring some parentheses: $(\neg P) \vee Q \supset (P \supset (\neg P) \vee Q)$.

COROLLARY. (Replacement rule, or replacement property of equivalence.) *If* $\models C_A$ *and* $\models A \sim B$, *then* $\models C_B$.

PROOF. By the hypothesis that $\models A \sim B$ with the theorem, $\models C_A \sim C_B$. So by Theorem 4 (b), C_A and C_B have the same table. By the hypothesis that $\models C_A$, this table has all t's.

EXERCISES. 4.1. In the manner of Example 2, reprove *31, *34, *49, *55a, *55c of Theorem 2.

4.2°. Similarly establish that:

(a) $\models (A \sim B) \sim (A \,\&\, B) \vee (\neg A \,\&\, \neg B)$.

(b) $\models \neg(A \sim B) \sim (A \,\&\, \neg B) \vee (\neg A \,\&\, B)$.

4.3. Illustrate the proof of Theorem 5 by computing the second line (for t f assigned to P Q) of the tables for $\neg P \vee Q \supset (P \supset \neg P \vee Q)$ and $\neg P \vee Q \supset (P \supset (P \supset Q))$. Underline the common parts (as in Tables (a), (b)).

4.4°. Use Theorem 5 with *55a in Theorem 2 to establish that $\models \neg\neg(\neg A \vee \neg B) \sim \neg(\neg\neg A \,\&\, \neg\neg B)$. (Observe that, whatever formulas constructed from P, Q, R, . . . , P_1, P_2, P_3, . . . "A" and "B" stand for here, *55a will hold when *its* A and B are *the present* $\neg A$ and $\neg B$.)

4.5. Using $\models \neg A \vee B \sim A \supset B$ (Example 2), infer *10a from 5a.

4.6. Give three proofs that: *If* $\models A$ *and* $\models A \sim B$, *then* $\models B$. (By Theorem 4 (b); by Corollary Theorem 5; using 10a and Theorem 3.)

4.7. Show by an example that Corollary Theorem 5 does not hold with "$\supset$" in place of "$\sim$".

4.8. Establish the following propositions, where A is a formula containing no occurrence of the symbol ¬, and B is any formula.
(a) The truth table of A has t in its first line.
(b) If ⊨ ¬B, then B contains at least one occurrence of ¬.
(c) If ⊨ B ∼ ¬A, then B contains at least one ¬.

§ 5. Model theory: chains of equivalences. It is often useful to know that two formulas A and B have the same truth table, or to transform a given formula A into a formula B of some specified sort which has the same table. By Theorem 4 (b), A and B have the same table exactly when ⊨ A ∼ B, i.e. when the formula asserting the (material) equivalence of A and B is valid. In this case, we may say that A and B are (*logically*) *equivalent* (*in the propositional calculus*). Of the 45 results in Theorem 2, 26 are thus assertions of equivalences holding in the propositional calculus.

The chain method which we present next is useful in establishing such equivalences.

First note that: **(α)** ⊨ A & B *if and only if both* ⊨ A *and* ⊨ B. This is immediate from the truth table for & (or we can infer it by Theorem 3 from 3, 4a and 4b in Theorem 2).

Next we observe that equivalence in the propositional calculus is reflexive, symmetric and transitive: **(β)** ⊨ A ∼ A. **(γ)** *If* ⊨ A ∼ B, *then* ⊨ B ∼ A. **(δ)** *If* ⊨ A ∼ B *and* ⊨ B ∼ C, *then* ⊨ A ∼ C. These three statements are immediate from Theorem 4 (b). (Alternatively, (β) is *19; (γ) follows from *20 by Exercise 4.6; and (δ) from *21 using (α) and Theorem 3.)

Using (β)–(γ): **(ε)** *If* ⊨ $A_0 \sim A_1$ *and* ⊨ $A_1 \sim A_2$ *and* ⊨ $A_2 \sim A_3$, *then* ⊨ $A_i \sim A_j$ *for each of the* 16 *pairs of subscripts* i, j ($i, j = 0, 1, 2, 3$); i.e. ⊨ $A_0 \sim A_0$, ⊨ $A_0 \sim A_1$, ⊨ $A_0 \sim A_2$, ⊨ $A_0 \sim A_3$, ⊨ $A_1 \sim A_0$, . . . , ⊨ $A_3 \sim A_0$, ⊨ $A_3 \sim A_1$, ⊨ $A_3 \sim A_2$, ⊨ $A_3 \sim A_3$. Thus (α) gives ⊨ $A_0 \sim A_0$; to get ⊨ $A_3 \sim A_1$, we use ⊨ $A_1 \sim A_2$ and ⊨ $A_2 \sim A_3$ with (δ), and the result with (γ); etc. Or (ε) can be recognized as true directly from Theorem 4 (b); for, the three hypotheses of (ε) say that in the list A_0, A_1, A_2, A_3 each successive formula has the same table as the preceding, and the conclusion says that any pair of formulas in the list have the same table.

Now we adopt "$A_0 \sim A_1 \sim A_2 \sim A_3$" as an abbreviation for $((A_0 \sim A_1) \& (A_1 \sim A_2)) \& (A_2 \sim A_3)$. Then by two applications of (α): **(ζ)**. *The hypothesis of* (ε) *is equivalent to* ⊨ $A_0 \sim A_1 \sim A_2 \sim A_3$.

We call $A_0 \sim A_1 \sim A_2 \sim A_3$ a "chain of (three) equivalences". It has the properties that we can establish its validity by establishing the validity of the (three) "links", and that once the chain is established as valid we can infer the equivalence of any pair of the formulas A_0, A_1, A_2, A_3 (joined by links) in the chain.

Everything beginning with (ϵ) said using A_0, A_1, A_2, A_3 applies similarly to $A_0, \ldots, A_n$ for any $n \geq 2$ (and trivially even for $n = 0, 1$).

Now we remark that, once *49, *55a and *55c in Theorem 2 are established (as proposed in § 3, or by Exercise 4.1), then all of *55b, *56–*61 follow by the chain method. For example:

*55b. By *49 (since the A of *49 can be any formula, e.g. the present $\neg A \vee \neg B$) and (γ): **(1)** $\vDash \neg A \vee \neg B \sim \neg\neg(\neg A \vee \neg B)$. By Theorem 5 with *55a (as in Exercise 4.4):

(2) $\vDash \neg\neg(\neg A \vee \neg B) \sim \neg(\neg\neg A \,\&\, \neg\neg B)$. By Theorem 5 with *49:

(3) $\vDash \neg(\neg\neg A \,\&\, \neg\neg B) \sim \neg(A \,\&\, \neg\neg B)$,

(4) $\vDash \neg(A \,\&\, \neg\neg B) \sim \neg(A \,\&\, B)$. From (1)–(4) by ($\epsilon$) (for $n = 4$), $\vDash \neg(A \,\&\, B) \sim \neg A \vee \neg B$, as was to be shown. — Using (ζ), we can put this proof in the following shorthand:

$\vDash \neg A \vee \neg B \sim \neg\neg(\neg A \vee \neg B)$ [*49] $\sim \neg(\neg\neg A \,\&\, \neg\neg B)$ [*55a] $\sim \neg(A \,\&\, B)$ [*49].

*57. $\vDash A \,\&\, B \sim \neg\neg(A \,\&\, B)$ [*49] $\sim \neg(\neg A \vee \neg B)$ [*55b].

*60. $\vDash A \,\&\, B \sim A \,\&\, \neg\neg B$ [*49] $\sim \neg(A \supset \neg B)$ [*55c].

Since in § 3 we could accept all the results of Theorem 2 as established by someone else's computation, the real point of these reproofs is to make it evident how, if we remember *49, and any two of *55a–*61 which together contain all three of the symbols $\supset$, &, $\vee$, we can quickly derive the others.

Now we use the chain method to get a new result.

THEOREM 6°. *Let* E *be any formula constructed from atoms* $P_1, \ldots, P_n$ *and their negations* $\neg P_1, \ldots, \neg P_n$ *using only* & *and* $\vee$. *Let* $E^\dagger$ *come from* E *by interchanging* & *with* $\vee$ *and each unnegated atom with its negation* (cf. the example in the proof).[19] *Then* $\vDash \neg E \sim E^\dagger$.

PROOF. Using *55a and *55b (with the chain method), we can move the initial $\neg$ of $\neg E$ progressively to the right (inward) across all the &'s and $\vee$'s, which interchanges them. Then we can use *49 to remove the resulting double negations, so that the atoms will be interchanged with their negations. The following example illustrates this proof.

$$
\begin{aligned}
\vDash\ & \neg(\neg Q \,\&\, (\neg P \vee Q)) \\
\sim\ & \neg\neg Q \vee \neg(\neg P \vee Q) \quad [*55b] \\
\sim\ & \neg\neg Q \vee (\neg\neg P \,\&\, \neg Q) \quad [*55a] \\
\sim\ & Q \vee (P \,\&\, \neg Q) \quad [*49].
\end{aligned}
$$

COROLLARY°. *Each formula* E *is equivalent to a formula* F (i.e. $\vDash E \sim F$) *in which* $\neg$ *occurs only applied directly to atoms.*

[19] If in writing E parentheses were omitted under our ranking of & ahead of $\vee$ (cf. § 1), they should be restored before performing the operation † here, or the operation ′ in Theorem 7.

PROOF. First, we can eliminate $\sim$ and $\supset$ from E by *63a, and *58 or *59 (or possibly *55c, *60 or *61). Next, we can suppress any double negations by *49. Finally, Theorem 6 can be used to eliminate successively each $\neg$ which does not apply directly to an atom; in doing so, we work each time on such a $\neg$ that is innermost (i.e. does not have another such $\neg$ within its scope, Example 1). This should become clear from the following illustration.

$\models \neg\{\neg P \supset \neg(\neg\neg P \vee \neg Q) \& R\}$

$\sim \neg\{\neg\neg P \vee (\neg(\neg\neg P \vee \neg Q) \& R)\}$ [eliminating $\supset$ by *59]

$\sim \neg\{P \vee (\neg(P \vee \neg Q) \& R)\}$ [suppressing the double negations by *49]

$\sim \neg\{P \vee (\neg P \& Q \& R)\}$ [applying Theorem 6 to the part $\neg(P \vee \neg Q)$]

$\sim \neg P \& (P \vee \neg Q \vee \neg R)$ [applying Theorem 6 to the whole].

As we eliminate the $\supset$, we supply a pair of parentheses, which before were superfluous since $\supset$ outranks &. In the fourth and final formulas, we omit a pair of parentheses (as mathematicians do in writing "$a+b+c$"), since by *31 and *32 it is immaterial for present purposes which way the triple conjunction and disjunction are associated.[20]

EXERCISES. 5.1. Use the chain method to derive *56, *58, *59, *61 (taking *49, *55a–*55c as already established).

5.2. Find equivalent formulas with $\neg$ applied only to atoms:
(a) $\neg((P \& \neg Q) \vee R \vee (S \& \neg P))$. (b) $\neg(P \vee \neg Q \supset (R \& \neg\neg S) \vee Q)$.
(c) $\neg(\neg(P \& Q) \sim P)$.

5.3°. Establish the following, using as far as possible recent results rather than new direct appeals to truth tables:
(a) $\models (A \sim B) \sim (A \supset B) \& (B \supset A) \sim (\neg A \vee B) \& (A \vee \neg B) \sim (A \& B) \vee (\neg A \& \neg B)$. (b) $\models \neg(A \sim B) \sim (A \& \neg B) \vee (\neg A \& B) \sim (A \vee B) \& (\neg A \vee \neg B) \sim (A \vee B) \& \neg(A \& B)$ [which expresses exclusive disjunction] $\sim (\neg A \supset B) \& (A \supset \neg B) \sim (\neg B \supset A) \& (B \supset \neg A) \sim (\neg A \sim B) \sim (A \sim \neg B)$.

★§ 6. Model theory: duality.[21]

THEOREM 7°. (Duality.) *Let* E *and* F *be formulas of the type described in Theorem* 6. *Let* E′, F′ *come from* E, F *by interchanging* & *with* V.[19] *Then*:

(a) *If* $\models \neg E$, *then* $\models E'$. (b) *If* $\models E$, *then* $\models \neg E'$.
(c) *If* $\models E \sim F$, *then* $\models E' \sim F'$. (d) *If* $\models E \supset F$, *then* $\models F' \supset E'$.

[20] Although by *31 the association is ordinarily immaterial to us, we can for definiteness regard "$A_1 \& \ldots \& A_m$" for $m \geq 3$ as an abbreviation for $(\ldots((A_1 \& A_2) \& A_3) \ldots \& A_{m-1}) \& A_m$. For $m = 1$, "$A_1 \& \ldots \& A_m$" means simply A_1. Similarly with *32 and V. (For $m = 0$, cf. Footnote 248.)

[21] Starred sections can be omitted without loss of continuity (references from unstarred sections will be incidental). This § 6 is not required for later starred sections (except for the end of § 19).

PROOF. (a) Assume $\models \neg E$. By Theorem 6 with Corollary Theorem 5 (or Exercise 4.6), $\models E^{\dagger}$. Thence by Theorem 1, $\models E^{\dagger *}$ where $*$ indicates the substitution of $\neg P_1, \ldots, \neg P_n$ simultaneously for the atoms $P_1, \ldots, P_n$. Finally, by *49 with Corollary Theorem 5, $\models E^{\dagger * \ddagger}$ where $\ddagger$ indicates the removal of a double negation before each prime part which in E was unnegated. But $E^{\dagger * \ddagger}$ is E', as the following example illustrates.

$\models \neg E.$	$\models \neg(\quad P \,\&\, \neg\ P).$
$\models E^{\dagger}.$	$\models \quad \neg\ P \vee \quad P.$
$\models E^{\dagger *}.$	$\models \quad \neg\neg P \vee \quad \neg P.$
$\models E^{\dagger * \ddagger}$, i.e. $\models E'$.	$\models \quad P \vee \quad \neg P.$

(b) Assume $\models E$. Then by *49 with Corollary Theorem 5, $\models \neg\neg E$. So by Theorem 6 and Corollary Theorem 5, $\models \neg E^{\dagger}$. Thence $\models \neg E^{\dagger *}$. Thence $\models \neg E^{\dagger * \ddagger}$, i.e. $\models \neg E'$.

(c) Assume $\models E \sim F$. Then by Theorem 5, $\models \neg E \sim \neg F$. So by Theorem 6, $\models E^{\dagger} \sim F^{\dagger}$. Thence $\models E^{\dagger *} \sim F^{\dagger *}$. Thence $\models E^{\dagger * \ddagger} \sim F^{\dagger * \ddagger}$, i.e. $\models E' \sim F'$.

If we had established 4a, 4b, *31, *33, *35, *37, *39, *50 and duality (Theorem 7), but not yet 5a, 5b, *32, *34, *36, *38, *40, *51, the latter would follow by duality and substitution (Theorem 1). For example, by *50 with P as the A, $\models \neg(P \,\&\, \neg P)$. Thence by duality (Theorem 7 (a)), $\models P \vee \neg P$. Thence by substitution, $\models A \vee \neg A$, which is *51.

The effect of using Theorem 1 (substitution) with Theorem 7 is to allow the duality transformation to be applied to a resolution of E (or of E, F) into components $A_1, \ldots, A_n$ not necessarily prime, which must then retain their identity (be "treated as prime") throughout the transformation. (Similarly with Theorem 6 and Corollary.) —

Suppose a visitor from Mars is confused by what he observes upon his arrival on Earth, and mistakes our true "t" for false "F", and our false "f" for true "T"; i.e. let F = t and T = f. Then our table for & would for him read as our table of V for us, and vice versa. To see this, let us view the tables in the square arrangement (available in the case of two components), in which properties of the tables are more easily visualized. Table (1) is our table for &; (2) is the same rewritten using F = t and T = f; (3) is (2) rearranged to the normal order of T first and F second

(1) A & B

A \ B	t	f
t	t	f
f	f	f

(2) A & B

A \ B	F	T
F	F	T
T	T	T

(3) A & B

A \ B	T	F
T	T	T
F	T	F

(4) A V B

A \ B	t	f
t	t	t
f	t	f

(according to the Martian's ideas). Now observe that Table (3) looks just like our table (4) for V, except that it is written in the capitals T, F instead of the small letters $\mathfrak{t}$, $\mathfrak{f}$. The table for $\neg$ written with T, F in the Martian's normal order will look just like our table for $\neg$ written with $\mathfrak{t}$, $\mathfrak{f}$, as the reader may verify.

These observations suggest new proofs of Theorems 6 and 7.[22] Also they suggest how to avoid excluding $\sim$ and $\supset$ from the formulas E, F (and our restriction on $\neg$ was inessential above). We need simply add to our symbolism two new propositional connectives $\nsim$ and $\not\subset$, choosing the tables for $A \nsim B$ and $A \not\subset B$ so that they will look to the Martian as the tables for $\sim$ and $\supset$, respectively, look to us. The reader may verify that this is accomplished if $A \nsim B$ has the table for $\neg(A \sim B)$ and $A \not\subset B$ has the table for $\neg(B \supset A)$. We may, if we wish, regard these as temporary additions to our symbolism, used while applying duality, and then eliminated by rewriting each part $A \nsim B$ as $\neg(A \sim B)$ (or $A \sim \neg B$, by Exercise 5.3 (b)) and $A \not\subset B$ as $\neg(B \supset A)$ (or $B \mathbin{\&} \neg A$).

We now prove THEOREM 6a° (= Theorem 6 when E may be any formula, even containing $\nsim$ and $\not\subset$, and † is the operation of interchanging $\sim$ with $\nsim$, and $\supset$ with $\not\subset$, and & with V, and of changing by one the number of $\neg$'s on each atom). By Theorem 4 (b) it will suffice to show that $\neg E$ and $E^\dagger$ have the same table. In computing (any given line for) $\neg E$, we first compute a value for E (from the $\mathfrak{t}$'s and $\mathfrak{f}$'s assigned to the atoms $P_1, \ldots P_n$) using our tables for $\sim$, $\nsim$, $\supset$, $\not\subset$, &, V, $\neg$, and then (by the $\neg$ of $\neg E$) we change the resulting $\mathfrak{t}$ or $\mathfrak{f}$ to T or F. In computing $E^\dagger$, we first change the $\mathfrak{t}$'s and $\mathfrak{f}$'s assigned to $P_1, \ldots, P_n$ to T's and F's (by the change by one in the number of $\neg$'s on $P_1, \ldots, P_n$), and then (because of the interchange of $\sim$ with $\nsim$, and $\supset$ with $\not\subset$, and & with V) we do the same computation using the Martian's tables with T, F as before we did using ours with $\mathfrak{t}$, $\mathfrak{f}$. Thus the two computations differ only in whether we change $\mathfrak{t}$, $\mathfrak{f}$ to T, F at the end or at the beginning.

We prove THEOREM 7a° (= Theorem 7 similarly extended), thus.

(a) $\{\models \neg E\} \equiv$ {all lines in the table for $\neg E$ have $\mathfrak{t}$} $\equiv$ {all lines in the table for $\neg E'$ have T} $\equiv$ {all lines in the table for $\neg E'$ have $\mathfrak{f}$} $\equiv$ {all lines in the table for E' have $\mathfrak{t}$} $\equiv \{\models E'\}$.[23]

[22] The preceding treatment is basically as in IM pp. 121–124, which was inspired by Hilbert and Ackermann 1928 Chapter 1 § 5. The following treatment is inspired by Church 1956 § 16 pp. 106–108. Duality in logic was first recognized by Schröder 1877.

[23] For brevity, we are using "$\equiv$" for "(is) equivalent (to)" or "if and only if", with the chain method, in the observer's language. We prefer the symbol "$\equiv$" different from "$\sim$" in order to keep clear the distinction between the two languages.[15]

In (c), we similarly use "$\rightarrow$" for "implies" or "only if" in the observer's language, with the chain method but lacking symmetry; thus $A_0 \rightarrow A_1 \rightarrow A_2 \rightarrow A_3$ implies $A_i \rightarrow A_j$ only for $i \leq j$.

(c) $\{\models E \sim F\} \rightarrow \{\not\models \neg(E \sim F)'\}$ [(b)] $\rightarrow \{\not\models \neg(E' \nsim F')\} \rightarrow \{\not\models \neg\neg(E' \sim F')\}$ [Theorem 4 with the table for $\nsim$, and Corollary Theorem 5] $\rightarrow \{\models E' \sim F'\}$ [8 in Theorem 2 with Theorem 3].

EXERCISES. 6.1. Prove Theorem 7 (d). (b) From the proof of Theorem 7a (a), infer (b). (c) Prove Theorem 7a (d).

6.2. By applying Theorem 7a (c) (with P, Q, for A, B), extend the list in Example 5.3 (a) of "equivalents" of $A \sim B$.

§ 7. Model theory: valid consequence. We started this chapter by saying that logic has the important function of saying what follows from what, and thus of saying what propositions are theorems for given axioms. Yet thus far we have dealt only with tautologies, i.e. valid formulas, which logic asserts to hold without regard to any extra-logical assumptions whatsoever.

Still keeping in mind that in the propositional calculus we do not look at the internal structure of the atoms (or will not know the propositions they express), let us suppose that we are given from outside the propositional calculus that a formula A is true by assumption or fact. That is, we may be told that it is an axiom of some abstract theory (like geometry or group theory), so it is true by fiat for the purpose of that theory. Or it may be a proposition which is true in physical fact or by intuitive mathematical reasoning. How does this alter our position with regard to what formulas we can assert to be true by use *otherwise* of only the propositional calculus?

Consider an example; say A is $(P \supset Q) \& (P \vee R)$ (Table (h)).

	P	Q	R	(h) $(P \supset Q) \& (P \vee R)$	(i) $Q \vee R$	(j) $P \supset R$	(k) $P \vee \neg Q \supset R$	(l) $P \& \neg Q$
1.	t	t	t	t	t	t	t	f
2.	t	t	f	t	t	f	f	f
3.	t	f	t	f	t	t	t	t
4.	t	f	f	f	f	f	f	t
5.	f	t	t	t	t	t	t	f
6.	f	t	f	f	t	t	t	f
7.	f	f	t	t	t	t	t	f
8.	f	f	f	f	f	t	f	f

Remember that what P, Q and R really are is top-secret information, and practitioners of the propositional calculus are not cleared for it. Nevertheless, if we are told that $(P \supset Q) \& (P \vee R)$ is true, we have been told something. Namely, we then know that the truth values of P, Q, R must form one of the four assignments (Lines 1, 2, 5, 7) which give t to (P ⊃ Q) & (P ∨ R) in Table (h). So now, in trying to decide what other formulas B are true on the basis of the propositional calculus *plus* the information that A is true, we need consider only these four assignments. Thus, upon being given that A is true, we know that Q ∨ R is true because its table (i) has only t's in Lines 1, 2, 5, 7; but we still do not have enough information to know whether P ⊃ R is true, because its table (j) has f in Line 2.

This leads us to the following definition. Consider two formulas A and B, and let $P_1, \ldots, P_n$ be the atoms occurring in A or in B. We say that B is a *valid consequence* of A (in, or by, the propositional calculus), or in symbols A ⊨ B, if, in truth tables for A and B entered from $P_1, \ldots, P_n$, the formula B has the value t in all those lines in which A has t.

Thus, as we have just observed, (P ⊃ Q) & (P ∨ R) ⊨ Q ∨ R but not (P ⊃ Q) & (P ∨ R) ⊨ P ⊃ R.

We note that "A ⊨ B" *is a stronger statement than* "If ⊨ A, then ⊨ B"; by this we mean that the first statement always implies the second, but the second may hold without the first holding.

To see that the first always implies the second, assume the first "A ⊨ B" and the hypothesis "⊨ A" of the second. Then (by A ⊨ B) B has t in all those lines in which A has t; and (by ⊨ A) these are all lines; so ⊨ B.

When A, B are (P ⊃ Q) & (P ∨ R), P ⊃ R, the second statement "If ⊨ A, then ⊨ B" holds as a material implication (§ 2) since "⊨ A" is false; but as we observed above "A ⊨ B" does not hold. The point is that, when not ⊨ A, so it is not the case that A is t in all lines, then "If ⊨ A, then ⊨ B" holds automatically, while "A ⊨ B" holds only if B is t in those lines if any in which A is t.

Now suppose m formulas $A_1, \ldots, A_m$ are given. Generalizing from the case $m = 1$, we define: B is a *valid consequence* of $A_1, \ldots A_m$ (in, or by, the propositional calculus), or in symbols $A_1, \ldots A_m$ ⊨ B, if, in truth tables entered from a list $P_1, \ldots, P_n$ of the atoms occurring in one or more of $A_1, \ldots A_m$, B, the formula B is t in all those lines in which $A_1, \ldots, A_m$ are simultaneously t. The symbol "⊨" may be read "entail(s)".

Not only is it obviously immaterial here in what order the atoms occurring in any of $A_1, \ldots, A_m$, B are listed as $P_1, \ldots, P_n$. But also by beginning § 4, the outcome will be the same if the tables for $A_1, \ldots A_m$, B are entered from a list $P_1, \ldots, P_n$ including still more atoms.

Inspection of the tables shows that (i), (j) ⊨ (k) (Lines 1, 3, 5, 6, 7); (i), (j), (l) ⊨ (k) (Line 3); (h), (l) ⊨ (i) (there are *no* lines in which it needs to be checked that (i) is t); but not (i), (j) ⊨ (h) (e.g. Line 3).

THEOREM 8. (a) $A \models B$ *if and only if* $\models A \supset B$. (b) *More generally, for* $m \geq 1$: $A_1, \ldots, A_{m-1}, A_m \models B$ *if and only if* $A_1, \ldots, A_{m-1} \models A_m \supset B$.

PROOF. (a) Consider tables for A, B, $A \supset B$ entered from a list $P_1, \ldots, P_n$ including all their atoms. Those lines in which A is f do not matter for whether $A \models B$, and in those lines $A \supset B$ is t anyway (by the table for $\supset$). So consider the remaining lines, i.e. the lines in which A is t. If $A \models B$, then B is t in these lines; so by the table for $\supset$, $A \supset B$ is t in these lines (as well as the others); so $\models A \supset B$. Conversely, if $\models A \supset B$, then $A \supset B$ is t in these lines (as well as the others); so by the table for $\supset$, B is t in these lines (the lines for which A is t); so $A \models B$.

(b) FOR $m \geq 2$. Consider tables for $A_1, \ldots, A_m$, B, $A \supset B$. We reason as before, with A_m as the A, except that we confine our attention throughout to only lines in which $A_1, \ldots, A_{m-1}$ are t.

COROLLARY. *For* $m \geq 1$: $A_1, \ldots, A_{m-1}, A_m \models B$ *if and only if* $\models A_1 \supset (\ldots (A_{m-1} \supset (A_m \supset B)) \ldots)$.

PROOF. By *m* successive applications of the theorem.

By Corollary Theorem 8, the problem of what formulas are valid consequences of given formulas $A_1, \ldots, A_m$ is reduced to the problem of what formulas are valid. This is one reason why tautologies are important.

One might reverse the argument and consider this as a reason why the valid consequence relationship is unimportant. However, the valid consequence relationship corresponds more directly to the way we ordinarily use logic. Many manipulations are easier to make in terms of valid consequence relationships than when these relationships are condensed by Corollary Theorem 8 into the validity of iterated implications.

For reasons which will appear later, we prefer to emphasize these manipulations in another context, that of "proof theory", which we will begin studying in § 9. Therefore we relegate the further development in terms of valid consequence to the exercises, which may help to make some of the manipulations more meaningful when we take them up in proof theory.

EXERCISES. 7.1. (a) Find all the true statements "(h) $\models$ B" and "(h), (j) $\models$ B" where B is one of (h)–(ł). (Counting trivial ones like "(h) $\models$ (h)", there are six.) (b) Prove that for *every* formula B: (h), (ł) $\models$ B. (c) Prove that for every formula B: (i), (j) $\models$ B if and only if (k) $\models$ B.

7.2. Verify by truth tables: (a) P, $P \supset Q \models Q$. (b) P, $Q \models P \& Q$. (c) $P \& Q \models P$. (d) $P \& Q \models Q$. (e) P, $\neg P \models Q$. (f) not $P \supset Q \models Q \supset P$.

7.3. Show that, with notation as in Theorem 1 (but with $m+1$ formulas): *If* $A_1, \ldots, A_m \models B$, *then* $A_1^*, \ldots, A_m^* \models B^*$. (HINT: use Corollary Theorem 8.)

7.4. (a) Apply Exercise 7.3 to generalize Exercise 7.2 (a)–(e) from P,

Q to A, B. (b) Thence by Theorem 8 and Corollary reprove 3, 4a, 4b and *10a of Theorem 2.

7.5. For $m \geq 1$, show that: (a) $A_1, \ldots, A_m \vDash B$ *if and only if* $A_1 \,\&\, \ldots \,\&\, A_m \vDash B$.[20] Thence by Theorem 8: (b) $A_1, \ldots, A_m \vDash B$ *if and only if* $\vDash A_1 \,\&\, \ldots \,\&\, A_m \supset B$. (This gives us an alternative to Corollary Theorem 8.)

7.6. Establish the following.

(i) *For* $m \geq 1$:

$A_1, \ldots, A_m \vDash A_1,$

$\ldots,$

$A_1, \ldots, A_m \vDash A_m.$

(ii) *For* $m, p \geq 0$:

If $A_1, \ldots, A_m \vDash B_1,$

$\ldots,$

$A_1, \ldots, A_m \vDash B_p$ *and* $B_1, \ldots, B_p \vDash C,$

then $A_1, \ldots, A_m \vDash C.$

7.7. Show directly from the definition of valid consequence:

(a) *If* $A \vDash B$ *and* $A \vDash \neg B$, *then* $\vDash \neg A$. (Reductio ad absurdum.)

(b) *If* $A \vDash C$ *and* $B \vDash C$, *then* $A \vee B \vDash C$. (Proof by cases.)

7.8. Do Exercise 7.7 instead from Theorems 2, 8 and 3.

7.9. Observe that the reasoning in §§ 4, 5 (for the chain method) holds good when we confine our attention to assignments (i.e. lines of the tables) for which a given list of formulas $A_1, \ldots, A_m$ are all t; so Theorem 3, Theorem 5 (and Corollary), and (α)–(ζ), and thus the chain method, hold good with "$\vDash$" replaced throughout by "$A_1, \ldots, A_m \vDash$". Now show that:

(a) $P \sim \neg Q \vDash P \,\&\, \neg Q \sim P$. (Hint: cf. Theorem 2.)

(b) $(P \vee Q) \sim \neg(P \,\&\, Q) \vDash (P \vee Q) \vee (\neg(P \,\&\, Q) \,\&\, P) \sim (P \vee Q)$.

★§ 8. Model theory: condensed truth tables. We used the idea of the truth tables to define when a formula E is valid (in symbols, $\vDash E$) and when a formula B is a valid consequence of formulas $A_1, \ldots, A_m$ (in symbols, $A_1, \ldots, A_m \vDash B$). The tables themselves have been used (often with shortcuts, as in Exercise 2.3) in illustrations and in the original proofs of the results in Theorem 2 and some other results. But hereafter, to establish that $\vDash E$ or that $A_1, \ldots, A_m \vDash B$, it will ordinarily be more efficient to employ theorems about validity and "valid consequence" than actually to compute truth tables. In §§ 4 and 5 we began, and we shall continue, to develop techniques for using such results systematically in lieu of truth tables. If we wish to show that not $\vDash E$ or not $A_1, \ldots, A_m \vDash B$, we need to compute only one suitably chosen line of the table(s). Often we can spot such a line with a little trial and error, without computing the full table(s).

A formula E is called *inconsistent* or *contradictory* or *identically false*, if it has a solid column of f's in its truth table; *contingent*, if it is neither valid nor inconsistent. Thus formulas fall into three classes with respect to the presence of t's and f's in their truth tables.

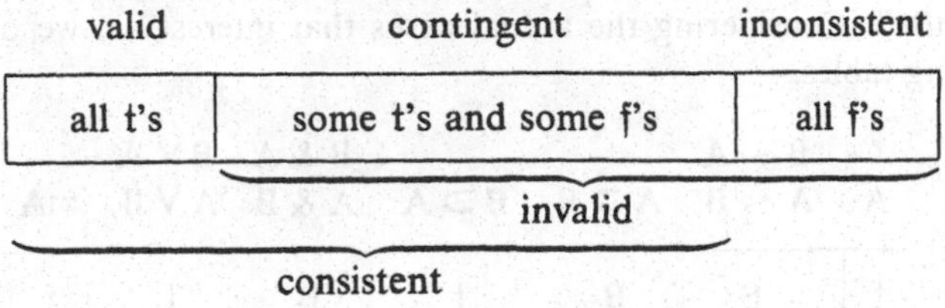

valid	contingent	inconsistent
all t's	some t's and some f's	all f's

invalid (contingent and inconsistent)

consistent (valid and contingent)

A formula E is inconsistent or consistent, according as ¬E is valid or invalid. To establish that a formula is contingent, computation of two suitable lines would suffice.

If, notwithstanding, we should find we need to do much truth-table computation, it will be worthwhile to seek further economies in the writing and computing of the tables.[24] We used 8 lines in writing the tables for formulas with 3 atoms; with a dozen atoms, 4096 lines would be required similarly.

Consider the table (1) in § 2 for P ⊃ (Q ∨ R ⊃ (R ⊃ ¬P)). Since the value t is common to the last four lines, those lines can be replaced by one. Likewise, the first and third, and also the second and fourth, lines can be combined. Thus (1) condenses to:

(2)

P	R	P ⊃ (Q ∨ R ⊃ (R ⊃ ¬P))
t	t	f
	f	t
f		t

In § 2, we observed a shortcut in the computation of the table for P ⊃ (Q ∨ R ⊃ (R ⊃ ¬P)) which gave the t's in the last four lines of (1) en masse. This shortcut is an instance of a technique that is advantageous generally. The technique consists in assigning a value t or f to just one of the letters and computing as far as we can with that, then repeating with another letter, etc. (instead of assigning values to all of the letters first, and computing second). Consider the basic table for any *binary* propositional connective ○. If we pick a value (t or f) for A, then, with that value fixed, A ○ B has the table of a unary connective applied to B. There are only four possible tables for a unary connective, with entries respectively: (i) t, t, (ii) t, f (the same as the values of B), (iii) f, t (the same as ¬B), (iv) f, f. So, after picking a value of A, A ○ B can be evaluated as one of

[24] If the number of tables or their complexity is very great, the use of modern high-speed computing machines should be considered. How best to put truth-value problems on the machines must take into account features of the machines, and belongs in the area of "computer sciences". Cf. Wang 1960.

t, B, ¬B, f. Considering the actual cases that interest us, we obtain the following tables.

(I)

A	B ∼ A A ∼ B	A ⊃ B	B ⊃ A	B & A A & B	B ∨ A A ∨ B	¬A
t	B	B	t	B	t	f
f	¬B	t	¬B	f	B	t

Now we use this technique on the previous example.

(3)

$$\begin{array}{ccc}
 & P \supset (Q \vee R \supset (R \supset \neg P)) & \\
t \supset (Q \vee R \supset (R \supset \neg t)) & & f \supset (Q \vee R \supset (R \supset \neg f)) \\
Q \vee R \supset (R \supset f) & & t \\
Q \vee R \supset \neg R & & \\
Q \vee t \supset \neg t \qquad Q \vee f \supset \neg f & & \\
t \supset f \qquad\qquad Q \supset t & & \\
f \qquad\qquad\qquad t & &
\end{array}$$

In the first line is the formula before values have been assigned to any of the letters. In the second line, P has been assigned t and f in the left and right columns, respectively. The tables in (I) are then used to simplify the resulting expressions to Q ∨ R ⊃ ¬R in the left column (by three steps) and to t in the right column (by one step). Continuing, in the fourth line of the left column, R is assigned t and f in the left and right subcolumns, respectively. The figure (3) is called a *truth-value analysis* by Quine 1950, who was apparently the first to emphasize the advantages of computing on one letter at a time. The table (2) can be read off the analysis (3) (and indeed Quine in 1950 never writes (2) at all).[25]

A good rule of thumb is to select each time, for assignment of t or f, a letter occurring frequently. This will tend to promote rapid simplification, as each binary combination in which it occurs must disappear. By looking ahead a bit, one may be able to recognize cases when a different choice would be better.

Steps can be saved by using *49 (with Theorems 4, 5) to suppress double negations, either present initially or introduced in using (I). Thus if B is ¬C, f ∼ B simplifies by (I) to ¬¬C, which further simplifies by *49 to C.

The method of (3) is not guaranteed to bring us directly to a most condensed table, like (2). If we treat Q second (instead of R), we come out with a 5-line table.

[25] The tables (I) appear in a different format as *40a–*48 in Theorem 22 § 24 below. (They appear as *41–*48 on p. 118 of IM, which was written without knowledge of Quine 1950. But their use, as on p. 474, was not emphasized in IM.)

The formula $\{P \supset (Q \vee R \supset (R \supset \neg P))\} \& \{((Q \supset P) \supset Q) \supset Q\}$ admits the table (2); but no order of treating the letters gives a 3-line table directly. Of course, afterwards we can combine lines, as we did to get (2) from (1). If we first notice that $((Q \supset P) \supset Q) \supset Q$ is valid (Exercise 2.3 (a)), we can pass to $P \supset (Q \vee R \supset (R \supset \neg P))$, and thence to (2).

Indeed, the method of (3) will never produce a table of less than two lines, since it requires an assignment of t or f to one of the letters P_1 to get started. But $((Q \supset P) \supset Q) \supset Q$, being valid, admits the 1-line table:

	$((Q \supset P) \supset Q) \supset Q$
	t

Full truth tables, preferably written in condensed form, can be used in simplifying formulas. Suppose a complicated formula E arises in some problem, and we wish to investigate valid consequences of E, or more generally to investigate whether various relationships $A_1, \ldots, A_m \models B$ hold in which E is a specified one (or a part of one) of $A_1, \ldots, A_m$, B. For these purposes, E can be replaced by any formula F with the same truth table as E (i.e. by Theorem 4, a formula F such that $\models E \sim F$), or as we said in § 5 a formula F *equivalent to* E (*in the propositional calculus*). So if we can find a formula F equivalent to E but simpler than E, our investigation will be furthered.

For example, say E is $P \supset (Q \vee R \supset (R \supset \neg P))$ or $\{P \supset (Q \vee R \supset (R \supset \neg P))\} \& \{((Q \supset P) \supset Q) \supset Q\}$, either of which has the truth table (2). By starting from (2), and asking what formulas have that truth table, we find ones simpler than E, namely $(P \& \neg R) \vee \neg P$ (which is picked to have exactly the two t's in (2)), $\neg(P \& R)$ (where $P \& R$ is picked to have exactly the one f in (2)), $\neg P \vee \neg R$ (thence by De Morgan's law *55b), and $P \supset \neg R$ (thence by *59). The formula $(P \& \neg R) \vee \neg P$ gives a resolution of the truth of E into two *disjoint* (i.e. non-overlapping) cases (namely: P, R are t, f; P is f), and $\neg P \vee \neg R$ condenses this by allowing the cases to overlap. Thus we could have found $\neg P \vee \neg R$ without going through $\neg(P \& R)$.

It is worth noting that consolidation and recombination of cases can be effected by chains of equivalences, since facility with such transformations is useful. Besides results in Theorem 2, we need the following.[20]

*52°. $\models A \& (B \vee \neg B) \sim A$. *53. $\models A \vee (B \& \neg B) \sim A$.

*54. $\models A \& B \& \neg B \sim B \& \neg B$. *55°. $\models A \vee B \vee \neg B \sim B \vee \neg B$.

We give an illustration, starting with an equivalent of $P \supset (Q \vee R \supset (R \supset \neg P))$ read from (1) with only the last four lines combined:[26]

$\models$ $(P \& Q \& \neg R) \vee (P \& \neg Q \& \neg R) \vee \neg P \sim$
$(P \& \neg R \& (Q \vee \neg Q)) \vee \neg P$ [*35 with *33 (and *31)] $\sim$
$(P \& \neg R) \vee \neg P$ [*52] $\sim$ $(P \vee \neg P) \& (\neg P \vee \neg R)$ [*36, *34] $\sim$
$\neg P \vee \neg R$ [*52, *33].

Say that for any formula E we have found a $p+q$-line truth table (condensed or not) with p t's and q f's. As illustrated, we can then write an equivalent formula D of the form $D_1 \vee \ldots \vee D_p$ (which reduces to simply D_1 if $p = 1$, and say to $P_1 \& \neg P_1$ if $p = 0$) where $D_1, \ldots, D_p$ correspond to the respective t's, and an equivalent formula C of the form $C_1 \& \ldots \& C_q$ (which reduces to C_1 if $q = 1$, and say to $P_1 \vee \neg P_1$ if $q = 0$) where $C_1, \ldots, C_q$ correspond to the respective f's. To give another example, if E is $P \sim Q$, then D is $(P \& Q) \vee (\neg P \& \neg Q)$ and C is $(\neg P \vee Q) \& (P \vee \neg Q)$ [by Theorem 6 from $\neg\{(P \& \neg Q) \vee (\neg P \& Q)\}$]. Such a formula D (a disjunction of conjunctions of negated and unnegated atoms) equivalent to E is called a *disjunctive normal form* of E, and such a formula C (vice versa) a *conjunctive normal form*.

By a formula F being "simpler" than E we mean in a pragmatic sense that it is easier to comprehend and use, even if it is not shorter or much shorter. What is simpler will then depend on what uses we have in view, or on what sorts of formulas we have become adept at handling. (Thus we scarcely need to simplify $P \sim Q$.) Disjunctive normal forms of E are useful if we are endeavoring to infer valid consequences from E (cf. Exercise 7.7 (b)) or from E and other formulas, and conjunctive normal forms of E if we are endeavoring to infer E as a valid consequence of other formulas (cf. Exercise 7.2 (b)).

EXERCISES. 8.1. Establish *52–*55 with P, Q (preparatory to substituting A, B by Theorem 1) by using (I) with *50, *51 (and Theorem 4).

8.2. Simplify:

(a) $(P \vee S \sim \neg Q) \& (R \vee Q) \& \neg S \supset \neg((S \supset R \vee Q) \vee P)$.

(b) $\neg\{(P \supset R) \& S \sim \neg R \vee \neg(R \supset Q)\} \& (Q \supset \neg P) \& R$. (Write a simple equivalent after treating only R and Q.)

(c) $(P \supset Q) \& (\neg(Q \& R) \vee P) \& \neg(R \& P)$. (Treat Q first; and use *55b and *36+*53 or *34+*39.)

[26] In fact, $P \supset (Q \vee R \supset (R \supset \neg P))$ reduces to $\neg P \vee \neg R$ simply by using *59 thrice, *55a, *38 and *40 (with *32–*34). In general, the reduction of formulas can be performed by first using *63a (or Exercise 5.3) to eliminate $\sim$ and *58 or *59 (or *55c, *60 or *61) to eliminate $\supset$, then Theorem 6, then "multiplying out" using *35 (or dually *36) as analogous to $a(b + c) = ab + ac$ in algebra, and finally consolidating and recombining as illustrated using *31–*40, *52–*55. This technique goes back in essentials to Boole 1847. Cf. IM pp. 135–136.

§ 9. Proof theory: provability and deducibility. The proof of theorems, or the deduction of consequences of assumptions, in mathematics typically proceeds à la Euclid, by putting sentences in a list called a "proof" or "deduction". We use the word "proof" (and call the assumptions "axioms") when the assumptions have a permanent status for a theory under consideration, "deduction" when we are not thinking of them as permanent. Each step from some sentences in the list to another is mediated by logic, as analyzed above for the case the logic is propositional logic. Thus one sentence follows from others if it is a valid consequence of those others; this relationship we defined by truth tables in § 7. A sentence can be put into the list without reference to earlier sentences, if it is one of the assumptions or is valid. In the definitions of "valid consequence" and validity, we stood outside the language of the sentences themselves (the object language), and observed in another language (the observer's language) how the sentences (or formulas) are composed from atoms. In the observer's language, we also developed various results concerning validity and the valid consequence relationship which often are more convenient to use than direct application of the truth tables. We call this treatment of logic "model theory", as in it we replace the atoms by truth values t and f in all possible combinations, to obtain what can be considered models or concrete replicas of what the sentences may express.

Now we shall take up another way of founding logic. This treatment, called "proof theory", arises from asking the question whether the proofs and deductions of logic itself cannot be given in an analogous way. But now, since it is logic itself that we would treat in the axiomatic-deductive manner, the inferences cannot be made by appealing to logical criteria but only to specifically stated axioms and rules. In proof theory, some sentences or formulas will be taken as logical "axioms", and some "rules of inference" will be established for making the inferences from some sentences to another sentence.

We now give such a formulation of the classical propositional calculus, both by itself and for its application to deduction from assumptions. Later we shall show that the two formulations, that of model theory and that of proof theory, give equivalent results.

As *axioms* for our (proof-theoretic) system of the (classical) propositional calculus, we take all formulas of any of the forms shown after the symbol "⊨" in 1a–10b of Theorem 2 (and in the "List of Postulates" p. 387). These forms themselves we call *axiom schemata*. Each axiom schema includes infinitely many axioms, one for each choice of the formulas denoted by "A", "B", "C". For example, corresponding to 1a in Theorem 2, we have as Axiom Schema 1a: $A \supset (B \supset A)$. Particular axioms by this schema are: $P \supset (P \supset P)$, $P \supset (Q \supset P)$, $Q \supset (P \supset Q)$, $\neg P \supset (Q \& R \supset \neg P)$, $(P \supset (\neg R \supset P)) \supset [R \supset (P \supset (\neg R \supset P))]$, etc.

As the sole *rule of inference*, called the $\supset$-*rule* or *modus ponens* or the *rule of detachment*, we take the operation of passing from two formulas of the respective forms A and $A \supset B$ to the formula B, for any choice of formulas A and B (cf. Theorem 3). In an *inference* by this rule, the formulas A and $A \supset B$ are the *premises*, and B is the *conclusion*.

We define a (*formal*) *proof* (in the propositional calculus) to be a finite list of (occurrences of) formulas $B_1, \ldots, B_l$ each of which either is an axiom of the propositional calculus or comes by the $\supset$-rule from a pair of formulas preceding it in the list. A proof is said to be a proof *of* its last formula B_l. If a proof of a given formula B exists, we say B is (*formally*) *provable*, or is a (*formal*) *theorem*, or in symbols $\vdash B$.

EXAMPLE 4. For each formula A, the following list of five formulas $B_1, \ldots, B_l$ is a proof of the formula $A \supset A$. (Here $l = 5$, and B_1 is $A \supset (A \supset A), \ldots, B_5$ is $A \supset A$.)[27]

1. $A \supset (A \supset A)$ — Axiom Schema 1a.
2. $\{A \supset (A \supset A)\} \supset \{[A \supset ((A \supset A) \supset A)] \supset [A \supset A]\}$ — Ax. Sch. 1b.
3. $[A \supset ((A \supset A) \supset A)] \supset [A \supset A]$ — modus ponens, 1, 2.
4. $A \supset ((A \supset A) \supset A)$ — Axiom Schema 1a.
5. $A \supset A$ — modus ponens, 4, 3.

Besides the proof itself $B_1, \ldots, B_l$, we give at the left numbers for reference, and at the right reasons which justify the inclusion of each of the formulas $B_1, \ldots, B_l$ in the proof (an "analysis" of the proof). Thus at Step 1, we have applied Axiom Schema 1a with A both as the A and as the B of the schema (i.e. with the formula presently denoted by "A" as the formula denoted by "A" and by "B" in the statement of the schema). In Step 2, A is the A, $A \supset A$ is the B, and A is the C, of Axiom Schema 1b. In applying modus ponens to Formulas 1 and 2 at Step 3, the A of the rule is $A \supset (A \supset A)$ (which is 1) and the B of the rule is 3. — Because 1–5 constitutes a proof (for any fixed formula A), we can say that $A \supset A$ is provable, or in symbols $\vdash A \supset A$. Similarly, $\vdash A \supset (A \supset A)$, etc. (Why?)

A few remarks may help to make this way of formulating the propositional calculus as an axiomatic-deductive theory seem reasonable. We have an infinite number of axioms from each axiom schema, as mentioned above. This could be avoided by requiring the language in which the prime formulas are constructed to include single letters as "proposition variables" and adding a second rule of inference, the "substitution rule", to say that E* can be inferred from E under the circumstances of Theorem 1 when

[27] What we actually exhibit is a "proof schema", which becomes a particular proof for each choice of the formula denoted by "A".

$P_1, \ldots, P_n$ are proposition variables. That procedure seems less in keeping with usual mathematical language than the one we have followed (due to von Neumann 1927).[28]

The list of thirteen axiom schemata may seem surprisingly long. However, each propositional connective must have axioms to characterize it, i.e. to provide the deductive properties we want it to have. We are getting along with only two or three axiom schemata for each of ⊃, &, V, ¬, ~, namely, one or two (left column in Theorem 2) which help us to prove formulas in which the symbol is used (i.e. to "introduce" the symbol), and (except for ⊃) one or two (right column) which help us to infer formulas not containing the symbol (or not containing it so often) from formulas containing it (i.e. to "eliminate" the symbol). In the case of ⊃, the ⊃-rule provides for "elimination". This (to our mind) elegant arrangement of axiom schemata is due essentially to Gentzen 1934–5.

We could get along with fewer axiom schemata by foregoing the use of some of the symbols &, V, ~ as an official part of the object language. For example, if each time we wrote "A ~ B" we understood it as an abbreviation for (A ⊃ B) & (B ⊃ A), then Axiom Schemata 9a, 10a, 10b could be omitted.[29]

For the propositional calculus applied to infer formulas from assumptions $A_1, \ldots, A_m$, the formulas $A_1, \ldots, A_m$ are in effect allowed to function as axioms also. However, we shall not call them "axioms" (for the propositional calculus), but (when we need a name) *assumption formulas*; and (for $m > 0$) we shall not call $B_1, \ldots, B_l$ a "proof", but a "deduction" from $A_1, \ldots, A_m$. That is, a finite list of (occurrences of) formulas $B_1, \ldots, B_l$ is a (*formal*) *deduction* (*of* B_l) *from* $A_1, \ldots, A_m$ (in, or by, the propositional calculus), if each formula in the list is one of $A_1, \ldots, A_m$, or one of the axioms (of the propositional calculus, i.e. by one of Axiom Schemata 1a–10b), or comes from two earlier formulas in the list by the ⊃-rule. If there is a deduction of a given formula B from $A_1, \ldots, A_m$, we say that B is *deducible* from $A_1, \ldots, A_m$, or in symbols $A_1, \ldots, A_m \vdash B$. The symbol "⊢" may be read "yield(s)". (In using this terminology when

[28] In either case, the rule of inference must have the character of a schema, with the Roman capital letters "A" and "B" standing for any formulas, in order to provide for infinitely many different applications of it. Our axiom schemata can be considered as rules of inference with zero premises. Hence Carnap 1934 called the axiom schemata and the rules of inference together "transformation rules". The rules defining the class of formulas (§ 1), analogous to the rules of syntax in grammar, are the "formation rules".

[29] This is done in IM. If we took "A & B", "A V B", "A ~ B" to be abbreviations for ¬(A ⊃ ¬B), ¬A ⊃ B, ¬((A ⊃ B) ⊃ ¬(B ⊃ A)), respectively, we would need only the four axiom schemata 1a, 1b, 7, 8. There are still other possibilities. Cf. Church 1956 pp. 119, 136–138.

$m \geq 0$, "deduction" and "deducible" include "proof" and "provable" as the case $m = 0$.)[30]

EXAMPLE 5. For each choice of formulas A, B, C, the following sequence of 8 formulas is a deduction of C from $A \supset (B \supset C)$, A & B.

1. A & B — 2nd assumption formula.
2. A & B $\supset$ A — Axiom Schema 4a.
3. A — modus ponens, 1, 2.
4. $A \supset (B \supset C)$ — 1st assumption formula.
5. $B \supset C$ — modus ponens, 3, 4.
6. A & B $\supset$ B — Axiom Schema 4b.
7. B — modus ponens, 1, 6.
8. C — modus ponens, 7, 5.

Thus we can say that C is deducible from $A \supset (B \supset C)$, A & B; or briefly, $A \supset (B \supset C), A \,\&\, B \vdash C$.

We said "formal proof" and "formal deduction" in the definitions above (though we shall usually omit the word "formal") to emphasize that these proofs and deductions are in the object language, which we are studying in the observer's language (§ 1). From the standpoint of the observer's language, we look only at the *form* of the formulas (in contrast to their meaning or content) in determining under the definitions just given whether a given sequence of formulas $B_1, \ldots, B_l$ is in fact a (formal) proof, or a (formal) deduction from given assumption formulas $A_1, \ldots, A_m$. A sequence of formulas $B_1, \ldots, B_l$ is a (formal) proof, or a (formal) deduction from $A_1, \ldots, A_m$, only when it exactly fits the above definition (as illustrated in Examples 4 and 5.)[31] This stereotyping of the operations that can be performed in constructing a formal proof or deduction makes the formal proofs and deductions (in the object language) definite enough in structure to serve as objects of our study.

In our study (in the observer's language), we shall also be proving theorems, deducing consequences of assumptions, etc. This will be as much the case here in proof theory as it was in model theory. In these *informal* proofs and deductions, we may operate flexibly, on the basis of the

[30] The symbol "$\vdash$" goes back to Frege 1879; the present use of it to Rosser 1935 and Kleene 1934. (Rosser proposed it to express deducibility by the rule(s) of inference, and Kleene suggested including also use of the axioms.) The parallel use of "$\models$" (§§ 2, 7) is perhaps original with Kleene 1956a.

[31] In particular, we do not simply allow a formula to be included just because it is valid, or to be inferred from preceding formulas just because it is a valid consequence of them. If we did so, no proof would need to be longer than one formula, and no deduction from $A_1, \ldots, A_m$ would need to be longer than $m+1$ formulas, but the one inference might be enormously complicated. There is an aim in proof theory to analyze the inferences into simple ones, as psychologically they must be in our actual reasoning. (More on this subject later.)

meanings of our statements, using any inferences that carry conviction. (It may indeed be that some of these inferences are the informal counterparts of operations available in the formal proofs, but they are not restricted to be such.)[32]

The student who keeps in mind that there are the two languages, where now (formal) proofs and deductions as well as formulas in the one language (the object language) are being studied, using (of necessity) informal proofs and deductions in another language (the observer's language), should have no trouble keeping matters straight. In model theory, the object language was dealt with only as an assemblage of formulas, whose truth tables we investigated; so there was not quite as much of a parallelism of terminology as we shall now have.

We conclude this section by giving in the observer's language two easy (informal) theorems (and their proofs), which have as their subject or object formal proofs and deductions in the object language.

THEOREM 9.

(i) *For* $m \geq 1$:

$$A_1, \ldots, A_m \vdash A_1,$$
$$\ldots,$$
$$A_1, \ldots, A_m \vdash A_m.$$

(ii) *For* $m, p \geq 0$:

If $A_1, \ldots, A_m \vdash B_1,$

$\ldots,$

$A_1, \ldots, A_m \vdash B_p$ *and* $B_1, \ldots, B_p \vdash C,$

then $A_1, \ldots, A_m \vdash C.$

PROOF. (i) The definition of a deduction $B_1, \ldots, B_l$ does not require that each of the assumption formulas $A_1, \ldots, A_m$ actually occur in the list $B_1, \ldots, B_l$. So, for each i from 1 to m, A_i standing by itself constitutes a deduction of A_i from $A_1, \ldots, A_m$. (ii) In a given deduction of C from $B_1, \ldots, B_p$, we can replace the occurrences of the assumption formulas $B_1, \ldots, B_p$ by deductions of $B_1, \ldots, B_p$ respectively from $A_1, \ldots, A_m$. Thereby we obtain a deduction of C from $A_1, \ldots, A_m$. —

Let $A_1, \ldots, A_m$ be a given list of formulas. Suppose we are exploring the class of the formulas B which are deducible from $A_1, \ldots, A_m$. Theorem 9 (i) tells us that $A_1, \ldots, A_m$ themselves are in this class. Theorem 9 (ii) tells us that any formula C is in this class which is deducible from any formulas $B_1, \ldots, B_p$ already known to be in this class.

From this standpoint, the role of Theorem 9 should be clear. However, we shall return to this in § 13, after obtaining some practice meanwhile with particular applications of Theorem 9. —

Before continuing, it may be well to contrast the meanings of four expressions.

[32] This is not to say that our informal proofs and deductions in the observer's language need not conform to any logical standards. But we are not now trying to regulate them or to study them as specimens of logic.

"$\models A \supset B$" means that the formula $A \supset B$ is valid, i.e. its truth table has a solid column of t's.

"$A \models B$" means that the formula B is a valid consequence of the formula A, i.e. B has t in each of those lines of its truth table in which A has t.

"$\vdash A \supset B$" means that the formula $A \supset B$ is provable, i.e. there is a finite sequence of formulas such that each formula of the sequence either is an axiom or comes from two preceding formulas of the sequence by modus ponens, and the last formula of the sequence is $A \supset B$.

"$A \vdash B$" means that the formula B is deducible from the formula A, i.e. there is a finite sequence of formulas such that each formula of the sequence either is A or is an axiom or comes from two preceding formulas of the sequence by modus ponens, and the last formula of the sequence is B.

By the end of § 12, we shall have found that these four expressions are equivalent, i.e. if any one is true so are the other three. (For the first two, we have this already in Theorem 8.)

Likewise, we shall find that "If $\models A$, then $\models B$" and "If $\vdash A$, then $\vdash B$" are equivalent. These two expressions are weaker than the preceding four, as we have already seen in § 7 for the ones with "$\models$".

THEOREM 10. (a) *If* $\vdash A \supset B$, *then* $A \vdash B$. (b) *More generally, for any* $m \geq 1$: *If* $A_1, \ldots, A_{m-1} \vdash A_m \supset B$, *then* $A_1, \ldots, A_{m-1}, A_m \vdash B$.

PROOF. (b) By hypothesis, there is a deduction of $A_m \supset B$ from $A_1, \ldots, A_{m-1}$ (say it has k formulas). Using this, we can construct a deduction of B from $A_1, \ldots, A_{m-1}, A_m$ as follows.

1.
. . .
k. $A_m \supset B$ } the deduction of $A_m \supset B$ from $A_1, \ldots, A_{m-1}$ given by the hypothesis of the theorem.

$k+1$. A_m — mth assumption formula.

$k+2$. B — modus ponens, $k+1$, k.

COROLLARY. *If* $\vdash A_1 \supset (\ldots (A_{m-1} \supset (A_m \supset B)) \ldots)$, *then* $A_1, \ldots, A_{m-1}, A_m \vdash B$.

EXERCISES. 9.1. Add to 1–5 in Example 4 to make a proof of $A \sim A$.

9.2. The following is a deduction of C from A, B, $A \supset (B \supset C)$. Supply the reasons (or "analysis"), and state the result using "$\vdash$".

1. A.
2. $A \supset (B \supset C)$.
3. $B \supset C$.
4. B.
5. C.

9.3. By constructing appropriate deductions, show that:
(a) $A, A \supset B \vdash B$. (b) $A, B \vdash A \& B$. (c) $A \& B \vdash A$. (d) $A \& B \vdash B$. (e) $A \vdash A \vee B$. (f) $B \vdash A \vee B$. (g) $\neg\neg A \vdash A$.
(h) $A \supset B, B \supset A \vdash A \sim B$. (i) $A \sim B \vdash A \supset B$. (j) $A \sim B \vdash B \supset A$.

9.4. Supply missing hypotheses (and justify them) or conclusion in the following applications of Theorem 9 (ii):
(a) $A, \neg A \vdash \neg\neg B$ and $\neg\neg B \vdash B$; therefore ____. ($m = 2, p = 1$.)
(b) ____ and $A \& B \vdash A$; therefore $A \supset (B \supset C), A \& B \vdash A$. ($m = 2, p = 1$.)
(c) $A \supset (B \supset C), A \& B \vdash A$ and $A \supset (B \supset C), A \& B \vdash B$ and ____ and $A, B, A \supset (B \supset C) \vdash C$; therefore $A \supset (B \supset C), A \& B \vdash C$. ($m = 2, p = 3$.)
(d) ____ and ____ and $A, A \supset B \vdash B$; therefore $A, \neg B, A \supset B \vdash B$. ($m = 3, p = 2$).

9.5. Corresponding to Exercise 9.4 (c), illustrate the proof of Theorem 9 (ii) by combining the deductions of Exercise 9.3 (c) and (d) (construed as deductions from $A \supset (B \supset C), A \& B$) with that of Exercise 9.2. Compare your result with Example 5.

9.6. Write out the $m = 0$ and $p = 0$ cases of Theorem 9 (ii) and adapt the proof to them.

9.7. Show how the result of Exercise 9.1 (i.e. $\vdash A \sim A$) comes by Theorem 9 from $\vdash A \supset A$ (Example 4) and Exercise 9.3 (h). (Take $m = 0$.)

9.8. Show that: *If* $\vdash A_1 \& \ldots \& A_m \supset B$, *then* $A_1, \ldots, A_m \vdash B$.[20]

§ 10. Proof theory: the deduction theorem. The property of deducibility expressed by the next theorem corresponds to a familiar method in our informal reasoning. To establish an implication "If A, then B", we often assume A "for the sake of the argument" and undertake to deduce B. This method is also available in the presence of other assumptions $A_1, \ldots, A_{m-1}$.

The proof is longer than that of the preceding theorems (except Theorem 2, if all the computations are included). But it has a simple plan, after which the rest of the work falls into four simple cases.

THEOREM 11. (The deduction theorem, Herbrand 1930.)[33] (a) *If* $A \vdash B$, *then* $\vdash A \supset B$. (b) *If* $A_1, \ldots, A_{m-1}, A_m \vdash B$, *then* $A_1, \ldots, A_{m-1} \vdash A_m \supset B$.

[33] The deduction theorem as an informal theorem proved about particular systems like the propositional calculus and the predicate calculus (Chapter II) first appears explicitly in Herbrand 1930 (and without proof in Herbrand 1928); and as a general methodological principle for axiomatic-deductive systems in Tarski 1930. According to Tarski 1956 footnote to p. 32, it was known and applied by Tarski since 1921.

PROOF. (b) The student should first review exactly what the hypothesis and conclusion mean under our definition of "$\vdash$". (Each asserts the existence of a certain kind of a finite list of formulas, say $B_1, \ldots, B_l$ and $B'_1, \ldots, B'_p$. There are two differences in the specifications for these two lists. In the first list but not in the second, a formula may be inserted on the ground that it is the mth assumption formula A_m. In the first list the last formula must be B, in the second $A_m \supset B$.)

We must show that, whenever we are given a deduction of B from $A_1, \ldots, A_{m-1}, A_m$ (the "given deduction"), we can find a deduction of $A_m \supset B$ from $A_1, \ldots, A_{m-1}$. There is in fact a uniform method by which, from any such given deduction, we can always find a deduction of $A_m \supset B$ from $A_1, \ldots, A_{m-1}$ (the "resulting deduction"). We describe this method in general terms now, and in Example 6 below we illustrate it.

Say the given deduction is

$$(\alpha) \qquad B_1, \ldots, B_l,$$

so B_l is B. (In Example 6, (α) is the left column, consisting of Formulas 1–5.) As the first step toward constructing the resulting deduction, we prefix to each formula of the given deduction (α) the symbols $A_m \supset$, supplying parentheses as appropriate. Thus we obtain

$$(\beta) \qquad A_m \supset B_1, \ldots, A_m \supset B_l,$$

which does have last the formula $A_m \supset B$ which must be last in the resulting deduction. (In Example 6, (β) is 3′, 8′, 11′, 14′, 17′ in the right column.) But this sequence (β) is not in general a deduction from $A_1, \ldots, A_{m-1}$. However we can insert some additional formulas into it before each one $A_m \supset B_i$ of its formulas ($i = 1, \ldots, l$) so that it will become a deduction (γ) from $A_1, \ldots, A_{m-1}$. For each i, the choice of the formulas to be inserted before $A_m \supset B_i$ depends on the reason given for the inclusion of B_i in the given deduction (α).[34]

CASE 1: B_i is one of the first $m-1$ assumption formulas $A_1, \ldots, A_{m-1}$, which are retained as assumption formulas for the resulting deduction; say B_i is A_j ($j < m$). Then we insert the first two of the following formulas before the third, which is $A_m \supset B_i$.

k'. A_j — jth assumption formula.
$k+1'$. $A_j \supset (A_m \supset A_j)$ — Axiom Schema 1a.
$k+2'$. $A_m \supset A_j$ — modus ponens, k', $k+1'$.

[34] Thus the uniform method applies to the given deduction $B_1, \ldots, B_l$ as a sequence of formulas, *together with* a reason for the inclusion of each formula in it. These reasons we call an *analysis* of a deduction. More than one analysis may be possible; e.g. in Example 6 (where A, B, C are understood to be *any* formulas, not necessarily distinct), if C is the same formula as A, another analysis would justify Formula 3 by Axiom Schema 1a.

(In Example 6, this is illustrated by 1′–3′ with $k' = 1'$, and again by 9′–11′ with $k' = 9'$.)

Case 2: B_i is the last assumption formula A_m, which will not be retained as an assumption formula for the resulting deduction (unless A_m happens to be the same formula as one of $A_1, \ldots, A_{m-1}$). We insert the first four of the formulas in the proof of $A \supset A$ in Example 4 for A_m as the A. (In Example 6, this is illustrated by 4′–8′, A_m being A so the insertions read exactly as in Example 4.)

Case 3: B_i is an axiom. Treated similarly to Case 1. (Not illustrated in Example 6.)

Case 4: B_i comes from two preceding formulas B_g and B_h $(g, h < i)$ by modus ponens. We leave it to the reader as an exercise to work out the treatment of this case, and to supply the insertions 12′, 13′ and 15′, 16′ in Example 6 (Exercise 10.1).

Example 6. To illustrate the *proof* of the deduction theorem, we give: (α) in the left column, a deduction of C from $A \supset (B \supset C)$, B, A; and (γ) in the right column, the deduction of $A \supset C$ from $A \supset (B \supset C)$, B resulting from that given deduction by the uniform method described in the proof of the theorem.

	1′. B — 2nd assumption formula.
	2′. $B \supset (A \supset B)$ — Axiom Schema 1a.
1. B — 2nd ass'n. formula.	3′. $A \supset B$ — modus ponens, 1′, 2′.
	4′. $A \supset (A \supset A)$ — Axiom Schema 1a.
	5′. $\{A \supset (A \supset A)\} \supset \{[A \supset ((A \supset A) \supset A)] \supset [A \supset A]\}$ — Axiom Schema 1b.
	6′. $[A \supset ((A \supset A) \supset A)] \supset [A \supset A]$ — modus ponens, 4′, 5′.
	7′. $A \supset ((A \supset A) \supset A)$ — Ax. Sch. 1a.
2. A — 3rd. ass'n formula.	8′. $A \supset A$ — modus ponens, 7′, 6′.
	9′. $A \supset (B \supset C)$ — 1st assumption formula.
	10′. $\{A \supset (B \supset C)\} \supset \{A \supset (A \supset (B \supset C))\}$ — Axiom Schema 1a.
3. $A \supset (B \supset C)$ — 1st. a. f.	11′. $A \supset (A \supset (B \supset C))$ — m.p., 9′, 10′.
	12′.
	13′.
4. $B \supset C$ — m. p., 2, 3.	14′. $A \supset (B \supset C)$ — m. p., 11′, 13′.
	15′.
	16′.
5. C — modus ponens, 1, 4.	17′. $A \supset C$ — modus ponens, 14′, 16′.

(A simplification applicable to this particular example leads to a shorter deduction of $A \supset C$ from $A \supset (B \supset C)$, B. Indeed, the 7 formulas 1′, 2′, 3′, 9′, 15′, 16′, 17′ suffice in place of the 17 which our uniform method gives.)

Applying the uniform method to 1′–17′ as the given deduction leads to a deduction 1″–53″ of $B \supset (A \supset C)$ from $A \supset (B \supset C)$; and application of the method to this in turn leads to a proof 1‴–161‴ of $(A \supset (B \supset C)) \supset (B \supset (A \supset C))$. (Cf. Exercise 10.2.)

We wrote out the 17-formula deduction of $A \supset C$ from $A \supset (B \supset C)$, B resulting by our method applied to 1–5 in order to help the reader visualize the proof of the deduction theorem. But now that the deduction theorem is proved, we shall be satisfied to use it to infer the *existence* of deductions and proofs without actually constructing them. Thus actual construction of the deduction 1–5 (left column) establishes that $A \supset (B \supset C), B, A \vdash C$. Thence it follows by three successive applications of Theorem 11 that $A \supset (B \supset C), B \vdash A \supset C$, that $A \supset (B \supset C) \vdash B \supset (A \supset C)$, and finally that $\vdash (A \supset (B \supset C)) \supset (B \supset (A \supset C))$, i.e. that there exists a proof of the formula $(A \supset (B \supset C)) \supset (B \supset (A \supset C))$. This satisfies us; we have no interest in actually seeing a proof of that formula, least of all the 161-formula proof resulting by three successive applications of our uniform method to the 5-formula deduction in the left column of Example 6. (There are shorter proofs of $(A \supset (B \supset C)) \supset (B \supset (A \supset C))$. Using the simplification of 1′–17′ noted above before reapplying the uniform method, we get one having 71 formulas.) But while our uniform method in the proof of Theorem 11 may be uneconomical for constructing deductions and proofs, it is efficient for proving Theorem 11; and Theorem 11 itself is very efficient for establishing their existence. We believe the reader right after Example 5 would have found it a fairly difficult exercise to construct (or show the existence of) a proof of $(A \supset (B \supset C)) \supset (B \supset (A \supset C))$.

Changing the letters in the preceding illustration (which we can do since A, B, C were arbitrary formulas, e.g. the B, A, C of the next), $\vdash (B \supset (A \supset C)) \supset (A \supset (B \supset C))$. Now using Axiom Schema 9a and modus ponens twice, $\vdash A \supset (B \supset C) \sim B \supset (A \supset C)$ (cf. *3).

EXAMPLE 7. Similarly, applying the deduction theorem to the result of Example 5, $A \supset (B \supset C) \vdash A \,\&\, B \supset C$ and $\vdash (A \supset (B \supset C)) \supset (A \,\&\, B \supset C)$.

COROLLARY. *If* $A_1, \ldots, A_{m-1}, A_m \vdash B$, *then* $\vdash A_1 \supset (\ldots (A_{m-1} \supset (A_m \supset B)) \ldots)$.

EXERCISES. 10.1. Treat Case 4, and supply 12′, 13′, 15′, 16′ in the right column of Example 6.

10.2. Show that, when the given deduction has *l* formulas, the resulting deduction has $3l + 2$ formulas if A_m is used as such in it; otherwise, $3l$ formulas.

10.3. Show that $A \& B \supset C \vdash A \supset (B \supset C)$ (start by constructing an appropriate deduction) and (using Example 7)
$\vdash A \supset (B \supset C) \sim A \& B \supset C$ (cf. *4a).

10.4. Show that $\vdash (A \supset B) \supset ((B \supset C) \supset (A \supset C))$ (cf. *2).

10.5. Show that: *If* $A_1, \ldots, A_m \vdash B$, *then* $\vdash A_1 \& \ldots \& A_m \supset B$.

§ 11. Proof theory: consistency, introduction and elimination rules. The corollaries to Theorems 10 and 11 accomplish the reduction of the deducibility notion "$A_1, \ldots, A_m \vdash B$" to the provability notion "$\vdash E$" in a manner parallel to the reduction of the notion of valid consequence "$A_1, \ldots, A_m \models B$" to the notion of validity "$\models E$" given in Corollary Theorem 8.

Hence, if we can show that "$\vdash E$" and "$\models E$" are equivalent, we shall then have completed the demonstration of the equivalence of proof theory and model theory for the propositional calculus, both as used in treating absolute logical truths and when applied under assumptions $A_1, \ldots, A_m$. We do this in Theorems 12 and 14.

THEOREM 12. *Each provable formula* E *is valid; using our symbols: if* $\vdash$ E, *then* $\models$ E.

PROOF. By 1a–10b in Theorem 2, each axiom of the propositional calculus is valid. By Theorem 3, given that the premises A and B for an application of modus ponens are valid, so is the conclusion B. Thus, as we construct a proof $B_1, \ldots, B_l$ of E, each of the formulas $B_1, B_2, B_3, \ldots$ successively introduced (either as an axiom, or as a consequence by modus ponens) is valid. Therefore, the last formula B_l, which is E, is valid.

COROLLARY. *For no formula* B *are both* B *and* $\neg$B *provable; using our symbols: for no formula* B *do both* $\vdash$ B *and* $\vdash \neg$B *hold.*

PROOF. Suppose $\vdash$ B and $\vdash \neg$B for some B. Then by the theorem, $\models$ B and $\models \neg$B; i.e. B has all t's in its truth table and so does $\neg$B. This is absurd, since, by the table for $\neg$, if B has all t's then $\neg$B has all f's. —

In general, by a "consistency property" of an axiomatic-deductive system (to be called a "formal system" in § 37) we mean a property that *at most* certain formulas are provable (e.g. only ones having some desired property, or lacking some undesired property). By a "completeness property", we mean a property that *at least* certain formulas are provable (e.g. all having a certain desired property).

Thus Theorem 12 establishes "consistency of the propositional calculus with respect to validity", and its corollary establishes the so-called "simple consistency".

To establish the converse of Theorem 12 (Theorem 14, giving the "completeness of the propositional calculus with respect to validity"). we shall need to develop the proof theory of the propositional calculus for a certain distance. In this development, the deduction theorem (Theorem 11) is a most helpful tool. We begin (in Theorem 13) with a collection of fourteen rules adapted from Gentzen 1934–5, which we call "introductions" and "eliminations" of logical symbols. For completeness, we include the deduction theorem itself as "⊃-introduction", and modus ponens restated using "⊢" as "⊃-elimination". The rest of the rules (with one exception, "weak ¬-elimination") amount essentially to restatements of the axiom schemata in the light of these two rules. To save space, we let "Γ" stand for a list of zero or more formulas, so we can write "Γ, A ⊢ B" for "$A_1, \ldots, A_{m-1}, A_m$ ⊢ B" with $A_m = A$(Γ empty for $m = 1$).

The rule of ¬-introduction (next to the bottom in the left column) corresponds to the informal method of "reductio ad absurdum": to prove "not A" or that A is false, we assume A "for the sake of the argument" and deduce a contradiction B and "not B". This argument can be carried out in the presence of prior assumptions Γ.

The rule of V-elimination corresponds to the informal method of "proof by cases". If we have established "A or B" or are assuming this, then to show that C follows it suffices to show it in two cases, the case that A holds and the case that B holds. (For the problem of deducing C from A V B, we can thus "eliminate" the V, and attempt the deduction from A and from B separately. It is in this sense that we can consider the rule as an elimination rule.)

THEOREM 13. *For any finite list of (zero or more) formulas* Γ, *and any formulas* A, B, C:

	Introduction	Elimination
⊃	*If* Γ, A ⊢ B, *then* Γ ⊢ A ⊃ B.	A, A ⊃ B ⊢ B.
&	A, B ⊢ A & B.	A & B ⊢ A. A & B ⊢ B.
V	A ⊢ A V B. B ⊢ A V B.	*If* Γ, A ⊢ C *and* Γ, B ⊢ C, *then* Γ, A V B ⊢ C. (Proof by cases.)
¬	*If* Γ, A ⊢ B *and* Γ, A ⊢ ¬B, *then* Γ ⊢ ¬A. (Reductio ad absurdum.)	¬¬A ⊢ A. ((Double) negation elimination.)° A, ¬A ⊢ B. (Weak negation elimination.)

$\sim$ A ⊃ B, B ⊃ A ⊢ A ~ B. A ~ B ⊢ A ⊃ B.

A ~ B ⊢ B ⊃ A.

PROOFS. We already have ⊃-introduction in Theorem 11, and ⊃-elim., &-introd., &-elim., V-introd., (double) ¬-elim. and the three ~-rules in Exercise 9.3.

V-elimination (proof by cases).

1. Γ, A ⊢ C – hypothesis.
2. Γ, B ⊢ C – hypothesis.
3. Γ ⊢ A ⊃ C – ⊃-introd. (the deduction theorem), 1.
4. Γ ⊢ B ⊃ C – ⊃-introd. (the deduction theorem), 2.
5. A ⊃ C, B ⊃ C, A V B ⊢ C – using Axiom Schema 6, and modus ponens thrice. (That is, we can construct the following deduction: 1′. A ⊃ C. 2′. (A ⊃ C) ⊃ ((B ⊃ C) ⊃ (A ⊃ B ⊃ C)).
   3′. (B ⊃ C) ⊃ (A V B ⊃ C). 4′. B ⊃ C. 5′. A V B ⊃ C. 6′. A V B.
   7′. C. The student should supply the reasons.)
6. Γ, A V B ⊢ C – combining 3, 4, 5 by Theorem 9. (For,
   Γ, A V B ⊢ A ⊃ C [using 3; cf. Exercise 11.1],
   Γ, A V B ⊢ B ⊃ C [using 4],
   Γ, A V B ⊢ A V B [Th. 9 (i)] and A ⊃ C, B ⊃ C, A V B ⊢ C [by 5];
   so Γ, A V B ⊢ C [by Theorem 9 (ii) with $m = 2, p = 3$.])

Weak ¬-elimination.

1. A, ¬A, ¬B ⊢ A – (Theorem 9 (i)).
2. A, ¬A, ¬B ⊢ ¬A – (Theorem 9 (i)).
3. A, ¬A ⊢ ¬¬B ⊢ B – ¬-introd., 1, 2; ¬-elim.[35]

By this rule, from a contradiction A, ¬A, any formula B can be deduced. The idea of the proof of it just given is to deduce a contradiction from A, ¬A, ¬B and blame the contradiction on ¬B.

EXERCISES. 11.1. Infer "Γ, A V B ⊢ A ⊃ C" from 3 in the proof of V-elimination by two methods: directly from the definition of "deduction"; by use of Theorem 9 (as in Exercise 9.4 (b) and (d)).

11.2. Prove the rule of ¬-introduction (reductio ad absurbum).

§ 12. Proof theory: completeness. We shall prove the completeness of the propositional calculus by a method due to Kalmár 1934–5. In preparation, we first establish four lemmas.

LEMMA 1. *To each entry (or line) in each of the five basic truth tables for the propositional calculus in* § 2, *a corresponding deducibility relationship*

[35] "Γ ⊢ A_1 ⊢ A_2" is an abbreviation for "Γ ⊢ A_1 and A_1 ⊢ A_2", from which "Γ ⊢ A_2" follows by Theorem 9 (ii). Cf. Exercise 9.4 (a). Similarly with longer chains. (We use only one formula *after* each "⊢".)

holds. For example, the tables for ⊃ and ¬ follow at the left, the corresponding deducibility relationships at the right.

A	B	A ⊃ B		
t	t	t	A, B ⊢ A ⊃ B	(1)
t	f	f	A, ¬B ⊢ ¬(A ⊃ B)	(2)
f	t	t	¬A, B ⊢ A ⊃ B	(3)
f	f	t	¬A, ¬B ⊢ A ⊃ B	(4)

A	¬A		
t	f	A ⊢ ¬¬A	(5)
f	t	¬A ⊢ ¬A	(6)

(*Altogether*, 18 *deducibility relationships are asserted by this lemma.*)

Proofs. For illustration we establish three of the four relationships above for A ⊃ B.

(1) 1. A, B, A ⊢ B — (Theorem 9 (i)).
2. A, B ⊢ A ⊃ B — ⊃-introd. (the deduction theorem), 1.

(2) 1. A, ¬B, A ⊃ B ⊢ B — ⊃-elim. (and Th. 9; cf. Ex. 9.4 (d)).
2. A, ¬B, A ⊃ B ⊢ ¬B — (Theorem 9 (i)).
3. A, ¬B ⊢ ¬(A ⊃ B) — ¬-introd., 1, 2.

(4) 1. ¬A, ¬B, A ⊢ B — weak ¬-elim. (and Theorem 9).
2. ¬A, ¬B ⊢ A ⊃ B — ⊃-introd. (the deduction theorem), 1.

Lemma 2. *Consider the truth table for any formula* E *containing* (*at most*) *the atoms* $P_1, \ldots, P_n$. *To each of the* 2^n *entries* (*or lines*) *in this table, a corresponding deducibility relationship holds.* For example, let E be P ⊃ (Q ∨ R ⊃ (R ⊃ ¬P)). Corresponding to the f in Line 3 of its truth table in § 2, the lemma asserts that

$$P, \neg Q, R \vdash \neg\{P \supset (Q \vee R \supset (R \supset \neg P))\}.$$

Proof. We explain the method using the illustration. Corresponding to the computation step from f, t for Q, R to t for Q ∨ R, Lemma 1 gives ¬Q, R ⊢ Q ∨ R, whence obviously (or by Theorem 9):

1. P, ¬Q, R ⊢ Q ∨ R.

Corresponding to the computation step from t for P to f for ¬P, (5) in Lemma 1 gives P ⊢ ¬¬P, whence:

2. P, ¬Q, R ⊢ ¬¬P.

Corresponding to the computation step from t, f for R, ¬P to f for

$R \supset \neg P$, (2) in Lemma 1 (with R, $\neg P$ as the A, B) gives $R, \neg\neg P \vdash \neg(R \supset \neg P)$; combining this with 2 by Theorem 9:
3. $P, \neg Q, R \vdash \neg(R \supset \neg P)$.
Continuing in this manner, we obtain successively $P, \neg Q, R \vdash D$ or $P, \neg Q, R \vdash \neg D$ for each formula part D of E, according as that part receives the value t or $\mathfrak{f}$ in the computation for the assignment of t, $\mathfrak{f}$, t to P, Q, R. At the end, since the whole E receives the value $\mathfrak{f}$, we thus have:
5. $P, \neg Q, R \vdash \neg\{P \supset (Q \vee R) \supset (R \supset \neg P))\}$.
This completes our illustration of the proof of Lemma 2.[36]

It may be instructive to view in a diagram with two "trees" how each computation step (horizontal line in the left tree) corresponds to a deducibility relationship of Lemma 1 (horizontal line in the right tree). The left tree is the computation given in § 2, reproduced omitting repetitions in the writing of values. By Theorem 9, it follows that in the right tree each formula is deducible from the distinct formulas occurring at the tops of branches over it (or any larger set of formulas).

$P \supset (Q \vee R \supset (R \supset \neg P))$

$$\dfrac{\text{t} \qquad \dfrac{\dfrac{\mathfrak{f} \quad \text{t}}{\text{t}} \qquad \dfrac{\text{t} \quad \dfrac{\text{t}}{\mathfrak{f}}}{\mathfrak{f}}}{\mathfrak{f}}}{\mathfrak{f}}$$

$P \supset (\ Q \vee R \supset \ (R \supset \neg P))$

$$\dfrac{P \qquad \dfrac{\dfrac{\neg Q \quad R}{Q \vee R} \qquad \dfrac{R \quad \dfrac{P}{\neg\neg P}}{\neg(R \supset \neg P)}}{\neg(\ Q \vee R \supset \ (R \supset \neg P))}}{\neg\{P \supset \ (\ Q \vee R \supset \ (R \supset \neg P))\}}$$

LEMMA 3. *If the formula* E *in Lemma* 2 *is valid* (*i.e.* $\models E$), *then* $P_1 \vee \neg P_1, \ldots, P_n \vee \neg P_n \vdash E$.

PROOF. E.g. take $n = 2$. Then by Lemma 2 with the present hypothesis:

$P_1, \quad P_2 \vdash E$.
$P_1, \neg P_2 \vdash E$.
$\neg P_1, \quad P_2 \vdash E$.
$\neg P_1, \neg P_2 \vdash E$.

By two applications of $\vee$-elimination:

$P_1, P_2 \vee \neg P_2 \vdash E$.
$\neg P_1, P_2 \vee \neg P_2 \vdash E$.

[36] Of course, the verification of one case of a general theorem (or lemma) does not prove the theorem (or lemma). The proof consists in the fact that the method used in the illustration is general, i.e. applies to all cases. The illustration illustrates a pattern of treatment, applicable to all cases. When this fact should be obvious, we may omit stating the proof in general terms, as we do now.

By a third application of ∨-elimination:

$P_1 \vee \neg P_1, P_2 \vee \neg P_2 \vdash E$.

LEMMA 4°. *For each formula* A: $\vdash A \vee \neg A$. (The law of the excluded middle; cf. *51 in Theorem 2.)

We leave the proof as an exercise (Exercise 12.2).[37]

THEOREM 14°. *Each valid formula* E *is provable; using our symbols: if* $\models E$, *then* $\vdash E$.

PROOF. By Lemma 3 with Lemma 4 and Theorem 9 (ii) for $m = 0$.

This completes the proof of the equivalence of proof theory and model theory for the propositional calculus. Some of the results obtained in model theory we had to develop independently in proof theory before establishing the equivalence. However, we can now take over with "⊢" replacing "⊨" any result we established in our model-theoretic treatment of the propositional calculus (§§ 2–8).[38] For example, all the results in Theorem 2 hold with "⊢" in place of "⊨"; before we had Theorem 14, we had this explicitly only for 1a–10b (because we took them as axiom schemata), *1 (by Example 4 in § 9), *2, *3, *4a (end § 10), *19 (Exercise 9.1) and *51 (Lemma 3).

Natural as the development of the propositional calculus by truth tables (model theory) seems now, it was actually the more recent approach to be fully exploited, by Post, who first proved Theorems 12 and 14 in 1921, and by Łukasiewicz in 1921, although some of the development goes back to Frege 1879 and Peirce 1885. Although an algebra of logic was initiated by Boole 1847 and De Morgan 1847, the proof theory of the propositional calculus properly appeared with Frege's "Begriffschrift" in 1879, and in Russell's work, especially in the "Principia Mathematica" of Whitehead and Russell 1910–13.[39]

OTHER PROPOSITIONAL CALCULI. As illustration of our remark in § 2 that there are different systems of logic, we mention that Łukasiewicz

[37] Although a proof can be given at this stage in six short lines (IM p. 120 *51), it is a bit tricky to find. Later (end § 13) it will be easy. We refrain from giving it here in order not to spoil the fun the student may have in trying it himself. (A proof is also implicit in our demonstration of *51 ($\models A \vee \neg A$) by duality, following Theorem 7 in § 6; the ingredients of that proof are easily developed in proof theory.)

[38] These replacements *in simple contexts* constitute applications *in the observer's language* of the replacement rule which as applied to the object language we established in Corollary Theorem 5. For now we have established (with $m \geq 0$) "$A_1, \ldots, A_m \models B$ if and only if $A_1, \ldots, A_m \vdash B$", and we can use this in the role of the $A \sim B$ of Corollary Theorem 5.

[39] See Church 1956 pp. 155 ff.

in 1920 introduced a *3-valued propositional calculus*, in which the model theory is given using three truth values instead of just the two t and f. Post in 1921 (independently of Łukasiewicz) generalized from the classical (= 2-valued) propositional calculus to *n-valued propositional calculus* for each positive integer $n \geq 2$. To what extent n-valued logics for $n > 2$ are a tour de force is moot.[40]

Modal propositional calculi deal with such notions as "A is necessary" (in symbols, $\Box$A) and "A is possible" (in symbols, $\Diamond$A or equivalently $\neg\Box\neg$A). These notions enter in domains of thinking where there are understood to be two different kinds of "truth", one more universal or compelling than the other. For example, it is impossible that $2+2 = 5$ (it is contrary to mathematical laws); but it is possible that there is a large continent in the middle of the Pacific ocean (it is contrary only to the geographical facts). A zoologist might declare that it is impossible that salamanders or any other living creatures can survive fire; but possible (though untrue) that unicorns exist, and possible (though improbable) that abominable snowmen exist. Modern treatments of modal logic begin with Lewis 1912, 1917 and Lewis and Langford 1932, with some anticipation by MacColl 1896–7.[41]

Another example of a nonclassical propositional calculus is the *intuitionistic propositional calculus*, in which the law of the excluded middle A $\vee$ $\neg$A and the law of double negation $\neg\neg$A $\supset$ A are not affirmed. The standpoint from which the intuitionistic system of logic arose and is of interest will be considered later (in § 36). We do not attempt here a model-theoretic description of the intuitionistic propositional calculus. A proof-theoretic formulation is obtained by replacing our Axiom Schema 8 ($\neg\neg$A $\supset$ A) by: 8^{I}. $\neg$A $\supset$ (A $\supset$ B). Since any axiom by this schema is provable in the classical propositional calculus (cf. *10a), the intuitionistic propositional calculus is a *subsystem* of the classical, i.e. all formulas provable in the intuitionistic system are provable in the classical. Those of our officially stated results involving "$\vdash$" (including ones we first stated with "$\models$") which are not readily established for the intuitionistic system also are identified by "°".[42]

EXERCISES. 12.1. Establish the first two deducibility relationships of Lemma 1 for & and the last two for $\vee$.

12.2*. Prove Lemma 4.[37]

[40] See Rosser and Turquette 1952.

[41] See von Wright 1951, Feys 1965.

[42] In fact, the results so marked in this book (and in IM) do not hold for the intuitionistic system; but this we are in no position to prove now. A method is suggested in § 54 below which suffices for all such results in this book except Theorem 27, and the actual proofs are in IM § 80.

12.3. True or false (and why)?
(a) "For each formula A: if $\vdash \neg$A, then not $\vdash$ A."
(b) "For each formula A: if not $\vdash$ A, then $\vdash \neg$A."
(c) "For each formulas A and B: if $\vdash$ A $\vee$ B, then $\vdash$ A or $\vdash$ B."
(d) "For each formulas A and B: if $\vdash$ A or $\vdash$ B, then $\vdash$ A $\vee$ B."

12.4°*. Consider any unprovable formula, e.g. P $\supset$ Q. Adjoin a corresponding axiom schema A $\supset$ B to our present list 1a–10b of axiom schemata. Show that in the resulting system, every formula is provable. (Post completeness, 1921; IM p. 134 Corollary 2.)

12.5°*. Show that exactly the same formulas are provable in the system described in Footnote 29 § 9 with four axiom schemata (after allowing for &, $\vee$, $\sim$ being symbols of abbreviation rather than "primitive" symbols) as in our system.

12.6. Establish weak $\neg$-elimination (in Theorem 13) for the intuitionistic propositional calculus. (The other twelve rules of Theorem 13 *without* (double) $\neg$-elimination hold for the intuitionistic propositional calculus by the same proofs as we gave above.)

12.7.[23] Show that: $\{\models \neg(A_1 \,\&\ldots\&\, A_m)\} \equiv \{A_1, \ldots, A_m$ are not simultaneously t for any assignment$\} \equiv \{$for some formula B, both $A_1, \ldots, A_m \vdash B$ and $A_1, \ldots, A_m \vdash \neg B\} \equiv \{A_1, \ldots, A_m \vdash B$, for every formula B$\}$. Hence, if **S** is the axiomatic-deductive system obtained from the propositional calculus by adding $A_1, \ldots, A_m$ as axioms: $\{A_1 \,\&\ldots\&\, A_m$ is consistent, § 8$\} \equiv \{$"$A_1, \ldots, A_m$ are consistent"$\} \equiv \{$**S** is simply consistent$\} \equiv \{$not every formula is provable in **S**$\}$.

§ 13. Proof theory: use of derived rules.

In model theory we gave one answer to the question "What formulas E shall hold in (classical) propositional logic?": those which have only t's in their truth tables, or in symbols for which $\models$ E. In proof theory we gave another answer: those for which there are formal proofs using just axioms by Axiom Schemata 1a–10b and modus ponens, or in symbols for which $\vdash$ E. By §§ 11, 12 we know the two answers are equivalent.

Similarly we gave two equivalent answers to the question "What formulas B shall follow by (classical) propositional logic from a given list of formulas $A_1, \ldots, A_m$?": those for which $A_1, \ldots, A_m \models B$ (model theory), those for which $A_1, \ldots, A_m \vdash B$ (proof theory).

Neither in model theory nor in proof theory did we stop with these answers. Instead we derived various properties of "$\models$" or "$\vdash$", which often are easier to use than direct applications of their definitions. In this section, we shall elaborate on the use of such results, especially the "introduction and elimination rules" of Theorem 13. We shall give the discussion in proof theory (with "$\vdash$"), though we could do so in model theory (cf. Exercises 7.4 (a), 7.6, 7.7).

We call the rules of Theorem 13 and similar results *derived rules*. For they are results which we have "derived" about the axiomatic-deductive system, after establishing it by choosing (or "postulating") the primitive or "postulated" rule of inference (modus ponens) and the axiom schemata (Axiom Schemata 1a–10b).

In using such rules, and generally in demonstrating the existence of deductions, we often find it convenient to write down lists of formulas that we can successively recognize as deducible from given assumption formulas $A_1, \ldots, A_m$. Apart from deductions themselves, one such list of formulas is the four formulas $A \supset C$, $B \supset C$, $A \vee B$, C written after "$\Gamma, A \vee B \vdash$" in the explanation of 6 in the proof of $\vee$-elimination for Theorem 13. Another is in:

EXAMPLE 8. With A & B as the assumption formula, we can use &-elimination and &-introduction to construct the following list of formulas 1–4 deducible from A & B (left column).

1. A & B — assumption formula.	1′. A & B — assumption formula.
2. A — &-elim., 1.	2′. $A \& B \supset A$ — Ax. Sch. 4a.
3. B — &-elim., 1.	3′. A — modus ponens, 1′, 2′.
4. B & A — &-introd., 3, 2.	4′. $A \& B \supset B$ — Ax. Sch. 4b.
	5′. B — modus ponens, 1′, 4′.
	6′. $B \supset (A \supset B \& A)$ — Ax. Sch. 3.
	7′. $A \supset B \& A$ — m. p., 5′, 6′.
	8′. B & A — modus ponens, 3′, 7′.

Thence we can conclude that $A \& B \vdash B \& A$, i.e. that there exists a deduction of B & A from A & B. This list of formulas 1–4 is not itself a deduction of B & A from A & B, since it does not *exactly* fit the definition of deduction in § 9. The list 1′–8′ in the right column is a deduction of B & A from A & B.

In Example 8 the use of &-elim. (two rules) and &-introd. provides a slight abbreviation compared to constructing the deduction itself; in effect these rules provide certain prefabricated units. These three rules and eight others in Theorem 13 have the form "$B \vdash C$" or "$B_1, B_2 \vdash C$". They assert that we can construct a deduction leading "directly" from B (or B_1, B_2) to C; so we call them *direct rules* (and likewise rules "$B_1, \ldots, B_p \vdash C$" for any $p \geq 0$).

The other three rules in Theorem 13 ($\supset$-introd., $\vee$-elim., $\neg$-introd.) enable us, *from* the existence of one or two "given deductions" or "subsidiary deductions", to infer the existence of another deduction (the "resulting deduction"); so we call them *subsidiary deduction rules* (and likewise with any number $s \geq 1$ of subsidiary deductions). Their use is illustrated for $\supset$-introd. by Examples 6 and 7 in § 10. Here the saving compared to actual construction of deductions is more impressive.

Theorem 9 gives two general principles concerning the construction of lists of formulas recognized successively as deducible from given assumption formulas $A_1, \ldots, A_m$. By (i), each of $A_1, \ldots, A_m$ itself can be put into the list. By (ii), if C is deducible from any formulas $B_1, \ldots, B_p$ already in the list, then C can be put into the list.

In Example 8 (the left column), we used (i) at Step 1, and (ii) with $p = 2$ at Step 4.

As a special case of (ii) (with (i)): If $B_1, \ldots, B_p \vdash C$, and each of $B_1, \ldots, B_p$ is one of $A_1, \ldots, A_m$, then $A_1, \ldots, A_m \vdash C$. (For then by (i), $A_1, \ldots, A_m \vdash B_i$ for $i = 1, \ldots, p$.) Thus any formula C deducible from a given list of assumption formulas $B_1, \ldots, B_p$ is deducible from any list $A_1, \ldots, A_m$ which includes all of $B_1, \ldots, B_p$ and maybe additional formulas. Using this for $p = 0$, any valid formula C is deducible from any list of assumption formulas $A_1, \ldots, A_m$. Using the same principle in both directions, whether $A_1, \ldots, A_m \vdash C$ holds depends only on what formulas occur in the list $A_1, \ldots, A_m$ with or without repetitions.

These consequences of Theorem 9 are also obvious enough directly from the definition of "$\vdash$". We call Theorem 9 and its consequences "*general* properties of $\vdash$", because these properties are independent of the particular list of "postulates" we chose (Axiom Schemata 1a–10b, the $\supset$-rule). The student should become facile in using them, either directly from the meaning of "$\vdash$", or by using Theorem 9.

In the subsidiary deduction rules of Theorem 13, the list of assumption formulas for the (or each) subsidiary deduction differs from that for the resulting deduction. So the use of these rules cannot simply consist in listing formulas successively recognized as deducible from one set of assumption formulas $A_1, \ldots, A_m$. We could construct several such lists, for different sets of assumption formulas. An alternative, which we shall illustrate presently, is to construct one list of deducible formulas, but alter from place to place in the list the set of the assumption formulas which are "in force", i.e. from which the formulas are being claimed to be deducible.

EXAMPLE 9. The following list (A) of 19 statements is a reproof of *55a of Theorem 2 with "$\vdash$" in place of "$\models$". (Of course we already had the result by Theorems 2 + 14; cf. end § 12.) The student should have no trouble checking (A) step by step. (Here we are no longer mentioning Theorem 9 for each use of general properties of $\vdash$, as we did in §§ 11, 12.)

1. $\neg(A \vee B), A \vdash A \vee B$ — $\vee$-introd.
2. $\neg(A \vee B), A \vdash \neg(A \vee B)$.
3. $\neg(A \vee B) \vdash \neg A$ — $\neg$-introd., 1, 2.
4. $\neg(A \vee B), B \vdash A \vee B$ — $\vee$-introd.
5. $\neg(A \vee B), B \vdash \neg(A \vee B)$.

6. $\neg(A \vee B) \vdash \neg B$ — $\neg$-introd., 4, 5.
7. $\neg(A \vee B) \vdash \neg A \,\&\, \neg B$ — &-introd., 3, 6.
8. $\vdash \neg(A \vee B) \supset \neg A \,\&\, \neg B$ — $\supset$-introd., 7.
9. $\neg A \,\&\, \neg B, A \vdash A$.

(A) 10. $\neg A \,\&\, \neg B, A \vdash \neg A$ — &-elim.
11. $\neg A \,\&\, \neg B, A \vdash \neg(A \vee B)$ — weak $\neg$-elim., 9, 10.
12. $\neg A \,\&\, \neg B, B \vdash B$.
13. $\neg A \,\&\, \neg B, B \vdash \neg B$ — &-elim.
14. $\neg A \,\&\, \neg B, B \vdash \neg(A \vee B)$ — weak $\neg$-elim., 12, 13.
15. $\neg A \,\&\, \neg B, A \vee B \vdash \neg(A \vee B)$ — $\vee$-elim., 11, 14.
16. $\neg A \,\&\, \neg B, A \vee B \vdash A \vee B$.
17. $\neg A \,\&\, \neg B \vdash \neg(A \vee B)$ — $\neg$-introd., 16, 15.
18. $\vdash \neg A \,\&\, \neg B \supset \neg(A \vee B)$ — $\supset$-introd., 17.
19. $\vdash \neg(A \vee B) \sim \neg A \,\&\, \neg B$ — $\sim$-introd., 8, 18.

However our purpose is not just that the student should be able to follow such (informal) proofs, but that he should be able to discover them for himself. So we now review how we are led to this proof. We are to establish an equivalence $\neg(A \vee B) \sim \neg A \,\&\, \neg B$ (Line 19). Looking at the rules provided in Theorem 13, the obvious choice for our purpose is $\sim$-introd. This requires us first to establish two implications (Lines 8 and 18). To establish the first of these $\neg(A \vee B) \supset \neg A \,\&\, \neg B$ (Line 8), the obvious method is $\supset$-introd. (the deduction theorem), which requires us to get Line 7, i.e. to show the deducibility of $\neg A \,\&\, \neg B$ from $\neg(A \vee B)$. By &-introd., it will suffice to deduce $\neg A$ and $\neg B$ separately. So this part of our problem becomes to get Lines 3 and 6. For Line 3, to deduce $\neg A$ the obvious method is $\neg$-introd. (reductio ad absurdum), for which we add A to the assumption $\neg(A \vee B)$ we already have, and attempt to deduce a contradiction, i.e. to obtain $\neg(A \vee B), A \vdash C$ and $\neg(A \vee B), A \vdash \neg C$ for some formula C. (These were written "$\Gamma, A \vdash B$" and "$\Gamma, A \vdash \neg B$" in the statement of $\neg$-introd. in Theorem 13; but the B of the rule need not be the present B, so we write it "C".) It isn't hard to see that we can do this, if we pick for C the formula $A \vee B$ (Lines 1 and 2). Similarly, Lines 4 and 5 suffice for Line 6. Now we must pick up the loose end at Line 18. The student should try to retrace how we discovered Lines 9–17 supporting Line 18. This isn't quite as straightforward. To get Line 17, obviously we aim to get $\neg A \,\&\, \neg B, A \vee B \vdash C$ and $\neg A \,\&\, \neg B, A \vee B \vdash \neg C$ for some C. It may not be immediately evident what to pick for C, but it should be evident that we must utilize $A \vee B$; and $\vee$-elim. (proof by cases) is the rule by which we can hope to do so. So we make two "cases" (corresponding to the "$\Gamma, A \vdash C$" and "$\Gamma, B \vdash C$" of $\vee$-elim.), i.e. we first replace $A \vee B$ as assumption formula by A (and later by B), and investigate what we can then deduce. Once we have a contradiction within either case, by weak ¬-elim. we can deduce anything else we please within that case.

We numbered (A) in the logical order 1–19, in which the statements are inferred (or checked); but as we have just seen, we discover them by working upward from the bottom more or less (say in the order 19, 8, 18, 7, 3, 6, 1, 2, 4, 5, 17, . . .).

The student should practice finding such informal proofs of results about formal provability and deducibility, working up from the bottom. After some practice, he should become fairly adept at looking ahead.

Now we give essentially the same proof in a more condensed format (B_1). Here we omit writing the symbol "⊢". When we wish to employ a formula A as assumption formula, we say simply "Assume A". This means that A is being added to the list of assumption formulas for the construction of deductions. In this example, all the assumption formulas are introduced preparatory to an application of a subsidiary deduction rule (⊃-introd., ∨-elim. or ¬-introd.). When that application occurs, the assumption is "discharged", i.e. no longer remains "in force". The student is expected to be sufficiently familiar with these rules to know when an assumption is thus discharged. To help in the discussion, we put small numbers before the formulas as they are successively introduced, i.e. either assumed or inferred. We number separately reintroductions of a formula under new sets of assumptions.

I. Assume for ⊃-introd., ${}_1\neg(A \vee B)$. Assume for ¬-introd., ${}_2A$. Thence by ∨-introd., ${}_3A \vee B$, contradicting $\neg(A \vee B)$. By the planned ¬-introd., ${}_4\neg A$. Assume for another ¬-introd., ${}_5B$. By ∨-introd., ${}_6A \vee B$, contradicting $\neg(A \vee B)$. By the ¬-introd., ${}_7\neg B$. By &-introd., ${}_8\neg A \& \neg B$. By the ⊃-introd., ${}_9\neg(A \vee B) \supset \neg A \& \neg B$.

II. Assume for ⊃-introd., ${}_{10}\neg A \& \neg B$. Assume for ¬-introd.,
(B_1) ${}_{11}A \vee B$. CASE 1: ${}_{12}A$. From $\neg A \& \neg B$ by &-elim., ${}_{13}\neg A$, contradicting the case hypothesis. By weak ¬-elim., ${}_{14}\neg(A \vee B)$. CASE 2: ${}_{15}B$. From $\neg A \& \neg B$ by &-elim., ${}_{16}\neg B$. Again by weak ¬-elim., ${}_{17}\neg(A \vee B)$. — By the cases (∨-elim.), ${}_{18}\neg(A \vee B)$, contradicting $A \vee B$. By the ¬-introd., ${}_{19}\neg(A \vee B)$. By the ⊃-introd., ${}_{20}\neg A \& \neg B \supset \neg(A \vee B)$.

From I and II by ∼-introd., ${}_{21}\neg(A \vee B) \sim \neg A \& \neg B$.

We discover (B_1) in essentially the same way as (A), but we have "written it up" differently. Thus, we know we have to prove two implications to get the desired equivalence (by ∼-introd.); and we number as "I" and "II" the work we do for these respective implications. For the first, we shall use ⊃-introd., so we assume its antecedent $\neg(A \vee B)$. We look ahead to deducing $\neg A \& \neg B$ by &-introd. from $\neg A$ and $\neg B$; and toward proving the first of these by $\neg$-introd. we assume A; etc. Naturally, one does not always succeed in writing out such an informal proof consecutively without a little exploratory scratch work.

We intend (B_1) as an abbreviated, and very convenient, way of presenting a series of applications of the rules of Theorem 13 (with Theorem 9).

To be sure there is no error in this, we should be able to translate (B_1) into explicit applications of the rules of Theorem 13. We now do this, using the idea of a list of formulas successively recognized as deducible, where from place to place in the list the set of assumption formulas from which the formulas are being claimed to be deducible may be changed.

First, we write the 21 numbered formulas in (B_1) in a list (B_2), as follows.

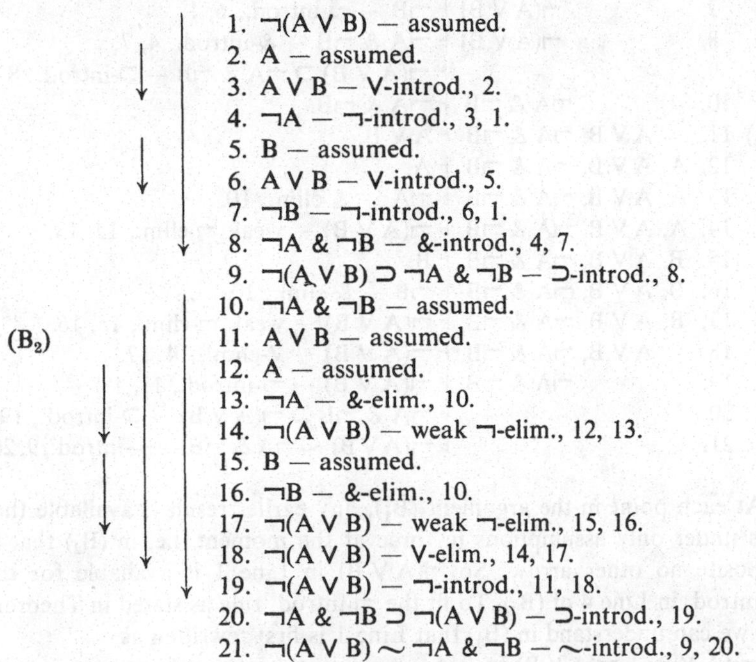

(B_2)

1. $\neg(A \vee B)$ — assumed.
2. A — assumed.
3. $A \vee B$ — $\vee$-introd., 2.
4. $\neg A$ — $\neg$-introd., 3, 1.
5. B — assumed.
6. $A \vee B$ — $\vee$-introd., 5.
7. $\neg B$ — $\neg$-introd., 6, 1.
8. $\neg A \;\&\; \neg B$ — &-introd., 4, 7.
9. $\neg(A \vee B) \supset \neg A \;\&\; \neg B$ — $\supset$-introd., 8.
10. $\neg A \;\&\; \neg B$ — assumed.
11. $A \vee B$ — assumed.
12. A — assumed.
13. $\neg A$ — &-elim., 10.
14. $\neg(A \vee B)$ — weak $\neg$-elim., 12, 13.
15. B — assumed.
16. $\neg B$ — &-elim., 10.
17. $\neg(A \vee B)$ — weak $\neg$-elim., 15, 16.
18. $\neg(A \vee B)$ — $\vee$-elim., 14, 17.
19. $\neg(A \vee B)$ — $\neg$-introd., 11, 18.
20. $\neg A \;\&\; \neg B \supset \neg(A \vee B)$ — $\supset$-introd., 19.
21. $\neg(A \vee B) \sim \neg A \;\&\; \neg B$ — $\sim$-introd., 9, 20.

The arrows at the left show exactly how long each assumption formula remains in force. Thus, since $_{1}\neg(A \vee B)$ is introduced in (B_1) in preparation for the $\supset$-elim. which takes place (discharging it) with result $_{9}\neg(A \vee B) \supset \neg A \;\&\; \neg B$, the arrow in (B_2) beginning at Formula 1 ends at Formula 8, the last one for which $_{1}\neg(A \vee B)$ is intended as an assumption.

Now, we replace the arrows by writing in the respective assumption formulas themselves followed by the symbol "$\vdash$". The numbers are moved out of the way to the left (and "— assumed" is omitted). Thus (B_2) becomes

(B_3), and we are back to essentially the same format as (A). Each line in (B_3) reads correctly as an application of a rule of Theorem 13 (with Theorem 9) or simply of (i) of Theorem 9. From the way we have arrived at (B_3) from (B_1) via (B_2), slight differences have resulted in the exact list of statements compared to (A). We note two of these differences.

(B_3)

1. ¬(A ∨ B) ⊢ ¬(A ∨ B).
2. A, ¬(A ∨ B) ⊢ A.
3. A, ¬(A ∨ B) ⊢ A ∨ B — ∨-introd., 2.
4. ¬(A ∨ B) ⊢ ¬A — ¬-introd., 3, 1.
5. B, ¬(A ∨ B) ⊢ B.
6. B, ¬(A ∨ B) ⊢ A ∨ B — ∨-introd., 5.
7. ¬(A ∨ B) ⊢ ¬B — ¬-introd., 6, 1.
8. ¬(A ∨ B) ⊢ ¬A & ¬B — &-introd., 4, 7.
9. ⊢ ¬(A ∨ B) ⊃ ¬A & ¬B — ⊃-introd., 8.
10. ¬A & ¬B ⊢ ¬A & ¬B.
11. A ∨ B, ¬A & ¬B ⊢ A ∨ B.
12. A, A ∨ B, ¬A & ¬B ⊢ A.
13. A, A ∨ B, ¬A & ¬B ⊢ ¬A — &-elim., 10.
14. A, A ∨ B, ¬A & ¬B ⊢ ¬(A ∨ B) — weak ¬-elim., 12, 13.
15. B, A ∨ B, ¬A & ¬B ⊢ B.
16. B, A ∨ B, ¬A & ¬B ⊢ ¬B — &-elim., 10.
17. B, A ∨ B, ¬A & ¬B ⊢ ¬(A ∨ B) — weak ¬-elim., 15, 16.
18. A ∨ B, ¬A & ¬B ⊢ ¬(A ∨ B) — ∨-elim., 14, 17.
19. ¬A & ¬B ⊢ ¬(A ∨ B) — ¬-introd., 11, 18.
20. ⊢ ¬A & ¬B ⊃ ¬(A ∨ B) — ⊃-introd., 19.
21. ⊢ ¬(A ∨ B) ∼ ¬A & ¬B — ∼-introd., 9, 20.

At each point in the argument (B_1), any earlier result is available that was under only assumptions in force at the moment, i.e. in (B_2) that is opposite no other arrow. So ¬(A ∨ B) in Line 1 is available for the ¬-introd. in Line 4 of (B_2). To fit the ¬-introd. rule as stated in Theorem 13, we can understand in (B_3) that Line 1 is first rewritten as A, ¬(A ∨ B) ⊢ ¬(A ∨ B) by general properties of ⊢ before the ¬-introd. (By general properties of ⊢, the changed order of assumptions compared to the statement in Theorem 13 is immaterial.)

In (B_1), A ∨ B naturally comes to mind as an assumption at Step 11, preparatory to the ¬-introd. at Step 19. We could suspend the assumption of A ∨ B during the two case arguments (Lines 12–14, and Lines 15–17, in $(B)_2$). However it is simpler (and does no harm) to keep it in force until it is finally discharged. Thereby we can represent the duration of each assumption in (B_2) by a single arrow. Then the ∨-elim. for Line 18 gives directly A V B, A V B, ¬A & ¬B ⊢ ¬(A V B); but this simplifies at once by general properties of ⊢ to A V B, ¬A & ¬B ⊢ ¬(A V B). —

This concludes Example 9. We believe the student, after he has a good grasp of Theorem 13 (and the general properties of ⊢) will find (B_1) used flexibly the easiest format. In using this he should have a clear mental picture of a (B_2), i.e. of just how long each of his assumptions remains in force (and of when formulas are reintroduced under new circumstances), whence automatically a (B_3) can be written that will consist step by step of correct applications of Theorem 13 (with general properties of ⊢). In any case of doubt, he should write out (B_3) or (A).

In tackling a new logical problem, the student should of course use any previously established results that are available (except here, where to give examples and exercises we are asking the student to forget for the moment that all the results of Theorem 2 with "⊢" were established at the end of § 12).

Not all of the results in Theorem 2 can be proved by straightforward use of the rules of Theorem 13 as illustrated in Example 9. All those not marked with "°" (and some which are) can be.

Example 10. For *55b with "⊢", the implication ¬A V ¬B ⊃ ¬(A & B) is easy. But ¬(A & B) ⊃ ¬A V ¬B requires a trick. After assuming ¬(A & B), we undertake to deduce ¬¬(¬A V ¬B), from which by (double) ¬-elim. we will be able to pass to ¬A V ¬B. So for ¬-introd. we further assume ¬(¬A V ¬B). Thence by the result (or the argument) in Example 9 (*55a), we can deduce ¬¬A & ¬¬B, and the rest is straightforward.

In contrast to the "direct methods" employed in Example 9, here at the crucial point we use the "indirect method" which consists in obtaining a contradiction from the negation of what we want.

Rather than handling *55b etc. as above (except for practice), we might as well now add the replacement theorem (Theorem 5) and the method of chains of equivalences to our tools in proof theory. Here we can do so on the basis of having them in model theory with "⊨" (§§ 4, 5), and the equivalence of ⊨ to ⊢ (§§ 11, 12). With these, as we have already seen in § 5, all the rest of *55a–*61 can be established, once we have *49, *55a and *55c.

In particular (supposing *49 was established meanwhile):
⊢ ¬A V ¬B ∼ ¬¬(¬A V ¬B) [*49] ∼ ¬(¬¬A & ¬¬B) [*55a] ∼ ¬(A & B) [*49]. —

Pragmatically, the student who is well versed in direct use of the rules of Theorem 13 (e.g. Example 9) and the chain of equivalences method (end Example 10), and is prepared if he runs into trouble to use the indirect method (beginning Example 10), will be well equipped to deal efficiently with problems of provability and deducibility in the classical propositional calculus.

In the intuitionistic propositional calculus (end § 12), we don't have double ¬-elimination, so the indirect method is unavailable; and for constructing chains of equivalences of course only equivalences holding in the intuitionistic system can be used in the links.

EXERCISES. 13.1. Use Theorem 9 (i) and (ii) to justify the general properties of ⊢ used in Example 9 (A) Steps 1, 2, 7.

13.2. Show that, if $\vdash A_{m+1}$, then $A_1, \ldots, A_m \vdash B$ if and only if $A_1, \ldots, A_m, A_{m+1} \vdash B$.

13.3. Show that $B \vdash C$ if and only if: for every list $A_1, \ldots, A_m$ of formulas, if $A_1, \ldots, A_m \vdash B$ then $A_1, \ldots, A_m \vdash C$.

13.4. The following proof of *59 uses results coming earlier in Theorem 2 (supposed already proved with "⊢"). Some obvious steps are tacit. Supply the tacit steps, convert the result successively to the formats (B_2) and (B_3) (as in Example 9), and check (B_3):

I. Assume A ⊃ B. By *51, A ∨ ¬A. CASE 1: A. Thence (by ⊃-elim. from A ⊃ B) B, whence by ∨-introd. ¬A ∨ B. CASE 2: ¬A. By ∨-introd., ¬A ∨ B.

II. Assume ¬A ∨ B. CASE 1: ¬A. By *10a, A ⊃ B. CASE 2: B. By Axiom Schema 1a, A ⊃ B.

13.5. Use the rules of Theorem 13 to establish with "⊢" in place of "⊨": *12a, *35, *40, *49, *55c.

13.6. Taking *55a with "⊢" as already established (by Example 9), use an indirect proof to establish *51 with "⊢".

13.7. True or false (and why)?

(a) "Q ∨ P, Q ⊃ ¬R ⊢ Q ∨ (¬R ∼ P)."

(b) "S ∼ T, T ⊃ (S ⊃ Q), ¬R ⊃ ¬P ⊢ P ∨ S ⊃ R ∨ Q."

13.8. Which of the four statements

"⊢ A ○ B ⊃ C ∼ (A ⊃ C) Δ (B ⊃ C)", where ○ is & or ∨, and Δ is & or ∨, hold for all formulas A, B, C? Similarly, which of

"⊢ C ⊃ A ○ B ∼ (C ⊃ A) Δ (C ⊃ B)"?

13.9°. Establish:

(a) $\Gamma \vdash A_1 \supset (\ldots (A_m \supset B) \ldots)$ if and only if, for some formula C, both $\Gamma, A_1, \ldots, A_m, \neg B \vdash C$ and $\Gamma, A_1, \ldots, A_m, \neg B \vdash \neg C$.

(b) A & B ⊃ C, ¬D ⊃ ¬(E ⊃ F), C ⊃ (E ⊃ F) ⊢ A ⊃ (B ⊃ D).

(c) A ⊃ B, C ⊃ ¬B, (D ⊃ ¬A) ⊃ C ⊢ ¬E ∨ A ⊃ (E ⊃ D).

★§ 14. Applications to ordinary language: analysis of arguments. In this section we shall treat some points concerning the application of the classical propositional calculus to reasoning in ordinary language (English).

A *full* procedure for solving logical problems arising in verbal form would be first to translate the sentences concerned into the symbolism of the propositional calculus,[43] and second to apply the theory and techniques of the calculus (as developed above) to the resulting formulas.

In *simple* reasoning, we can apply the calculus without first explicity translating. We have been reading & as "and", ⊃ as "implies" or "if . . . then . . ." or "only if", and ¬ as "not". So we can, almost without conscious effect, apply simple properties of ∼, ⊃, &, V, ¬ directly in verbal form. Many of those properties we must already have been applying in verbal form, since all of us have been using propositional calculus from when we first learned to talk. However, our seeing logical principles stated succinctly with the aid of symbols may help to fix them as part of our mental apparatus. Thus the formal study of logic may reinforce and extend our native facility.

EXAMPLE 11. Consider the following argument in words. Letters are suggested in brackets to symbolize prime components of the composite sentences. "I will pay them for fixing our T.V. [P] only if it works [W]. But our T.V. still doesn't work. Therefore I won't pay them." This argument can be symbolized thus:

(1) $$P \supset W, \neg W \therefore \neg P.$$

To say that this is correct *reasoning* (disregarding whether or not the premises P ⊃ W and ¬W are both true *in fact*) should mean that whenever P ⊃ W and ¬W are both true, then ¬P is also true. This is what we expressed exactly in § 7 by saying that ¬P is a valid consequence of P ⊃ W, ¬W, or in symbols by

(2) $$P \supset W, \neg W \models \neg P.$$

Accordingly, just in this case we will say the argument (1) is *valid*. By §§ 11 and 12, (2) is equivalent to

(3) $$P \supset W, \neg W \vdash \neg P.$$

In establishing validity, we shall generally use this latter form, tacitly employing the consistency theorem (Theorem 12). (In establishing invalidity, we usually deal directly with (2).) We now establish (3) by writing down in succession formulas deducible from P ⊃ W, ¬W, beginning with those formulas themselves, until we reach ¬P (as in Example 8, etc.): 1. P ⊃ W. 2. ¬W. 3. ¬W ⊃ ¬P [from 1 by contraposition *12a (with Corollary Theorem 5, or ∼- and ⊃-elim.)]. 4. ¬P [from 2 and 3 by modus ponens (⊃-elim.)].

[43] Or to construe them as formulas in our symbolism, if the object language is verbal (cf. § 1 top p. 6).

EXAMPLE 12. "If he doesn't tell her [$\neg T$], she'll never find out [$\neg F$]. If she doesn't ask [$\neg A$], he won't tell. She did find out. So she must have asked." We put this in symbols as follows.

(1) $\neg T \supset \neg F,\ \neg A \supset \neg T,\ F \therefore A.$

Now we establish that the argument is valid, or equivalently that

(3) $\neg T \supset \neg F,\ \neg A \supset \neg T,\ F \vdash A.$

1. $\neg T \supset \neg F$. 2. $\neg A \supset \neg T$. 3. F. 4. $F \supset T$ [contraposition *12a, 1]. 5. $T \supset A$ [*12a, 2]. 6. T [modus ponens, 3, 4]. 7. A [modus ponens, 6, 5]. — Alternatively: 4. $\neg A \supset \neg F$ [chain inference *2, 2, 1 (and modus ponens twice)].[44] 5. $F \supset A$ [*12a, 4]. 6. A [modus ponens, 3, 5].

EXAMPLE 13. "The Governor will retain the support of labor [L] only if he signs the bill [S]. He will keep the farm vote [F] only if he vetoes it [V]. Obviously, he must either not sign the bill or not veto it. Therefore the Governor will lose either the labor vote or the farm vote."

(1) $L \supset S,\ F \supset V,\ \neg S \vee \neg V \therefore \neg L \vee \neg F.$

To establish

(3) $L \supset S,\ F \supset V,\ \neg S \vee \neg V \vdash \neg L \vee \neg F,$

by cases ($\vee$-elim. in Theorem 13), it will suffice to establish (3a) and (3b) below, each of which we do directly in parentheses (by listing successive consequences of the underlined assumptions).

(3a) $L \supset S, F \supset V,\ \neg S \vdash \neg L \vee \neg F$

($\underline{L \supset S}$, $\underline{\neg S}$, $\neg S \supset \neg L$, $\neg L$, $\neg L \vee \neg F$).

(3b) $L \supset S,\ F \supset V,\ \neg V \vdash \neg L \vee \neg F$

($\underline{F \supset V}$, $\underline{\neg V}$, $\neg V \supset \neg F$, $\neg F$, $\neg L \vee \neg F$).

The demonstrations of (3) in Examples 11 and 12 may appear long because we wrote out the reasons; we omitted the reasons in demonstrating (3a) and (3b) of Example 13. We hope the reader is gaining enough facility to write down without hesitation lists of formulas demonstrating deducibility relationships directly (when simple direct demonstrations exist).

We hesitate to say whether a person's having seen contraposition (*12a) stated as a law of logic promotes his using it with more fluency and sureness than before, in examples like the three foregoing. That will depend on how proficient he already was. But it seems likely that a

[44] Chain inference ($A \supset B,\ B \supset C \vdash A \supset C$) is sometimes called "hypothetical syllogism" in traditional logic.

person who hadn't previously studied logic will find in the collection of results in Theorems 2 and 13 (with all the supporting material) *some* useful logical principles which he wasn't already using effectively.

Of course our lists of principles could be extended. The result of Example 11 constitutes a principle of inference ($A \supset B, \neg B \vdash \neg A$) called "modus (tollendo) tollens" in traditional logic. The inference $A \vee B, \neg A \vdash B$ (which we get by first using *61 to change $A \vee B$ to $\neg A \supset B$ and then modus (ponendo) ponens) is called "disjunctive syllogism". The whole result of Example 13
($A \supset C, B \supset D, \neg C \vee \neg D \vdash \neg A \vee \neg B$) is called "destructive dilemma", as the Governor can appreciate. We omitted these and others from our compilations above only because they are such immediate consequences of principles we did include. The student may add them and any other correct principles he wishes to his working lists of principles and rules. The foregoing theory (especially § 13) should make it easy to prove such additional results as he desires. Different persons will have different habits and preferences as to which principles they use directly and which they assemble out of others when needed, and these habits may change with time and with the problems at hand.

Now, whether or not logic studied formally actually augments our native capacity to *discover* correct arguments, it certainly is of value in *checking* the correctness of proposed arguments. For, it provides an analysis of the basis of reasoning with both models (model theory) and norms for correct reasoning (proof theory). Thus we can appeal to formal logic to confirm the correctness of our reasoning or to detect errors in it, whenever we find ourselves in risk of becoming confused. And if we have not found ourselves prone to confusions in reasoning, we have no doubt noticed that others sometimes are, especially our adversaries in arguments.

EXAMPLE 14. "He said he would come [C] if it doesn't rain [$\neg R$] (and we can depend on what he says). It's raining. Therefore he won't come". In symbols,

$$(1) \qquad \neg R \supset C, R \therefore \neg C.$$

To try to confirm this by establishing $\neg R \supset C, R \vdash \neg C$, we naturally try modus ponens. But we would need $\neg R$ to use that directly. If we first contrapose $\neg R \supset C$ (by *12a, simplifying $\neg\neg R$ to R by *49), we obtain $\neg C \supset R$, and we're no better off. Thus what was spoken as though it were an obvious one-step inference ("Therefore") doesn't follow by any obvious principle. If we hadn't studied truth tables, we would already be able to say to the speaker that we don't see how her conclusion follows, and strongly suspect that it doesn't follow. But we have studied truth tables. Giving R, C the values t, t, both the premises (assumption formulas)

$\neg R \supset C$, R become t, while the proposed conclusion $\neg C$ becomes f. Thus not $\neg R \supset C, R \models \neg C$. So the conclusion really doesn't follow. "Maybe he'll come anyway. Or, did he rather say, 'I'll come only if it doesn't rain'?".[45]

In examples like these, we admit to laboring the obvious. But it is not so easy in *long* chains of deductive reasoning to be sure one has nowhere gone astray. In polemical arguments, the speaker may be deliberately attempting to lead his audience to a conclusion which his assumptions do not strictly justify.

To check any separate step in a chain of reasoning, if it is correct, should be quite easy, *once* the reasoning has been clearly formulated. Translation into logical symbolism (carried out partially or fully as the circumstances may require) forces a resolution of any ambiguities or obscurities there may be in the verbal presentation. It is outside the scope of this book to treat exhaustively the topic of verbal argumentation.[46] But we shall note a few aspects of it.

The principles we are noting here will come up again in the predicate calculus (Chapter II), where there are more ways of going astray if one is not explicit and careful.

An example of a simple argument which we cannot adequately analyze now, but will be able to in the predicate calculus, is the following: "All men are mortal. Socrates is a man. Therefore Socrates is mortal." As we recall from § 1, the *propositional* calculus is concerned only with logical relationships which result from the way sentences are built from certain sentences (called by us *prime formulas* or *atoms*) which enter as unanalyzed wholes. In this example, all we can say now is that the foregoing argument has the form $P, Q \therefore R$; and $P, Q \models R$ does not hold. We give this example as a warning of what not to attempt in the way of logical analyses at this stage.

We recall that, for the *classical* propositional calculus, it is assumed that each unanalyzed sentence (or atom) is either true or false but not both; and we are to use no more than this (§ 2). In *classical* mathematics, this assumption is considered as holding strictly (§ 36). In daily life, it is notoriously the case that propositions do not always fall neatly into two classes, the true and the false. We may tell the three dozen people we are inviting to a picnic, "If the weather is good [G], the picnic will be this

[45] In verbal examples, we suspend our convention of using only $P, Q, R, \ldots, P_1, P_2, P_3, \ldots$ for prime formulas (atoms), to allow the letters chosen for the prime formulas to match the words (as here, C for "he will come"). We already did this in Examples 11–13; but those three applications of propositional calculus did not depend on the formulas symbolized being prime.

[46] Cf. Clark and Welsh 1962.

Sunday [S]; otherwise, the next good Sunday [N]". Then when Sunday comes, the weather is not good, and everybody knows not to show up. $(G \supset S) \& (\neg G \supset N), \neg G \vdash N$. Only, often it isn't this simple. Sunday morning the weather outlook is very ambiguous. We have to decide whether to go through with the picnic (risking that everyone will be miserable); or to postpone it (and have the family spend half the week eating up the picnic hamburger and rolls).

Nevertheless, the classical propositional calculus is very useful in daily life. It helps us to be precise in our logic, even when we must recognize there is an element of guesswork in the assumptions. Thus we may phrase a number of statements, none of which we are absolutely sure is true, but which represent our best appraisal of the facts. Then we would like to know exactly what will follow. Or we may wish to consider several alternative sets of assumptions, perhaps attaching probabilities to them; then, before deciding upon some course of action, we deduce by two-valued logic what consequences follow exactly from each set of assumptions.

These rather obvious remarks are to emphasize that one should not lose sight of the role of his logic in the total intellectual situation. Our allusions to modal and intuitionistic propositional calculi (end § 12) illustrate that there are situations which call for other kinds of logic than the classical. —

In case of doubt about a piece of verbal propositional reasoning, translate it into the symbolism of the propositional calculus.

Here is a list (not exhaustive) of expressions on the right which can (or often can) be translated by the symbols on the left.

$A \sim B$.	A if and only if B.	A iff B [an abbreviation].	
	A if B, and B if A.	If A then B, and conversely.	
	A exactly if B.	A exactly when B.	A just in case B.
	A is (a) necessary and sufficient (condition) for B.		
	A is materially equivalent to B.		
	A is equivalent to B [sometimes].		
$A \supset B$.	If A, then B.	B if A.	A only if B.
	When A, then B.	B when A.	A only when B.
	In case A, B.	B in case A.	A only in case B.
	B provided that A.		
	A is (a) sufficient (condition) for B.		
	B is (a) necessary (condition) for A.		
	A materially implies B.	A implies B [sometimes].	
A & B.	A and B.	Both A and B.	
	A but B.	Not only A but B.	
	A although B.	A despite B.	A yet B. A while B.

$A \vee B$.	A or B or both.	A and/or B [in legal documents].
	A or B [usually].	Either A or B [usually].
	A unless B [usually].	A except when B [usually].
$(A \vee B) \& \neg(A \& B)$	A or B but not both.	A or else B [usually].
(equivalently by	A or B [sometimes].	
Exercise 5.3 (b):	Either A or B [sometimes].	
$\neg(A \sim B)$, $\neg A \sim B$,	A unless B [sometimes].	
$A \sim \neg B$).	A except when B [sometimes].	
$\neg(A \vee B)$ (equivalently by *55a, $\neg A \& \neg B$).		Neither A nor B.
$\neg A$.	Not A [or the result of transforming A to put "not" just after the verb or an auxiliary verb].	
	A doesn't hold.	A isn't so.
	It is not the case that A.	

In translating English words by our propositional connectives with two-valued truth tables, shades of meaning that are present in the English words may be lost, at the same time that precision for the purpose of logic is gained.

Although in the propositional calculus, A & B is equivalent to B & A, the report "Jane had a baby and got married" will strike Jane's friends differently from "Jane got married and had a baby".[47] In this example, the order of the conjunctands suggests a temporal (or causal) succession. The temporal succession can be rendered in classical logic, if we wait until we have the symbolism of the predicate calculus. But the translation A & B is simpler, and suffices for logical analyses in which the temporal (or causal) component doesn't matter.

Expressions like "A but B", "A although B" and "A despite B" have nuances of meaning not possessed by "A and B", and lost in the translation A & B. The young man may respond differently if his girl friend tells him "I love you and I love your brother almost as well" than if she says "I love you but I love your brother almost as well".[48]

Although we propose to translate "A unless B" ordinarily by $A \vee B$, which is equivalent to $B \vee A$, one would say "I won't go unless she apologizes" but hardly "She apologizes unless I won't go".[49] This is another example in which the English suggests a temporal or causal succession.

A further difficulty in translating is that there may be ambiguity as to

[47] Example from Strawson 1952 p. 80.
[48] Example from Suppes 1957 p. 4.
[49] Example from Clark and Welsh 1962 p. 45.

what meaning reduced to terms of two-valued truth tables is intended. When the dinner menu says, "Tea or coffee included", we are not surprised to be charged extra if we order both. If it is announced that donations of books will be received at the school or at the church, we won't expect our books to be refused at the church because we have already left some at the school. Since the inclusive "or", symbolized by $\vee$, is the more useful one, we personally are accustomed to using "or" inclusively (not bothering to add "or both"); and always adding "but not both" or the like in any cases of doubt when we do want the exclusive "or". If A and B are such that $\neg(A \& B)$ is known or assumed anyway, then the inclusive and the exclusive "A or B" are equivalent, and one might as well use the simpler translation $A \vee B$. (Indeed, $\neg(A \& B) \vdash (A \vee B) \& \neg(A \& B) \sim A \vee B$.) Thus in a lecture to a mathematical audience, when I say "*n* is even [A] or *n* is an odd prime [B]", it is immaterial whether the sentence be understood as $A \vee B$ or as $(A \vee B) \& \neg(A \& B)$. But if I were speaking to an audience of people who did not know that a number cannot be both even and an odd prime, it could make a difference which way my "or" was understood. Although it is farfetched to suppose such an audience would have invited me to speak about *n*, a similar situation is less so. We might be analyzing an argument involving the sentence "*n* is even [A] or *n* is an odd prime [B]". Then if the premises so not include enough facts about the number system so that $\neg(A \& B)$ is deducible from them, it could be necessary for the argument to be valid either to use the translation $(A \vee B) \& \neg(A \& B)$ instead of simply $A \vee B$, or to add $\neg(A \& B)$ or something that entails it to the premises (whereupon $(A \vee B) \& \neg(A \& B) \sim A \vee B$ becomes deducible from them).

The meaning of "A unless B" and "A except when B" which we prefer and use ourselves is that B is an "escape clause" which lets us off from our assertion of A in the case of B; i.e. in saying "A unless B" we intend to claim that A when $\neg B$, and to make no claim when B. Then $\neg B \supset A$, or equivalently $A \vee B$, is the translation. But one must be on guard to recognize when a speaker intends to claim that $\neg A$ when B; then the translation is $(\neg B \supset A) \& (B \supset \neg A)$, or equivalently (by Exercise 5.3 (b)) $(A \vee B) \& \neg(A \& B)$, so "unless" and "except when" in this usage amount truth-functionally to the exclusive "or".

In ordinary language, we don't use parentheses to indicate how the parts of a composite sentence are to be associated, and often fairly subtle clues serve instead. "If Jones is present [J] or Williams speaks up for our proposal [W] and Stark doesn't come out against it [$\neg S$], it will be adopted [A]." Should we translate this as **(a)** $(J \vee W) \& \neg S \supset A$ or as **(b)** $J \vee (W \& \neg S) \supset A$? In written language a comma before "and" would resolve the ambiguity in favor of (a); in spoken language, we can indicate

(a) by putting "both" before "Jones" or repeating "if" before "Stark", and accenting "and" (after a slight pause).

To summarize, the process of translating from English to symbols is not automatic. The translator must first seek to understand clearly the passage to be translated. If it is his own words, he must choose the interpretation he intends. If it is the words of another, he must undertake to divine that person's intentions when the words are ambiguous, using any clues the context may provide; and he may even need to try out different translations to see which makes the best sense.

In translating from symbols to English words, if we choose from the tops of our lists of translations we can hardly go wrong in rendering the connectives, but care will still be needed to be sure that the scopes of the connectives are unambiguously rendered. The result should then be unambiguous, but may be far from being idiomatic or in good style.

EXERCISES. 14.1. Translate each of the following arguments into logical symbolism, and analyze the result in the manner of Examples 11–14.

(a) If he belongs to our group [B], he is virtuous and trustworthy [V & T]. He does not belong to our group. Therefore, either he is not virtuous or he is not trustworthy.

(b) Unless taxes are raised [R], there will be a deficit [D]. If there is a deficit, state services will be curtailed [C]. Therefore, if taxes are raised, state services will not be curtailed.

(c) If he is responsible for this rumor, he must be either stupid or unprincipled. He is neither stupid nor unprincipled. Therefore he is not responsible for the rumor.

(d) If the suspect committed the robbery, either it was planned very carefully or he had an insider as confederate. If it was planned very carefully, then, if he had an insider as confederate, more loot would have been taken than was taken. Therefore the suspect is innocent.

(e) If peace breaks out, then there will be a depression unless the country rearms with new weapons or carries out a massive domestic program in education, conservation, antipoverty, etc. It will not be possible to get agreement on what a massive domestic program should be. Therefore, if peace breaks out and there is not to be a depression, the country must rearm.

(f) The proposed attack will succeed, only if the enemy is taken by surprise or the position is weakly defended. The enemy will not be taken by surprise, unless he is overconfident. He will not be overconfident, if the position is weakly defended. Hence the attack will not succeed.

(g) Unless we continue price supports, we will lose the farm vote. Overproduction will continue if we continue supports, unless we institute production controls. Without the farm vote, we cannot be reelected. Therefore, if we are reelected and do not institute production controls, over-production will continue.

(h) If **(1)** $x + 3 = \sqrt{3 - x}$, then $x^2 + 6x + 9 = 3 - x$. But $x^2 + 6x + 9 = 3 - x$ if and only if $(x + 6)(x + 1) = 0$, which is the case if and only if $x = -6$ or $x = -1$. Therefore only -6 and -1 can be roots of the equation (1); i.e. $x + 3 = \sqrt{3 - x}$ implies that $x = -6$ or $x = -1$.

(i) Like (h), except changing the conclusion to: Therefore -6 and -1 are roots of (1); i.e. $x = -6$ implies $x + 3 = \sqrt{3 - x}$, and $x = -1$ implies $x + 3 = \sqrt{3 - x}$.

14.2. (Keisler.) Brown, Jones and Smith are suspected of income tax evasion. They testify under oath as follows.

BROWN: Jones is guilty and Smith is innocent.

JONES: If Brown is guilty, then so is Smith.

SMITH: I'm innocent, but at least one of the others is guilty.

Let B, J, S be the statements "Brown is innocent", "Jones is innocent", "Smith is innocent", respectively. Express the testimony of each suspect by a formula in our logical symbolism, and write out the truth tables for these three formulas (in parallel columns, like (h)–(ł) § 7). Now answer the following questions.

(a) Are the testimonies of the three suspects consistent? (Exercise 11.3.)

(b) The testimony of one of the suspects follows from that of another. Which from which?

(c) Assuming everybody is innocent, who committed purjury?

(d) Assuming everyone's testimony is true, who is innocent and who is guilty?

(e) Assuming that the innocent told the truth and the guilty told lies, who is innocent and who is guilty?

★§ 15. Applications to ordinary language: incompletely stated arguments. In daily life and public affairs, it is common for arguments to be given in which the intended premises (or assumption formulas $A_1, \ldots, A_m$) are not all of them explicitly stated. It would be beside the point to chide a speaker by saying that his argument is not valid because not $A_1, \ldots, A_p \models B$ where $A_1, \ldots, A_p$ express the premises he stated, when it would be fair to assume that he intended further premises $A_{p+1}, \ldots, A_m$ to be understood. Arguments that are intended to have such tacit assumptions $A_{p+1}, \ldots, A_m$ may be called *enthymemes*. Traditionally, the term related to the inferential patterns (or syllogisms) of traditional logic. As logic has now become more flexible, it is natural to extend the term to cover any argument in which one or more premises or the conclusion is tacit.

Enthymemes have a proper and well-nigh indispensible role. Without them communication would become exceedingly slow and tedious. We can properly omit what should be obvious; we'll quickly lose our audience if we don't. A premise may be obvious for an argument, because it's well known and universally accepted, or because we have recently been talking about it. But also, if it's to be omitted without obscurity, what is left of the argument should more or less clearly indicate that it's called for as a premise. Indeed this alone can be sufficient license for leaving it tacit. Thus I could say to a hostess who didn't already know that I planned to retire early, "If I drink coffee [C], I can't get to sleep early [¬S]. So please don't pour me any". The enthymematic argument (before supplying the missing premise) is

$$C \supset \neg S \therefore \neg C$$

It's clear enough that this is an abbreviation for

$$C \supset \neg S, S \therefore \neg C.$$

In the black art of persuasion, enthymemes may be employed to detract attention from a premise whose truth the hearer might doubt.

As an enthymeme with the conclusion tacit, if I have just been offered a cup of coffee, simply $C \supset \neg S$ would be clearly enough an abbreviation for $C \supset \neg S, S \therefore \neg C$. If we communicate statements that (perhaps with others that are obvious) would constitute the premises for the inference of a conclusion that we prefer not to state baldly, we are engaging in *innuendo*.

Thus logical analysis comes to include attempting to supply premises (or conclusion) for what is ostensibly an incompletely stated argument. In some cases, the form of the argument may leave no room for doubt about what is to be supplied. In other cases, we may have to experiment with different trials of unstated premises $A_{p+1}, \ldots, A_m$ in the attempt to find a set that will make the argument valid; and we may find more than one such set.

Hence it is appropriate to pause and consider more carefully what an argument "$A_1, \ldots, A_m \therefore B$" is. When someone says "$A_1, \ldots, A_m \therefore B$", he doesn't simply mean that B is a valid consequence of $A_1, \ldots, A_m$ (in symbols, $A_1, \ldots, A_m \models B$). He also intends to claim that $A_1, \ldots, A_m$ are true (or at least available as though true). Thus the full meaning of "$A_1, \ldots, A_m \therefore B$" is "(i) $A_1, \ldots, A_m$ are true, and (ii) $A_1, \ldots, A_m \models B$; and therefore B is true". The purpose of the argument is to persuade the hearer of the truth of B on the grounds (i) and (ii). When *both* (i) and (ii) hold, we call the argument not simply "valid" but *sound*.

Whether $A_1, \ldots, A_m$ are true or not may be a matter of empirical fact, or of belief, or may rest on earlier assumptions under which the argument is being pursued and which make $A_1, \ldots, A_m$ available for the purpose of the argument. Soundness is thus relative to whatever criteria or standards are being presupposed in the claim of the argument that $A_1, \ldots, A_m$ are available; and a full statement on the matter of soundness would include such reference. Also it seems convenient to recognize graduations by calling an argument simply *plausible* when it is valid but we can only say that $A_1, \ldots, A_m$ are plausible.

Now if there are different choices of unstated premises that will make an enthymematic argument valid, it will be of interest to see if a choice can be found for which they are all true (or at least all plausible). Whether or not the proposed premises $A_1, \ldots, A_m$ are all true is of course ordinarily not a question of logic.

If $\models \neg(A_1 \,\&\ldots\&\, A_m)$ ($A_1 \,\&\ldots\&\, A_m$ is *inconsistent*, beginning § 8), then "$A_1, \ldots, A_m \therefore B$" is unsound simply by logic (cf. Exercise 12.7). An opponent's argument will be destroyed, if his premises are shown to be inconsistent, or to be "inconsistent with the facts", i.e. to become inconsistent on adding to them other statements known to be true.

Similarly, if $\models A_1 \,\&\ldots\&\, A_m$ ($A_1 \,\&\ldots\&\, A_m$ is *valid*), $A_1, \ldots, A_m$ are all true on logical grounds; so then, if "$A_1, \ldots, A_m \therefore B$" is valid, $\models B$.

It is in the remaining case, that neither $\models \neg(A_1 \,\&\ldots\&\, A_m)$ nor $\models A_1 \,\&\ldots\&\, A_m$ ($A_1 \,\&\ldots\&\, A_m$ is *contingent*, beginning § 8), that the soundness of a valid argument "$A_1, \ldots, A_m \therefore B$" depends on considerations outside of logic.

In discourse, it may not always be clear when an argument (perhaps enthymematic) is intended. Here are some phrases which when inserted for the dots in "$A_1, \ldots, A_m; \ldots B$" indicate an argument "$A_1, \ldots, A_m \therefore B$":

therefore, hence, whence, thence, so, it follows that, we infer that, we conclude that, consequently, but then, these imply, thus.

The following phrases have the same effect when inserted for the dots in "$B; \ldots A_1, \ldots, A_m$":

this follows from, in consequence of, for, this is implied by, because, since, in as much as.

The word "implies" requires special comment. According to the context and the user, "A implies B" may mean (I) "if A then B" or in symbols $A \supset B$, or (II) "B follows by logic from A" or in symbols "$A \models B$", which is equivalent to "$A \vdash B$", to "$\models A \supset B$", and to "$\vdash A \supset B$". In brief, "implies" may be translatable by (I) $\supset$ or (II) either of "$\models$" and "$\vdash$". Clearly, an implication in Meaning (I) is a statement in the object language; in Meaning (II), a statement in the observer's language.[11]

Under (I), the truth of "A implies B" will depend ordinarily on circumstances outside of logic, e.g. on matters of empirical fact; hence the name "material implication" for $A \supset B$. Under (II), "A implies B" is a "logical implication", being true exactly when $A \supset B$ is t for all assignments of t's and f's to the prime components of $A \supset B$; in later chapters and elsewhere, this will generalize to "$\vdash A \supset B$" in whatever system (in place of propositional calculus) is under consideration there. Some writers (beginning with Quine 1940) avoid (I), and call $A \supset B$ a "conditional" rather than an "implication".

Similarly, "A is equivalent to B" may be translated by (I) $A \sim B$ ("A is materially equivalent to B") or (II) either of "$\models A \sim B$" and "$\vdash A \sim B$" ("A is equivalent to B *in* the system to which $\models$ and $\vdash$ refer"). Those who call $A \supset B$ a conditional call $A \sim B$ a "biconditional".

In this book, we generally avoid Meaning (II) of "implies" (finding "$\models$" and "$\vdash$" more convenient and explicit). But sometimes we use Meaning (II) of "equivalent", indicating explicitly or by the context the reference to a system (cf. beginning § 5, end § 8).

It is more or less customary in definitions to write simply "if" or "when" or "in this case" meaning "if and only if" or "equivalent". (We did so in the definitions of "valid consequence" § 7, of "inconsistent" and "contingent" in § 8, of "provable", "deduction" and "deducible" in § 9, and in effect in the definition of "valid" in § 2 by saying "Such" meaning "Such and only such".) —

The verbal examples above all led to quite trivial logical problems when they were translated into symbols.

It is easy enough to make up problems in which the logical part would be more challenging. All we need to do is to take some chain of arguments (in which conclusions of earlier arguments become premises of later ones), and give only the original premises (i.e. those not conclusions of earlier "links") and the final conclusion. Such problems that are both ingenious and entertaining have been constructed by a number of logicians, above all by C. L. Dodgson (= Lewis Carroll) 1887, 1897. Whether they belong to the serious study of logic, or only to the department of mathematical puzzles and recreations, is a matter of one's educational philosophy.

We offer some reflections, without claiming definitely to answer the question.

First, consider the role of logic in *checking* the validity of arguments already constructed. (In § 14, we took this second.) The person giving an argument, whether in ordinary conversation, or in a debate or a trial, or in mathematical exposition, is supposed when he says "therefore" to be making a step which his hearers or readers can reasonably be expected to follow. Thus it is in the nature of an *individual argument* that it should be simple (or if it is fallacious, that it should pretend to be simple). It is the arguer's job, if he is seriously claiming to demonstrate something, to break his demonstration into a series of chunks that can each be swallowed. If the individual arguments or inferences (chunks) get too complicated, we can accuse him of not having given us a demonstration. Of course, how big a step can reasonably be justified by "therefore" is a function of the audience's proficiency in logic, and familiarity with the subject matter. Individual arguments in the chain may be expressed enthymematically; and in an extended piece of exposition (like a text book) successions of individual arguments that recur may be combined into individual arguments after the audience has had time to become familiar with them. We conclude, however, that "arguments" consisting of a list of premises, followed by a conclusion (with "therefore" prefixed) that can only be inferred by a *long* succession of steps of kinds with which the audience can be expected to be familiar, are rather artificial. Such "arguments" do not naturally occur.

Of course, the arguer may elect not to give his demonstration in full on a given occasion. He may omit some steps that the audience is not supposed to be able without pausing to supply (which omissions we would not consider enthymematic), saying "It can be proved that . . .". (It is poor style to substitute the phrase "Obviously . . .".) Then we have the question whether it is the reader's (or hearer's) job to try to fill in the missing steps (which could constitute an extended logical problem of the sort we are discussing).

A similar situation arises if one of the allegedly simple arguments in a chain of arguments offered to us is fallacious. Refuting *that* individual argument by a choice of truth values doesn't settle whether the whole demonstration, taken as an argument from the original premises to the final conclusion, is valid or not. Is it our job to make a repair, or to demonstrate the impossibility of repair? Of course, if a trifling change will fix matters, we would feel silly not to have noticed it.

But how much work we should be expected to do in either of these situations would depend on our stake in the matter. Primarily, it is the responsibility of the person who would demonstrate something to supply a demonstration, broken into reasonably sized pieces, each individually valid.

We now come to the other role of logic, which is to discover demonstrable results and demonstrations of them. We aren't just looking for valid formulas and valid arguments. We are interested in the fact that $P \supset P$ is valid; but much less so in the facts of the validity of each of the formulas of the infinite collection of which

$$\neg\neg\neg\neg(P \supset P) \;\&\; (P \supset P) \;\&\; \neg\neg(P \supset P) \;\&\; \neg\neg\neg\neg\neg\neg(P \supset P)$$

is another member. A catalog simply listing all the valid formulas found among the formulas up to some given length would not sell. Whether the validity of a formula or an argument interests us depends on whether it has some bearing on actual affairs, or may play a role in some logical or mathematical theory, which may in turn interest us for practical or theoretical reasons. So a question whether $\models$ E or whether $A_1, \ldots, A_m \models B$ is only likely to interest us if we have some reason for wanting the result that the conjecture does (or does not) hold. This reason may be the possible applicability of the result to questions that previously interested us, or it may be a reason suggested by the form of the conjecture itself, or it may be that in settling the conjecture we anticipate developing interesting methods. Moreover such a reason often provides hunches why the conjecture may hold, and thus clues as to how to go about trying to establish it.

For these reasons we are rather skeptical that complicated logical problems, given in vacuo (with motivation or clues), are very relevant to the use of logic, even to its use in discovering quite complicated proofs. But a very firm grasp on the individual rules of logic, the simple principles, repeated applications of which are used to construct complicated proofs, is essential. Practice with these principles can of course be obtained in using them to derive still other logical principles, which in turn may be useful, and in applying them in the course of deducing interesting results in mathematics, science and daily life, instead of in working extended made-up logic problems. So we regard the logical theory and applications in the following chapters as in part a substitute for more problem work here.

We conclude this chapter with one more example, to illustrate a domain of logic in which complicated problems may naturally arise. This is in simplifying logical expressions. We refer to the theory given in § 8.

EXAMPLE 15. (Venn 1881 p. 261.) A certain club has the following rules: **(1)** The Financial Committee shall be chosen from amongst the General Committee. **(2)** No one shall be a member both of the General and Library Committees unless he be also on the Financial Committee. **(3)** No member of the Library Committee shall be on the Financial Committee. Simplify these rules. SOLUTION. Let x be any person (member of the club, presumably). Let P be "x is on the Financial Committee", Q be "x is on the General Committee", R be "x is on the Library Committee". Then (1)–(3) are expressed by $(P \supset Q) \,\&\, (\neg(Q \,\&\, R) \vee P) \,\&\, \neg(R \,\&\, P)$ (Exercise 8.2 (c)). This has the conjunctive normal form (§ 8)
$(\neg Q \vee \neg R) \,\&\, (Q \vee \neg P)$, which is equivalent to $(Q \supset \neg R) \,\&\, (P \supset Q)$. Thus a simpler set of rules is (1) with: **(2′)** No member of the General Committee shall be on the Library Committee.

EXERCISE. 15.1. Premises (or conclusion) may be missing in the following. If so, attempt to supply them to produce a valid argument. Do you consider the (completed) argument sound (or at least plausible)?

(a) The accused could be guilty of the crime [G] only if he was in New York at 6 P.M. on January 1 [N]. But it has been established that he was in Washington at that time [W]. Therefore he is not guilty.

(b) We have no proof that he committed the crime. Therefore he must be acquitted.

(c) We have no proof that he committed the crime. Therefore he is innocent.

(e) If it has snowed, it will be poor driving. If it is poor driving, I will be late unless I start early. Indeed, it has snowed. Therefore I must start early.

(f) If you use our "Gallant Tailor Spray", you will not be troubled by insects.

CHAPTER II

THE PREDICATE CALCULUS

§ 16. Linguistic considerations: formulas, free and bound occurrences of variables. In the propositional calculus, we studied those logical relationships which depend on how some propositions are composed from other propositions by operations (expressed by the symbols $\sim$, $\supset$, &, V, $\neg$) in which the latter propositions enter as unanalyzed wholes. In the *predicate calculus*, we carry the analysis a step deeper to take into account also what in grammar is called "subject-predicate structure", and we use two further operations, $\forall$ ("for all") and $\exists$ ("for some" or "there exists") which depend on that structure. (The predicate calculus includes the propositional calculus.)

Consider the proposition (expressed by the sentence) "Socrates is a man". The part of this proposition (expressed by) "— is a man" or "x is a man" is a *predicate;* "Socrates" is a *subject.* Read "x is a man", using the mathematical notation of a variable, the predicate is seen to be a *propositional function*, i.e. for each value of the (independent) variable "x", it becomes (or takes as value) a proposition, true for example when x is Socrates, false in Greek mythology when x is Chiron, and in the Kleene household when x is Fleck. To take another example, "John loves Jane" is a proposition, which can be thought of as a value of any one of three propositional functions, "x loves Jane", "John loves y" and "x loves y". In grammar, "x loves Jane" is a *predicate*, but not "John loves y" or "x loves y"; and a value of "x" is a *subject*, of "y" an *object*. Mathematically, these distinctions are unimportant. We shall simply adopt the term *predicate* as short for the more cumbersome *propositional function* $P(x_1, \ldots, x_n)$, for any number $n \geq 0$ of (independent) variables;[50] and the term *object* or *individual* for a value of any one of the variables. For $n = 0$, we have a *proposition* as a special case of a predicate; for $n = 1$ a *property;* for $n = 2$, a (*binary*) *relation;* for $n = 3$, a *ternary relation;* etc.

This explains the name *predicate calculus* for the logic of propositional

[50] The term "*independent* variable" arises from the mathematical practice with functions like $x^2 + 3x + 1$ or $\sin x$ of writing "$y = x^2 + 3x + 1$" or "$y = \sin x$" where "y" is another variable (the *dependent* variable) assuming the values of the function. We shall not do thus here.

functions. A fully descriptive (but cumbersome) name is *calculus of propositional functions.*[51]

The notation "x is a man" is a bit more concise than "— is a man". The advantage of variables over blanks to show the "open places" in a predicate is increasingly apparent in examples like "$—_1$ loves $—_2$", "$—_1$ loves $—_1$" (synonymous with "— loves himself"), and "$—_1$ is father of $—_2$, or $—_1$ is mother of $—_2$" (synonymous with "$—_1$ is a parent of $—_2$").

Such expressions as "$—_1$ loves $—_2$" or "x loves y" are not used directly in ordinary language.[52] To relate them to ordinary language, we may begin by regarding the blanks or variables which we have introduced as "place holders" for words naming objects. Of course, the words to be supplied do not need to be proper nouns, like "John" and "Jane". We can have, for example: (a_1) "Somebody loves Jane", (a_2) "There is someone who loves Jane", (b) "Nobody loves Jane", (c) "Everybody loves Jane", (d) "Everybody loves someone", (e) "Someone is loved by everybody", (f) "Everybody loves himself", (g) "There is no one who does not love himself". Using "L(x, y)" as short for "x loves y", and supposing for the moment that "bodies" or persons are the range of the variables "x" and "y", these can be written using $\forall$ and $\exists$, thus: (a′) $\exists x$L(x, Jane), (b′) $\neg\exists x$L(x, Jane), (c′) $\forall x$L(x, Jane), (d′) $\forall x \exists y$L(x, y), (e′) $\exists y \forall x$L(x, y), (f′) $\forall x$L(x, x), (g′) $\neg\exists x\neg$L(x, x). In this symbolism we have formed sentences expressing propositions, without displacing or completely displacing the variables "x" and "y". Note particularly that in (a′) as expressing (a_2) (which is synonymous with (a_1)), "someone" and "who" are each represented by an (occurrence of) "x"; similarly, three different words of (g) are each represented in (g′) by an "x".

Here, rather than thinking of variables as simply (or always) "place holders", they may be thought of as a stock of names (nouns or pronouns), which are available to us to name various objects. What objects are named

[51] The name *functional calculus* is often used. This name has seemed to us a little unhappy (though historical precedence can be claimed for it), as it leaves out of account the most descriptive part, that the functions are propositional. Thus it might suggest functions from numbers to numbers (like $x^2 + 3x + 1$ and sin x, where for each number as value of "x", $x^2 + 3x + 1$ and sin x become numbers); these functions are commonly called simply "functions" when propositional functions are being called "predicates". A further possibility of confusion is with *functionals*, in the sense of functions from functions of the last sort to numbers (e.g. $\int_1^2 f(x)\,dx$, which, for each suitable function like $x^2 + 3x + 1$ or sin x as value of f, becomes a number); the branch of mathematics called "functional analysis" deals with these.

[52] Except as the mathematical notation of variables has come to be adopted in such examples as "A loves B", "If A does this, B will do that", with "A", "B", "C", ... as variables for persons.

may depend on how they are incorporated into sentences, or on the context in which those sentences appear.

The use of variables is thus not basically so very different from constructions that are used in ordinary language. "Somebody" and "everybody" serve as names for unspecified persons; and even the proper name "Jane" isn't specific, unless we have explained that we mean Jane Austen, or Jane Grey, or Jane Addams, or Jane who lives down the street. The legal "John Doe" is the equivalent of a variable (ranging over men) whose value is being left unspecified.

In the propositional calculus we studied logical relationships between propositions without taking into account what propositions are expressed by the prime formulas. This comes to the same thing as saying that, for the logical relationships we dealt with there, the propositions expressed by "P", "Q", "R", . . . could be any propositions. Likewise, here we shall not identify the objects or individuals which may constitute values of the variables. To keep the symbolism as simple as we can in this chapter, we shall also not assume any other list of names for individuals to be provided than the one list of variables. So, if we wish to symbolize (a_1) "Somebody loves Jane", we shall have to write e.g. $\exists x L(x, y)$ (or $\exists x L(x, j)$) and agree that "y" (or "j") is a name for the person Jane. This is not too great a sacrifice here, since we shall only be concerned in this chapter with general logical relationships, in which the special charms of Jane won't figure (or at least only to the extent they are enumerated, and the relationships will then apply to any other lady possessing all those same charms). Later (§ 28), some symbols different from variables may be provided as names for particular individuals.

We return to the matter of notation for predicates. To take a mathematical example now, "$x < y$" can be used to name a predicate. Then when (x, y) become or take as values (2, 7) or (5, 100), $x < y$ becomes or takes as value a true proposition. When (x, y) become (7, 7) or (100, 5), $x < y$ becomes a false proposition. Now if we say (h) "For every (real) number x, there exists a number y such that $x < y$" (or in our new symbolism, (h′) $\forall x \exists y\, x < y$), the "$x < y$" as part of (h) or (h′) is not being used to name a predicate, but to say something about two numbers, the first an arbitrary (i.e. completely unspecified) number named by "x", and the second a suitably chosen number named "y" (the choice depending on the number named "x"). This example illustrates that it is necessary (but we think, easy) to distinguish between the use of an expression like "$x < y$" to name a predicate, and its use to express a proposition which is the value of that predicate when "x", "y" are being thought of as naming objects.[53] It is

[53] If those objects are unspecified, the proposition expressed then by "$x < y$" may be called the "ambiguous value" of the predicate. Cf. IM pp. 33, 227.

only necessary to keep clearly in mind that a predicate $P(x, y)$ is not a proposition, but a *correspondence* (or *correlation*) by which, from the various choices of pairs of objects as values of "x", "y", respective propositions arise.

So long as "$x < y$" is used just to name a predicate, it is quite satisfactory to think of "x" and "y" as place holders, not (or not *yet*) naming anything. But mathematicians, without changing "x" and "y" to anything else, then often go over to thinking of "x" and "y" as naming objects; so by a transformation in interpretation, the expression "$x < y$" for a predicate becomes an expression for a proposition, as we have illustrated. It is the ease with which this transformation can be made that accounts in part for the popularity of variables as a notational device.

The predicate (= propositional function) named by "$x < y$" can also be named simply "$<$". Similarly, the function named "sin x" can be named simply "sin" or "sine"; but there is no commonly used name for the function $x^2 + 3x + 1$ that does not contain "x" (or some other variable instead). The vast majority of predicates we shall wish to name will not have commonly used names without variables.

A notation for a predicate using variables we may call a *name form* for the predicate. The variables used as part of such a notation constitute the *name form variables* in that notation, and have the *predicate interpretation* or the *name form interpretation*. We postpone further discussion of the use of variables in expressing propositions.

We began our study of the propositional calculus in § 1 by assuming that we are dealing with one or another object language, and that in this object language there are some (declarative) sentences (expressing propositions) which are to retain their identity throughout any particular investigation in the propositional calculus but are not to be analyzed in the investigation.

Now, to get started on our study of predicate logic, we need to assume that the object language contains some expressions or linguistic constructions for predicates (of given numbers of variables), which expressions shall retain their identity throughout any particular investigation in the predicate calculus but shall not be analyzed. These expressions we call *prime predicate expressions* or *ions*, and we denote them by "P", "P(—)", "P(—, —)", "P(—, —, —)", . . . , "Q", "Q(—)", "Q(—, —), "Q(—, —, —)", . . . , "R", . . . , also using subscripts when convenient.[54] Each capital

[54] There is some analogy with chemistry, where positive one-atom ions are atoms with some electrons missing. When the places for those electrons are filled by electrons, the ions become atoms, just as when the places for names in prime predicate expressions are filled by names of objects they become prime formulas. (For the present, our only names are variables.)

A chemist friend suggested "neucleons" as more appropriate; but, since the analogy with chemistry will be imperfect anyway, we prefer the shorter word "ions".

Roman letter from the latter part of the alphabet will be used as a name for a different *n-place* prime predicate expression or ion for each number $n \geq 0$ of variables; thus P, P(—), P(—, —), P(—, —, —) are four different ions (expressing respectively a 0-, 1-, 2- and 3-place predicate), and Q, Q(—), Q(—, —), Q(—, —, —) are four other ions. By including $n = 0$, we allow P, Q, R, . . . , expressing propositions, as in the propositional calculus; i.e. any atoms we had there which we do not now analyze further are allowed as the $n = 0$ case of n-place ions.

As in § 1, we are remaining silent here about what exactly the underlying object language is, both because we do not wish to be drawn into details about it now, and because we wish to leave the way open to various applications. The object language may be a symbolic language constructed by logicians using logical symbols, and perhaps also mathematical symbols; or it may be a suitably restricted and regulated part of English or some other natural language, without or with mathematical symbols added. Now *we* need to operate with variables; and we now agree to use the small Roman letters "a", "b", "c", . . . , "x", "y", "z", "a_1", "a_2", "a_3", . . . , "x_1", "x_2", "x_3", . . . as names for variables (or their equivalent as in (a_1)–(g) above) in the object language. Our convention here is that distinct small Roman letters will be names for *distinct* variables (or their equivalent) in the object language, *except when* we say they need not be distinct.[55]

Sometimes it will be more convenient to consider the predicates expressed by the ions (with blanks as place holders) as named by expressions with variables (as above we had a choice between "$—_1$ loves $—_2$" and "x loves y"). We call P(x, y, z), P(y, z, x), P(u, v, w), etc. different *name forms* for the same 3-place prime predicate expression or ion P(—, —, —); the variables x, y, z or y, z, x or u, v, w are the *name form variables* in these name forms, and they are to be distinct (as in the three examples). We also call P(x, y, z), P(y, z, x), P(u, v, w) *prime predicate expressions* (or *ions*) *with attached* (*name form*) *variables.* In a particular logical analysis, one name form for a given ion is the most we will need; however we may need to choose the name form variables in it to avoid conflict with variables already being used in other ways (e.g. in Exercise 19.1 below). We do not allow P(x, x, y), P(x, y, x) etc. as name forms for a 2-place prime predicate expression; for these are not prime, but exhibit that the predicate named arises by identifying two of the variables of a 3-place predicate expressed

[55] In fact, we shall reserve "r", "r_1", "r_2", "r_3", . . . as names for variables that need not be distinct from each other and the other variables present. Under our convention in this chapter, however, we shall repeat this each time they are used.

Later we shall be obliged to use some of the Roman lower case letters for other purposes than naming variables (§§ 28, 38, 57).

by the ion P(—, —, —). Because in a 3-place prime predicate expression we exclude considering any further structure (otherwise it wouldn't be prime for us), the blanks in P(—, —, —) are to be filled independently, so we don't need subscripts thus: P($—_1$, $—_2$, $—_3$). Similarly, as a *prime* predicate expression, "— loves —" must be "$—_1$ loves $—_2$"; and "— < —" (or simply "<")[56] must be "$—_1 < —_2$".

Now we are ready to describe a class of sentences assumed to exist in the object language called *formulas*, just as we were in § 1 after introducing the atoms there. But here we are starting further down in the structure of the object language (or assuming the object language has more than the minimum structure assumed there); i.e. we must now start with the ions.

For each n-place ion P(—, . . . , —) and each choice of variables $r_1, \ldots, r_n$ *not necessarily distinct*, P($r_1, \ldots, r_n$) shall be a *prime formula* or *atom*. For example, from the ion P(—, —, —), we obtain as atoms P(x, y, z), P(y, z, x), P(u, v, w), P(x, x, y), P(x, y, x), P(u, u, u), etc. From the ion P, we get just P as atom. From Q(—) we get Q(x), Q(y), Q(u), etc. (With $n = 0$, the atoms of the propositional calculus are again atoms here. With $n > 1$, the atoms are a more extensive class than the name forms of ions, since in the atoms the variables replacing the blanks need not be distinct.)[57]

The *formulas* shall comprise exactly the prime formulas or atoms, and the additional formulas (*composite formulas* or *molecules*) constructible from atoms by repeated introductions of the *logical symbols* ∼, ⊃, &, V, ¬, ∀, ∃, thus. If A and B are any *formulas* (either prime formulas, or composite formulas already constructed), then A ∼ B, A ⊃ B, A & B, A V B and ¬A are (*composite*) *formulas*. If A is any *formula*, and x is any variable, then ∀xA (read "for all x, A") and ∃xA (read "for some x, A" or "(there) exists (an) x (such that) A") are (*composite*) *formulas*.

[56] Since "<" is a permanent notation for a binary predicate. Likewise, in a discussion where P(—, —) but not P, P(—), P(—, —, —), . . . occur, we can write simply "P" for "P(—, —)" without ambiguity; etc.

[57] So long as one is working only in the predicate calculus, as treated in this chapter, "P(x)", "P(x, y)", "Q(x, y, z)", . . . , and the similar expressions "A(x)", "A(x, y)", "B(x, y, z)", . . . with capitals from the beginning of the alphabet (introduced below), can be simplified by omitting parentheses, or commas, or both thus: "Px", "Pxy", "Qxyz", "Ax", "Axy", "Bxyz". We should do so if we planned to dwell on the predicate calculus for a long time. However, we plan to advance rapidly to more complicated systems, where the "arguments" may not be simply variables x, y, z, but also for example 5, 12, xy (= x · y) etc., which would render the notations without commas ambiguous, or even ones that would render the notations without parentheses difficult to read if not actually ambiguous. So we prefer to keep the parentheses and commas that are usually (but not invariably) taken as part of the notation for functions in mathematics. (The student is welcome to omit them in this chapter, if he won't have trouble putting them back later.)

∀x is called a *universal quantifier*, and ∃x an *existential quantifier*. The quantifiers act as unary operators in building formulas, and with our other unary operator ¬ are ranked last under the convention for omitting parentheses. Thus, "∀xA ⊃ B" means (∀xA) ⊃ B, not ∀x(A ⊃ B).[58]

So that there will be no ambiguity as to what the atoms are, we stipulate that none of them be of any of the seven forms A ∼ B, A ⊃ B, A & B, A ∨ B, ¬A, ∀xA, ∃xA which the molecules have. Also each atom shall come unambiguously from just one ion. In brief, the internal structure of the ions (whatever it may be) shall be such that it cannot get mixed up with what is added in building formulas and shown explicitly in our symbolism for formulas.

For example, "— is a man", "— loves —", "— = —", "— < —", "— + — = —", "2 · 2 = 4" could be ions (with 1, 2, 2, 2, 3, 0 places, respectively). Then "x is a man", "y is a man", "x loves y", "x loves x", "$x = y$", "$y = y$", "$x + y = z$", "$x + x = y$", "2 · 2 = 4", etc. will be atoms; and these can include "Socrates is a man" and "Chiron is a man", or "John loves Jane", if we interpret "x" and "y" as names for Socrates or Chiron, or John and Jane, etc. Examples of molecules would then be "x is a man and x loves y", "x loves y or x loves z", "for some x, x loves y" or $\exists x L(x, y)$, etc. as in (a′)–(h′).

In § 1 we emphasized the distinction between our use of P, Q, R, . . . for distinct prime formulas and of A, B, C, . . . for any formulas not necessarily distinct or prime. Here we are again so using A, B, C, . . . ; and presently we shall so use A(x), A(y), B(x, y), etc. These capitals from the beginning of the alphabet, with or without variables, will be names for formulas built up from P, Q, R, . . . , variables, parentheses, commas, and the logical symbols ∼, ⊃, &, ∨, ¬, ∀, ∃; and they can be names for the same or different such formulas.

In integral calculus, $\int_0^y x^2y\,dx$ is not a quantity that depends on x, though it does depend on y. Similarly, $\sum_{n=0}^{\infty} x^n/n!$ does not depend on n, though it does on x. We can express this by saying that in the first expression "x" is a *bound* variable and "y" is *free;* in the second, "n" is bound and "x" is free. In "$3x + \int_0^y x^2y\,dx$", the first occurrence of "x" is free and the other two are bound, while "y" is free in both occurrences. (The notation in the third example is unambiguous, though some people would

[58] The predicate calculus (or what it adds to the propositional calculus) is sometimes called "quantification theory". In the literature, "(∀x)", "(x)", "Λ_x", "Π_x" are often used for our "∀x"; "(∃x)", "(Ex)", "V_x", "Σ_x" for "∃x".

prefer "$3x + \int_0^y t^2y\,dt$".) An example not presupposing calculus is "the least y such that $2y \geq x$"; here "x" is free and "y" is bound.

Similarly, we have *bound* and *free* variables or occurrences of variables in the predicate calculus, where the two operators which bind variables are the quantifiers ∀x and ∃x (rather than $\int \ldots dx$ or $\sum_{n=0}^{\infty}$ or "the least y such that").[59]

Consider the formula

(1) $\forall x(P(x) \,\&\, \exists x Q(x, z) \supset \exists y R(x, y)) \vee Q(z, x)$.

In the part ∃xQ(x, z), each x is bound by the ∃x, which we can indicate by attaching a subscript $_1$ to these two x's to show that they belong together. Similarly we can indicate by subscripts $_2$ and $_3$ the variable occurrences bound by ∃y and ∀x, respectively. Note that, since the x in Q(x, z) is already bound by ∃x, it is not free in the part $P(x) \,\&\, \exists x Q(x, z) \supset \exists y R(x, y)$ on which the ∀x operates (call that part the *scope* of that ∀x; cf. Example 1 in § 1), so the ∀x cannot bind it. In supplying the subscripts we therefore always work from the inside out, following the order of the steps by which the formula is built up from its atoms P(x), Q(x, z), R(x, y), Q(z, x). To standardize the method of numbering, we can further agree to work at each stage on the leftmost "eligible" quantifier, i.e. on the leftmost quantifier whose scope contains no other quantifier not yet treated. In this manner we obtain

(1a) $\forall x_3(P(x_3) \,\&\, \exists x_1 Q(x_1, z) \supset \exists y_2 R(x_3, y_2)) \vee Q(z, x)$.

The variable occurrences not thus receiving subscripts (two of z and one of x) are free. As another example, consider

(2) $\forall y(P(y) \,\&\, \exists x Q(x, z) \supset \exists z R(y, z)) \vee Q(z, x)$.

Supplying subscripts, we get

(2a) $\forall y_3(P(y_3) \,\&\, \exists x_1 Q(x_1, z) \supset \exists z_2 R(y_3, z_2)) \vee Q(z, x)$.

Erasing the bound (occurrences of) variables in (1a) and (2a) gives the same expression from both, namely

(1b), (2b) $\forall\ _3(P(\ _3) \,\&\, \exists\ _1 Q(\ _1, z) \supset \exists\ _2 R(\ _3,\ _2)) \vee Q(z, x)$.

[59] Hilbert and Bernays 1934, 1939 and some other authors use different letters for free and for bound variables, say "a", "b", "c", . . . for free occurrences only and "x", "y", "z", . . . for bound occurrences only (contrary to the practice in informal mathematics). The writer did so during a decade in teaching the material of Part II of IM, before changing in 1946; he is now convinced that using one list of variables for both free and bound occurrences has a slight but definite advantage.

This illustrates that the two formulas (1) and (2) are *congruent*. They are not congruent to (3), (4) or (5) (below); for, upon supplying subscripts and erasing the bound variable occurrences, we would obtain expressions (3b), (4b) and (5b) each differing from (1b). For formulas of moderate length, it may be easier to use lines (instead of subscripts) to indicate which quantifiers bind which variable occurrences, thus:

(1) ∀x(P(x) & ∃xQ(x, z) ⊃ ∃yR(x, y)) ∨ Q(z, x).

(2) ∀y(P(y) & ∃xQ(x, z) ⊃ ∃zR(y, z)) ∨ Q(z, x).

(3) ∀x(P(x) & ∃xQ(x, z) ⊃ ∃xR(x, x)) ∨ Q(z, x).

(4) ∀z(P(z) & ∃xQ(x, z) ⊃ ∃yR(z, y)) ∨ Q(z, x).

(5) ∀x(P(x) & ∃xQ(x, z) ⊃ ∃yR(x, y)) ∨ Q(z, y).

The student can compare these figures to determine congruence or incongruence, disregarding the bound variable occurrences or imagining them erased. In (3) the seventh variable occurrence (an occurrence of x) is bound by the third quantifier (the second ∃x), whereas in (1) the seventh variable occurrence (an occurrence of x) is bound by the first quantifier (the ∀x). In (4) the fifth variable occurrence (an occurrence of z) is bound by the first quantifier (the ∀z), whereas in (1) the fifth variable occurrence is free. In (5) there are no such differences from (1) in the bound variable occurrences, but the last variable occurrence is a free occurrence of y, whereas in (1) it is a free occurrence of x.

Of course, if two formulas are not the same to within the choices of variables, they are incongruent anyway.

Each formula expresses a predicate of the number of variables which occur free in it (those variables serving as name form variables). For example, "x is a man", "$x+x = x$", L(x, x), ∃xL(x, y) express predicates of one variable; "$x < y$", "$x < y \vee x = y$", L(x, y), ∀x(P(x) & ∃xQ(x, z) ⊃ ∃yR(x, y)) ∨ Q(z, x) express predicates of two variables; and "$2 \cdot 2=4$" expresses a predicate of 0 variables, i.e. a proposition. (Formulas can also be considered as expressing predicates of more variables. Thus "$2 \cdot 2 = 4$" also expresses a constant predicate of one variable, or of two variables, etc.; and L(x, x) expresses a predicate of two variables x, y which is constant in the second variable y.)

But as we remarked above in our discussion of notation for predicates, by interpreting the variables as standing for particular objects the formulas come to express not the predicates but propositions taken as their values. (There are other ways in which a formula with free variables can be interpreted as expressing a proposition, as will be discussed in §§ 20, 38.)

The formulas (1) and (2) express the same predicate, with z, x as the name form variables, while (3) and (4) express two other predicates, again with z, x as the name form variables. (These three predicates depend on what predicates the ions P(—), Q(—, —), R(—, —) express.) This should be evident now from the words proposed for reading the quantifiers; and in § 17 it will be further emphasized. Formula (5) expresses the same predicate as (1) and (2), but using instead z, y as the name form variables. If we should prefix ∀x to (1), (2) and (5) after enclosing them in parentheses (to give ∀x their wholes as scope), the third resulting formula would not be congruent to the first two; indeed, the first two would express one-variable predicates, the third still a predicate of two variables. If we interpret x, y, z as names for objects, then (5) will express a different proposition than (1) and (2) if the objects named by x and y are different.

Exercise. 16.1. Show by subscripts (or lines) which quantifiers bind which variable occurrences. Which pairs of formulas are congruent?

(a) ∀z∃y(P(z, y) & ∀zQ(z, x) ⊃ R(z)). (f) ∃z∀x(P(z, x) ∨ ∀zQ(x, y, z)).
(b) ∀x∃y(P(x, y) & ∀yQ(y, x) ⊃ R(x)). (g) ∃z∀x(P(z, x) ∨ ∀xQ(x, y, z)).
(c) ∀y∃z(P(y, z) & ∀zQ(z, x) ⊃ R(y)). (h) ∃y∀x(P(y, x) ∨ ∀xQ(x, y, z)).
(d) ∀z∃x(P(z, x) & ∀zQ(z, y) ⊃ R(z)). (i) ∃y∀x(P(z, x) ∨ ∀xQ(x, u, y)).
(e) ∀y∃z(P(z, y) & ∀zQ(z, x) ⊃ R(y)). (j) ∃x∀z(P(x, z) ∨ ∀uQ(u, y, x)).

§ 17. Model theory: domains, validity. We are now at the stage corresponding to the beginning of § 2 in Chapter I. There we said that, for the classical propositional calculus, each atom (or prime formula) is assumed to express a proposition that is either true or false but not both (but which is the case is not a datum for the propositional calculus).

Now, for the *classical* predicate calculus, we shall wish to make a corresponding assumption about each ion (or prime predicate expression). But the first step toward properly talking about the *n*-place predicate (= propositional function of *n* variables) expressed by an *n*-place ion is to consider what objects may be values of the variables, or in mathematical terminology what are the *ranges* of the variables. In examples like "*x* is a man" or "*x* loves *y*", a meaningful statement, and thus one expressing a proposition which classically we can regard as true or false, will not necessarily result whatever noun is substituted for "*x*" or whatever nouns for "*x*" and "*y*". It is debatable whether "*x* loves *y*" becomes false or simply meaningless, when we substitute for "*x*" and "*y*" names of vegetables. Furthermore, in ordinary language one can find borderline cases when it is not clear whether anything is named by an expression ostensibly serving as a noun. We shut off debate on these issues, at least for now, by assuming that there is a particular nonempty set or collection of objects, called the *domain D*, over which each of the (independent) variables of our propositional functions ranges; i.e. the members of D are the objects to be allowed as values of the variables.

This is not at all a trivial assumption, since it is not always clearly satisfied in ordinary discourse. In mathematics likewise, logic can become pretty slippery when no D has been specified explicitly or implicitly, or the specification of a D is too vague.

In saying that D shall be the range of each variable of our propositional functions, we mean that the predicate expressed by the ion P(—, —), or by its name form P(x, y), becomes a proposition, or as mathematicians say is "defined", for each pair of values chosen for x and y from the set D; and similarly with $P(x_1, \ldots, x_n)$ for any $n > 0$. (For $n = 0$, D is not involved.) For example, the predicate $x < y$ can't be used when D is the set of the complex numbers $a + bi$, since $x < y$ isn't defined (meaningful) for every pair x, y in that D. The predicate $x < y$ can be used when D is the real numbers, or the natural numbers 0, 1, 2, ... But then $\sqrt{x} = y$ is not "always defined". In none of the three D's just mentioned is $x \div y = z$ always defined (it isn't defined when $y = 0$).[60]

For our study of the predicate calculus in general (i.e. without a stated further restriction), we shall consider that we do not know what nonempty set D is. In other words, we undertake to develop the predicate logic applicable to any nonempty D whatsoever. Thus we are excluding from the sets as possible choices of D only the empty set; i.e. there shall be at least one object in the range of our variables. (E.g. D cannot be the real roots of the equation $x^2 + 1 = 0$.)[61]

In mathematics, often different variables are employed with different ranges, like $x, y, z, \ldots$ ranging over real numbers and $m, n, p, \ldots$ over natural numbers. To keep matters as simple as possible, we are not providing for this now; all our variables are to have the same range D (though what this D is can be varied in our applications). It is not difficult, after the predicate calculus has been treated thus (as *one-sorted*) to proceed to predicate calculus with two sorts of variables, some with one range D_1 and some with another D_2, called *two-sorted* predicate calculus. Similarly, with more than two sorts.[62]

[60] In such situations, mathematicians often get around the difficulty, if they need to, by extending the D or extending the definition of the predicate on the original D.

[61] The predicate calculus with the empty domain allowed presents some differences (that on the whole are not advantages) compared to the treatment here. Cf. Mostowski 1951a, Jaśkowski 1934.

[62] See IM pp. 179–180, and references given there (or, for higher-order predicate calculi, Hilbert and Ackermann 1949, Church 1956).

Another form of predicate calculus treats the ions as variables which may also be quantified, so ∀P, ∀Q, ∃P, ∃Q, . . . become part of the symbolism. This gives a more considerable extension of the predicate calculus (as we study it), called *second-order* predicate calculus; and on iteration, *higher-order* predicate calculi.[62] To distinguish the form of the predicate calculus which we treat from these, it may be called *restricted* or *lower* or *first-order* predicate calculus.[63]

Now we make one more assumption, for the *classical* predicate calculus, as we intimated we would do paralleling the treatment in § 2. This assumption is that, for each pair of values of x, y taken from *D*, the proposition which results as value of P(x, y) is either true (t) or false (f) but not both. (However we are not told which is the case.) Considering that this happens for each x, y in *D*, it comes to the same thing to say that there is correlated to the ion P(—, —) or its name form P(x, y) a function I(x, y) which, for each pair of values of x, y in *D*, takes either t or f as value (in mathematical terminology, a function I(x, y) from $D \times D$ to {t, f}).[64] Such a function I(x, y) we call a (*2-place*) *logical function*. Similarly, to each ion $P(x_1, \ldots, x_n)$ with *n* attached variables, an *n*-place logical function $I(x_1, \ldots, x_n)$ is correlated. In the case $n = 0$, i.e. simply P, the logical function $I(x_1, \ldots, x_n)$ is simply a t or f, as in the propositional calculus.

The truth tables given for ∼, ⊃, &, ∨, ¬ in the propositional calculus (§ 2) shall again apply. We also define now the process of evaluating ∀xA and ∃xA. We shall have occasion to evaluate these only when we are already in a position to evaluate A for each choice of a member of *D* as value of x at its free occurrences in A, or briefly when we can evaluate A by a logical function of x. We define ∀xA to be true (t) if this logical

[63] Our one kind of variables may be called "individual variables" or "object variables" to emphasize that they range over the domain *D* of individuals or objects. This is in contrast to e.g. "predicate variables" in second-order predicate calculus, or in first-order predicate calculus with a postulated substitution rule for such variables.

[64] Here we are deviating from our usual notational conventions, under which we should (and occasionally will) write instead: a function I(x, y) which, for each x, y in *D* (or for each pair of values of "x" and "y" in *D*) takes either t or f as value.

For, it is convenient to use the same variables x, y in our name "I(x, y)" for the logical function as in the name form P(x, y) for the ion P(—, —) to which it is correlated or assigned as value. So here, where the actual variables in the formula P(x, y) are unspecified, we use our names "x" and "y" for them in our name "I(x, y)" for the logical function. Similarly, with $P(x_1, \ldots, x_n)$ and "$I(x_1, \ldots, x_n)$" (and rarely in naming predicates expressed by formulas).

Then, in the computations or evaluations (below) we shall simply substitute our names for logical functions ("$I_1(x)$", "$I_2(x)$", "$I_1(x, y)$", etc.), for truth values ("t", "f") and for members of *D* ("1", "2", etc.) into the formulas of the object language (or into our names for them). So there we manipulate an object language augmented by "$I_1(x)$", "t", "1", etc. (We did this already with "t", "f" in Chapter I; and we shall do the like, with also functions having values in *D*, in Chapter III §§ 28, 29, etc.)

function has 𝔱 for all its values, otherwise false (𝔣); and $\exists x A$ to be 𝔱 if this logical function has at least one 𝔱 among its values, otherwise 𝔣.

Now, can we compute a truth table for any formula E? To begin with, *D*, though supposed fixed, is unknown. Actually, only the number $\overline{\overline{D}}$ (> 0) of members of *D* (still unknown) matters.[65]

EXAMPLE 1. For illustration, however, let us suppose *D* is a domain of two objects, which for convenience we write simply "1" and "2"; i.e. $D = \{1, 2\}$. Take as E the formula $P(y) \vee \forall x(P(x) \supset Q)$. To compute a truth value for this, we must start from an *assignment* consisting of a logical function of one variable ranging over *D* as value of the ion P(—) or its name form P(x), a truth value (or logical function of zero variables) as value of Q, and a member of *D* as value of the free variable y; i.e. we shall compute a table to be entered from these three quantities. Before computing this table, we list the 4 ($= 2^2$) possible logical functions of one variable over *D* ($= \{1, 2\}$), as follows:

x	$\mathfrak{l}_1(x)$	$\mathfrak{l}_2(x)$	$\mathfrak{l}_3(x)$	$\mathfrak{l}_4(x)$
1	𝔱	𝔱	𝔣	𝔣
2	𝔱	𝔣	𝔱	𝔣

Here is the table for $P(y) \vee \forall x(P(x) \supset Q)$:

		P(x)	Q	y	$P(y) \vee \forall x(P(x) \supset Q)$
	1.	$\mathfrak{l}_1(x)$	𝔱	1	𝔱
	2.	$\mathfrak{l}_1(x)$	𝔱	2	𝔱
	3.	$\mathfrak{l}_1(x)$	𝔣	1	𝔱
	4.	$\mathfrak{l}_1(x)$	𝔣	2	𝔱
	5.	$\mathfrak{l}_2(x)$	𝔱	1	𝔱
	6.	$\mathfrak{l}_2(x)$	𝔱	2	𝔱
	7.	$\mathfrak{l}_2(x)$	𝔣	1	𝔱
(1)	8.	$\mathfrak{l}_2(x)$	𝔣	2	𝔣
	9.	$\mathfrak{l}_3(x)$	𝔱	1	𝔱
	10.	$\mathfrak{l}_3(x)$	𝔱	2	𝔱
	11.	$\mathfrak{l}_3(x)$	𝔣	1	𝔣
	12.	$\mathfrak{l}_3(x)$	𝔣	2	𝔱
	13.	$\mathfrak{l}_4(x)$	𝔱	1	𝔱
	14.	$\mathfrak{l}_4(x)$	𝔱	2	𝔱
	15.	$\mathfrak{l}_4(x)$	𝔣	1	𝔱
	16.	$\mathfrak{l}_4(x)$	𝔣	2	𝔱

[65] As will appear below (§ 34), we can talk about the number $\overline{\overline{D}}$ even when *D* is infinite.

Here is the computation for the entry in Line 8 (explanation follows):

	$P(y) \vee \forall x(P(x) \supset Q)$
(i)	$I_2(2) \vee \forall x(I_2(x) \supset \mathfrak{f})$
(ii)	$\mathfrak{f} \quad \vee \quad \mathfrak{f}$
(iii)	$\mathfrak{f}$

The first step is to substitute the assignment represented by Line 8 into the formula to be computed; this gives (i). Toward (ii), we get $\mathfrak{f}$ as the value of $I_2(2)$ by the table for $I_2(x)$ above. But before we can evaluate the other part $\forall x(I_2(x) \supset \mathfrak{f})$ of (i), we need to compute $I_2(x) \supset \mathfrak{f}$ as a logical function of x; the result is shown in the supplementary table at the left below; and the computations of its two lines are at the right.

x	$I_2(x) \supset \mathfrak{f}$
1	$\mathfrak{f}$
2	t

$I_2(x) \supset \mathfrak{f}$	$I_2(x) \supset \mathfrak{f}$
$I_2(1) \supset \mathfrak{f}$	$I_2(2) \supset \mathfrak{f}$
$t \supset \mathfrak{f}$	$\mathfrak{f} \supset \mathfrak{f}$
$\mathfrak{f}$	t

Continuing the main computation, since the supplementary table does not have all t's, $\forall x(I_2(x) \supset \mathfrak{f})$ is evaluated as $\mathfrak{f}$; so we get (ii). Finally we get (iii) by the table for V.

This illustrates the definition of the table for a formula E, for the given *D*. As before, shortcuts are possible. In our example, the observation that $A \supset B$ is t whenever B is t shows that, whenever Q is t, $P(x) \supset Q$ will have a supplementary table of all t's, so by our prescription for evaluating $\forall xA$ the value of $\forall x(P(x) \supset Q)$ will be t, so by the table for V the whole will be t. Thus we can write t in Lines 1, 2, 5, 6, 9, 10, 13, 14 without further ado.[66]

EXAMPLE 2. Again with $D = \{1, 2\}$, we give several lines of the table for $\forall x(\exists xP(x) \supset P(x))$ & P(x).

	P(x)	x	$\forall x(\exists xP(x) \supset P(x))$ & P(x)
1.	$I_1(x)$	1	t
2.	$I_1(x)$	2	t
3.	$I_2(x)$	1	$\mathfrak{f}$
	. . .		. . .
8.	$I_4(x)$	2	$\mathfrak{f}$

[66] The method of truth-value analysis, illustrated by (3) in § 8, can be applied successively to each proposition letter in E (i.e. in this example, just to Q).

Here is the computation for the entry in Line 3.

$$\forall x(\exists x P(x) \supset P(x)) \;\&\; P(x)$$

(i)	$\forall x(\exists x I_2(x) \supset I_2(x)) \;\&\; I_2(1)$
(ii)	f t
(iii)	f

In (i), we use the value 1 of x only at the free occurrence of x (the last one) in $\forall x(\exists x P(x) \supset P(x)) \;\&\; P(x)$. To get (ii) we need a supplementary table for $\exists x I_2(x) \supset I_2(x)$ as a function of x; in setting this up we ignore the value 1 previously assigned to x for the whole formula $\forall x(\exists x P(x) \supset P(x)) \;\&\; P(x)$. The supplementary table follows at the left, with the computations of its two lines at the right.

x	$\exists x I_2(x) \supset I_2(x)$
1	t
2	f

$$\begin{array}{ccc} \exists x I_2(x) \supset I_2(x) & & \exists x I_2(x) \supset I_2(x) \\ \exists x I_2(x) \supset I_2(1) & & \exists x I_2(x) \supset I_2(2) \\ t \quad \supset t & & t \quad \supset f \\ t & & f \end{array}$$

For the two computations at the right, we need a further supplementary table, namely for $I_2(x)$ (ignoring the values already given to x for the respective lines of the supplementary table at the left which we are engaged in computing); however this is simply the table for $I_2(x)$ as given preceding (1) in Example 1.

It should be clear from these examples (and the others below) that, provided a domain D has been selected and is finite, a table can be computed for any given formula E, at least in theory (i.e. disregarding practical limitations). Of course, for large finite $\bar{\bar{D}}$ (or even for a small $\bar{\bar{D}}$, if E is complicated), if no shortcuts are used, the computation may be of impractical length. If D is infinite, the table is no longer a finite object which in theory can be computed; but what is meant by the table should be clear enough (from the standpoint of classical mathematics), and we may be able to reason about it. When D may be infinite, we shall avoid the word "compute" and say instead "evaluate" or "determine".

When can a formula E be said to be true on the basis of only the predicate calculus? Considering that both the D (or $\bar{\bar{D}}$), and the logical functions over D as values of the ions in E (or truth values in the case of zero variables) and the members of D as values of the free variables of E, will be unavailable, the answer must be as follows: The formula E is true on the basis of the predicate calculus, exactly if, for each choice of D (or of the number $\bar{\bar{D}}$ of its elements), the resulting truth table has only t's in its value column. In this case we say E is *valid* (*in the predicate calculus*) and write $\models$ E. (In this chapter, it will be understood that "valid" and "$\models$" refer to the predicate calculus, unless the contrary is stated.)

It is also often of interest to consider the predicate calculus supplemented by a choice of D (or of the number $\overline{\overline{D}}$ of members of D); we then say E is *valid in the domain D* or is *$\overline{\overline{D}}$-valid*, and write $\overline{\overline{D}}$-⊨ E, exactly if the truth table of E for the chosen D has all t's. Interesting cases are $\overline{\overline{D}} = k$ (a positive integer), and D is the set of the natural numbers {0, 1, 2, . . .}.

There is a vast difference now from the situation we had in the propositional calculus. There each question as to the validity of a formula E could (in theory) be settled mechanically by computing the truth table. Now the definition of validity refers to a whole infinite family of truth tables, one for each $\overline{\overline{D}}$, and for an infinite $\overline{\overline{D}}$ we cannot (even theoretically) compute the table. For validity, every one of these tables should give all t's. Despite this difficulty, we shall see that logical theory has gone quite far in solving problems of the predicate calculus.

For a demonstration of invalidity, it suffices to find just one D and one line of the table for this D which gives f. Thus we already know that P(y) ∨ ∀x(P(x) ⊃ Q) is not valid, because we found an f in Line 8 of its table (1) for $\overline{\overline{D}} = 2$.

The notion of validity, e.g. as applied to the E of Example 1, arose by considering E as expressing a proposition with its free variable y naming some member of D. Since we were in ignorance, not only of D and the values of P(—) and Q, but also of the member of D named by y, we could know (using the predicate calculus and nothing else) that E is true when and only when, for every D, the table for E has all t's.

But formulas also serve as names for predicates, as we stressed in § 16. Indeed, we constructed the supplementary table for $\mathfrak{l}_2$(x) ⊃ f from this point of view, and not because we were interested in whether P(x) ⊃ Q expresses a true or false proposition. For the chosen D, and the assigned values of P(x) and Q, the supplementary table gives the logical function which then evaluates P(x) ⊃ Q interpreted as standing for a predicate of x.[67,64]

If we are interested in the whole formula E as expressing a predicate

[67] We can think of truth values and logical functions as X-ray pictures of propositions and predicates. Thus "1 < 2", "$\int_0^{\pi/2} \sin x\, dx = 1$", "Socrates is a man" and "Madison is an inland city" are four propositions, with quite different meanings; but under the X-ray the flesh becomes invisible, and only t shows. Likewise, "2 < 2", "$\int_0^{\pi/2} \sin x\, dx = 2$", "Chiron is a man" and "New York is an inland city" are four other propositions, which appear under the X-ray as simply f. For $D = \{1, 2\}$: "$x < 2$" and "$\int_0^{\pi/2} \sin x\, dx = x$" when 1 and 2 are themselves, "x is a man" when 1 is Socrates and 2 is Chiron, and "x is an inland city" when 1 is Madison and 2 is New York, are four different predicates (their respective values, listed above, are eight different propositions); but under the X-ray all four of them appear as the logical function $\mathfrak{l}_2(x)$.

rather than a proposition, the distribution of 𝔱's and 𝔣's in its truth table will interest us, not just whether they are all 𝔱's or not all 𝔱's. This we can illustrate by going one step further to construct the formula ∀yE or the formula ∃yE from the formula E of Example 1. For $D = \{1, 2\}$, ∀yE and ∃yE each have an 8-line truth table, since their tables are entered from a value of P(x) and of Q, but not of y. The supplementary tables that we need in the next to the last step of computation appear as subtables of (1); indeed, the pairs of lines 1–2, 3–4, . . . , 15–16 give these supplementary tables. Thus inspection of (1) enables us at a glance to write the tables for ∀yE and ∃yE:

		P(x)	Q	∀y(P(y) ∨ ∀x(P(x) ⊃ Q))	∃y(P(y) ∨ ∀x(P(x) ⊃ Q))
	1.	$\mathfrak{l}_1(x)$	𝔱	𝔱	𝔱
	2.	$\mathfrak{l}_1(x)$	𝔣	𝔱	𝔱
	3.	$\mathfrak{l}_2(x)$	𝔱	𝔱	𝔱
(2)	4.	$\mathfrak{l}_2(x)$	𝔣	𝔣	𝔱
	5.	$\mathfrak{l}_3(x)$	𝔱	𝔱	𝔱
	6.	$\mathfrak{l}_3(x)$	𝔣	𝔣	𝔱
	7.	$\mathfrak{l}_4(x)$	𝔱	𝔱	𝔱
	8.	$\mathfrak{l}_4(x)$	𝔣	𝔱	𝔱

From (2), we see that ∀y(P(y) ∨ ∀x(P(x) ⊃ Q)) is not valid, but ∃y(P(y) ∨ ∀x(P(x) ⊃ Q)) is at least 2-valid.

EXAMPLE 3. As another illustration, we shall show that ∀x∃yP(x, y) ⊃ ∃y∀xP(x, y) is not valid, by computing one suitable line of its table for the domain $D = \{1, 2\}$. This table must be entered from a logical function as value of P(x, y). We first list the 16 (= 2^4) possible such logical functions, as follows:

x	y	$\mathfrak{l}_1(x, y)$	$\mathfrak{l}_2$	$\mathfrak{l}_3$	$\mathfrak{l}_4$	$\mathfrak{l}_5$	$\mathfrak{l}_6$	$\mathfrak{l}_7$	$\mathfrak{l}_8$	$\mathfrak{l}_9$	$\mathfrak{l}_{10}$	$\mathfrak{l}_{11}$	$\mathfrak{l}_{12}$	$\mathfrak{l}_{13}$	$\mathfrak{l}_{14}$	$\mathfrak{l}_{15}$	$\mathfrak{l}_{16}$
1	1	𝔱	𝔱	𝔱	𝔱	𝔱	𝔱	𝔱	𝔱	𝔣	𝔣	𝔣	𝔣	𝔣	𝔣	𝔣	𝔣
1	2	𝔱	𝔱	𝔱	𝔱	𝔣	𝔣	𝔣	𝔣	𝔱	𝔱	𝔱	𝔱	𝔣	𝔣	𝔣	𝔣
2	1	𝔱	𝔱	𝔣	𝔣	𝔱	𝔱	𝔣	𝔣	𝔱	𝔱	𝔣	𝔣	𝔱	𝔱	𝔣	𝔣
2	2	𝔱	𝔣	𝔱	𝔣	𝔱	𝔣	𝔱	𝔣	𝔱	𝔣	𝔱	𝔣	𝔱	𝔣	𝔱	𝔣

Here is the truth table for our formula showing the entry for Line 10; the fact this entry is 𝔣 proves the invalidity of the formula.

	P(x, y)	∀x∃yP(x, y) ⊃ ∃y∀xP(x, y)
	. . .	. . .
10.	$\mathfrak{l}_{10}(x, y)$	𝔣
	. . .	. . .

Here is the computation for the entry in Line 10 (explanation follows):

	$\forall x\exists yP(x, y) \supset \exists y\forall xP(x, y)$
(i)	$\forall x\exists yl_{10}(x, y) \supset \exists y\forall xl_{10}(x, y)$
(ii)	t $\supset$ f
(iii)	f

The first step is to substitute the assignment represented by Line 10 into the formula to be computed; this gives (i).

Before we can evaluate the part $\forall x\exists yl_{10}(x, y)$, we need to compute $\exists yl_{10}(x, y)$ as a logical function of x. We construct the following table (a), in which we tabulate this logical function (explanation follows).

(a)

x	$\exists yl_{10}(x, y)$
1	t
2	t

To compute the truth values in (a), we must in turn construct subsidiary tables as follows, (b) for the case in which the value of x is 1, and (c) for the case in which x is 2.

(b)

y	$l_{10}(1, y)$
1	f
2	t

(c)

y	$l_{10}(2, y)$
1	t
2	f

The values in (b) and (c) we of course obtain from our table of the possible logical functions of two variables over *D*. In (b) a t appears, so that by the evaluation rule for the existential quantifier $\exists y$ we get t in Line 1 of (a); and similarly in Line 2.

In Table (a) we now have all t's, so by the evaluation rule for $\forall$ we get t to use as the value of $\forall x\exists yl_{10}(x, y)$ in (ii).

Proceeding similarly to evaluate $\exists y\forall xl_{10}(x, y)$, we need to compute $\forall xl_{10}(x, y)$ as a logical function of y. We construct the following table (a′), in which we tabulate the truth values of this function.

(a′)

y	$\forall xl_{10}(x, y)$
1	f
2	f

To determine the truth values in (a′), we must again construct two subsidiary tables (b′) and (c′).

(b′)

x	$l_{10}(x, 1)$
1	f
2	t

(c′)

x	$l_{10}(x, 2)$
1	t
2	f

In each of (b′) and (c′) we fail to have a solid column of $\mathfrak{t}$'s, so by the rule for evaluating $\forall$ the two values in (a′) are $\mathfrak{f}$. Since we do not have any $\mathfrak{t}$ in the value column of (a′), the rule for evaluating $\exists$ gives us $\mathfrak{f}$ as the value of $\exists y \forall x \mathfrak{l}_{10}(x, y)$ to use in (ii).

Thus we reach the situation depicted in (ii) of the computation of the entry in Line 10, and can proceed to (iii) to get the value $\mathfrak{f}$ in Line 10 for the whole formula.

EXAMPLE 4. We shall now show the formula $P(y) \supset \exists x P(x)$ to be valid. In doing so, we cannot help using some general reasoning, as we must show that we get all $\mathfrak{t}$'s in the table for any $\bar{D}$. However, to help ourselves picture the situation, we begin by taking $D = \{1, 2, 3\}$. The 1-place logical functions are now 8 ($= 2^3$) in number, as follows:

x	$\mathfrak{l}_1(x)$	$\mathfrak{l}_2(x)$	$\mathfrak{l}_3(x)$	$\mathfrak{l}_4(x)$	$\mathfrak{l}_5(x)$	$\mathfrak{l}_6(x)$	$\mathfrak{l}_7(x)$	$\mathfrak{l}_8(x)$
1	$\mathfrak{t}$	$\mathfrak{t}$	$\mathfrak{t}$	$\mathfrak{t}$	$\mathfrak{f}$	$\mathfrak{f}$	$\mathfrak{f}$	$\mathfrak{f}$
2	$\mathfrak{t}$	$\mathfrak{t}$	$\mathfrak{f}$	$\mathfrak{f}$	$\mathfrak{t}$	$\mathfrak{t}$	$\mathfrak{f}$	$\mathfrak{f}$
3	$\mathfrak{t}$	$\mathfrak{f}$	$\mathfrak{t}$	$\mathfrak{f}$	$\mathfrak{t}$	$\mathfrak{f}$	$\mathfrak{t}$	$\mathfrak{f}$

The table for $P(y) \supset \exists x P(x)$ will have 24 ($= 8 \cdot 3$) lines, since all 8 logical functions have to be listed under $P(x)$, each with each of the 3 members of D as y. We show two lines as a sample.

	P(x)	y	$P(y) \supset \exists x P(x)$
14.	$\mathfrak{l}_5(x)$	2	$\mathfrak{t}$
22.	$\mathfrak{l}_8(x)$	1	$\mathfrak{t}$

For Line 14, note that $\mathfrak{l}_5(x)$ has a $\mathfrak{t}$ in its table, e.g. for $x = 2$. Hence by the rule for evaluating $\exists x$, the part $\exists x P(x)$ takes the value $\mathfrak{t}$; so by the table for $\supset$, the whole is $\mathfrak{t}$. This consideration suffices for the first 21 lines of the table, in each of which the $\mathfrak{l}(x)$ has a $\mathfrak{t}$ in its table; i.e. it suffices for every $\mathfrak{l}(x)$ except $\mathfrak{l}_8(x)$. For Line 22 on the other hand, $\mathfrak{l}_8(1)$ is $\mathfrak{f}$; so by the table for $\supset$, the whole is $\mathfrak{t}$. This consideration suffices for the last three lines, in which, since $\mathfrak{l}_8(x)$ has only $\mathfrak{f}$'s in its table, $\mathfrak{l}_8(y)$ and thus $P(y)$ will be $\mathfrak{f}$ whatever y is. It should be clear now that for any D, even an infinite one, the table for $P(y) \supset \exists x P(x)$ will have all $\mathfrak{t}$'s. The demonstration is given by classifying the assignments to $P(x)$, y (or the "lines"). First, consider any assignment with an $\mathfrak{l}(x)$ as value of $P(x)$ other than the logical function with all $\mathfrak{f}$'s; then $\exists x P(x)$ is $\mathfrak{t}$, so the whole is $\mathfrak{t}$. Second, consider any line with the $\mathfrak{l}(x)$ whose table is all $\mathfrak{f}$'s as value of $P(x)$; then $P(y)$ is $\mathfrak{f}$ whatever the value assigned to y, so again the whole is $\mathfrak{t}$.

The same reasoning shows that $P(x) \supset \exists x P(x)$ is valid; for any assignment, the value assigned to x for the whole formula $P(x) \supset \exists x P(x)$ is ignored in evaluating $\exists x P(x)$ from the logical function assigned as value to P(—). (Cf. Example 2.)

Altogether, we can thus say that $\models P(r) \supset \exists x P(x)$, for any variables x and r, where r is not necessarily distinct from x.

EXERCISES. 17.1. How many lines are there in the truth table for $D = \{1, 2\}$? Compute in full detail the line indicated.
(a) $\forall z(P(x) \supset \neg Q \vee P(z))$ when P(x), Q, x are $l_3(x)$, t, 2.
(b) $P(x, y) \supset \forall x(P(x, y) \supset \exists x P(x, x))$ when P(x, y), x, y are $l_{14}(x, y)$, 2, 1.

17.2. Show that $\forall x \exists y P(x, y) \supset \exists y \forall x P(x, y)$ is 1-valid, by computing in full its table for $D = \{1\}$. (By Example 3, it is not 2-valid.)

17.3. Show that each of the following formulas is invalid.
(a) $\neg[\forall x \exists y P(x, y) \supset \exists y \forall x P(x, y)]$. (b) $\exists x \exists y P(x, y) \supset \exists x P(x, x)$.
(c) $\exists x P(x) \ \& \ \exists x Q(x) \supset \exists x(P(x) \ \& \ Q(x))$.

17.4. Give a demonstration by cases (classifying the assignments) that $\forall x P(x) \supset P(y)$ (and $\forall x P(x) \supset P(x)$) is valid.

17.5. Is the formula valid? (Show why or why not.)
(a) $P(x) \supset \forall x P(x)$. (b) $\exists x P(x) \supset P(x)$.
(c) $\forall x P(x) \supset \exists x P(x)$. (d) $\exists x P(x) \supset \forall x P(x)$.
(e) $\exists y(P(y) \vee \forall x(P(x) \supset Q))$. (Cf. the right column in (2).)

17.6. Show that, for any variable x and formula A: $\models$ A if and only if $\models \forall x$A. Similarly, with more variables.

17.7*. Find (a) a formula which is 1-valid and 2-valid but not 3-valid, and (b) a formula which is 1-, 2- and 3-valid but not 4-valid.

§ 18. Model theory: basic results on validity. It is a little fussy to extend Theorem 1 § 3 to the predicate calculus, so we postpone that to § 19 (Theorem 17). However, in the special case that the formula E into which we substitute is a formula of the propositional calculus, the reasoning we used in § 2 suffices. Thus: THEOREM 1 holds (with "$\models$" referring to the predicate calculus) when E is any formula of the propositional calculus containing only the atoms (i.e. 0-place ions) $P_1, \ldots, P_n$ but $A_1, \ldots, A_n$ are any formulas of the predicate calculus. Consequently: THEOREM 2 holds when A, B, C are any formulas of the predicate calculus. Also by the same reasoning as before: THEOREM 3 holds when A, B are any formulas of the predicate calculus. We defer the extensions of Theorems 4–7a to § 19.

In the next theorem (Theorem 15), we generalize the results of Example 4 and Exercise 17.4. This will come under the extension of Theorem 1 to the predicate calculus (Theorem 17), but again the special case in question is simpler.

To generalize the result that ⊨ P(r) ⊃ ∃xP(x) (Example 4), we shall replace the two atoms P(x) and P(r) by any two formulas suitably related to each other, i.e. so related that the reasoning in Example 4 will still apply. Toward this end, consider any formula whatsoever, which we shall call "A(x)" rather than simply "A". Now we denote by "A(r)" the result of substituting r for the free occurrences of x in A(x). For example, if A(x) is ∀z(Q(x) ∨ ∀xP(x, y) ∨ P(z, x)) and r is y, then A(r) is ∀z(Q(y) ∨ ∀xP(x, y) ∨ P(z, y)). With this notation, our proposed generalization of "⊨ P(r) ⊃ ∃xP(x)" will read "⊨ A(r) ⊃ ∃xA(x)", as though it simply consists in changing "P" to "A"; but we must not forget what is behind the notation.

Will the reasoning which in Example 4 established that ⊨ P(r) ⊃ ∃xP(x) now give that ⊨ A(r) ⊃ ∃xA(x)? It will *if* A(r) *differs from* A(x) *exactly by* A(r) *having a free occurrence of* r *in each position where* A(x) *has a free occurrence of* x. For then, whatever domain *D* we consider and whatever assignment (or line in the table) for that *D*, the value of A(r) will be among the values in the supplementary table for A(x) used in evaluating ∃xA(x); namely, the value of A(r) will be the same as the value of A(x) when x has the value of r. This is what was essential to our reasoning in Example 4. (A concrete example will follow.)

By the way we obtained A(r) from A(x), A(r) differs from A(x) exactly by A(r) having an occurrence of r wherever A(x) has a free occurrence of x. But are *these* occurrences of r in A(r) all free? It depends on what variables and formula x, r and A(x) are. If these occurrences (i.e. the occurrences of r in A(r) *resulting* from the substitution of r for the free occurrences of x in A(x)) are all free, we say that r is *free for* x *in* A(x) or that the substitution of r for x in A(x) (with result A(r)) is *free*. In this case, as we have said, the former reasoning carries over and establishes that A(r) ⊃ ∃xA(x).

Thus we obtain (b) of the theorem. Similarly, (a) generalizes Exercise 17.4. First, we repeat the key notational convention and definition. Whenever we introduce a notation like "A(x)" for a formula showing a variable x in the notation, we shall thereafter understand by "A(r)" for any variable r the result of substituting r for the free occurrences of x in A(x). (It is not required that the formula denoted by "A(x)" actually contain x free; if A(x) does not contain x free, then A(r) is simply A(x) itself. Also it is not excluded that A(x) may contain free other variables than x.) We call r *free for* x *in* A(x), or say the substitution is *free*, if the resulting occurrences of r in A(r) are free.

THEOREM 15. *Let* x *be any variable*, A(x) *be any formula*, r *be any variable not necessarily distinct from* x, *and* A(r) *be the result of substituting* r *for the free occurrences of* x *in* A(x). *If* r *is free for* x *in* A(x), *then:*

(a) $\models \forall x A(x) \supset A(r)$. (b) $\models A(r) \supset \exists x A(x)$.

The proof has already been indicated; but we shall illustrate it in an example. Also we shall show how the reasoning (and also the conclusion) fails in another example that does not satisfy the final hypothesis.

EXAMPLE 5. Let A(x) be $\forall z(Q(x) \vee \forall x P(x, y) \vee P(z, x))$ and r be y. Then A(r) is $\forall z(Q(y) \vee \forall x P(x, y) \vee P(z, y))$, where exactly the first and third occurrences of y result from the substitution for the free occurrences of x in A(x). These occurrences of y are both free. Thus r is free for x in A(x). So the theorem applies, and tells us that $A(r) \supset \exists x A(x)$, i.e.

$$\forall z(Q(y) \vee \forall x P(x, y) \vee P(z, y)) \supset \exists x \forall z(Q(x) \vee \forall x P(x, y) \vee P(z, x)),$$

is valid. To illustrate how the proof of the theorem applies to this case, consider for example the domain $D = \{1, 2\}$ and the assignment of $\mathfrak{l}_4(x)$, $\mathfrak{l}_7(x, y)$, 2 to Q(x), P(x, y), y respectively (cf. Examples 1 and 3). We get the two entries in the supplementary table for A(x) entered from x (as required to evaluate $\exists x A(x)$) by evaluating the two expressions

A(1): $\forall z(\mathfrak{l}_4(1) \vee \forall x \mathfrak{l}_7(x, 2) \vee \mathfrak{l}_7(z, 1))$,

A(2): $\forall z(\mathfrak{l}_4(2) \vee \forall x \mathfrak{l}_7(x, 2) \vee \mathfrak{l}_7(z, 2))$,

while the value of A(r) is that of the expression

A(r): $\forall z(\mathfrak{l}_4(2) \vee \forall x \mathfrak{l}_7(x, 2) \vee \mathfrak{l}_7(z, 2))$;

the latter is the second of the two expressions to be evaluated for the supplementary table. If fact, the supplementary table has only $\mathfrak{f}$'s (as in the second case in Example 4); and since the value of A(r) is one of the values in the supplementary table (the second in fact), it is $\mathfrak{f}$, so $A(r) \supset \exists x A(x)$ is $\mathfrak{t}$. If we change the example to use $\mathfrak{l}_6(x, y)$ instead of $\mathfrak{l}_7(x, y)$, then the supplementary table has a $\mathfrak{t}$ (in the first line, for A(1)); so $\exists x A(x)$ is $\mathfrak{t}$, and $A(r) \supset \exists x A(x)$ is $\mathfrak{t}$ anyway. Similarly whatever the domain D and the assignment, $A(r) \supset \exists x A(x)$ will be $\mathfrak{t}$ in one of the two ways illustrated.

EXAMPLE 6. Let A(x) be as in Example 5, but let r be z. Then A(r) is $\forall z(Q(z) \vee \forall x P(x, y) \vee P(z, z))$. Exactly the second and fourth occurrences of z result from the substitution; and these are not both free (in fact, neither is). So the substitution is not free, and the theorem does not apply. To see how the reasoning which establishes the theorem fails to apply here, consider for example the domain $D = \{1, 2\}$ and the assignment $\mathfrak{l}_4(x)$, $\mathfrak{l}_7(x, y)$, 2. The values in the supplementary table for A(x) are those of A(1) and A(2) as in Example 5, but the value of A(r) is that of

A(r): $\forall z(\mathfrak{l}_4(z) \vee \forall x \mathfrak{l}_7(x, 2) \vee \mathfrak{l}_7(z, z))$.

In this example, the latter expression is not one of the former two. In fact, the latter is $\mathfrak{t}$, while as before both the former are $\mathfrak{f}$ and hence $\exists x A(x)$ is $\mathfrak{f}$; so $A(r) \supset \exists x A(x)$ is $\mathfrak{f}$. Therefore $A(r) \supset \exists x A(x)$ is not valid. —

In the next theorem, we again denote a formula by a notation "A(x)" showing a variable x. (The formula denoted by "A(x)" is not required to contain x free; it may contain other variables free.) In this case, we do so to contrast A(x) with another formula, denoted by "C", which shall not contain x free. (Usually, when we simultaneously denote some formulas by notations showing a variable x and others by notations not showing x, this is to help us remember that the latter formulas are not to contain x free, while the former are allowed but not required to contain x free.)

THEOREM 16. *Let* x *be any variable,* A(x) *be any formula, and* C *be any formula not containing any free occurrence of* x. *Then:*

(a) *If* $\models$ C $\supset$ A(x), *then* $\models$ C $\supset$ $\forall$xA(x).

(b) *If* $\models$ A(x) $\supset$ C, *then* $\models$ $\exists$xA(x) $\supset$ C.

PROOF. (a) Suppose $\models$ C $\supset$ A(x). We must show that $\models$ C $\supset$ $\forall$xA(x). Choose any domain *D*. For this *D*, consider any assignment of logical functions and members of *D* to exactly the ions and free variables of C $\supset$ $\forall$xA(x) (i.e. consider any line of the table entered from exactly these); call this the "given assignment". Since x does not occur free in C, this does not include an assignment of a member of *D* to x. CASE 1: for the given assignment, C is f. Then, by the table for $\supset$, C $\supset$ $\forall$xA(x) is t. CASE 2: for the given assignment, C is t. Then, for the given assignment supplemented by *any* assignment to x, C is still t (cf. the discussion preceding Theorem 3), so, since C $\supset$ A(x) is t (by the hypothesis $\models$ C $\supset$ A(x)), A(x) is t (by the table for $\supset$). As this was for *any* assignment to x together with the given assignment, $\forall$xA(x) is t for the given assignment, by the rule for evaluating $\forall$. Hence, by the table for $\supset$, C $\supset$ $\forall$xA(x) is t for the given assignment. (b) By a similar argument by cases (Exercise 18.2).

EXERCISES. 18.1. Does Theorem 15 justify concluding that the formula is valid? (If not say why Theorem 15 does not apply; and try to settle by other methods whether the formula is valid or not.)

(a) $\forall$x$\exists$yP(x, y) $\supset$ $\exists$yP(y, y).
(b) $\forall$x$\exists$zP(x, z) $\supset$ $\exists$zP(y, z).
(c) $\exists$yP(y, y) $\supset$ $\exists$x$\exists$yP(x, y).
(d) P(x, x) $\supset$ $\exists$yP(x, y).
(e) P(x, x) $\supset$ $\exists$yP(y, y).
(f) $\forall$yP(x, y) $\supset$ P(y, y).

18.2. Prove Theorem 16 (b).

18.3. Show that Theorem 16 would not hold in general without the restriction that the C not contain the x free. (HINT: try P(x) as both A(x) and C.)

★§ 19. Model theory: further results on validity.[68] We shall now examine under what conditions the reasoning which gave us Theorem 1

[68] Some of the results in this section will not be used later, and the others will be obtained independently in proof theory.

in § 3 will hold good in the predicate calculus. (The student should review the proof of Theorem 1.)

First, we consider the mechanics of substituting for an ion. Since an ion, e.g. P(—), stands for a propositional function, the process of substitution for P(—) is similar to that for a function variable in mathematics rather than simply for a number variable.

For example, consider the statement

(a) $$\forall x[f(-x) + f(x) = 2f(0)],$$

which is true for some functions f and false for others. (So long as we are concerned only with the mechanics of substitution, the truth or falsity of the formulas is beside the point.) In analyzing substitutions into (a), it will be convenient to pick a name form, say "$f(w)$" for the function f. Now if we substitute "cos w" or "$w^3 - w$" for "$f(w)$" in (a), we obtain respectively:

(b) $$\forall x[\cos -x + \cos x = 2 \cos 0],$$

(c) $$\forall x[((-x)^3 - (-x)) + (x^3 - x) = 2(0^3 - 0)].$$

Notice that in performing these substitutions, the arguments "$-x$", "x" and "0" of "f" in (a) get substituted for the name form variable "w" in "cos w" or "$w^3 - w$". A simpler alternative analysis is available in the case of (b); namely, we can regard the substitution as simply of "cos" for "f". In the case of (c), this alternative is not available, because our only permanent name for the function to be substituted is "$w^3 - w$", which has "w" in it. Of course, we could introduce a name for $w^3 - w$ without "w" in the name, say "g", by putting $g(w) = w^3 - w$. Then, simply substituting "g" for "f" in (a), we get:

(c′) $$\forall x[g(-x) + g(x) = 2\, g(0)].$$

But we don't want to tie up "g" permanently as a name for the function $w^3 - w$; so we have to evaluate "$g(-x)$", "$g(x)$" and "$g(0)$" in (c′). This brings us back to (c) with "$-x$", "x" and "0" substituted for the "w" of "$w^3 - w$".

The second example is like those we will usually have in the predicate calculus, and we deal with them in the same manner. For illustration, say we are to substitute for P(w) in P(y) ⊃ ∃xP(x). (We are slightly altering the example preceding Theorem 15 to fit our present aims better.) We shall substitute a formula ordinarily (but not necessarily) containing w free; and for this formula we adopt "A(w)" as a temporary notation. Then the result of the substitution can be written A(y) ⊃ ∃xA(x) (simply changing "P" to "A"); but in getting rid of the temporary notation "A(w)", the arguments y and x of P in P(y) ⊃ ∃xP(x) are substituted for the *free occurrences* of w in the formula abbreviated by "A(w)". Here we are using the notational convention introduced just before Theorem 15, but with w in the role of the x there, and y and x successively as the r there.

Having now described the mechanics of substituting for P(w) in $P(y) \supset \exists x P(x)$, we show the result for five choices of A(w):

	A(w)	$A(y) \supset \exists x A(x)$
I.	$\forall z Q(w, z, w)$	$\forall z Q(\underline{y}, z, \underline{y}) \underline{\supset \exists x} \forall z Q(\underline{x}, z, \underline{x})$
II.	$\forall y Q(w, y, w)$	$\forall y Q(\underline{y}, y, \underline{y}) \underline{\supset \exists x} \forall y Q(\underline{x}, y, \underline{x})$
III.	$Q(w, u, w)$	$Q(\underline{y}, u, \underline{y}) \underline{\supset \exists x} Q(\underline{x}, u, \underline{x})$
IV.	$Q(w, x, w)$	$Q(\underline{y}, x, \underline{y}) \underline{\supset \exists x} Q(\underline{x}, x, \underline{x})$
V.	$\forall w P(w) \vee Q(w)$	$\forall w P(w) \vee Q(\underline{y}) \underline{\supset \exists x}(\forall w P(w) \vee Q(\underline{x}))$

Of these five substitutions, we shall regard only I, III and V as "free". In II and IV the evaluation of a line for the truth table for $A(y) \supset \exists x A(x)$ (with a given domain *D*) will not always split into two parts, the first being the determination of a logical function (described by a supplementary table entered from w) as value of A(w), and the second coinciding with the evaluation of $P(y) \supset \exists x P(x)$ from that logical function as value of P(w). Such a split is necessary to the reasoning to generalize Theorem 1. Briefly stated, the trouble in II is that the free y in P(y) becomes bound by the ∀y of the A(w), and in IV that the free x in the A(w) becomes bound by the ∃x in ∃xP(x). We shall say this more fully.

In the examples I–V of $A(y) \supset \exists x A(x)$, the parts which originate from $P(y) \supset \exists x P(x)$ have been underlined, leaving nonunderlined the parts which originate from the respective A(w)'s. Thus in II, the 2nd and 4th y's originate from the y of $P(y) \supset \exists x P(x)$, while the 1st, 3rd, 4th and 5th originate from the y's of the A(w), i.e. of ∀yQ(w, y, w). The trouble in II is that the first ∀y (originating from the A(w)) binds not only the 1st and 3rd y's (as it should) but also the 2nd and 4th. Similarly in IV, the ∃x (originating from $P(y) \supset \exists x P(x)$) binds not only the 2nd, 3rd and 5th x's (as it should) but also the 4th (originating from the A(w)).

For a "free" substitution we wish to avoid such mix-ups in the way the variables are bound after the substitution. So we say a substitution is *free*, if in the resulting formula (A) no nonunderlined quantifier binds an underlined variable, and (B) no underlined quantifier binds a nonunderlined variable.

Now we describe the mechanics of substitution for ions in the general case. Say E is a formula which contains only the distinct ions having the respective name forms

$$(1) \qquad P_1(w_1, \ldots, w_{p_1}), \ldots, P_n(w_1, \ldots, w_{p_n})$$

(where $n \geq 1$ and $p_1, \ldots, p_n \geq 0$), i.e. each prime component of E is of the form $P_i(r_1, \ldots, r_{p_i})$ where $1 \leq i \leq n$ and $r_1, \ldots, r_{p_i}$ is a list of variables not necessarily distinct from each other or from the variables $w_1, w_2, w_3 \ldots$. *The substitution of* formulas

$$(2) \qquad A_1(w_1, \ldots, w_{p_1}), \ldots, A_n(w_1, \ldots, w_{p_n})$$

for (1) *in* E with *result* E* is effected by replacing simultaneously each prime component $P_i(r_1, \ldots, r_{p_i})$ of E by $A_i(r_1, \ldots, r_{p_i})$, where (extending the notation introduced in § 18) $A_i(r_1, \ldots, r_{p_i})$ is the result of substituting simultaneously $r_1, \ldots, r_{p_i}$ for the free occurrences (if any) of $w_1, \ldots, w_{p_i}$ respectively in $A_i(w_1, \ldots, w_{p_i})$.

The rule for underlining, in terms of which we have expressed our definition of *free*, is the following: First, underline all of E outside the parts $P_i(r_1, \ldots, r_{p_i})$ to be replaced in passing to E*. Second, in the replacements $A_i(r_1, \ldots, r_{p_i})$ for those parts, underline those occurrences of $r_1, \ldots, r_{p_i}$ which take the places of the free occurrences of $w_1, \ldots, w_{p_i}$ in $A_i(w_1, \ldots, w_{p_i})$ (i.e. the occurrences of $r_1, \ldots, r_{p_i}$ in $A(r_1, \ldots, r_{p_i})$ resulting from the substitution for $w_1, \ldots, w_{p_i}$ in $A(w_1, \ldots, w_{p_i})$).[69]

When E* comes from E by a free substitution, the evaluation of any line of the truth table for E* (for a given *D*) can be analyzed into, first, the determination of logical functions

$$(3) \qquad \mathfrak{l}_1(w_1, \ldots, w_{p_1}), \ldots, \mathfrak{l}_n(w_1, \ldots, w_{p_n})$$

as values of (2), and second, further steps corresponding to the underlined operators in E* (the underlined ∃x and ⊃ in our illustrations). These further steps coincide with the evaluation of E from (3) as values of (1) and the same values of the free variables of E as in that line for E*. So, if ⊨ E, then ⊨ E*. Thus:

THEOREM 17. (Substitution for ions, extending Theorem 1.) *Let* E* *come from* E *by substituting* (2) *for* (1) *as described above. If this substitution is free, then*: *If* ⊨ E, *then* ⊨ E*.

Considering only the part of the preceding discussion which relates to $A(w_1, \ldots, w_p)$ and $A(r_1, \ldots, r_p)$ (taking $n = 1$ and omitting "i"), we have the next theorem. Indeed, under the stated condition, any line in the table for $A(r_1, \ldots, r_p)$ receives the value we obtain by first determining a logical function $\mathfrak{l}(w_1, \ldots, w_p)$ as value of $A(w_1, \ldots, w_p)$, and then using

[69] In testing for freedom in the substitution of (2) for (1), we only need to underline variable occurrences. (The rest of the underlining we did was to cite in the proof below.) Of course, the definition of "free substitution" can be stated without recourse to underlining. Cf. IM p. 156 with p. 79.

the value of this function when $w_1, \ldots, w_p$ have the values assigned to $r_1, \ldots, r_p$.

THEOREM 18. (Substitution for individual variables.)[63] *Let* $w_1, \ldots, w_p$ *be any (distinct) variables,* $A(w_1, \ldots, w_p)$ *be any formula,* $r_1, \ldots, r_p$ *be any variables not necessarily distinct from each other or from the variables* $w_1, \ldots, w_p$, *and* $A(r_1, \ldots, r_p)$ *be the result of substituting* $r_1, \ldots, r_p$ *simultaneously for the free occurrences of* $w_1, \ldots, w_p$ *respectively in* $A(w_1, \ldots, w_p)$. *If this substitution is free* (i.e. the occurrences of $r_1, \ldots, r_p$ in $A(r_1, \ldots, r_p)$ resulting from it are free), *then: If* $\models A(w_1, \ldots, w_p)$, *then* $\models A(r_1, \ldots, r_p)$.

We resume our quest for results in the validity theory of the propositional calculus (through § 6) which we can take over for the predicate calculus.

THEOREM 19. (Theorem 4 adapted to the predicate calculus.) (a) *For each domain D and assignment,* $A \sim B$ *is* t *if and only if* A *and* B *have the same truth value.* Hence: (b)° $\models A \sim B$ *if and only if, for each domain D,* A *and* B *have the same truth table.*

THEOREM 5 and COROLLARY, and (α)–(ζ) in § 5 (and thence the chain method) extend to the predicate calculus, no change being necessary in their wordings. (Also, for each particular *D*, these results hold reading "$\overline{\overline{D}}$-$\models$" in place of "$\models$".)

Two formulas which are congruent (end § 16) have the same truth tables, for any given domain *D*. For, the alphabetical differences in the bound variables will make no difference in the process of determining the tables. Hence, by Theorem 19 (b):

THEOREM 20. *If* A *is congruent to* B, $\models A \sim B$.

We could now give a list of valid forms of formulas, similar to Theorem 2 for the propositional calculus. However we shall wait till we are ready to establish the results in proof theory (Theorem 26 § 25). We give just two of them now, for the purpose of extending Theorems 6 and 7.

*82a. $\models \neg\exists x A(x) \sim \forall x \neg A(x)$. *82b°. $\models \neg\forall x A(x) \sim \exists x \neg A(x)$.

PROOFS. *82a. For any *D*, and any assignment to the ions and free variables of $\neg\exists x A(x) \sim \forall x \neg A(x)$, consider the supplementary table for A(x) as a logical function of x. CASE 1: this table has a t in some line. Then $\exists x A(x)$ receives the value t, and $\neg\exists x A(x)$ the value f. The supplementary table for $\neg A(x)$ then has an f in some line (each one where A(x) has t); so $\forall x \neg A(x)$ also has the value f. So by the table for $\sim$ (or Theorem 19 (a)), $\neg\exists x A(x) \sim \forall x \neg A(x)$ is t. CASE 2: the supplementary table for A(x) has only f's. Then similarly, $\neg\exists x A(x)$ and $\forall x \neg A(x)$ both have the value t, and hence again $\neg\exists x A(x) \sim \forall x \neg A(x)$ is t.

*82b. Using the chain method: $\models \exists x \neg A(x) \sim \neg\neg\exists x \neg A(x)$ [*49] $\sim \neg\forall x \neg\neg A(x)$ [*82a] $\sim \neg\forall x A(x)$ [*49].

THEOREMS 6 (and COROLLARY), 7, 6a and 7a all extend now to the predicate calculus, when the operations † and $'$ are extended to include the interchange of $\forall$ with $\exists$. To establish Theorem 6 thus extended, *82a and *82b take care of moving a $\neg$ rightward across a quantifier. For Theorem 7, the substitution rule is again available (as Theorem 17), and and indeed the substitution simultaneously of $\neg P_i(w_1, \ldots, w_{p_i})$ for each ion $P_i(w_1, \ldots, w_{p_i})$ ($i = 1, \ldots, n$) is free (quite trivially). For Theorems 6a and 7a, we observe that the evaluation rule for $\forall$ when written in terms of the Martian's T, F reads just like ours for $\exists$ in terms of t, f; and vice versa. In the compilation of results to be given later (Theorem 26), it will be possible to recognize many pairs in which (with $\models$) one entails the other by duality (the extended Theorems 7, 7a).

EXERCISES. 19.1. Perform each of the following substitutions. For each, say whether the substitution is free or not; if not free, say why.

(a) $\exists zP(z, w, y)$, Q for P(w), Q in $P(z) \supset Q$.
(b) $\exists xP(x, w, y)$, Q(w) for P(w), Q(w) in $\forall y(P(z) \supset Q(y))$.
(c) $\exists zP(z, w, y)$, Q(w) for P(w), Q(w) in $\forall x(P(x) \supset Q(x))$.
(d) P(w, v, x), Q for P(v, w), Q(w) in $\forall xP(x, y) \vee Q(x) \supset P(y, x))$.
(e) $\exists zQ(z, w, w, y)$, $\forall xP(v, w, x)$ for P(w), Q(v, w) in $\forall x(P(x) \supset Q(x, x))$.
(f) $\exists zQ(z, w, w, y)$, $\forall xP(v, w, x)$ for P(w), Q(v, w) in $\forall xP(x) \supset \forall yQ(y, y)$.
(g) $\exists z\forall wP(z, w, y)$, $Q(z, w) \;\&\; \exists wR(w)$ for P(w), Q(w) in $\exists zP(z) \supset Q(z)$.

19.2. Analyze Example 5 as an application of Theorem 17 to $\models P(r) \supset \exists xP(x)$, and show why Theorem 17 does not apply in Example 6.

19.3. (a) Give an argument (by classifying the assignments) that $\models P \;\&\; \exists xQ(x) \sim \exists x(P \;\&\; Q(x))$. (b) Under what conditions can you thence infer that $\models A \;\&\; \exists xB(x) \sim \exists x(A \;\&\; B(x))$ where A and B(x) are formulas? (c) Show that $\models A \;\&\; \exists xB(x) \sim \exists x(A \;\&\; B(x))$ does not hold for all choices of formulas A and B(x).

19.4. Find an equivalent formula containing $\neg$ only applied directly to atoms.

(a) $\neg\forall x\{(P(x) \vee \exists y \neg Q(x, y)) \;\&\; \forall yR(y)\}$
(b) $\neg\{\neg(\exists xP(x) \supset \forall xQ(x, y)) \vee \forall x \neg P(x)\}$.

§ 20. Model theory: valid consequence. In § 7, we defined "valid consequence" in the propositional calculus. Adapting this definition to the predicate calculus in the obvious way, we say that (for $m \geq 1$) B is a *valid consequence* of $A_1, \ldots, A_m$ (in, or by, the predicate calculus), or in symbols $A_1, \ldots, A_m \models B$, under exactly the following circumstances: For each domain D, in truth tables entered from a list including all the ions and free variables of $A_1, \ldots, A_m$, B, the formula B is t in all those lines in which $A_1, \ldots, A_m$ are simultaneously t. As before, it is immaterial which such list of ions and variables is used. (We define "$A_1, \ldots, A_m$ $\bar{\bar{D}}$-$\models$ B" similarly, using a domain D with $\bar{\bar{D}}$ elements; cf. § 17.)

EXAMPLE 7. Which of the four formulas $\neg P(x)$, $\neg P(y)$, $\forall x \neg P(x)$, $\neg \forall x P(x)$ are valid consequences of $\neg P(x)$; or in symbols, which of the four statements "$\neg P(x) \models \neg P(x)$", "$\neg P(x) \models \neg P(y)$", "$\neg P(x) \models \forall x \neg P(x)$", "$\neg P(x) \models \neg \forall x P(x)$" hold? We must compare the truth tables for the four formulas with the truth table for the assumption formula $\neg P(x)$ (which is the first formula). We begin by constructing the truth tables with $D = \{1, 2\}$, using $\mathfrak{I}_1, \mathfrak{I}_2, \mathfrak{I}_3, \mathfrak{I}_4$ as before, namely:

x	$\mathfrak{I}_1(x)$	$\mathfrak{I}_2(x)$	$\mathfrak{I}_3(x)$	$\mathfrak{I}_4(x)$
1	t	t	f	f
2	t	f	t	f

For convenience, we enter all four tables (a)–(d) from P(x), x, y, although y is necessary only in investigating "$\neg P(x) \models \neg P(y)$".

	P(x)	x	y	(a) $\neg P(x)$	(b) $\neg P(y)$	(c) $\forall x \neg P(x)$	(d) $\neg \forall x P(x)$
1.	$\mathfrak{I}_1(x)$	1	1	f	f	f	f
2.	$\mathfrak{I}_1(x)$	1	2	f	f	f	f
3.	$\mathfrak{I}_1(x)$	2	1	f	f	f	f
4.	$\mathfrak{I}_1(x)$	2	2	f	f	f	f
5.	$\mathfrak{I}_2(x)$	1	1	f	f	f	t
6.	$\mathfrak{I}_2(x)$	1	2	f	t	f	t
7.	$\mathfrak{I}_2(x)$	2	1	t	f	f	t
8.	$\mathfrak{I}_2(x)$	2	2	t	t	f	t
9.	$\mathfrak{I}_3(x)$	1	1	t	t	f	t
10.	$\mathfrak{I}_3(x)$	1	2	t	f	f	t
11.	$\mathfrak{I}_3(x)$	2	1	f	t	f	t
12.	$\mathfrak{I}_3(x)$	2	2	f	f	f	t
13.	$\mathfrak{I}_4(x)$	1	1	t	t	t	t
14.	$\mathfrak{I}_4(x)$	1	2	t	t	t	t
15.	$\mathfrak{I}_4(x)$	2	1	t	t	t	t
16.	$\mathfrak{I}_4(x)$	2	2	t	t	t	t

We must consider those lines in which (a) has t, namely Lines 7, 8, 9, 10, 13, 14, 15, 16. In each of these lines (a) and (d) have t. Thus "$\neg P(x) \models \neg P(x)$" and "$\neg P(x) \models \neg \forall x P(x)$" both hold so far as the test by $D = \{1, 2\}$ goes. To prove that they hold (without restriction to $\bar{\bar{D}} = 2$), we need to reason that we would get the same result no matter what (nonempty) D is used. We leave this to the student (Exercise 20.1). In Line 7, (b) has f and so does (c). Thus (without having to consider other D's), we find that "$\neg P(x) \models \neg P(y)$" and "$\neg P(x) \models \forall x \neg P(x)$" do not hold.

As before (§ 7), "A ⊨ B" is stronger than "If ⊨ A, then ⊨ B".

With the definition of "valid consequence" just given, THEOREM 8 and COROLLARY extend to the predicate calculus. The proofs are essentially as before.

Now we must notice that the foregoing definition of "valid consequence" (call it "(I)") is not the only one that arises in the predicate calculus (and in mathematics).

In that definition, we treated the ions and the free variables of $A_1, \ldots, A_m$ both in the same manner as in § 7 we treated the atoms (now 0-place ions). Thus, we considered each ion as standing for a predicate, and each variable in its free occurrences as standing for a member of D. That predicate, or member of D, is to be the same throughout any discussion in the predicate calculus, though what it is, is to be unknown to us or not to be taken into account in the predicate calculus. So we entered tables from all the ions and free variables of $A_1, \ldots, A_m$, B in testing whether B is t whenever $A_1, \ldots, A_m$ are all t.

In the alternative definition of "valid consequence" which we shall presently give (call it "(II)"), we shall treat the free variables or some of them differently. Specifically, we shall not require the member of D which a free variable names to be the same throughout the discussion encompassing both $A_1, \ldots, A_m$ and B, but shall allow it to be different in different formulas or different applications of the same formula.

To see how these two possible treatments of the free variables arise in practice, we consider some examples in the language of informal mathematics. The story is really a familiar one to students of elementary algebra from the difference between (I) a *conditional equation* and (II) an *identical equation*. Examples of conditional equations are (1) $x^2 - 2x - 3 = 0$ and (2) $y = x + 1$; of identical equations, (3) $x + y = y + x$ and (4) $(x + 1)^2 = x^2 + 2x + 1$. From (1) we should not infer that $2^2 - 2 \cdot 2 - 3 = 0$ or $y^2 - 2y - 3 = 0$, though we can infer $(x - 3)(x + 1) = 0$, thence $x - 3 = 0 \vee x + 1 = 0$, and thence $x = 3 \vee x = -1$. From (3) we can infer $3 + 1 = 1 + 3$ and $(x + z) + 2x = 2x + (x + z)$.

We say that in (1) and (2) the variables have the *conditional interpretation* (I), in (3) and (4) the *generality interpretation* (II). Under the conditional interpretation of x in an assumption formula A(x) containing x free, any conclusions we draw containing x free should refer to the same member of D as in the assumption A(x), which expresses a "condition on x"; so the conclusion applies to any member of D satisfying the condition. Under the generality interpretation, we are justified in inferring whatever follows from A(x) being true for all x or true "identically" or "true in general". In the inferences from (1) we can say the variable "x" is "held constant" since it stands for the same number throughout the deductions. In the inferences from (3), "x" and "y" are "allowed to vary", since their values may be changed.[70]

There would not be any problem of distinguishing these two interpretations, if only bound variables were used. But the use of free variables is very convenient, as is shown by their frequence in mathematical writings.

The result of the above inferences from (1) can be written with x bound, after discharging the assumption (1), thus:

(a) $$\forall x(x^2 - 2x - 3 = 0 \supset x = 3 \vee x = -1).$$

In (3) we could have used bound variables in the assumption itself (thus: $\forall x \forall y(x + y = y + x)$); and the first result can be written, after discharging the assumption, thus:

(b) $$\forall x \forall y(x + y = y + x) \supset 3 + 1 = 1 + 3.$$

Note that in (a) the parentheses close *after* the ⊃, in (b) *before* the ⊃.

Which interpretation (I) or (II) to use is for us to choose in each case we propose to infer consequences from assumptions, depending on the role we intend the assumptions to have. The choice can be made independently for each free variable of each assumption.

An inappropriate choice will not result in error, unless we subsequently

[70] In mathematics, letters are often classified as "constants" and "variables". Close inspection shows that the distinction *in use* is relative to a context (whatever the terminology; cf. IM p. 150). A given letter is used as name for some object, and throughout some context every occurrence of it is as a name for the same object. From outside this context, it is indicated that the object may be any one (some one, etc.) of the members of some collection or set D. Thus the letter is "held constant" within the context, while from outside it may be "varied".

It is important to know exactly the duration of the context, irrespective of how often the letter occurs in it; e.g. $\forall x(A(x) \supset B)$ and $\forall x A(x) \supset B$ have different meanings even when x does not occur free in B. As we have illustrated, in our logical formulas and likewise in formulas as written in mathematics, when a variable appears free, it names any member of D, but there are two possibilities as to the duration of the context: (I) the context is an extended discussion, specifically here an entire deduction, (II) the context is just the whole formula.

The universal quantifier $\forall x$ is necessary in logic as a device by which the context throughout which a variable x names the same arbitrary member of D can be made less than the whole formula.

misstate what we have done. For example, using the generality interpretation of x in (1) (which is inappropriate), we can infer $2^2 - 2 \cdot 2 - 3 = 0$, and thus establish

(c) $$\forall x(x^2 - 2x - 3 = 0) \supset 2^2 - 2 \cdot 2 - 3 = 0,$$

which is a correct but uninteresting result. Only if we claim that $2^2 - 2 \cdot 2 - 3 = 0$ follows from (1) under the conditional interpretation, and thus claim to establish

(d) $$\forall x(x^2 - 2x - 3 = 0 \supset 2^2 - 2 \cdot 2 - 3 = 0),$$

do we make an error. From (d) by specializing to $x = 3$ and simplifying, $0 = 0 \supset -3 = 0$, and thence $-3 = 0$.

The valid consequence relationship (I) formulated at the beginning of this section and illustrated by Example 7 corresponds to using the conditional interpretation of all the free variables of the assumption formulas $A_1, \ldots, A_m$. The alternative one (II) to be formulated next corresponds to using instead the generality interpretation of some or all $x_1, \ldots, x_q$ of those variables. To keep matters as simple as possible, we shall suppose here that the variables $x_1, \ldots, x_q$ having the generality interpretation in any of $A_1, \ldots, A_m$ have the same interpretation in all of $A_1, \ldots, A_m$ which contain them free.[71]

The intervening discussion should have made it clear that to get (II) from (I), all we need to do is to prefix universal quantifiers $\forall x_1 \ldots \forall x_q$ to the assumption formulas $A_1, \ldots, A_m$ (with each whole one of $A_1, \ldots, A_m$ as the scope of the prefix $\forall x_1 \ldots \forall x_q$) before applying (I), i.e. before constructing the truth tables called for under (I).

In formulating (II) "officially", we shall use a concise notation for the prefixed universal quantifiers.

We call a formula A *closed*, if it contains no variables free (i.e. it has at most bound occurrences of variables); otherwise, *open*. By the *closure* of A, abbreviated "$\forall$A", we mean the closed formula $\forall z_1 \ldots \forall z_p A$ (i.e. $\forall z_1 \ldots \forall z_p(A)$) where $z_1, \ldots, z_p$ are the distinct free variables of A, taken for definiteness in order of first free occurrence in A. If A is closed (i.e. $p = 0$), $\forall$A is simply A. "$\forall A \supset B$" means $(\forall A) \supset B$, etc.

Now let $\forall' (\simeq \forall x_1 \ldots \forall x_q)$ be the operation of *closure with respect to* (*just*) $x_1, \ldots, x_q$, which we get from $\forall$ by omitting from $\forall z_1 \ldots \forall z_p$ the quantifiers other than on $x_1, \ldots, x_q$. For example, if $x_1, \ldots, x_q$ is x_1, x_2, x_3 ($q = 3$), and A contains in order of first free occurrence exactly the distinct variables y_1, x_3, y_2, y_3, x_1, then $\forall$A is $\forall y_1 \forall x_3 \forall y_2 \forall y_3 \forall x_1 A$ and $\forall' A$ is $\forall x_3 \forall x_1 A$. Thus $\forall'$ is $\forall x_1 \ldots \forall x_q$ apart from the order of quantifiers and omissions of quantifiers on variables not free in the scope. Where it

[71] In IM §§ 22–24, (II) is treated in proof theory without this restriction.[75]

does not matter for our purposes (as is usually the case), we may not bother to distinguish carefully between $\forall'$ and $\forall x_1 \ldots \forall x_q$.

Now we say B is a *valid consequence of* $A_1, \ldots, A_m$ (in, or by, the predicate calculus) *holding constant all variables except* $x_1, \ldots, x_q$ or *allowing* $x_1, \ldots, x_q$ *to vary* or *with* $x_1, \ldots, x_q$ *general*, or in symbols $A_1, \ldots, A_m \models^{x_1 \ldots x_q} B$, exactly if $\forall'A_1, \ldots, \forall'A_m \models B$. Thus $A_1, \ldots, A_m \models^{x_1 \ldots x_q} B$ holds exactly if, for each domain D, in truth tables entered from a list including all the ions and free variables of $\forall'A_1, \ldots, \forall'A_m$, B, the formula B is t in all those lines in which $\forall'A_1, \ldots, \forall'A_m$ are simultaneously t.[72]

This definition and notation need not be restricted to the case each of $x_1, \ldots, x_q$ occurs free in some of $A_1, \ldots, A_m$; but any variable not so occurring can be added or removed without affecting the meaning. The order of listing the superscripts, and the presence or absence of repetitions, is likewise immaterial.

EXAMPLE 8. Which of the formulas $\neg P(x)$, $\neg P(y)$, $\forall x \neg P(x)$, $\neg \forall x P(x)$ are valid consequences of $\neg P(x)$ with x general; or in symbols, which of the four statements "$\neg P(x) \models^x \neg P(x)$", "$\neg P(x) \models^x \neg P(y)$", "$\neg P(x) \models^x \forall x \neg P(x)$", "$\neg P(x) \models^x \neg \forall x P(x)$" hold? We must now compare the tables (a)–(d) with the table for the closure $\forall x \neg P(x)$ of the assumption formula $\neg P(x)$ (which closure is the third formula). Thus we must now consider those lines in which (c) has t, namely Lines 13, 14, 15, 16. In each of these lines (a), (b), (c) and (d) have t. This is for $D = \{1, 2\}$; but we easily convince ourselves of the like for any D. So "$\neg P(x) \models^x \neg P(x)$", "$\neg P(x) \models^x \neg P(y)$", "$\neg P(x) \models^x \forall x \neg P(x)$" and "$\neg P(x) \models^x \neg \forall x P(x)$" all hold.

Say A contains only x free. *The statement* "$A \models B$" *is stronger than* "$A \models^x B$"; i.e. if $A \models B$ then $A \models^x B$, but not in general conversely. To establish this, consider any D, and take tables entered from x and the other quantities required to evaluate A and B. For $A \models^x B$ to hold, it is only required that B be t in those lines in which the other quantities cause $\forall x A$ to be t, i.e. cause A to be t for *all* values of x (in Example 8, where A is $\neg P(x)$, Lines 13, 14, 15, 16). For $A \models B$ to hold, it is required that B be t in all lines, entered from a value of x and the other quantities, in which A is t for *that* value of x. So we have to consider all the lines that had to be considered for $A \models^x B$ and in general others as well (in Example 7, Lines 7, 8, 9, 10 as well).

Generalizing this to any number of assumption formulas and variables:

[72] The phrase "except $x_1, \ldots, x_q$" merely removes the variables $x_1, \ldots, x_q$ from the assertion about all variables being held constant. It does not exclude the possibility that B is a valid consequence of A with all variables held constant or intermediate possibilities.

If $A_1, \ldots, A_m \models^{x_1 \ldots x_q} B$, *then* $A_1, \ldots, A_m \models^{x_1 \ldots x_q x_{q+1} \ldots x_r} B$; *but not in general conversely*. Of course, *the converse does hold when* $x_{q+1}, \ldots, x_r$ *do not occur free in any of* $A_1, \ldots, A_m$.

The valid consequence relationship (I) ("$A_1, \ldots, A_m \models B$") is the one which will have the principal role in this book. When we say "B is a valid consequence of $A_1, \ldots, A_m$" we shall mean that one, though for emphasis we may say more fully "B is a valid consequence of $A_1, \ldots, A_m$ holding all (free) variables (of $A_1, \ldots, A_m$) constant".

EXERCISES. 20.1. Supply the reasoning for Example 7 (a) and (d).

20.2. Show that:
(a) Not $R \supset P(x) \models R \supset \forall x P(x)$. (b) $R \supset P(x) \models^x R \supset \forall x P(x)$.
(c) Not $P(x) \supset R \models \exists x P(x) \supset R$. (d) $P(x) \supset R \models^x \exists x P(x) \supset R$.

20.3. Show that: $\forall' A_1, \ldots, \forall' A_m \models B$ if and only if $\forall' A_1, \ldots, \forall' A_m \models \forall' B$.

20.4. Note that Exercise 7.6 holds for the predicate calculus. Now use Exercise 20.3 to infer that it also holds with "$\models^{x_1 \ldots x_q}$" in place of "$\models$".

20.5. Show that for any variable x and formula A: (a) $\forall x A \models A$. (b) $A \models^x \forall x A$. (c) "$A \models \forall x A$" does not hold in general.

20.6. Show that: $A \models^x B$ if and only if $\models \forall x A \supset B$, and also if and only if $\models \forall x(\forall x A \supset B)$, and also if and only if $\models \forall x A \supset \forall x B$; but $A \models B$ if and only if $\models A \supset B$, and if and only if $\models \forall x(A \supset B)$. (Cf. Exercise 17.6.)

20.7. Show that, when C does not contain x free: $A(x) \models^x C$ if and only if $\models \forall x A(x) \supset C$; but $A(x) \models C$ if and only if $\models \exists x A(x) \supset C$.

20.8*. Does Theorem 5 § 5 hold in the predicate calculus with "$A_1, \ldots, A_m \models$" in place of "$\models$"? (For $m = 0$, cf. end § 19.)

§ 21. Proof theory: provability and deducibility. To establish the proof theory of the predicate calculus, we begin with the axiom schemata and rule of inference of the propositional calculus (i.e. Axiom Schemata 1a–10b, and modus ponens or the $\supset$-rule; cf. § 9). Of course, these are now to be used with the present sense of formula (§ 16).

To these, we add two new axiom schemata, $\forall x A(x) \supset A(r)$ (the $\forall$-*schema*) and $A(r) \supset \exists x A(x)$ (the $\exists$-*schema*), where r is free for x in A(x) etc. (cf. Theorem 15 § 18). That is, the axioms are now to include also any formulas having either of these forms.

We also add two new rules of inference: the $\forall$-*rule* allows us to pass from $C \supset A(x)$ to $C \supset \forall x A(x)$, and the $\exists$-*rule* from $A(x) \supset C$ to $\exists x A(x) \supset C$, when C does not contain x free (cf. Theorem 16).[73]

[73] These four simple postulates for the quantifiers are due to Bernays (according to Hilbert and Ackermann 1928 p. 54). But as in § 9 (following von Neumann 1927) we are using axiom schemata, instead of particular axioms with a postulated substitution rule. (There were defects in all the early formulations of the substitution rule for the predicate calculus; cf. Church 1956 pp. 289–290.)

The definition of a *proof* $B_1, \ldots, B_l$ (*of* B_l), and of B *is provable* or in symbols $\vdash$ B, are as before (§ 9), allowing that now we have two more axiom schemata and two more rules of inference.

The definition of a *deduction* $B_1, \ldots, B_l$ (*of* B_l) *from* $A_1, \ldots, A_m$ is likewise as before.

Moreover, we say that in such a deduction *all (free) variables (of* $A_1, \ldots, A_m$*) are held constant*, if the following is the case: the $\forall$-rule and the $\exists$-rule are not applied with respect to a variable (as the x of the rule) occurring free in $A_1, \ldots, A_m$ except preceding the first occurrence of $A_1, \ldots, A_m$ (as such) in the deduction. (We can disregard occurrences of $A_1, \ldots, A_m$ in the deduction $B_1, \ldots, B_l$ which are justified otherwise than as assumption formulas.)[74]

Thus, if we say that in the deduction $B_1, \ldots, B_l$ from $A_1, \ldots, A_m$ all variables are held constant, we mean the following: Let B_k be the first of $B_1, \ldots, B_l$ which is justified as an assumption formula, if any of $B_1, \ldots, B_l$ is so justified; otherwise, there is no B_k and B_{k-1} shall be B_l. In the part $B_1, \ldots, B_{k-1}$ of the deduction (if it exists), any variable may be the x of an inference (with conclusion in that part) by the $\forall$-rule or the $\exists$-rule. In the part $B_{k+1}, \ldots, B_l$ (if it exists), only a variable not occurring free in $A_1, \ldots, A_m$ may be the x of an inference (with conclusion in that part) by the $\forall$-rule or the $\exists$-rule.

If there is a deduction of B from $A_1, \ldots, A_m$ holding all variables constant, we say that B *is deducible from* $A_1, \ldots, A_m$ (*holding all variables constant*), and write $A_1, \ldots, A_m \vdash$ B. In this book we usually omit the words "holding all variables constant".

We say that B is *deducible from* $A_1, \ldots, A_m$ *holding constant all variables except* $x_1, \ldots, x_q$ or *allowing* $x_1, \ldots, x_q$ *to vary* or *with* $x_1, \ldots, x_q$ *general*, and write $A_1, \ldots, A_m \vdash^{x_1 \ldots x_q}$ B, if $\forall' A_1, \ldots, \forall' A_m \vdash$ B in the foregoing sense, where $\forall'$ ($\simeq \forall x_1 \ldots \forall x_q$) indicates closure with respect to $x_1, \ldots, x_q$. The significance of this notion should be evident from the parallelism of "$\vdash$" here to "$\models$" in § 20. In this book, we shall scarcely use "$\vdash^{x_1 \ldots x_q}$" more than as a shorthand for prefixing $\forall x_1 \ldots \forall x_q$ to each of the assumption formulas.[75]

[74] Thus this definition applies to the deduction $B_1, \ldots, B_l$ together with an analysis of it, i.e. a reason for the inclusion of each formula in it.[34]

[75] In IM §§ 22 ff. the derived rules are formulated so as to provide for manipulations of statements using "$\vdash$" with superscripts allowed. We are foregoing that here, in order to simplify the present treatment.

In IM, "$\vdash$" is used in some passages with a "systematic ambiguity" to stand for a notion equivalent to what we write here as "$\vdash^{x_1 \ldots x_q}$", with various lists of variables $x_1, \ldots, x_q$ ($q \geq 0$), and the word "deducible" is used there similarly. Thus rules with "$\vdash$" in hypothesis and conclusion can be applied there introducing superscripts on "$\vdash$" in the hypothesis, if then the superscripts are carried forward to the conclusion;

We can take over automatically most of the results of the proof theory of the propositional calculus. For, all proofs and deductions (with formulas in the present sense) which we can construct or show to exist *in the propositional calculus* (i.e. using only Axiom Schemata 1a–10b and the $\supset$-rule) are proofs and deductions (with all variables held constant) in the predicate calculus, since in the predicate calculus all the same reasons are still available to justify each formula in them (and the $\forall$-rule and $\exists$-rule are not used). In particular, if $\vdash B$ in the propositional calculus, then $\vdash B$ in the predicate calculus. If $A_1, \ldots, A_m \vdash B$ in the propositional calculus, then $A_1, \ldots, A_m \vdash B$ in the predicate calculus. (In the terminology of § 13, direct rules established for the propositional calculus hold ipso facto in the predicate calculus.)

But from "If $A \vdash B$, then $\vdash A \supset B$" in the propositional calculus (Theorem 11 (a)), we can not immediately infer the like in the predicate calculus, but only "If $A \vdash B$ in the propositional calculus, then $\vdash A \supset B$ in the predicate calculus". (Subsidiary deduction rules do not automatically extend.)

EXAMPLE 9. The following is a deduction of $R \supset \forall x P(x)$ from $\forall y(R \supset P(y))$.

1. $\forall y P(y) \supset P(x)$ — $\forall$-schema.
2. $\forall y P(y) \supset \forall x P(x)$ — $\forall$-rule, 1.
3. $\forall y(R \supset P(y))$ — assumption formula.
4. $\forall y(R \supset P(y)) \supset (R \supset P(y))$ — $\forall$-schema.
5. $R \supset P(y)$ — $\supset$-rule, 3, 4.
6. $R \supset \forall y P(y)$ — $\forall$-rule, 5.
7. $(\forall y P(y) \supset \forall x P(x)) \supset \{R \supset (\forall y P(y) \supset \forall x P(x))\}$ — Axiom Schema 1a.
8. $R \supset (\forall y P(y) \supset \forall x P(x))$ — $\supset$-rule, 2, 7.
9. $\{R \supset \forall y P(y)\} \supset \{\{R \supset (\forall y P(y) \supset \forall x P(x))\} \supset \{R \supset \forall x P(x)\}\}$ — Axiom Schema 1b.
10. $\{R \supset (\forall y P(y) \supset \forall x P(x))\} \supset \{R \supset \forall x P(x)\}$ — $\supset$-rule, 6, 9.
11. $R \supset \forall x P(x)$ — $\supset$-rule, 8, 10.

Since no variables are free in $\forall y(R \supset P(y))$, all variables are held constant in this deduction. Thus $\forall y(R \supset P(y)) \vdash R \supset \forall x P(x)$.

---

cf. IM p. 103. In $\vdash$-statements standing by themselves, the superscripts are generally shown (when applicable) in IM, as here.

In IM, instead of using just one list $x_1, \ldots, x_q$ of variables with the generality interpretation in all the assumption formulas $A_1, \ldots, A_m$, a different list $x_{i1}, \ldots, x_{iq_i}$ can be selected to have the generality interpretation in each respective assumption formula A_i ($i = 1, \ldots, m$). To do this here, we can simply employ $\forall'_1 A_1, \ldots, \forall'_m A_m$ (where $\forall'_i \simeq \forall x_{i1} \ldots \forall x_{iq_i}$) instead of $\forall' A_1, \ldots, \forall' A_m$.

The equivalence of the present treatment of variables having the generality interpretation to that in IM follows from IM Lemma 8a p. 104.

If we omit from 1–11 the two formulas 3 and 4, and justify 5 instead by "assumption formula", we obtain a deduction of $R \supset \forall xP(x)$ from $R \supset P(y)$. But we cannot say all variables are held constant. For, the $\forall$-rule is used at 6 with respect to the variable y; and the variable y occurs free in the assumption formula $R \supset P(y)$ used at 5 (above 6). Thus we are not justified in saying "$R \supset P(y) \vdash R \supset \forall xP(x)$". In fact, this is not true, since by the theory to follow (§§ 22, 23) it could only be true if "$R \supset P(y) \models R \supset \forall xP(x)$" were true; and that is not the case by Example 20.2 (a) (where it is immaterial whether the free variable is x or y).

Example 10. For any formula A, the following is a deduction of A from $\forall x\forall yA$. (By our convention in § 16, x and y are distinct variables.)

1. $\forall x\forall yA$ — assumption formula.
2. $\forall x\forall yA \supset \forall yA$ — $\forall$-schema, with (the present) x, $\forall yA$, x as the x, A(x), r of the schema.
3. $\forall yA$ — $\supset$-rule, 1, 2.
4. $\forall yA \supset A$ — $\forall$-schema with y, A, y as the x, A(x), r.
5. A — $\supset$-rule, 3, 4.

There are no applications of the $\forall$- or $\exists$-rule, so all variables are held constant. Thus $\forall x\forall yA \vdash A$. More generally, for any list of variables $x_1, \ldots, x_q$, $\forall x_1 \ldots \forall x_q A \vdash A$.

In the predicate calculus, especial care is necessary to be sure that the r is free for the x in the A(x) each time we use the $\forall$-schema or the $\exists$-schema (cf. Theorem 15), and that the C does not contain the x free each time we use the $\forall$-rule or the $\exists$-rule (cf. Theorem 16).

Example 10 (concluded). In the two applications of the $\forall$-schema, the r is free for the x in the A(x), since each time the r is the x itself: the resulting occurrences of the r in the A(r) are simply the original free occurrences of the x in the A(x). Here the r and x are x in Step 2, and y in Step 4.

Example 11. In the left column below, we give a proof of $\forall x\exists wP(x, w, z) \sim \forall y\exists wP(y, w, z)$ (so $\vdash \forall x\exists wP(x, w, z) \sim \forall y\exists wP(y, w, z)$). In the right column, we generalize this to a proof of $\forall xA(x) \sim \forall yA(y)$ under suitable conditions on the formula A(x) and the variables x and y (to be stated in a moment). The reasons given are the same for both columns. The reader should first check their correctness for the left column (Exercise 21.1).

1. $\forall x\exists wP(x, w, z) \supset \exists wP(y, w, z)$ — $\forall$-schema.	1. $\forall xA(x) \supset A(y)$.
2. $\forall x\exists wP(x, w, z) \supset \forall y\exists wP(y, w, z)$ — $\forall$-rule, 1.	2. $\forall xA(x) \supset \forall yA(y)$.
3. $\forall y\exists wP(y, w, z) \supset \exists wP(x, w, z)$ — $\forall$-schema.	3. $\forall yA(y) \supset A(x)$.
4. $\forall y\exists wP(y, w, z) \supset \forall x\exists wP(x, w, z)$ — $\forall$-rule, 3.	4. $\forall yA(y) \supset \forall xA(x)$.
. . .	. . .
7. $\forall x\exists wP(x, w, z) \sim \forall y\exists wP(y, w, z)$ — using Axiom Schema 9a with 2 and 4 and the $\supset$-rule.	7. $\forall xA(x) \sim \forall yA(y)$.

For the right column, we shall suppose that x is any variable, A(x) is any formula, y is any variable (not necessarily distinct from x) such that

(i) y is free for x in A(x),
(ii) y does not occur free in A(x) (unless y is x),

and A(y) is the result of substituting y for the free occurrences of x in A(x). (Recall that (i) means that none of the occurrences of y resulting from this substitution is bound.)

Let us check the requirements that 1–7 be a proof in the right column. They are easily verified when y is x. Now take the case y is not x. For 1, y is the r, and A(y) is the A(r), for the application of the ∀-schema; and using (i) the requirements of the ∀-schema are met. For 2, ∀xA(x) is the C, and y is the x, of the ∀-rule; and by (ii) the x does not occur free in the C; so again the requirements are met. For 3, we need to see that ∀yA(y) ⊃ A(x) is of the form ∀yB(y) ⊃ B(r), where B(r) comes by substituting r for the free occurrences of y in a formula B(y) and r is free for y in B(y) (then y and B(y) will be the x and A(x) of the ∀-schema). We got A(y) by substituting y for the free occurrences of x in A(x); and by (i) each of the occurrences of y in A(y) resulting from this substitution is free, while by (ii) there are no free occurrences of y in A(x); thus A(y) has free y's exactly where A(x) had free x's, and hence A(x) does come by substituting x (as the r) for the free occurrences of y in A(y) (as the B(y)). Furthermore none of the occurrences of x which enter A(x) by this substitution are bound, as they are simply the original free occurrences of x in A(x). So 3 is justified. For 4, we need merely verify that ∀yA(y) contains no free occurrences of x. This is so because A(y) comes by substituting y for *all* the free occurrences of x in A(x).

In this example, the left column illustrates the right for the case the A(x) is ∃wP(x, w, z).

EXAMPLE 12. Now suppose instead that the A(x) of the right column is ∃yP(x, y, z) (with its x and y as the x and y). Then (i) would fail, and the ∀-schema would not be applicable at Line 1. At Line 1 ∀xA(x) ⊃ A(y) would be ∀x∃yP(x, y, z) ⊃ ∃yP(y, y, z). In fact, this formula is unprovable. For, by its truth table for $D = \{1, 2\}$, it is not valid, and hence by the theory below (Theorem 12 extended to the predicate calculus in § 23), it is unprovable.

Similar examples were given in § 18, where the question whether the conditions of Theorem 15 are met is now synonymous with whether the ∀-schema or the ∃-schema applies.

EXAMPLE 13. Suppose instead that the A(x) is ∃wP(x, w, y). Then (ii) would fail. Line 1 is correct (i.e. the ∀-schema applies). But at Line 2, the application of the ∀-rule with result ∀x∃wP(x, w, y) ⊃ ∀y∃wP(y, w, y) would be incorrect, since the x and C of the application would be y and $\forall x \exists w P(x, w, y)$, with the C containing the x free. Indeed, $\forall x \exists w P(x, w, y) \supset \forall y \exists w P(y, w, y)$ is unprovable.

THEOREM 10 and COROLLARY in § 9 again hold, with no change in wording or proof.

EXERCISES. 21.1. Check the left column of Example 11.

21.2. Is the list of formulas a deduction from its first formula? If not, find and explain the mistake.

(a) 1. $\exists z \forall x P(y, x, z)$ — assumption formula.
2. $\forall x P(y, x, z) \supset \exists w \forall x P(w, x, z)$ — $\exists$-schema.
3. $\exists z \forall x P(y, x, z) \supset \exists w \forall x P(w, x, z)$ — $\exists$-rule, 2.
4. $\exists w \forall x P(w, x, z)$ — $\supset$-rule, 1, 3.

(b) 1. $\exists x \forall y P(x, y, z)$ — assumption formula.
2. $\forall y P(x, y, z) \supset \exists w \forall y P(w, y, z)$ — $\exists$-schema.
3. $\exists x \forall y P(x, y, z) \supset \exists w \forall y P(w, y, z)$ — $\exists$-rule, 2.
4. $\exists w \forall y P(w, y, z)$ — $\supset$-rule, 1, 3.

(c) 1. $\exists y \forall y P(y, y, x)$ — assumption formula.
2. $\forall y P(y, y, x) \supset \exists z \forall y P(z, y, x)$ — $\exists$-schema.
3. $\exists y \forall y P(y, y, x) \supset \exists z \forall y P(z, y, x)$ — $\exists$-rule, 2.
4. $\exists z \forall y P(z, y, x)$ — $\supset$-rule, 1, 3.

21.3. State conditions on x, A(x), y, A(y), C under which the result of Example 9 generalizes to $\forall y(C \supset A(y)) \vdash C \supset \forall x A(x)$.

21.4. Add to the following proof to obtain a deduction of $\exists x P(x) \supset \exists x Q(x)$ from $\forall x(P(x) \supset Q(x))$ (check all requirements).

. . .
k. $(P(x) \supset Q(x)) \supset [(Q(x) \supset \exists x Q(x)) \supset (P(x) \supset \exists x Q(x))]$ } proof given by *2 in Theorem 2 with Theorem 14.

State the result which this construction establishes, using the symbol "$\vdash$". Does the result hold good when P(x), Q(x) are changed to any formulas A(x), B(x)?

21.5. Show that, under the conditions of Theorem 15:
(a) $\forall x A(x) \vdash A(r)$. (b) $A(r) \vdash \exists x A(x)$.

§ 22. Proof theory: the deduction theorem. We now reprove the deduction theorem (THEOREM 11 § 10) for the predicate calculus. Essential use is made of the condition, now incorporated into the meaning of the hypothesis "$A_1, \ldots, A_{m-1}, A_m \vdash B$", that in the given deduction

$$(\alpha) \qquad B_1, \ldots, B_l$$

of B from $A_1, \ldots, A_{m-1}, A_m$ all variables are held constant.

We use the former proof (§ 10) with some modifications and additions. (The treatment is illustrated by Example 14 below.)

If the last assumption formula A_m is not used (as such) in the given deduction (α), we shall now construct the resulting deduction simply by adding the following at the end of the given deduction:
$l+1'$. $B_l \supset (A_m \supset B_l)$ — Axiom Schema 1a.
$l+2'$. $A_m \supset B_l$ — modus ponens ($\supset$-rule), l', $l+1'$.

Now suppose the last assumption formula A_m is used (as such) in the given deduction (α); say the first use of it is B_n, i.e. this is the first of the formulas $B_1, \ldots, B_l$ which is justified in (α) on the ground of its being the last assumption formula A_m. We now begin by prefixing "$A_m \supset$" only to the formulas $B_n, \ldots, B_l$ in the given deduction, to obtain

$$(\beta) \qquad B_1, \ldots, B_{n-1}, A_m \supset B_n, \ldots, A_m \supset B_l.$$

Now insertions will be required preceding $A_m \supset B_i$ only for $i = n, \ldots, l$, instead of for $i = 1, \ldots, l$. We again specify the method of making these insertions by cases, depending on the reason given for the inclusion of B_i in (α).

In Cases 1–3 the former treatment applies unchanged. In Case 4, that B_i with $i > n$ comes from B_g and B_h ($g, h < i$) by modus ponens, we formerly had to provide steps to lead from $A_m \supset B_g$ and $A_m \supset B_h$ to $A_m \supset B_i$, where B_g, B_h, B_i are of the respective forms A, A $\supset$ B, B. (In § 10, these steps were left to the student in Exercise 10.1, but here they are illustrated in Example 14 by 12′–14′ and by 17′–19′.) Now, if $g < n$ we must first supply the following two steps to get $A_m \supset B_g$:

$B_g \supset (A_m \supset B_g)$ — Axiom Schema 1a.
$A_m \supset B_g$ — modus ponens, g', —.

Similarly, if $h < n$. (This is illustrated by 15′, 16′ in Example 14.) After these preliminaries, the former treatment of Case 4 applies.[76]

Case 5: B_i with $i > n$ comes from a preceding formula B_g ($g < i$) by the $\forall$-rule. So B_g, B_i are of the respective forms C $\supset$ A(x), C $\supset \forall$xA(x), where C does not contain x free. If $g < n$, we first supply two steps to get $A_m \supset B_g$, as in Case 4. Remember that $i > n$, i.e. the application of the $\forall$-rule comes after the first use B_n of the assumption formula A_m (as such). Hence x does not occur free in A_m; this is by the hypothesis that all variables are held constant in (α), with the definition in § 21 of a deduction in which all variables are held constant. By the stipulation for the $\forall$-rule, C does not contain x free. Hence A_m & C does not contain x free. This fact is used in justifying the new application of the $\forall$-rule at k_2+1' below (top p. 114).

[76] The changes so far described from the plan in § 10 could have been used there, and would often have been made the resulting deductions shorter.

k_1'. $A_m \supset B_g$, i.e. $A_m \supset (C \supset A(x))$
. . .
k_2'. $A_m \,\&\, C \supset A(x)$
} deduction given by Example 7 § 10

k_2+1'. $A_m \,\&\, C \supset \forall xA(x)$ — $\forall$-rule, k_2'.
. . .
k_3'. $A_m \supset (C \supset \forall xA(x))$, i.e. $A_m \supset B_i$
} deduction given by Exercise 10.3.

CASE 6: B_i with $i > n$ comes from a preceding formula B_g $(g < i)$ by the $\exists$-rule. Left to the student (Exercise 22.1).

We have now seen how to construct a particular deduction of $A_m \supset B$ from $A_1, \ldots, A_{m-1}$. In this "resulting deduction", all the applications of the $\forall$-rule and the $\exists$-rule are on the same respective variables, and occur in the same order relative to the uses of $A_1, \ldots, A_{m-1}$, as the applications in the given deduction. Hence, since all variables are held constant in the given deduction, all are held constant in the resulting deduction. Therefore $A_1, \ldots, A_{m-1} \vdash A_m \supset B$, as was to be proved.

EXAMPLE 14. As the left column illustrates, $\forall x(P(x) \supset Q(x))$, $\forall xP(x) \vdash \forall xQ(x)$. So by the theorem, $\forall x(P(x) \supset Q(x)) \vdash \forall xP(x) \supset \forall xQ(x)$. The right column shows the deduction of $\forall xP(x) \supset \forall xQ(x)$ from $\forall x(P(x) \supset Q(x))$ resulting by our uniform method from the deduction of $\forall xQ(x)$ from $\forall x(P(x) \supset Q(x))$, $\forall xP(x)$ given in the left column. Because the second assumption formula enters only at 4, the prefixing of "$\forall xP(x) \supset$" begins there.

1. $\forall x(P(x) \supset Q(x))$ — 1st ass'n. f.	1′. $\forall x(P(x) \supset Q(x))$ — same reason.
2. $\forall x(P(x) \supset Q(x)) \supset$ $(P(x) \supset Q(x))$ — $\forall$-schema.	2′. $\forall x(P(x) \supset Q(x)) \supset$ $(P(x) \supset Q(x))$ — same reason.
3. $P(x) \supset Q(x)$ — m.p., 1, 2.	3′. $P(x) \supset Q(x)$ — same reason.
	. . .
4. $\forall xP(x)$ — 2nd ass'n. f.	8′. $\forall xP(x) \supset \forall xP(x)$ } Example 4 § 9.
	9′. $\forall xP(x) \supset P(x)$ — $\forall$-schema.
	10′. $\{\forall xP(x) \supset P(x)\} \supset$ $\{\forall xP(x) \supset (\forall xP(x) \supset P(x))\}$ — Axiom Schema 1a.
5. $\forall xP(x) \supset P(x)$ — $\forall$-schema.	11′. $\forall xP(x) \supset (\forall xP(x) \supset P(x))$ — modus ponens, 9′, 10′.
	12′. $\{\forall xP(x) \supset \forall xP(x)\} \supset$ $\{\{\forall xP(x) \supset (\forall xP(x) \supset P(x))\} \supset$ $\{\forall xP(x) \supset P(x)\}\}$ — Ax. Sch. 1b.
	13′. $\{\forall xP(x) \supset (\forall xP(x) \supset P(x))\} \supset$ $\{\forall xP(x) \supset P(x)\}$ — m.p., 8′, 12′.

6. $P(x)$ — modus ponens, 4, 5.	14′. $\forall x P(x) \supset P(x)$ — m.p., 11′, 13′.	
	15′. $\{P(x) \supset Q(x)\} \supset \{\forall x P(x) \supset (P(x) \supset Q(x))\}$ — Ax. Sch. 1a.	
	16′. $\forall x P(x) \supset (P(x) \supset Q(x))$ — modus ponens, 3′, 15′.	
	17′. $\{\forall x P(x) \supset P(x)\} \supset \{\{\forall x P(x) \supset (P(x) \supset Q(x))\} \supset \{\forall x P(x) \supset Q(x)\}\}$ — Ax. Sch. 1b.	
	18′. $\{\forall x P(x) \supset (P(x) \supset Q(x))\} \supset \{\forall x P(x) \supset Q(x)\}$ — m.p., 14′, 17′.	
7. $Q(x)$ — modus ponens, 6, 3.	19′. $\forall x P(x) \supset Q(x)$ — m.p., 16′, 18′.	
	. . .	like
8. $Q(x) \supset ((P \supset P \lor P) \supset Q(x))$ — Axiom Schema 1a.	22′. $\forall x P(x) \supset \{Q(x) \supset ((P \supset P \lor P) \supset Q(x))\}$	9′–11′.
	. . .	like
9. $(P \supset P \lor P) \supset Q(x)$ — modus ponens, 7, 8.	25′. $\forall x P(x) \supset \{(P \supset P \lor P) \supset Q(x)\}$	12′–14′.
	. . .	Example
	k_1'. $\forall x P(x) \,\&\, (P \supset P \lor P) \supset Q(x)$	7 § 10.
	k_1+1'. $\forall x P(x) \,\&\, (P \supset P \lor P) \supset \forall x Q(x)$ — $\forall$-rule, k_1'.	Exercise
	. . .	10.3.
10. $(P \supset P \lor P) \supset \forall x Q(x)$ — $\forall$-rule, 9.	k_2'. $\forall x P(x) \supset \{(P \supset P \lor P) \supset \forall x Q(x)\}$	
	. . .	like
11. $P \supset P \lor P$ — Ax. Sch. 5a.	k_2+3'. $\forall x P(x) \supset (P \supset P \lor P)$	9′–11′.
	. . .	like
12. $\forall x Q(x)$ — m. p., 11, 10.	k_2+6'. $\forall x P(x) \supset \forall x Q(x)$	12′–14′.

Corollary Theorem 11 follows from the theorem as before.

We now observe that Theorem 9 (i) and (ii) hold for the predicate calculus with no change in statement.

However (ii) requires a new method of proof. If we simply combined the given deductions as in § 9, it could happen that applications of the $\forall$-rule or the $\exists$-rule to variables which occur free in $A_1, \ldots, A_m$ would come under the first use of $A_1, \ldots, A_m$ in the resulting deduction. By an obvious rearrangement we could mend this with respect to such applications coming from the p given deductions of $B_1, \ldots, B_p$ respectively from $A_1, \ldots, A_m$. A more serious difficulty is that such applications, with respect to variables occurring free in $A_1, \ldots, A_m$ but not in $B_1, \ldots, B_p$, might come from below the first use of the assumption formulas $B_1, \dots, B_p$ (as such) in the given deduction of C. This difficulty we shall overcome by use of the deduction theorem.

As the hypotheses of (ii) we have: **(1)** $A_1, \dots, A_m \vdash B_i$ $(i = 1, \dots, p)$ and **(2)** $B_1, \dots, B_p \vdash C$. Applying Corollary Theorem 11_{Pd} to (2):[77] **(3)** $\vdash B_1 \supset (\dots(B_p \supset C)\dots)$. Now we build a deduction of C from $A_1, \dots, A_m$ as follows. Take the p deductions of $B_1, \dots, B_p$ respectively from $A_1, \dots, A_m$ (which exist by (1)); write them consecutively; and then move to the front of the combination the part of each which precedes the first use in it of an assumption formula (as such). To the sequence of formulas thus obtained, prefix the proof of $B_1 \supset (\dots(B_p \supset C)\dots)$ (which exists by (3))). Finally, suffix applications of modus ponens which lead from $B_1 \supset (\dots(B_p \supset C)\dots), B_1, \dots, B_p$ to C.

In the deduction of C from $A_1, \dots, A_m$ thus constructed, there are no applications of the $\forall$-rule or the $\exists$-rule coming below uses of the assumption formulas (as such) except ones which came in from such applications in the deductions given by (1). By the hypothesis (implicit in our using "$\vdash$" without superscripts) that all variables are held constant in those deductions, the variables for these applications do not occur free in $A_1, \dots, A_m$. Thus in our deduction all variables are held constant, so $A_1, \dots, A_m \vdash C$. —

Having reinstated Theorem 9, we now have all the general properties of $\vdash$ (written *without* superscripts) that were noted in § 13.

EXERCISES. 22.1. Treat Case 6 in the proof of the deduction theorem.

22.2. Is the deduction theorem applicable to 3–12 of Example 14 as the "given deduction" (with 3 justified by "assumption formula")?

§ 23. Proof theory: consistency, introduction and elimination rules. The corollaries to Theorems 10_{Pd} and 11_{Pd} in §§ 21, 22 reduce the notion of deducibility "$A_1, \dots, A_m \vdash B$" to that of provability "$\vdash E$" in a manner parallel to the reduction by Corollary Theorem 8_{Pd} in § 20 of valid consequence "$A_1, \dots, A_m \models B$" to validity "$\models E$".[77] (Likewise, using $\forall'$, we have parallel reductions of "$A_1, \dots, A_m \vdash^{x_1 \dots x_q} B$" and "$A_1, \dots, A_m \models^{x_1 \dots x_q} B$".) So now to show the equivalence of model theory and proof theory for the predicate calculus, it remains to show that $\models E$ *if and only if* $\vdash E$.

The "if" part is easy. THEOREM 12 and COROLLARY of § 11 hold again. The proofs are as before, using now 1a–10b in Theorem 2_{Pd}, and Theorems

[77] In citing theorems taken over, or slightly adapted, from the propositional calculus, to the predicate calculus without being fully restated, we may use the old numbers with "$_{Pd}$" suffixed (e.g. "Theorem 11_{Pd}" here). We may also suffix "$\vdash$" if at the same time the "$\models$" is changed to "$\vdash$", etc. (e.g. "Theorem 6_{Pd}" end § 19, "Theorem $6_{Pd\vdash}$" end § 25, "Theorem $12_{Pd=}$" § 29).

3_{Pd}, 15 and 16, in § 18. The present corollary does not require the full force of the theorem, which refers to any D. It can be inferred from the theorem with "$\models$" replaced by "1-$\models$", i.e. using only $D = \{1\}$. This is the way the simple consistency of the predicate calculus was first proved by Hilbert and Ackermann in 1928.

The "only if" part, $\models$ E *only if* $\vdash$ E (extending Theorem 14 to the predicate calculus), is not as elementary as the theorems we have thus far given. This proposition is included in Gödel's completeness theorem 1930, which we shall postpone to Chapter VI (Theorem 37 § 51).

In *classical* logic (with which we are primarily concerned), we are interested in the valid formulas because they express universal logical truths, and in the provable ones because they are valid.

In the case of the (classical) *propositional* calculus, constructions of truth tables (or truth-value analyses, § 8) constitute a sure and unimpeachable method of demonstrating validity, directly from its definition in model theory. However, this method isn't always very practical, and it usually doesn't resemble actual thought processes, in which the utmost economy is essential. This is partly why we introduced also proof theory. Of course, we didn't stop with the postulated form of proof theory, but we proceeded thence to derived rules (§ 13). The postulated form of proof theory can be regarded there as a very convenient way station between model theory (validity, etc.) and the derived rules of proof theory.[78]

Now, in the (classical) *predicate* calculus, direct application of the definition of validity is no longer a simple matter. For, demonstrations of validity by considering the truth tables are no longer purely mechanical, but require general reasoning about the truth tables for all (non-empty) domains D; and for infinite D's, the truth tables are infinite objects.

Hence, in the predicate calculus, proof theory has the advantage over model theory, not only of convenience, but of greater concreteness. We are on firmer ground in establishing provability from its definition in § 21, with the help of elementary theorems such as the deduction theorem

[78] Our form of proof theory of the propositional calculus, so far as it consists in rules for establishing logical truths (§ 3 Paragraph 2), could be avoided by developing our *derived* rules entirely as propositions about model theory, beginning with 1a–10b in Theorem 2, and Theorem 3 (cf. § 13 Paragraph 3).

Then in the proof theory of the predicate calculus, one could simply take all tautologies (valid formulas) of the propositional calculus as axioms en masse (instead of postulating our Axiom Schemata 1a–10b or something similar).

There are some good reasons for not thus leaving out our postulated form of proof theory in the propositional calculus. Some of the results in the proof theory of the predicate calculus and more complicated systems are most easily obtained by beginning with a treatment in the propositional calculus. In any case, it is good to practice with proof theory in the simpler situation first. In the intuitionistic propositional calculus (end § 12), a simple truth-table method is not available.

and results based on the proof theory of the propositional calculus, than in demonstrating validity from its definition in § 17. By Theorem 12_{Pd}, demonstrations of provability do establish validity, for anyone who accepts the reasoning in its proof.[79]

Gödel's completeness theorem for the predicate calculus has the significance that, for all true statements of validity, this method of demonstration suffices. In using this method, we escape from having to reason further in terms of the validity notion, after handling the few cases of it that we needed for the proof of Theorem 12_{Pd}. Demonstrations of all other true statements of validity can be put together out of repetitions of the reasoning used in handling those few cases.

Because of our preference for the proof-theoretic approach to the predicate calculus on the ground of its greater concreteness, we now develop some of the further results proof-theoretically, including some which in Chapter I or § 19 we did model-theoretically.

THEOREM 13 again holds. For, as we remarked early in § 21, the direct rules carry over automatically from the propositional to the predicate calculus; and the three subsidiary deduction rules in Theorem 13 are established now by the same proofs from Theorem 11_{Pd} in § 22 as before from Theorem 11. We now add four more derived rules.

THEOREM 21. *Let* x *be any variable,* A(x) *be any formula,* r *be any variable not necessarily distinct from* x, *and* A(r) *be the result of substituting* r *for the free occurrences of* x *in* A(x). *Also let* Γ *be any list of (zero or more) formulas, and* C *be any formula. Then the following rules hold, provided*:

(A) *For* ∀*-elimination and* ∃*-introduction,* r *is free for* x *in* A(x).

(B) *For* ∀*-introduction and* ∃*-elimination,* Γ *does not contain* x *free.*

(C) *For* ∃*-elimination,* C *does not contain* x *free.*

	Introduction	Elimination
∀	*If* Γ ⊢ A(x), *then* Γ ⊢ ∀xA(x).	∀xA(x) ⊢ A(r).
∃	A(r) ⊢ ∃xA(x).	*If* Γ, A(x) ⊢ C, *then* Γ, ∃xA(x) ⊢ C.

[79] A person has the option of taking the postulates as the starting point of logic, without attempting to put behind them an argument like the proofs of Theorem 2_{Pd}, 3_{Pd}, 15, 16. Something very similar to those postulates must go into the intuitive reasoning in those proofs anyway. On the other hand, giving those proofs explicitly does help to guard against mistakes, such as were made in formulating the postulated substitution rule in the early systems of the predicate calculus.[73]

There is some arbitrariness whether our particular set of postulates or some other is accepted. The problem of completeness (solved by Gödel, e.g. for our postulates) only appears to a person who is not merely accepting a set of postulates as the starting point of logic.

PROOFS. ∀-elimination, ∃-introduction. Cf. Exercise 21.5.

∀-introduction. Let C be an axiom not containing x free.

1. $\Gamma \vdash A(x)$ — by hypothesis.
2. $\Gamma, C \vdash A(x)$ — Exercise 13.2, 1.
3. $\Gamma \vdash C \supset A(x)$ — Theorem 11_{Pd}, 2.
4. $\Gamma \vdash C \supset \forall x A(x)$ — from 3 using the ∀-rule, or in more detail thus: Consider a given deduction $B_1, \ldots, B_l$ of $C \supset A(x)$ from Γ with all variables held constant. To this we can add as B_{l+1} the formula $C \supset \forall x A(x)$, justified by the ∀-rule with B_l as premise, since C does not contain x free. In the resulting deduction from Γ, all variables are held constant, since the new application of the ∀-rule is on x, which by Proviso (B) does not occur free in Γ.
5. $\Gamma, C \vdash \forall x A(x)$ — Theorem 10_{Pd} (or modus ponens), 4.
6. $\Gamma \vdash \forall x A(x)$ — Exercise 13.2, 5.

∃-elimination.

1. $\Gamma, A(x) \vdash C$ — by hypothesis.
3. $\Gamma \vdash A(x) \supset C$ — Theorem 11_{Pd}, 1.
3. $\Gamma \vdash \exists x A(x) \supset C$ — from 2, using the ∃-rule (and Provisos (B), (C)).
4. $\Gamma, \exists x A(x) \vdash C$ — Theorem 10_{Pd}, 3.

In Exercise 23.1 below we shall illustrate the importance of watching Provisos (B) and (C) in using these rules. The importance of (A) should appear from examples in §§ 21 and 18 which deal with essentially the same matter.

The subsidiary deduction rule of ∀-introduction is unlike our other subsidiary deduction rules in having the same assumption formulas in the given or subsidiary deduction and the resulting deduction. It is stated as a subsidiary deduction rule because of Proviso (B). Except for this proviso, Example 13.3 would apply and give "$A(x) \vdash \forall x A(x)$". That however does not hold; for then, by Theorem 12_{Pd} (with Theorems 11_{Pd} and 8_{Pd}) we would have "$A(x) \models \forall x A(x)$"; and this is not the case e.g. when A(x) is P(x) or (as we saw in Example 7 § 20) $\neg P(x)$.

COROLLARY 1. *Let* $w_1, \ldots, w_p$ *be distinct variables,* $A(w_1, \ldots, w_p)$ *be a formula,* $r_1, \ldots, r_p$ *be variables not necessarily distinct from each other or from* $w_1, \ldots, w_p$, *and* $A(r_1, \ldots, r_p)$ *be the result of substituting* $r_1, \ldots, r_p$ *simultaneously for the free occurrences of* $w_1, \ldots, w_p$ *respectively in* $A(w_1, \ldots, w_p)$. *If this substitution is free* (i.e. if the resulting occurrences of $r_1, \ldots, r_p$ in $A(r_1, \ldots, r_p)$ are free), *then*:

(a) $\forall w_1 \ldots \forall w_p A(w_1, \ldots, w_p) \vdash A(r_1, \ldots, r_p)$. (*p*-fold ∀-elimination.)

(b) $A(r_1, \ldots, r_p) \vdash \exists w_1 \ldots \exists w_p A(r_1, \ldots, r_p)$. (*p*-fold ∃-introduction.)

PROOFS, for $p > 1$. (For $p = 0$, (a) and (b) are simply Theorem 9_{Pd} (i) for $m = 1$. For $p = 1$, they are in the theorem.) (a) For illustration, suppose $p = 2$, and w_1, w_2 are x, y, and $A(w_1, w_2)$ is $\exists wP(w, x, y)$, and r_1, r_2 are y, z. Then the substitution is free. (It would not be if r_1, r_2 were w, z.) We must show that $\forall x \forall y \exists wP(w, x, y) \vdash \exists wP(w, y, z)$. Evidently we should use $\forall$-eliminations, but a little care is necessary. We cannot first eliminate $\forall x$ with y as the r to obtain $\forall y \exists wP(w, y, y)$, as the resulting occurrence of y (the second) would not be free. To use an easily-described uniform method, let x_1, x_2 be two distinct new variables, i.e. variables distinct from each other and all the others in the illustration. By two successive $\forall$-elims. (with Theorem 9_{Pd} (ii)): **(1)** $\forall x \forall y \exists wP(w, x, y) \vdash \exists wP(w, x_1, x_2)$. Thence by two successive $\forall$-introds.: **(2)** $\forall x \forall y \exists wP(w, x, y) \vdash \forall x_1 \forall x_2 \exists wP(w, x_1, x_2)$. By two successive $\forall$-elims.: **(3)** $\forall x_1 \forall x_2 \exists wP(w, x_1, x_2) \vdash \exists wP(w, y, z)$. Combining (2) and (3) by Theorem 9_{Pd} (ii): **(4)** $\forall x \forall y \exists wP(w, x, y) \vdash \exists wP(w, y, z)$.

(b) Left to the student (Exercise 23.2 (b)).

COROLLARY 2. (Substitution for individual variables.)[63] *Under the circumstances of the theorem or Corollary* 1, *and provided* Γ *does not contain free occurrences of* x *or* $w_1, \ldots, w_p$, *respectively*:

(c) *If* $\Gamma \vdash A(x)$, *then* $\Gamma \vdash A(r)$.

(d) *If* $\Gamma \vdash A(w_1, \ldots, w_p)$, *then* $\Gamma \vdash A(r_1, \ldots, r_p)$.

PROOF is Exercise 23.3.

EXERCISES. 23.1. In each of the following, the formula in the last statement is not valid (by Exercises 17.3, 17.5) and therefore (by Theorem 12_{Pd}) not provable, contrary to what is asserted. Locate the fallacy.

I. "1. $P(x) \vdash P(x)$ — (i) (in Theorem 9_{Pd}).
  2. $P(x) \vdash \forall xP(x)$ — $\forall$-introd., 1.
  3. $\exists xP(x) \vdash \forall xP(x)$ — $\exists$-elim., 2.
  4. $\vdash \exists xP(x) \supset \forall xP(x)$ — $\supset$-introd., 3."

II. "1. $P(x), Q(x) \vdash \exists xP((x) \& Q(x))$ — &- and $\exists$-introd.
  2. $P(x), \exists xQ(x) \vdash \exists x(P(x) \& Q(x))$ — $\exists$-elim., 1.
  3. $\exists xP(x), \exists xQ(x) \vdash \exists x(P(x) \& Q(x))$ — $\exists$-elim., 2.
  4. $\exists xP(x) \& \exists xQ(x) \vdash \exists x(P(x) \& Q(x))$ — &-elims., 3 (with (ii)).
  5. $\vdash \exists xP(x) \& \exists xQ(x) \supset \exists x(P(x) \& Q(x))$ — $\supset$-introd., 4."

III. "1. $P(x) \vdash P(x)$ — (i).
  2. $\exists xP(x) \vdash P(x)$ — $\exists$-elim., 1.
  3. $\vdash \exists xP(x) \supset P(x)$ — $\supset$-introd., 2."

23.2. (a) Write out the six $\forall$-elims. and $\forall$-introds. in the illustration of the proof of Corollary 1 (a), and verify that the provisos are satisfied. (b) Give the proof of Corollary 1 (b) with the same illustration.

23.3. Prove Corollary 2. Also simplify the statement for the case Γ is empty.

23.4. Show that: For each variable x and formula A, ⊢ A if and only if ⊢ ∀xA; and hence, ⊢ A if and only if ⊢ ∀A (cf. § 20 for "∀").

23.5. Show that: $\forall' A_1, \ldots, \forall' A_m \vdash B$ if and only if $\forall' A_1, \ldots, \forall' A_m \vdash \forall' B$ (where $\forall' \simeq \forall x_1 \ldots \forall x_q$ as in § 20).

23.6. True or false (and why)?

(a) "∀z∃y∀xP(x, y, z) ⊢ ∀x∃y∀zP(x, y, z)."

(b) "∃x∀y∀zP(x, y, z) ⊢ ∀z∃xP(x, x, z)."

23.7. For each of the following modifications of the predicate calculus, say which of the two implications {⊢ E} ⇄ {⊨ E} (Theorem 12_{Pd} and Chapter VI) continue to hold for all formulas E, and why.

(a) Add as a new axiom schema ∀xA(x) ⊃ ∃xA(x).

(b) Add as a new axiom schema ∃xA(x) ⊃ ∀xA(x).

(c) Omit the ∀-schema, ∀-rule, ∃-schema and ∃-rule.

§ 24. Proof theory: replacement, chains of equivalences.

THEOREM 22. *Let* x *be any variable, and* A, B, C, A(x), B(x) *be any formulas, and for* *69a–*72a *let* Γ *be any list of (zero or more) formulas not containing* x *free.*[80]

*6.	A ⊃ B ⊢ (B ⊃ C) ⊃ (A ⊃ C).	*7.	A ⊃ B ⊢ (C ⊃ A) ⊃ (C ⊃ B).
*8a.	A ⊃ B ⊢ A & C ⊃ B & C.	*8b.	A ⊃ B ⊢ C & A ⊃ C & B.
*9a.	A ⊃ B ⊢ A ∨ C ⊃ B ∨ C.	*9b.	A ⊃ B ⊢ C ∨ A ⊃ C ∨ B.
*12.	A ⊃ B ⊢ ¬B ⊃ ¬A.	*13.	A ⊃ ¬B ⊢ B ⊃ ¬A.
*14°.	¬A ⊃ B ⊢ ¬B ⊃ A.	*15°.	¬A ⊃ ¬B ⊢ B ⊃ A.

*69a. *If* Γ ⊢ A(x) ⊃ B(x), *then* Γ ⊢ ∀xA(x) ⊃ ∀xB(x).

*70a. *If* Γ ⊢ A(x) ⊃ B(x), *then* Γ ⊢ ∃xA(x) ⊃ ∃xB(x).

(Introduction of a logical symbol into an implication, including contraposition with double negations suppressed.)

*25a.	A ∼ B ⊢ (A ∼ C) ∼ (B ∼ C).	*25b.	A ∼ B ⊢ (C ∼ A) ∼ (C ∼ B).
*26.	A ∼ B ⊢ A ⊃ C ∼ B ⊃ C.	*27.	A ∼ B ⊢ C ⊃ A ∼ C ⊃ B.
*28a.	A ∼ B ⊢ A & C ∼ B & C.	*28b.	A ∼ B ⊢ C & A ∼ C & B.
*29a.	A ∼ B ⊢ A ∨ C ∼ B ∨ C.	*29b.	A ∼ B ⊢ C ∨ A ∼ C ∨ B.

*30. A ∼ B ⊢ ¬A ∼ ¬B.

*71a. *If* Γ ⊢ A(x) ∼ B(x), *then* Γ ⊢ ∀xA(x) ∼ ∀xB(x).

*72a. *If* Γ ⊢ A(x) ∼ B(x), *then* Γ ⊢ ∃xA(x) ∼ ∃xB(x).

(Introduction of a logical symbol into an equivalence.)

[80] As before (Theorem 2), we are numbering results in agreement with IM, as far as possible. (*69a–*72a are modifications of IM *69–72. *25a, *25b, *40a, *40b do not occur as numbered results in IM.)[29]

*40a. $A \vdash (A \sim B) \sim (B \sim A) \sim B$. *40b. $\neg A \vdash (A \sim B) \sim (B \sim A) \sim \neg B$.
*41. $A \vdash A \supset B \sim B$. *43. $\neg A \vdash A \supset B \sim \neg A$.
*42. $A \vdash B \supset A \sim A$. *44. $\neg A \vdash B \supset A \sim \neg B$.
*45. $A \vdash A \& B \sim B \& A \sim B$. *47. $\neg A \vdash A \& B \sim B \& A \sim A$.
*46. $A \vdash A \vee B \sim B \vee A \sim A$. *48. $\neg A \vdash A \vee B \sim B \vee A \sim B$.

(Special cases of the binary propositional connectives.)

PROOFS. All of the results except the four involving quantifiers (*69a–*72a) can be established in the propositional calculus. This can be done automatically with $\models$, by truth table computation with P, Q, R, followed by substitution in the form of Exercise 7.3;[81] and then $\models$ can be changed to $\vdash$ by completeness (Theorem 14 with Theorems 8, 10); or we can use the rules of Theorem 13, employing *6–*12 in the proofs of *26–*30. Then they can be taken over into the predicate calculus, by beginning § 21. (Exercise 24.1.)

*69a, *70a. *70a is the result of Exercise 21.4, when we employ a deduction of $A(x) \supset B(x)$ from Γ instead of from $\forall x(A(x) \supset B(x))$. *69a is obtainable similarly, or by generalizing Example 14 from P(x), Q(x) to A(x), B(x) and replacing the first three steps by the assumed deduction of $A(x) \supset B(x)$. Or we can use the rules of Theorems 13 and 21 (Exercise 24.2).

*71a, *72a. *71a is established as follows, and *72a similarly.

1. $\Gamma \vdash A(x) \sim B(x) \vdash A(x) \supset B(x)$ — hypothesis; $\sim$-elim.[35]
2. $\Gamma \vdash \forall x A(x) \supset \forall x B(x)$ — *69a, 1.
3. $\Gamma \vdash \forall x B(x) \supset \forall x A(x)$ — similarly.
4. $\Gamma \vdash \forall x A(x) \sim \forall x B(x)$ — $\sim$-introd., 2, 3.

Now we can extend Theorem 5 to the predicate calculus with "$\Gamma \vdash$" in place of "$\models$". (The extension with "$\models$" is in § 19.)

THEOREM 23. (Replacement theorem.) *Let* C_A *be a formula containing a formula* A *as a specified (consecutive) part, and let* C_B *come from* C_A *by replacing that part by a formula* B. *Let* $x_1, \ldots, x_q$ *be the free variables of* A *or* B *which belong to quantifiers of* C_A *having the specified part within their scopes (i.e.* $x_1, \ldots, x_q$ *are the variables having free occurrences in* A *or* B *which become bound occurrences in* C_A *or* C_B*). Let* Γ *be a list of (zero or more) formulas not containing any of* $x_1, \ldots, x_q$ *free. Then:*

$$\textit{If } \Gamma \vdash A \sim B, \textit{ then } \Gamma \vdash C_A \sim C_B.$$

PROOF. We illustrate the proof by four examples.

[81] The tables (I) in § 8 provide shortcuts to *40a–*48.

EXAMPLE 15. Let C_A be $R \supset \forall z(\underline{\neg\exists x P(x, y, z)} \vee Q(z)$ with the specified part A underlined. Let "$\Gamma \vdash A \sim B$" be "$\vdash \neg\exists x P(x, y, z) \sim \forall x \neg P(x, y, z)$" (with Γ empty), which holds by *82a in Theorem 26; i.e. the formula 1 below is provable. Using *29a, *71a and *27 we infer successively that 2–4 are also provable.

$$\begin{array}{lrcl}
1. & \neg\exists x P(x, y, z) & \sim & \forall x \neg P(x, y, z) \\
2. & \neg\exists x P(x, y, z) \vee Q(z) & \sim & \forall x \neg P(x, y, z) \vee Q(z) \\
3. & \forall z(\neg\exists x P(x, y, z) \vee Q(z)) & \sim & \forall z(\forall x \neg P(x, y, z) \vee Q(z)) \\
4. & R \supset \forall z(\neg\exists x P(x, y, z) \vee Q(z)) & \sim & R \supset \forall z(\forall x \neg P(x, y, z) \vee Q(z))
\end{array}$$

Thus $\vdash R \supset \forall z(\neg\exists x P(x, y, z) \vee Q(z)) \sim R \supset \forall z(\forall x \neg P(x, y, z) \vee Q(z))$, which is what the theorem asserts in this example.

EXAMPLE 16. Let C_A be $P(x) \vee \forall x \exists y(\underline{P(x)} \supset Q(y))$ with the specified part A underlined. Let "$\Gamma \vdash A \sim B$" be "$\forall x \forall y[P(x) \sim \exists z R(x, y, z)] \vdash P(x) \sim \exists z R(x, y, z)$", which holds by double $\forall$-elim.; i.e. the formula 1 below is deducible from the Γ. Using *26, *72a (since the Γ does not contain y free), *71a (since x isn't free in Γ), *29b, we infer successively that 2–5 are also deducible from the Γ.

$$\begin{array}{lrcl}
1. & P(x) & \sim & \exists z R(x, y, z) \\
2. & P(x) \supset Q(y) & \sim & \exists z R(x, y, z) \supset Q(y) \\
3. & \exists y(P(x) \supset Q(y)) & \sim & \exists y(\exists z R(x, y, z) \supset Q(y)) \\
4. & \forall x \exists y(P(x) \supset Q(y)) & \sim & \forall x \exists y(\exists z R(x, y, z) \supset Q(y)) \\
5. & P(x) \vee \forall x \exists y(P(x) \supset Q(y)) & \sim & P(x) \vee \forall x \exists y(\exists z R(x, y, z) \supset Q(y))
\end{array}$$

Thus $\forall x \forall y[P(x) \sim \exists z R(x, y, z)] \vdash$

$$P(x) \vee \forall x \exists y(P(x) \supset Q(y)) \sim P(x) \vee \forall x \exists y(\exists z R(x, y, z) \supset Q(y)),$$

which is what the theorem asserts in this example.

EXAMPLE 17. In Example 16 change B to $\exists z R(x, w, z)$. Since w does not become bound in replacing the specified $P(x)$ by $\exists z R(x, w, z)$, it is permitted to be free in the Γ. Thus $\forall x[P(x) \sim \exists z R(x, w, z)] \vdash$

$$P(x) \vee \forall x \exists y(P(x) \supset Q(y)) \sim P(x) \vee \forall x \exists y(\exists z R(x, w, z) \supset Q(y)).$$

EXAMPLE 18. Changing the specified occurrence of $P(x)$ in Example 16 to be the first one, $C_A \sim C_B$ is deducible from $P(x) \sim B$ (using simply *29a), whatever the B.

COROLLARY 1. *For* C_A, C_B, $x_1, \ldots, x_q$ *as in the theorem:*

$$A \sim B \vdash^{x_1 \ldots x_q} C_A \sim C_B.$$

PROOF. Apply the theorem with $\forall x_1 \ldots \forall x_q(A \sim B)$ as the Γ, as in Examples 16, 17. Remember that "$A \sim B \vdash^{x_1 \ldots x_q}$" means "$\forall x_1 \ldots \forall x_q(A \sim B) \vdash$".

COROLLARY 2. (Replacement property of equivalence.) *Similarly:*

(a) *If* $\Gamma \vdash A \sim B$, *then* $\Gamma, C_A \vdash C_B$. (b) $A \sim B, C_A \vdash^{x_1 \ldots x_q} C_B$.

Now we can return to (α)–(ζ) in § 5, and verify that they hold with "$\models$" replaced by "$\Gamma \vdash$" (Exercise 24.3). So with Theorem 23 (or its Corollary 1), we have all the ingredients for using chains of equivalences in proof theory, with "$\Gamma \vdash$" standing before the chains (instead of "$\models$" as in §§ 5, 19). The formulas Γ must not contain free any of the variables $x_1, \ldots, x_q$ of quantifiers within the scopes of which replacements are performed (or such variables which they do contain free must be written as superscripts on "$\vdash$"). In most of our applications (e.g. those in the proofs of Theorems 25 and 26), Γ is empty (as in Example 15); i.e. the equivalences are provable, and we need not worry about $x_1, \ldots, x_q$.

Suppose that in C_A (as in Theorem 23) the specified part A is not in the scope of any $\sim$ (i.e. it is not in the D of any part D $\sim$ E or E $\sim$ D). We then say the specified part A is *positive* or *negative* according as it is in the D of an even or odd number of parts of the form $\neg$D or the form D $\supset$ E. For example, in $\overline{\underline{\neg}}\exists x((P(x) \underline{\supset} Q) \vee R) \overline{\underline{\supset}} \forall y P(y)$ the part P(x) is odd (count the underlined operators), and the part Q is even (count the overlined operators).

THEOREM 24. *Under the circumstances of Theorem 23, and supposing further that the specific part* A *is not in the scope of any* $\sim$:

If $\Gamma \vdash A \supset B$, *then* $\Gamma \vdash \begin{Bmatrix} C_A \supset C_B \\ C_B \supset C_A \end{Bmatrix}$ *if the specified part* A *is* $\begin{Bmatrix} \textit{positive} \\ \textit{negative} \end{Bmatrix}$.

COROLLARY. $A \supset B \vdash^{x_1 \ldots x_q} \begin{Bmatrix} C_A \supset C_B \\ C_B \supset C_A \end{Bmatrix}$ *if that part* A *is* $\begin{Bmatrix} \textit{positive} \\ \textit{negative} \end{Bmatrix}$.

PROOFS is Exercise 24.4.

LEMMA 5. *Let* x *be any variable,* A(x) *be any formula,* y *be any variable* (*not necessarily distinct from* x) *such that*

(i) y *is free for* x *in* A(x),

(ii) y *does not occur free in* A(x) (*unless* y *is* x),

and A(y) *be the result of substituting* y *for the free occurrences of* x *in* A(x). *Then:*

*73. $\vdash \forall x A(x) \sim \forall y A(y)$. *74. $\vdash \exists x A(x) \sim \exists y A(y)$.

PROOFS. *73. By the right column in Example 11. *74. Similarly.

THEOREM 25. *If* A *is congruent to* B, *then* $\vdash A \sim B$.

PROOF. We illustrate this by the example (1) and (2) of congruent formulas end § 16. In order to employ a uniform method applicable in all cases (though short cuts will generally be available in particular cases), we pick a set of new variables x_1, x_2, x_3 to correspond to the three quantifiers of either. By successive replacements of the variable-occurrences bound by one quantifier at a time, we can transform 1 (= (1)) into 4 (cf. (1b), (2b)) and that into 7 (= (2)), thus:

1. $\forall x\, (P(x) \,\&\, \exists x\, Q(x, z) \supset \exists y\, R(x, y)) \vee Q(z, x)$.
2. $\forall x\, (P(x) \,\&\, \exists x_1 Q(x_1, z) \supset \exists y\, R(x, y)) \vee Q(z, x)$.
3. $\underline{\forall x\, (P(x) \,\&\, \exists x_1 Q(x_1, z) \supset \exists x_2 R(x, x_2))} \vee Q(z, x)$.
4. $\underline{\forall x_3 (P(x_3) \,\&\, \exists x_1 Q(x_1, z) \supset \exists x_2 R(x_3, x_2))} \vee Q(z, x)$.
5. $\forall y\, (P(y) \,\&\, \exists x_1 Q(x_1, z) \supset \exists x_2 R(y, x_2)) \vee Q(z, x)$.
6. $\forall y\, (P(y) \,\&\, \exists x_1 Q(x_1, z) \supset \exists z\, R(y, z)) \vee Q(z, x)$.
7. $\forall y\, (P(y) \,\&\, \exists x\, Q(x, z) \supset \exists z\, R(y, z)) \vee Q(z, x)$.

The proof that $\vdash A \sim B$ (here A is (1) and B is (2)) is given by the chain method. E.g. for the "third link", the hypotheses of Lemma 5 are satisfied by x, $P(x) \,\&\, \exists x_1 Q(x_1, z) \supset \exists x_2 R(x, x_2)$, x_3 as the x, A(x), y; so by *73 the underlined parts of 3 and 4 are equivalent (i.e. the formula $\forall x A(x) \sim \forall y A(y)$ expressing their equivalence is provable); so by Theorem 23, 3 is equivalent to 4.

Exercises. 24.1. Run through all steps in the proofs of *6, *7 and *26 by each of the suggested methods.

24.2. Establish *69a and *70a by the rules of Theorems 13 and 21.

24.3. Verify that (α)–(ζ) in § 5 hold for the predicate calculus with "$\Gamma \vdash$". (For (β)–(δ), cf. *19–*21 in Theorem 2.)

24.4. Prove Theorem 24 and Corollary.

24.5. Justify the following. Format (B_1) § 13 is used in (b).

(a) 1. $\forall x(P(x) \supset M(x)) \vdash P(x) \supset M(x)$.

2. $\forall x(P(x) \supset M(x)) \vdash \exists x(S(x) \,\&\, \neg M(x) \supset \exists x(S(x) \,\&\, \neg P(x))$.

3. $\forall x(P(x) \supset M(x)), \exists x(S(x) \,\&\, \neg M(x)) \vdash \exists x(S(x) \,\&\, \neg P(x))$.

This expresses the traditional Aristotelian syllogism "Baroco": All P are M. Some S are not M. Therefore some S are not P.

(b) Assume $\forall x(P(x) \supset \neg M(x))$. Thence successively $P(x) \supset \neg M(x)$, $M(x) \supset \neg P(x)$, $\exists x(S(x) \,\&\, M(x)) \supset \exists x(S(x) \,\&\, \neg P(x))$. — Thus $\forall x(P(x) \supset \neg M(x)), \exists x(S(x) \,\&\, M(x)) \vdash \exists x(S(x) \,\&\, \neg P(x))$ ("Festino").[82]

§ 25. Proof theory: alterations of quantifiers, prenex form.

We shall use the proofs for Theorem 26 below to provide further illustrative and practice material in the technique of § 13 for using our derived rules. Formats "(A)", "(B_1)", "(B_2)", "(B_3)" are used as there. Now we have in addition the four introduction and elimination rules for the quantifiers (Theorem 21). The provisos for these rules must be observed carefully, as always.

[82] The translation into our symbolism is explained in § 26 below. Thirteen others of the Aristotelian syllogisms can be established similarly, while four fail to hold when similarly translated. Cf. Example 23 § 27, Hilbert and Ackermann 1928 Chapter 3 § 3.

The rule of ∀-introduction was stated in Theorem 21 as a subsidiary deduction rule in which the assumption formulas are not changed from subsidiary to resulting deduction. The effect is that, when we are listing formulas deducible from assumptions Γ *not containing* x *free*, we can proceed by ∀-introduction from A(x) to ∀xA(x).

The rule of ∃-elimination serves to eliminate the ∃x from ∃xA(x) preparatory to inferring from ∃xA(x) (and possibly other assumptions Γ *not containing* x *free*) a consequence C *not containing* x *free*. The procedure corresponds to the following familiar form of reasoning in mathematics: "There is an x such that $A(x)$; now pick such an x . . ." or "There is a number with the property A; let x be such a number . . .". If through a chain of reasoning using x a conclusion is reached not containing x, the conclusion can be taken to be established without the assumption about x.

One of the purposes of studying logic formally is to improve our reliability in the use of logic. This improvement is obtained by our being able, in cases of doubt, to trace how our proposed reasoning (or the reasoning of others) can be effected by the principles we have studied, or in brief how the reasoning can be "formalized". Also, as will appear in Chapter IV, we may sometimes wish to do so to make clear the bases of the reasoning, whether or not we have doubts about its validity.

Therefore, it is an objective of the formal study of logic to bring the methods of recognizing that formulas are provable in the strict sense of proof theory (§§ 9, 21) as close as possible to the methods which are used informally.

This we do not claim can be done once and for all, for all kinds of reasoning. However, in our opinion, the rules for introducing and eliminating logical symbols do this very successfully for the ordinary use of predicate logic, not just for proving results within logic (as in our next illustrations) but also in the application of logic in mathematics and elsewhere. The remarks at the end of § 13 on supplementing these rules appropriately by other devices should be kept in mind.

The introduction and elimination rules are due essentially to Gentzen 1932, 1934–5, 1936 § 5 (who profited from earlier work of Hertz 1929) and Jaśkowski 1934.[83] These authors were primarily interested in formulations of logic with such rules as postulated rules, while we have them as derived rules; cf. § 13. Our format (B_2) corresponds closely to Gentzen's "natural deduction" and Jaśkowski 1934.

It is controversial whether one does better to start with a postulated formulation of logic based on modus ponens (a "Hilbert-type system",

[83] These rules include the deduction theorem, originating in 1930 and before with Herbrand and Tarski.[33]

§§ 50, 54) and derive the introduction and elimination rules, as we have done, or to start out with a version of the latter as the postulated rules (a "Gentzen-type system").

No one formulation is most convenient for all purposes. One formulation with postulated axioms and rules is necessary to give a firm starting point for proof theory. But it is our view that, no matter what formulation one begins with, he will not want simply to use it, but rather he will soon wish to escape its limitations by devising short cuts, proving theorems that introduce new methods, etc. The history of mathematics is one of continually building on previous discoveries, and of devising methods which telescope or reorganize steps previously carried out separately into new processes of reasoning.

This being the case, we have preferred to start from the structurally simpler Hilbert-type systems in §§ 9, 21. These also lend themselves well to the addition of mathematical axioms and axiom schemata to establish systems of logic and mathematics, of which we shall have more to say in Chapter IV.[84]

THEOREM 26. *Let* x, y *be any (distinct) variables,* A(x), B(x), A(x, y) *be any formulas, and* A, B *be any formulas not containing* x *free. For* *79 *and* *80, *further let* A(x, x) *be the result of substituting* x *for the free occurrences of* y *in* A(x, y), *and suppose that* y *is free for* x *in* A(x, y) (i.e. the occurrences of x in A(x, x) resulting from this substitution are all free).[85]

*75. $\vdash \forall x A \sim A$.	*76. $\vdash \exists x A \sim A$.
*77. $\vdash \forall x \forall y A(x, y) \sim \forall y \forall x A(x, y)$.	*78. $\vdash \exists x \exists y A(x, y) \sim \exists y \exists x A(x, y)$.
*79. $\vdash \forall x \forall y A(x, y) \supset \forall x A(x, x)$.	*80. $\vdash \exists x A(x, x) \supset \exists x \exists y A(x, y)$.

*81. $\vdash \forall x A(x) \supset \exists x A(x)$.

*82. $\vdash \exists x \forall y A(x, y) \supset \forall y \exists x A(x, y)$.

(Alterations of quantifiers.)

*82a. $\vdash \neg \exists x A(x) \sim \forall x \neg A(x)$.	*82b°. $\vdash \neg \forall x A(x) \sim \exists x \neg A(x)$.
*83°. $\vdash \exists x A(x) \sim \neg \forall x \neg A(x)$.	*84°. $\vdash \forall x A(x) \sim \neg \exists x \neg A(x)$.

(Negation and quantifiers.)

[84] For purposes which will arise in Chapter VI, we shall then also want postulated Gentzen-type systems, proved equivalent to our present Hilbert-type system.

[85] The numbering follows IM. (*82a, *82b are the *86, *85 of IM renumbered to come earlier. *99b is not stated in IM.)

*87. $\vdash \forall x A(x) \,\&\, \forall x B(x) \sim \forall x(A(x) \,\&\, B(x))$.

*88. $\vdash \exists x A(x) \vee \exists x B(x) \sim \exists x(A(x) \vee B(x))$.

*89. $\vdash A \,\&\, \forall x B(x) \sim \forall x(A \,\&\, B(x))$.

*90. $\vdash A \vee \exists x B(x) \sim \exists x(A \vee B(x))$.

*91. $\vdash A \,\&\, \exists x B(x) \sim \exists x(A \,\&\, B(x))$.

*92°. $\vdash A \vee \forall x B(x) \sim \forall x(A \vee B(x))$.

*93. $\vdash \exists x(A(x) \,\&\, B(x)) \supset \exists x A(x) \,\&\, \exists x B(x)$.

*94. $\vdash \forall x A(x) \vee \forall x B(x) \supset \forall x(A(x) \vee B(x))$.

(Conjunction or disjunction, and quantifiers.)

*95. $\vdash A \supset \forall x B(x) \sim \forall x(A \supset B(x))$.

*97°. $\vdash A \supset \exists x B(x) \sim \exists x(A \supset B(x))$.

*98°. $\vdash \forall x A(x) \supset B \sim \exists x(A(x) \supset B)$.

*96. $\vdash \exists x A(x) \supset B \sim \forall x(A(x) \supset B)$.

*99°. $\vdash \forall x A(x) \supset \exists x B(x) \sim \exists x(A(x) \supset B(x))$.

*99b. $\vdash (\exists x A(x) \supset \forall x B(x)) \supset \forall x(A(x) \supset B(x))$.

(Implication and quantifiers.)

PROOFS.

*75. 1. $\forall x A \vdash A$ — $\forall$-elim. (with x as both the x and the r and A as the A(x), so trivially the r is free for the x in the A(x)).

2. $\vdash \forall x A \supset A$ — $\supset$-introd., 1.

(A) 3. $A \vdash A$.

4. $A \vdash \forall x A$ — $\forall$-introd., 3 (which is legitimate since A does not contain x free).

5. $\vdash A \supset \forall x A$ — $\supset$-introd., 4.

6. $\vdash \forall x A \sim A$ — $\sim$-introd., 2, 5.

I. Assume ${}_1\forall x A$ (preparatory to $\supset$-introd.). By $\forall$-elim., ${}_2A$.

(B_1) II. Assume ${}_4A$, which does not contain x free. So by $\forall$-introd., ${}_5\forall x A$.

1. $\forall x A$ — assumed.

2. A — $\forall$-elim., 1.

3. $\forall x A \supset A$ — $\supset$-introd., 2.

(B_2) 4. A — assumed.

5. $\forall x A$ — $\forall$-introd., 4.

6. $A \supset \forall x A$ — $\supset$-introd., 5.

7. $\forall x A \sim A$ — $\sim$-introd., 3, 6.

1. ∀xA ⊢ ∀xA.
2. ∀xA ⊢ A — ∀-elim., 1.
3. ⊢ ∀xA ⊃ A — ⊃-introd., 2.

(B_3) 4. A ⊢ A.
5. A ⊢ ∀xA — ∀-introd., 4.
6. ⊢ A ⊃ ∀xA — ⊃-introd., 5.
7. ⊢ ∀xA ∼ A — ∼-introd., 3, 6.

*76. 1. A ⊢ A.
2. ∃xA ⊢ A — ∃-elim., 1 (noting that A as the C does not contain x free).

(A) 3. ⊢ ∃xA ⊃ A — ⊃-introd., 2.
4. A ⊢ ∃xA — ∃-introd. (with x as the r).
5. ⊢ A ⊃ ∃xA — ⊃-introd., 4.
6. ⊢ ∃xA ∼ A — ∼-introd., 3, 5.

*80. 1. A(x, x) ⊢ ∃x∃yA(x, y) — ∃-introd. twice (first with y as the x and x as the r, using the hypothesis that x is free for y in A(x, y); then with x as both the x and the r); or by Corollary 1 (b)

(A) Theorem 21.
2. ∃xA(x, x) ⊢ ∃x∃yÂ(x, y) — ∃-elim., 1 (noting that ∃x∃yA(x, y) as the C does not contain x free).
3. ⊢ ∃xA(x, x) ⊃ ∃x∃yA(x, y) — ⊃-introd., 2.

*82a. 1. ¬∃xA(x), A(x) ⊢ ¬∃xA(x).
2. ¬∃xA(x), A(x) ⊢ ∃xA(x) — ∃-introd. (with x as the r).
3. ¬∃xA(x) ⊢ ¬A(x) — ¬-introd., 2, 1.
4. ¬∃xA(x) ⊢ ∀x¬A(x) — ∀-introd., 3 (noting that ¬∃xA(x) as the Γ does not contain x free).
5. ⊢ ¬∃xA(x) ⊃ ∀x¬A(x) — ⊃-introd., 4.
6. ∀x¬A(x), A(x) ⊢ A(x).

(A) 7. ∀x¬A(x), A(x) ⊢ ¬A(x) — ∀-elim.
8. ∀x¬A(x), A(x) ⊢ ¬∀x¬A(x) — weak ¬-elim., 6, 7.
9. ∀x¬A(x), ∃xA(x) ⊢ ¬∀x¬A(x) — ∃-elim., 8 (noting that neither ¬∀x¬A(x) as the C nor ∀x¬A(x) as the Γ contains x free).
10. ∀x¬A(x), ∃xAx) ⊢ ∀x¬A(x).
11. ∀x¬A(x) ⊢ ¬∃xA(x) — ¬-introd., 10, 9.
12. ⊢ ∀x¬A(x) ⊃ ¬∃xA(x) — ⊃-introd., 11.
13. ⊢ ¬∃xA(x) ∼ ∀x¬A(x) — ∼-introd., 5, 12.

(B₁) I. Assume ${}_1\neg\exists xA(x)$. Preparatory to reductio ad absurdum ($\neg$-introd.), assume ${}_2A(x)$. By $\exists$-introd., ${}_3\exists xA(x)$, contradicting $\neg\exists xA(x)$. So by the $\neg$-introd. (discharging the assumption $A(x)$), ${}_4\neg A(x)$. By $\forall$-introd. (as the remaining assumption $\neg\exists xA(x)$ doesn't contain x free), ${}_5\forall x\neg A(x)$. II. Assume ${}_7\forall x\neg A(x)$. For $\neg$-introd., assume ${}_8\exists xA(x)$. Preparatory to $\exists$-elim., assume ${}_9A(x)$. From $\forall x\neg A(x)$ by $\forall$-elim., ${}_{10}\neg A(x)$, contradicting $A(x)$. By weak $\neg$-elim., ${}_{11}\neg\forall x\neg A(x)$. Since neither this formula nor the assumption formulas $\forall x\neg A(x)$, $\exists xA(x)$ other than $A(x)$ contain x free, we can now complete the $\exists$-elim., and then (noting the contradiction to $\forall x\neg A(x)$) the $\neg$-introd., to obtain ${}_{13}\neg\exists xA(x)$.

(B₂)

1. $\neg\exists xA(x)$ — assumed.
2. $A(x)$ — assumed.
3. $\exists xA(x)$ — $\exists$-introd., 2.
4. $\neg A(x)$ — $\neg$-introd., 3, 1.
5. $\forall x\neg A(x)$ — $\forall$-introd., 4.
6. $\neg\exists xA(x) \supset \forall x\neg A(x)$ — $\supset$-introd., 5.
7. $\forall x\neg A(x)$ — assumed.
8. $\exists xA(x)$ — assumed.
9. $A(x)$ — assumed.
10. $\neg A(x)$ — $\forall$-elim., 7.
11. $\neg\forall x\neg A(x)$ — weak $\neg$-elim., 9, 10.
12. $\neg\forall x\neg A(x)$ — $\exists$-elim., 11.
13. $\neg\exists xA(x)$ — $\neg$-introd., 7, 12.
14. $\forall x\neg A(x) \supset \neg\exists xA(x)$ — $\supset$-introd., 13.
15. $\neg\exists xA(x) \sim \forall x\neg A(x)$ — $\sim$-introd., 6, 14.

(B₃)

1. $\neg\exists xA(x) \vdash \neg\exists xA(x)$.
2. $A(x), \neg\exists xA(x) \vdash A(x)$.
3. $A(x), \neg\exists xA(x) \vdash \exists xA(x)$ — $\exists$-introd., 2.
4. $\neg\exists xA(x) \vdash \neg A(x)$ — $\neg$-introd., 3, 1.
5. $\neg\exists xA(x) \vdash \forall x\neg A(x)$ — $\forall$-introd., 4.
6. $\vdash \neg\exists xA(x) \supset \forall x\neg A(x)$ — $\supset$-introd., 5.
7. $\forall x\neg A(x) \vdash \forall x\neg A(x)$.
8. $\exists xA(x), \forall x\neg A(x) \vdash \exists xA(x)$.
9. $A(x), \exists xA(x), \forall x\neg A(x) \vdash A(x)$.
10. $A(x), \exists xA(x), \forall x\neg A(x) \vdash \neg A(x)$ — $\forall$-elim., 7.
11. $A(x), \exists xA(x), \forall x\neg A(x) \vdash \neg\forall x\neg A(x)$ — weak $\neg$-elim., 9, 10.
12. $\exists xA(x), \forall x\neg A(x) \vdash \neg\forall x\neg A(x)$ — $\exists$-elim., 11.
13. $\forall x\neg A(x) \vdash \neg\exists xA(x)$ — $\neg$-introd., 7, 12.
14. $\vdash \forall x\neg A(x) \supset \neg\exists xA(x)$ — $\supset$-introd., 13.
15. $\vdash \neg\exists xA(x) \sim \forall x\neg A(x)$ — $\sim$-introd., 6, 14.

At each step the results in (B_2) which are opposite no other arrows are available. Thus for 4, we can use 1 as one of the two "given deductions" for the ¬-introd., since from ¬∃xA(x) ⊢ ¬∃xA(x) we have (by general properties of ⊢) A(x), ¬∃xA(x) ⊢ ¬∃xA(x). Though we could suspend the assumption ∃xA(x) for Lines 9–11, we prefer to keep it in force throughout Lines 8–12, so the immediate result of the ∃-elim. applied to 11 is ∃xA(x), ∃xA(x), ∀x¬A(x) ⊢ ¬∀x¬A(x). (Cf. Lines 12–17 in (B_3) of Example 9 § 13.)

*82b. We use the chain method (available by § 24). ⊢ ∃x¬A(x) ∼ ¬¬∃x¬A(x) [*49] ∼ ¬∀x¬¬A(x) [*82a] ∼ ¬∀xA(x) [*49].

*87–*94. All of these can be established in pairs, as we illustrate next for *91 and *92. (Also *88, *90, *94 can be established by direct methods, like *91.)

*91. I. Assume for ⊃-introd., ${}_1$A & ∃xB(x). Thence by &-elims., ${}_2$A and ${}_3$∃xB(x). Assume for ∃-elim., ${}_4$B(x). By &-introd., ${}_5$A & B(x). By ∃-introd., ${}_6$∃x(A & B(x)). Now complete the ∃-elim. and ⊃-introd.

(B_1) II. Assume ${}_9$∃x(A & B(x)). Assume for ∃-elim., ${}_{10}$A & B(x). By &-elims., ${}_{11}$A and ${}_{12}$B(x). By ∃-introd., ${}_{13}$∃xB(x). By &-introd., ${}_{14}$A & ∃xB(x). Complete the ∃-elim. and ⊃-introd.

*92. ⊢ A ∨ ∀xB(x) ∼ ¬(¬A & ¬∀xB(x)) [*56] ∼ ¬(¬A & ∃x¬B(x)) [*82b] ∼ ¬∃x(¬A & ¬B(x)) [*91] ∼ ∀x¬(¬A & ¬B(x)) [*82a] ∼ ∀x(A ∨ B(x)) [*56].

*95–*99b. These can all be obtained by the chain method from *88 *90, *92, *94 via *59 (⊢ A ⊃ B ∼ ¬A ∨ B), *82a, *82b (and *34). The transformations can be performed mentally (with a little practice), so this can be used to remember them from *88–*94. (Also some of them can be established by direct methods.)

It can be remembered that *87 and *88 hold rather than the similar formulas with ∀ and ∨ or with ∃ and & (cf. *94, *93), since ∀ and & are kindred operations (∀ can be thought of as a conjunction extended over D; e.g. if $D = \{1, 2, 3\}$, then ∀xP(x) is synonymous with P(1) & P(2) & P(3) in the extension of the predicate calculus having the symbols "1", "2", "3" as names for the members of D), and so are ∃ and ∨.

THEOREM 6° and COROLLARY° now extend to the predicate calculus with ⊢ (instead of ⊨), where the operation † now includes the interchange of ∀ with ∃. The proof is as before, now using also *82a and *82b.[86]

A *prenex* formula is one with all its quantifiers (if any) at the front, with all the rest of the formula as scope of the prefixed quantifiers. Thus ∀x∃y(P(x) ∨ Q(y)) is prenex but not ∀x(∃yP(x) ∨ Q(y)). If E is equivalent to F (i.e. ⊢ E ∼ F) and F is prenex, F is a *prenex form of* E.

THEOREM 27°. *Each formula E has a prenex form.*

PROOF. Say for example E is ¬∃xP(x) ∨ ∀xQ(x). We can transform this E to a prenex formula F by a chain of equivalences, thus:

$$
\begin{array}{lll}
\vdash & \neg\exists x P(x) \vee \forall x Q(x) & \\
& \sim \forall x \neg P(x) \vee \forall x Q(x) & [{*}82a] \\
& \sim \forall x(\neg P(x) \vee \forall x Q(x)) & [{*}92 \text{ with } {*}34] \\
& \sim \forall x(\neg P(x) \vee \forall y Q(y)) & [\text{Theorem 25, or } {*}73] \\
& \sim \forall x \forall y(\neg P(x) \vee Q(y)) & [{*}92].
\end{array}
$$

In the first application of *92 (with *34), ∀xQ(x) (which does not contain x free) is the A and ¬P(x) is the B(x). Then we use Theorem 25 to replace ∀xQ(x) by ∀yQ(y), so that in the final application of *92 (with y as the x) the A will not contain the x free.

THE INTUITIONISTIC PREDICATE CALCULUS. (Cf. end § 12.) Simply by using Axiom Schema 8^I in place of Axiom Schema 8 (i.e. basing the predicate calculus on the intuitionistic propositional calculus instead of on the classical propositional calculus), we obtain the *intuitionistic predicate*

[86] All that is lacking to extend Theorem 7 (duality) to the predicate calculus with ⊢, where ′ now includes the interchange of ∀ with ∃, is an easy special case of the substitution rule (Theorems 1, 17) in the predicate calculus with ⊢ (to give ⊢ E†* from ⊢ E†).

The idea of the proof-theoretic treatment of substitution for atoms in the propositional calculus and for ions in the predicate calculus (and of some other subsidiary-deduction substitution rules) is to perform throughout an entire proof the substitution desired in the proved formula, and then to argue that the result is still a proof. This is straightforward in the propositional calculus (except that, if the substitution involves a change in the underlying language, it must extend to the atoms if any in the proof that are not in the proved formula); cf. IM Theorem 3 pp. 109 ff. In the predicate calculus, there are the risks that the substitution into a proof might introduce a free x into the C of an inference by the ∀-rule or the ∃-rule (invalidating the inference), or introduce quantifiers into the A(x) of an axiom by the ∀-schema or the ∃-schema so that the r would no longer be free for the x (invalidating the axiom); cf. IM pp. 159 ff. (on p. 159 line 3 from below, add "or no free variables"). Neither of these two complications can arise in the very simple cases of substitution for ions that are needed in this book (as here, the substitution of negations of the ions for the ions). Substitutions into deducibility relationships can be treated similarly, or via the deduction theorem etc. as in Exercise 7.3.

calculus (proof-theoretically). Results established using only the derived rules of Theorems 13 and 21, except double negation elimination, hold good for it.[87]

EXERCISES. 25.1. Translate the following from the format (B_1) into (B_2) and (B_3), and check that all steps in (B_3) are correct.

(a) The proof of *91 given above.

(b) The following proof of *96: I. Assume (preparatory to ⊃-introd.) ∃xA(x) ⊃ B. Assume (for ⊃-introd.) A(x). By ∃-introd. ∃xA(x), and by ⊃-elim. B. By ⊃-introd. (discharging the assumption A(x)), A(x) ⊃ B. By ∀-introd., ∀x(A(x) ⊃ B). II. Assume ∀x(A(x) ⊃ B) and ∃xA(x). By ∀-elim. from the former, A(x) ⊃ B. Assume preparatory to ∃-elim. from the latter, A(x). By ⊃-elim., B.

25.2. Give proofs of *82 and *90 by direct methods in the format (B_1), and rewrite them in the formats (B_2) and (B_3).

25.3. Attempt similar treatment, and locate the fallacy. (Reduce the step in question to a proposed explicit application of one of the introduction and elimination rules, and show what requirement would be violated.)

(a) The converse of *82 (∀y∃xA(x, y) ⊃ ∃x∀yA(x, y)). (Cf. Example 3.)

(b) *92.[88]

25.4. Prove *83, *95, *98 from earlier results by the chain method.

25.5. Prove *99b from *94, and also give a direct proof in the format (B_1).

25.6. Give proofs in the format (B_1) of:

(a) ⊢ ∃x¬A(x) ⊃ ¬∀xA(x) (cf. *82b).

(b) ⊢ ∀xA(x) & ∃xB(x) ⊃ ∃x(A(x) & B(x)).

25.7°. Establish ⊢ ∀x(A(x) ∨ B(x)) ⊃ ∃xA(x) ∨ ∀xB(x).

25.8*. Show that *80 does not hold in general without the restriction stated in the second sentence of the theorem.

25.9. Show that *91 does not hold in general without the restriction that A not contain x free.

25.10. Do Exercise 19.4 with "A is equivalent to B" now meaning ⊢ A ∼ B.

25.11. Reduce to prenex form:

(a) (¬∃xP(x) ∨ ∀xQ(x)) & (R ⊃ ∀xS(x)).

(b) ¬{(¬∃xP(x) ∨ ∀xQ(x)) & (R ⊃ ∀xS(x))}. Save steps by applying Theorem $6_{Pd\vdash}$ to the result of (a).

[87] Fuller lists of results like those in Theorems 2 (but with "⊢"), 22 and 26 that hold intuitionistically (written without "°") are given in IM in Theorems 5, 7, 17 and Corollary on pp. 114, 119, 163, 165–166.

[88] P ∨ ∀xQ(x) ∼ ∀x(P ∨ Q(x)) is interesting as an example of a formula not containing ¬, but which (by IM § 80), cannot be proved without using Axiom Schema 8.

(c) $(\neg\forall x P(x) \lor \exists x Q(x)) \;\&\; (R \supset \exists x S(x))$. Come out with only two quantifiers.

(d) $\forall x P(x) \sim \exists x Q(x)$. Use *63a first.

25.12°*. (a) Prove the theorem:[66] *For each formula* E *and* 0-*place ion* P: $\vdash E \sim (P \,\&\, F_1) \lor (\neg P \,\&\, F_2)$ *where* F_i *is* P or ¬P *or a formula not containing* P ($i = 1, 2$). HINT: Use *49, *40a–*48 (in § 24), *75, *76, to obtain F_1, F_2 as described such that $P \vdash E \sim F_1$ and $\neg P \vdash E \sim F_2$.

(b) The theorem comprises 9 ($= 3 \cdot 3$) cases, since each of F_1 and F_2 can have one of three forms. Simplify $(P \,\&\, F_1) \lor (\neg P \,\&\, F_2)$ in 8 of these cases.

(c) State and establish a similar theorem with two 0-place ions P_1 and P_2 in place of one P.

★§ 26. Applications to ordinary language: sets, Aristotelian categorical forms. We continue from §§ 14, 15. Now a richer system is available for our logical analyses: the predicate calculus, or more specifically the restricted, one-sorted, classical predicate calculus, §§ 16, 17.

In translating verbal expressions into logical symbolism, we must as before (§ 14) set clearly before ourselves the interpretation, or possible models, of the symbolism.

So, when a translation using only the propositional calculus will not suffice, we must be able to regard all the sentences for the given application of the predicate calculus as speaking about the members of some one nonempty set *D*. After translating, we will test the translated argument for validity without assuming that we know what nonempty set *D* is. Therefore, we are not compelled before translating to be specific about what *D* is, so long as we agree that there is some nonempty set *D* to which all the objects under discussion belong, and which can serve as the range of the variables in the translations of statements using "all", "some", etc.

Having envisaged a *D*, we must next be able to select a list of predicates, i.e. of propositional functions (each of some number $n \geq 0$ of variables), as the independent basic predicates which the sentences concern; no one of these predicates is to be regarded (for the purpose of our analysis) as composed out of other predicates. Each of these predicates, when each of its variables becomes (or takes as value) a member of *D*, is to become (or take as value) a proposition, which is either true or false (but not both). We then establish the symbolism for our application of the predicate calculus by picking prime predicate expressions or ions, which we usually write from the beginning with attached name form variables, to express these predicates.

In these two steps, of picking a *D* and of picking predicates over *D* which take values that are each either true or false, we may be representing what we believe to be the actual situation, or we may be making simplifying assumptions to enable us to reason exactly.

Let us digress for a moment to introduce a little standard terminology about sets. We do not undertake now to criticize the notion of a "set" or "collection" or "aggregate" or "totality" or "class" C, as constituted by objects thought of together and said to "belong" to C or to be "members" of C or "elements" of C. We assume the reader has a sufficient understanding of this for present purposes. We write "$x \in C$" to say that x is a member of C; and "$x \notin C$" to say that x is not a member of C (i.e. "$x \notin C$" abbreviates $\neg\, x \in C$).

A set is to be completely determined by what objects are members of it. Thus C_1 and C_2 are the same set exactly if the same objects belong to each; in symbols $C_1 = C_2$ if and only if $\forall x[x \in C_1 \sim x \in C_2]$.

A set C is *included* in a set D or C is a *part* of D or C is a *subset* of D, or in symbols $C \subseteq D$, exactly if each member of C is a member of D. In symbols, $C \subseteq D$ if and only if $\forall x[x \in C \supset x \in D]$. (Then $C = D$ if and only if $C \subseteq D \;\&\; D \subseteq C$.) For any set D, this definition makes D itself a subset of D; we call D the *improper subset* of D. Any other subset C of D, i.e. any set C such that $C \subseteq D \;\&\; C \neq D$ (i.e. every member of C is a member of D, but some member of D is not a member of C), we call a *proper subset* of D, and write $C \subset D$. The set $\varnothing$ containing no objects as members, which we call *the empty set*, is a subset of every set D.[89]

For example, suppose D is the set of three elements a, b, c; in symbols, $D = \{a, b, c\}$. Then D has eight subsets:

$$\{\,\}, \{a\}, \{b\}, \{c\}, \{a, b\}, \{a, c\}, \{b, c\}, \{a, b, c\}.$$

The first $\{\}$ is the empty set; the last $\{a, b, c\}$ the improper subset of D; and the second, third and fourth $\{a\}$, $\{b\}$, $\{c\}$ are *unit sets*, i.e. sets with just one element each.

The domain D for an application of the predicate calculus may also be called the *universal set* or the *universe* or *universe of discourse*. It is to be simply a set, fixed for the given application, which contains as members all the objects which we are considering *in that application*. Thus, if a given piece of reasoning concerns natural numbers 0, 1, 2, . . . and no other objects, then D can be the set of all the natural numbers, but need not include shoes, ships, sealing wax, cabbages and kings. In the following example, which we didn't successfully analyze in § 14, it wouldn't do to take D to be just the natural numbers.

EXAMPLE 19. "All men are mortal. Socrates is a man. Therefore Socrates is mortal." If D, including Socrates, whom clearly we are talking about, were just all men, the second sentence would be unnecessary to the argument. If the set D of all the objects we are talking about, including

[89] Some books, including IM, use "$\subset$" as we are using "$\subseteq$". Our "$\varnothing$" is written O in IM.

Socrates, were *by definition* exactly the mortal objects, the argument itself would be redundant. So we had better think of D as consisting of, say, the "beings", which includes men, mortal objects and maybe more (as in Greek mythology, gods). *D could* include natural numbers and minerals, if "x is mortal" is construed as meaningful when x is a natural number or a mineral; but this seems far fetched. Using H(x) to express "x is a man" (human being), M(x) to express "x is mortal", and s to express "Socrates", the argument can be symbolized thus:[90]

(1) $$\forall \mathrm{x}[\mathrm{H}(\mathrm{x}) \supset \mathrm{M}(\mathrm{x})],\ \mathrm{H}(\mathrm{s}) \therefore \mathrm{M}(\mathrm{s}).$$

To say that this is correct reasoning would mean that

(2) $$\forall \mathrm{x}[\mathrm{H}(\mathrm{x}) \supset \mathrm{M}(\mathrm{x})],\ \mathrm{H}(\mathrm{s}) \models \mathrm{M}(\mathrm{s})$$

(cf. § 20). We have chosen (2) rather than

$$\forall \mathrm{x}[\mathrm{H}(\mathrm{x}) \supset \mathrm{M}(\mathrm{x})],\ \mathrm{H}(\mathrm{s}) \models^{\mathrm{s}} \mathrm{M}(\mathrm{s}),$$

because in (1) as a rendering of the verbal argument the variable s is supposed in both H(s) and M(s) to name the same member of D, i.e. Socrates. The reader will have no trouble in establishing (2) directly, or (by beginning § 23) via

(3) $$\forall \mathrm{x}[\mathrm{H}(\mathrm{x}) \supset \mathrm{M}(\mathrm{x})],\ \mathrm{H}(\mathrm{s}) \vdash \mathrm{M}(\mathrm{s}).$$

Since thus far we have been doing without "constants" as symbols separate from variables, it was important here to give s in (1) and thence in (2) the conditional interpretation ((I) in § 20). We could avoid any uncertainty about the status of free variables in premises by writing the premises in the form $\forall' \mathrm{A}_1, \ldots, \forall' \mathrm{A}_m$ where $\forall'$ is closure with respect to all variables not intended to be held constant. This probably corresponds best to usage in ordinary language (*outside* of mathematics), where constructions equivalent to the use of free variables not held constant scarcely occur. However, in mathematics premises are very often stated with free variables having the generality interpretation ((II) of § 20), e.g. $x+y=y+x$ and $x<y \mathbin{\&} y<z \supset x<z$. Hence we see no reason not to allow this practice here. We can avoid ambiguity then by writing the variables $\mathrm{x}_1, \ldots, \mathrm{x}_q$

[90] Our convention is to use Roman letters (capitals and small letters) to name linguistic objects (like formulas and variables), while using italic letters to name mathematical or empirical entities (like predicates, sets, and members of the domain D).

It seems preferable to us here to write "H(x)" as a name for "x is a man" (which makes "x" simply a name for "x") than to mix the Roman and italic letters by writing "H(x)". But we might also want to use "$H(x)$" as a shorter notation for the predicate "x is a man", whereupon "H(x)" can be a name for either of the linguistic expressions "$H(x)$" and "x is a man" we choose.

which are not necessarily held constant as superscripts on "∴", just as we did on "⊨" beginning in § 20, and on "⊢" in § 21.

The requirement that all the objects considered belong to one nonempty set D, called the domain, does not prevent our talking about the members of other sets $C_1, C_2, C_3, \ldots$, provided those other sets are included in D (i.e. provided $C_1 \subseteq D$, $C_2 \subseteq D$, $C_3 \subseteq D, \ldots$). Thus we can regard Example 19 as dealing with three sets $D =$ {the beings}, $H =$ {the humans} = {the x's in D such that $H(x)$} $= \hat{x}H(x)$, $M =$ {the mortals} = {the x's in D such that $M(x)$} $= \hat{x}M(x)$.[90] As this illustrates, to each one-place predicate $C(x)$, where the variable x ranges over D, we can define a subset C of D as containing just those x's for which $C(x)$ is a true proposition; in symbols, $C = \hat{x}C(x)$.[91]

Inversely, if we have first introduced some subset C of D, we can define a predicate $C(x)$ by $C(x) \equiv x \in C$. In particular, if the sets H of humans and M of mortals (as subsets of D) were first introduced, then "H(x)" and "M(x)" could be considered as standing for $x \in H$ and $x \in M$ respectively.[92]

Notice that in (1)–(3) the subsets H and M of D are mentioned via the ions H(—) and M(—). But D itself is not mentioned explicitly; D enters implicitly, as the range of the variable x, and also of s in the validity treatment, where we disregard its interpretation as naming Socrates.

Example 20. "Every odd natural number is a difference of two squares. 5 is an odd natural number. Therefore 5 is a difference of two squares". For this argument, no objects need to be considered which are not natural numbers. So let us take the domain $D =$ {the natural numbers}. Using O(x) for "x is odd", and D(x) for "x is a difference of two squares", and f to express 5, we obtain the symbolization:

(1) $\forall x[O(x) \supset D(x)], O(f) \therefore D(f)$.

Apart from the letters used, this is the same as (1) in Example 19; so the

[91] Other notations for $\hat{x}C(x)$ are "$\{x : C(x)\}$" and "$\{x \mid C(x)\}$".

[92] We have been considering two predicates $C_1(x)$ and $C_2(x)$ to be the same if and only if, for each x, $C_1(x)$ and $C_2(x)$ are the same propositions, i.e. have the same meaning. So it could happen that $C_1(x)$ and $C_2(x)$ are *different predicates*, but, for each x in D, $C_1(x)$ is true if and only if $C_2(x)$ is true, which would make $\hat{x}C_1(x)$ and $\hat{x}C_2(x)$ the *same set*.

If we are to maintain this point of view while defining a predicate $C(x)$ by $C(x) \equiv x \in C$, we must think of the set C as given together with some particular way of describing it which determines the meaning of $x \in C$.

This distinction makes no difference in *classical* logic, where in considering whether a logical relationship holds the predicates are in effect subjected to the X-ray of Footnote 67 § 17, leaving only logical functions.

The set $C = \hat{x}C(x)$ can be called the *extension* of the predicate $C(x)$. A predicate as we have been considering it is an *intensional* concept, since the meaning or "intension" matters, while a logical function is *extensional*.

argument is valid. (The demonstration of validity there encompasses both the argument about Socrates and the argument about five.) The *D* of this application is quite precise (or so mathematicians usually think), while that of Example 19 is a rather vaguely defined totality.

This example is similar to the second illustration [2] for material interpretation early in § 2. We weren't ready there to talk explicitly about a domain *D* (say the positive integers > 1) or to use a quantifier ∀x. So there we let *n* be a number which was fixed by being written on a paper in your pocket, but was unknown to me. So my assertion could be considered as simply O ⊃ F, where O is "*n* is odd" and F is "$x^n + y^n$ is factorable" for that fixed *n*. But since I was in ignorance of the particular number *n* in your pocket, I had to consider all the possibilities. Using our present notation, I can now say that, before risking a wager with you, I convinced myself of ∀x[O(x) ⊃ F(x)], where *D* = {all positive integers > 1}, O(x) is "*x* is odd" and F(x) is "$x^n + y^n$ is factorable". Thence I passed by ∀-elimination to O(x) ⊃ F(x) (or briefly O ⊃ F) with x referring now to the number *n* in your pocket. In this illustration, material implication ⊃, used in combination with a universal quantifier, enables us to make the statement ∀x(O(x) ⊃ F(x)) about just the odd positive integers > 1, while using a variable x that ranges over the set of all positive integers > 1. We are able to do this because we made the truth table for A ⊃ B give t whenever A is f. The relationship between propositional functions *O*(*x*) and *F*(*x*) (rather than just propositions) expressed by ∀x[O(x) ⊃ F(x)] is called "formal implication". Similarly, ∀x[P(x) ∼ Q(x)] expresses "formal equivalence" between propositional functions *P*(*x*) and *Q*(*x*). —

In the logic of Aristotle (384–322 B.C.) and his followers, down into the nineteenth century, a paramount role was given to four forms of statement called "categorical". We show these forms below at the right, and our symbolizations of them at the left. Here S(x) stands for "*x* possesses the property *S*" and P(x) for "*x* possesses the property *P*". The first premises in Examples 19 and 20 illustrate the A form.[93] In these forms, when S is expressed in the singular, "are" becomes "is". "Every" or "each" can replace "all".

A	∀x[S(x) ⊃ P(x)].	All S are P. (Only P are S.)
E	¬∃x[S(x) & P(x)] (equivalently by *82a, *55b, *59: ∀x[S(x) ⊃ ¬P(x)]).	No S are P. (All S are not-P.)
I	∃x[S(x) & P(x)].	Some S are P.
O	∃x[S(x) & ¬P(x)].	Some S are not P.

We have stipulated that the range of our variables, here x, be some non-empty set D. But, in the symbolizations of the first two forms, the effect of "S(x) ⊃" is to remove from our attention the part of D outside the set S, so that the assertion that $P(x)$ holds (first form) or fails to hold (second form) is made only for $x \in S$.

If for the moment we allow a second sort of variables, $\forall x[S(x) \supset P(x)]$ where x ranges over D is synonymous with $\forall \xi P(\xi)$ where ξ ranges over S (with $S \subseteq D$); and $\forall x[S(x) \supset \neg P(x)]$ is synonymous with $\forall \xi \neg P(\xi)$. Similarly, in the third form $\exists x[S(x) \,\&\, P(x)]$ is synonymous with $\exists \xi P(\xi)$; and in the fourth $\exists x[S(x) \,\&\, \neg P(x)]$ is synonymous with $\exists \xi \neg P(\xi)$.

It would not be worth while to complicate the symbolism by introducing a second sort of variables, unless we were to make a great deal of use of them. (For simplicity in this book, we are avoiding them in any case.) So we now revert to the one-sorted predicate calculus, in which we can get the *effect* of a variable over a subset $S = \hat{x}S(x)$ of the domain D by prefixing S(x) ⊃ to the scope of a universal quantifier and S(x) & to the scope of an existential quantifier.

We excluded the empty set as a possible choice of D. This only means that the (complete) range of our variables is not to be empty; there is to be at least one value of x. But when we confine our attention to a subset $S = \hat{x}S(x)$ of D by using $\forall x[S(x) \supset \ldots]$ or $\exists x[S(x) \,\&\, \ldots]$, this subset S may be empty. I may develop the theory of angels in our form of the predicate calculus; but if I am in doubt whether angels exist, I shall first have to embed the set S of angels in some totality D that I am sure is not empty.

Under our translations, when S is empty, "All S are P" and "All S are not-P" are true, while "Some S are P" and "Some S are not P" are false. For example, on the supposition that there are no angels, we can truthfully assert "All angels have wings" and also "All angels are wingless" or equivalently "No angels have wings".

This is a point on which ordinary language is ambiguous, just as in everyday speaking "or" is sometimes intended inclusively and sometimes exclusively. People wouldn't ordinarily say "All S are P" if they already knew that S is empty; but they might say it when they were in ignorance that S is empty or had not considered that possibility. Then if confronted with the fact or possibility that S is empty, they might disagree on whether to regard "All S are P" as true in that case (as we do) or as false. Indeed, Aristotle took the latter interpretation; i.e. he took "All S are P" to

[93] The letters "A", "E", "I", "O" come from the italicized vowels in the Latin "*a*ff*i*rmo" (affirm) and "n*e*g*o*" (negate). The A and I forms are "universal affirmative" and "particular affirmative"; and the E and O forms are "universal negative" and "particular negative". The letters "S" and "P" correspond to the two parts of the sentence considered respectively as "subject" and "predicate".

include the assertion that there are some S. Under Aristotle's interpretation, the translation of "All S are P" into our symbolism is $\exists x S(x) \,\&\, \forall x[S(x) \supset P(x)]$.[94] *When S is nonempty*, i.e. when $\exists x S(x)$ is true, this is equivalent to our $\forall x[S(x) \supset P(x)]$. (By *45 in § 24, $\exists x S(x) \vdash \exists x S(x) \,\&\, \forall x[S(x) \supset P(x)] \sim \forall x[S(x) \supset P(x)]$.)

Our reason for preferring to use "All S are P" and "All S are not-P" in the respective senses $\forall x[S(x) \supset P(x)]$ and $\forall x[S(x) \supset \neg P(x)]$ is that these senses are simpler and thus more useful. (In translating someone else's English, we must be ready to prefix $\exists x S(x)$ & if that is what he meant.) This simplicity is emphasized by writing them in the notation of set theory. Indeed, $\forall x[S(x) \supset P(x)]$ can be expressed as "$S \subseteq P$"; and $\forall x[S(x) \supset \neg P(x)]$ as "$S \subseteq \bar{P}$" where $\bar{P} = \hat{x}\neg P(x)$. We call $\bar{P}$ the *complement of P* (*with respect to our universal set D*).

Let us introduce the further set-theoretic notations $S \cap P = \hat{x}(S(x) \,\&\, P(x))$ (the *intersection* of S and P), and $S \cup P = \hat{x}(S(x) \vee P(x))$ (the *union* of S and P).[95] Recalling that $\varnothing = \{\text{the empty set}\} = \hat{x}[S(x) \,\&\, \neg S(x)]$, we can express $\exists x[S(x) \,\&\, P(x)]$ as "$S \cap P \neq \varnothing$" and $\exists x[S(x) \,\&\, \neg P(x)]$ as "$S \cap \bar{P} \neq \varnothing$".

Aristotle and we and the man on the street usually agree in rendering "Some S is P" by $\exists x[S(x) \,\&\, P(x)]$, and "Some S is not P" by $\exists x[S(x) \,\&\, \neg P(x)]$. If there are no angels, "Some angels have wings" and "Some angels are wingless" are both false in everyone's usage.

However, in everyday language, "some" (especially if accented) is sometimes used to mean "some but not all". When a politician says "*Some* politicians are crooks", he means "Not every politician is a crook, although some are", i.e. $\neg\forall x[P(x) \supset C(x)] \,\&\, \exists x[P(x) \,\&\, C(x)]$, where the first conjunctand is the part he is primarily interested in communicating.

In everyday language, "all" or "some" may be omitted. "Men are mortal" means "All men are mortal". But "Men have climbed Mt. Everest" means "Some men have climbed Mt. Everest".

★§ 27. Applications to ordinary language: more on translating words into symbols. As has been amply illustrated in §§ 16 ff., the universal and existential quantifiers can be combined with each other and with propositional connectives in all sorts of ways; and modern logic has

[94] If a person on the street has said, "All S are P", and we then raise a doubt about whether $\exists x S(x)$, he could respond in two ways: (1) "What I said was contingent on there being some S; i.e. $\exists x S(x) \supset$ was intended to be prefixed to my statement." (2) "I was meaning to claim that there are some S; i.e. $\exists x S(x)$ & was intended to be prefixed." These lead to the respective translations (1′) $\exists x S(x) \supset \forall x[S(x) \supset P(x)]$ and (2′) $\exists x S(x) \,\&\, \forall x[S(x) \supset P(x)]$. But (1′) is equivalent to our translation $\forall x[S(x) \supset P(x)]$.

[95] Sometimes $S \cap P$ is called the "meet", and $S \cup P$ the "join", of S and P.

gone far beyond the analysis of statements of just the four categorical forms A, E, I, O of Aristotle.

As in § 14, we give a list of expressions at the right usually translatable as shown at the left. In the translation of expressions using pronouns like "everybody", "somebody", etc., a variable is to be introduced. This is illustrated by our examples (a)–(g) at the beginning of § 16.

$\forall xA(x)$	For all x, A(x). For every x, A(x). For each x, A(x). For arbitrary x, A(x). Whatever x is, A(x). A(x) always holds. Everyone is A. Everybody is A. Everything is A. Each one is A. Each person is A. Each thing is A.
$\exists xA(x)$	For some x, A(x). For suitable x, A(x). There exists an x such that A(x). There is an x such that A(x). There is some x such that A(x). Someone is A. Somebody is A. Something is A. For at least one x, A(x). At least one is A.[96]

When negations and quantifiers appear together, care is necessary to get them in the intended order. A "not" before or after the quantifier seems unambiguous in examples like the following. (The list could be prolonged by using others of the verbal expressions for the quantifiers.)

$\neg\forall xA(x)$	Not for all x, A(x). A(x) does not hold for all x. A(x) does not always hold. Not everyone is A.
$\forall x\neg A(x)$	For all x, not A(x). A(x) always fails. Everyone is not A.
$\neg\exists xA(x)$	There does not exist an x such that A(x). There does not exist any x such that A(x). There exists no x such that A(x). There is no x such that A(x). There isn't any x such that A(x). There isn't anyone who is A.
$\exists x\neg A(x)$	For some x, not A(x). Someone is not A.

[96] To translate "At most one is A", "Exactly one is A", "At least two are A", etc., we must wait till Chapter III.

Now consider "All S are not P". The word order here, which contrasts with "Not all S are P", would seem to call unambiguously for the translation $\forall x[S(x) \supset \neg P(x)]$. However in examples like (a) "All that glisters is not gold" (Shakespeare, Merchant of Venice (1596–7), II, 7) and (b) "*All* women are not gold diggers",[97] the intended meaning is evidently $\neg\forall x[S(x) \supset P(x)]$ or equivalently $\exists x[S(x) \& \neg P(x)]$.[98] It would seem that in these examples the clear logical structure of the language has been corrupted. In (a) Shakespeare, and Chaucer before him, needed their meter.[99] In (b), the speaker could have avoided the ambiguity by saying "Not all women are gold diggers" if she meant what we think she did, or "Each woman is not a gold digger" (or equivalently "No woman is a gold digger") if she intended the meaning which is even more favorable to the fair sex. (Here we have a context in which "all" and "each" do not behave alike.) Of course, our job in translating is to take the language as it is used or misused, and attempt to supply the logical expression that fits best.

Because the meaning of "any" depends on the context, we left "any" out of the preceding lists (except for $\neg\exists xA(x)$ when "There exists" or "There is" indicates clearly the existential quantifier). When an any-expression stands by itself, "any" has the same logical force as "all".

$\forall xA(x)$	For any x, A(x).
	Anyone is A. Anybody is A. Anything is A.

But when an any-expression D is put into either of the contexts $\neg D$ or $D \supset E$ (cf. preceding Theorem 24), the meaning of "any" normally alters from "all" to "some", thus.

$\neg\exists xA(x)$	Not for any x, A(x).
	For no x, A(x) [no ≡ not any].
	No one is A. Nobody is A. Nothing is A.
$\exists xS(x) \supset P$	If S(x) for any x, then P.
	If anyone is S, then P.
$S \supset \forall xP(x)$	If S, then, for any x, P.

Here are a few specific examples. "I would do that for anyone": $\forall xA(x)$. But "I wouldn't do that to anyone": $\neg\exists xA(x)$. "If any man is godfearing, he is just": $\forall x[S(x) \supset P(x)]$. But "If any man is just, Aristides is just":

[97] Example from Ambrose and Lazerowitz 1948 p. 236.

[98] To avoid this ambiguity, we are hyphenating "not-P" when we write "All S are not-P" intending $\forall x[S(x) \supset \neg P(x)]$.

[99] Bartlett, "Familiar Quotations", 13th ed. 1955, p. 77b says that Tyrwhitt says that Chaucer took it from the "Parabolae" of Alanus de Insulis (d. 1294): "Non teneas aurum totum quod splendet ut aurum (Do not hold everything as gold which shines as gold)". Here it is unambiguously $\neg\forall x[S(x) \supset P(x)]$.

$\exists xS(x) \supset P$ (here the any-expression is just the antecedent of the implication, not the whole implication). "If Vidkun Quisling was a patriot, then any man is a patriot": $S \supset \forall xP(x)$ (here the any-expression is the consequent of the implication, which does not give "any" the force of "some"). There is a simple explanation of the rule by which "any" alters from "all" to "some" in the D of $\neg D$ or $D \supset E$ but not of $E \supset D$. "Any" before "x", "one", "body", "thing", etc. shall indicate that the "x", "one", . . . is chosen arbitrarily (in "any" way) from the domain *D*. But this choice is to be made for the sentence as a whole, i.e. treating the "x", "one", . . . as a free variable under the generality interpretation ((II) in § 20). This amounts to using "any" like "all", "every" and "each" except that for "any" the quantifier $\forall x$ is placed at the beginning of a composite sentence, rather than where the "any" actually appears, thus:

$\forall xA(x)$	For any x, A(x).
$\forall x[S(x) \supset P(x)]$	Any S is P.
$\forall x \neg A(x)$	Not for any x, A(x).
$\forall x[S(x) \supset P]$	If S(x) for any x, then P.
$\forall x[S \supset P(x)]$	If S, then, for any x, P(x).

Now when we move the $\forall x$ inward if possible in the composite expressions by *82a, *96 or *95, we get the results given by the rule. It may appear to be an exception that "Any S is P" is $\forall x[S(x) \supset P(x)]$; but "Not any S is P" and "No S is P" are $\neg\exists x[S(x) \& P(x)]$ rather than $\neg\exists x[S(x) \supset P(x)]$. But if we use a variable ξ over *S* to write "Not any S is P" as $\forall\xi\neg P(\xi)$, and then use the rule, we get $\neg\exists x[S(x) \& P(x)]$ upon changing $\exists\xi$ to $\exists x[S(x) \& \ldots]$. Clearly "any" is tricky. Not every user can be counted on to follow the above principles invariably, so the translator must be on guard.

The indefinite article "a" or "an" sometimes has the force of "all", sometimes of "some". "A child needs affection": $\forall x[C(x) \supset A(x)]$. "A man was here": $\exists x[M(x) \& H(x)]$. (Cf. end § 26.)

The scopes of quantifiers in relation to implications are usually clear. "For all x, if S(x), then P(x)" and "For all x, S(x) only if P(x)" are $\forall x[S(x) \supset P(x)]$. "If, for all x, S(x), then P(x)" and "P(x) if, for all x, S(x)" are $\forall xS(x) \supset P(x)$.

Here are a few examples of translating quantifiers with & or V.

$\exists x[S(x) \& P(x)]$	Someone is S and P.	
$\exists xS(x) \& \exists xP(x)$	Someone is S and someone is P.	
$\forall x[S(x) \vee P(x)]$	Everyone is S or P.	All except S are P.
$\forall xS(x) \vee \forall xP(x)$	Everyone is S or everyone is P.	

"No cats or dogs are allowed in apartments" is clearly $\neg\exists x[(C(x) \vee D(x)) \,\&\, A(x)]$ or equivalently $\forall x[C(x) \vee D(x) \supset \neg A(x)]$. But "All cats and dogs must be licensed" would be mistranslated as $\forall x[C(x) \,\&\, D(x) \supset L(x)]$ (the set $\hat{x}[C(x) \,\&\, D(x)]$ is empty); the correct translation is $\forall x[C(x) \supset L(x)] \,\&\, \forall x[D(x) \supset L(x)]$, or equivalently $\forall x[C(x) \vee D(x) \supset L(x)]$ (*87 and Exercise 13.8).

Sometimes "all" is used not as a quantifier but to describe a set. "All the kings horses and all the kings men could not put Humpty Dumpty together again" presumably does not mean simply that no one of the horses and men working alone could effect the repair, as the translation $\forall x[H(x) \vee M(x) \supset \neg P(x)]$ or equivalently $\forall x[H(x) \supset \neg P(x)] \,\&\, \forall x[M(x) \supset \neg P(x)]$ would say. Similar usage occurs with "and".[100] "John and William couldn't push the car out of the ditch" can be translated $\neg P(\{j, w\})$ where P(x) is "x could push the car out", or in strictly predicate calculus symbolism $\neg P(t)$ where t stands for the team John + William. But "John and William will pass this course" is $P(j) \,\&\, P(w)$ where P(x) is "x will pass this course"; John and William are not supposed to collaborate in writing the final examination. —

Our discussion early in this section of picking the domain D led us naturally to the consideration of subsets of D such as $\hat{x}S(x)$ and $\hat{x}P(x)$, and thence to the Aristotelian categorical forms and how to translate into them. But after picking D, and before introducing quantifiers and propositional connectives, we must pick the basic predicates and select ions or prime predicate expressions to stand for them.

EXAMPLE 21. "When I am tired and hungry, I want to go home. Now I am tired and hungry. So I want to go home now." Let E(x) be "at time x, I am tired and hungry", H(x) be "at time x, I want to go home" and n be "now".

(1) $\forall x[E(x) \supset H(x)],\ E(n) \therefore H(n).$

Valid, as already noted in Examples 19 and 20.

EXAMPLE 22. "When I am tired, I want to go home. When I am hungry, I want to go home or to a restaurant. Now I am tired and hungry. So I want to go home now."

(1) $\forall x[T(x) \supset H(x)],\ \forall x[Hu(x) \supset H(x) \vee R(x)],\ T(n) \,\&\, Hu(n) \therefore H(n).$

Again valid. The second premise is *redundant*, i.e. the argument is valid without it.

[100] The preceding anomaly about "cats and dogs" is connected with the use (sometimes) of "and" to describe a union of sets: {cats and dogs} $= \hat{x}(C(x) \vee D(x)) = C \cup D$. In the next example (John and William) we could use instead the translation P(j, w) with a binary predicate, but then we would need a ternary predicate for "John, William and Henry couldn't push the car out".

In Example 21, it sufficed to represent the compound notion "tired and hungry" simply by an ion E(x), since the argument was valid without further analysis. In Example 22, this wouldn't have sufficed. Example 21 illustrates that to confirm the validity of an argument, it may not be necessary to analyze it to the maximum possible extent. On the other hand, to establish the invalidity of an argument in a given system (in Chapter I, the propositional calculus; here, the predicate calculus), we must carry out the fullest analysis possible in the system. (We first remarked this in § 3 when we proved Theorem 1 and refuted its converse.)

In retrospect, Example 15 at the end of § 15 might now seem to call for predicate letters P(x), Q(x), R(x) and universal quantifiers, with the conclusion becoming
$\models \forall x[P(x) \supset Q(x)] \ \&\ \forall x[\neg(Q(x) \ \&\ R(x)) \vee P(x)] \ \&\ \forall x \neg[R(x) \ \&\ P(x)] \sim \forall x[P(x) \supset Q(x)] \ \&\ \forall x[Q(x) \supset \neg R(x)]$. However, we analyzed it successfully there by thinking of x as naming a fixed (but unspecified) member of the club. (Cf. Exercise 17.6.)

EXAMPLE 23. Consider the Aristotelian syllogism "Barbara": "All M are P. All S are M. Therefore all S are P." Using the Aristotelian A form, this becomes:

(1) $\forall x[M(x) \supset P(x)], \forall x[S(x) \supset M(x)] \therefore \forall x[S(x) \supset P(x)]$.

Its validity means

(2) $\forall x[M(x) \supset P(x)], \forall x[S(x) \supset M(x)] \models \forall x[S(x) \supset P(x)]$.

But simply by *2 in Theorem $2_{\vdash}$, and modus ponens,

(3a) $M(x) \supset P(x), S(x) \supset M(x) \vdash S(x) \supset P(x)$

in the propositional calculus. Thence using $\forall$-elims. and $\forall$-introd.,

(3) $\forall x[M(x) \supset P(x)], \forall x[S(x) \supset M(x)] \vdash \forall x[S(x) \supset P(x)]$.

More essential use is made of the predicate calculus with quantification of one-place predicates in Example 19 and the syllogisms Baroco, Festino, etc. (Exercise 24.5).

Now we give an example in which two-place predicates are essential. This example is beyond the resources of the traditional Aristotelian logic, which concerned itself with the syllogisms and related matters, which we have expressed in terms of one-place predicates.

EXAMPLE 24. "The relation $x < y$ is both transitive and irreflexive. Therefore it is asymmetric." In our symbols,

(1) $\forall x \forall y \forall z(x<y \ \&\ y<z \supset x<z), \forall x \neg x<x \therefore \forall x \forall y(x<y \supset \neg y<x)$,

or equivalently with free variables in the formulas (imitating the notation of mathematics text books, but making the generality interpretation of x, y, z explicit by our superscripts)

$$(1') \qquad x<y \,\&\, y<z \supset x<z,\ \neg x<x \ \therefore^{xyz}\ x<y \supset \neg y<x.$$

To establish the validity of (1) and (1′), it will suffice (by Theorem 12_{Pd} and beginning § 23) to establish

$$(3') \qquad x<y \,\&\, y<z \supset x<z,\ \neg x<x \ \vdash^{xyz}\ x<y \supset \neg y<x$$

(where "$x<y \,\&\, y<z \supset x<z,\ \neg x<x \vdash^{xyz}$" means "$\forall x \forall y \forall z(x<y \,\&\, y<z \supset x<z), \forall x \neg x<x \vdash$"). We do this in the format (B_1) of §§ 13, 25. Assume $\forall x \forall y \forall z(x<y \,\&\, y<z \supset x<z), \forall x \neg x<x$. By $\forall$-elims., $x<y \,\&\, y<x \supset x<x$ and $\neg x<x$, whence by propositional calculus $x<y \supset \neg y<x$. In this example the original notation "$<$" is concise enough so we use it directly.

Exercises. 27.1. Translate each of the following arguments into logical symbolism, and establish the validity or invalidity of the result.

(a) Everyone loves himself. Therefore someone is loved by somebody.

(b) Everybody loves Jane. Therefore everybody is loved by someone.

(c) No animals are immortal. All cats are animals. Therefore some cats are not immortal.

(d) Only birds have feathers. No mammal is a bird. Therefore each mammal is featherless.

(e) Some students are studious. No student is unqualified. Therefore some unqualified students are not studious.

(f) Each politician is a showman. Some showmen are insincere. Therefore some politicians are insincere.

(g) Nothing effective is easy. Something easy is popular. Therefore something popular is not effective.

(h) All of his friends are devoted. Some of his friends are insincere. No insincere person can be devoted. Therefore all of his friends are crooks.

(i) There isn't anyone who would believe that. Therefore the judge wouldn't believe it.

(j) Any fool could do that. I cannot do that. Therefore I am not a fool.

(k) If anyone can solve this problem, some mathematician can solve it. Cabot is a mathematician and cannot solve the problem. Therefore the problem cannot be solved.

(ł) Any mathematician can solve this problem if anyone can. Cabot is a mathematician and cannot solve the problem. Therefore the problem cannot be solved.

(m) Anyone who can solve this problem is a mathematician. Cabot cannot solve this problem. Therefore Cabot is not a mathematician.

(n) Anyone who can solve this problem is a mathematician. No mathematician can solve this problem. Therefore the problem cannot be solved.
(o) If any of the numbers (strictly) between 1 and 101 divides 101, then a prime number less than 11 divides 101. Each prime number less than 11 does not divide 101. Therefore no number between 1 and 101 divides 101.
(p) The person responsible for this rumor must be both clever and unprincipled. Cabot is not clever. Lowell is not unprincipled. Therefore neither Lowell nor Cabot is responsible.
(q) No one can interpret this message unless someone can supply the code. Therefore there is someone who can interpret this message only if he can supply the code.
(r) Every liberal advocates changes. Some conservatives favor no one who advocates changes. Therefore some conservatives favor no liberal.
(s) A man is an animal. Therefore the head of a man is the head of an animal. (Let M(x), A(x), H(y, x) be "x is a man", "x is an animal", "y is the head of x", respectively.)[101]
(t) One person being the father of another is an asymmetric relation; i.e. for any persons x and y, if x is the father of y, then y is not the father of x. Therefore, no person is his own father.
(u)* If each of two persons is related to a third person, the first person is related to the second. Everyone is related to someone. Therefore, if John is related to William, and William is related to Edith, then John is related to Edith.[102]
(v) For any persons x and y, x is brother of y if and only if x and y are male and x is a different person than y and x and y have the same two parents. Therefore, if x is brother of y, then y is brother of x.
(w) Hope is not lost. Therefore all is not lost.

27.2. In the following enthymemes (premise(s) or conclusion missing, § 15), attempt to supply what is missing to make a valid argument. Is the result sound?
(a) Only the brave deserve the fair. She is fair. He is not brave.
(b) Adults are admitted only if accompanied by a child. I am admitted. Therefore I am either a child or accompanied by a child.
(c) John is taller than Susan. Susan is taller than Peter. Therefore John is taller than Peter.
(d) San Francisco is west of New York. Tokyo is west of San Francisco. London is west of Tokyo. Therefore London is west of New York.

[101] Given by DeMorgan 1847 p. 114 as an inference not possible by Aristotelian syllogistic.
[102] The solution is essentially the same as of Exercise 29.3 (f).

CHAPTER III

THE PREDICATE CALCULUS WITH EQUALITY[103]

★**§ 28. Functions, terms.** In this chapter we extend the object language (or increase the part of it we take into account) in two ways. These extensions can be made separately or simultaneously.

First (in this section), we add expressions for functions whose values are in the same domain *D* as the *arguments*, i.e. values of the independent variable(s), in contrast to the predicates (i.e. propositional functions), whose values are propositions (cf. Footnote 51 § 16). Such "functions" play an important role in mathematics. For example, when *D* is the natural numbers {0, 1, 2, . . .}, we use as such functions $x+1$, $x+y$, xy, x^y, $x!$, etc. When *D* is the real numbers, we use also $x-y$, $\sqrt{x^2+y^2}$, $\sin x$, e^x, etc. We include individuals (i.e. members of *D*) as functions of zero variables (0-place functions), such as 0 and 1 when *D* is the natural numbers, and also π and *e* when *D* is the real numbers (just as we included propositions as 0-place predicates).

To make this extension, we assume that the object language has expressions for such functions, which expressions shall retain their identity but not be analyzed. These expressions we call *prime function expressions* or *mesons*,[104] and we denote them by "f", "f(—)", "f(—, —)", "f(—, —, —)", . . . , "g", "g(—)", "g(—, —)", "g(—, —, —)", ... Putting variables into these, we have *name forms* for them, or *prime function expressions* (or *mesons*) *with attached variables*. We can be brief here, as the treatment parallels that of prime predicate expressions in § 16.

Now we describe the larger class of expressions, called "terms", which name functions (using variables, except in the case of individuals).[105] The *terms* shall comprise exactly the variables and the additional *terms* obtainable thus: For each $n \geq 0$, each *n*-place meson f(—, . . . , —), and each

[103] The predicate calculus with equality is an important chapter of logic. But what will be needed in Part II of the book will be repeated or cited piecemeal. So the student who has not time for both can skip this chapter.

[104] We pick the short name "meson" (akin to "atom", "molecule", "ion"), without making any pretense that there is an analogy with chemistry or physics.

[105] When the variables are interpreted as expressing members of *D*, the formulas express propositions,[53] and the terms express members of *D*. However, we have nothing here for terms analogous to the generality interpretation of free variables $x_1, \ldots, x_q$ in formulas, by which a formula A can instead express the proposition expressed by $\forall$'A (§ 20).

list $r_1, \ldots, r_n$ of n *terms* $r_1, \ldots, r_n$ already constructed, $f(r_1, \ldots, r_n)$ shall also be a *term*.

Now we take over the definition of *formula* in § 16, but allowing the $r_1, \ldots, r_n$ in the definition of "prime formula" or "atom" to be any terms, instead of simply any variables. (If there are no mesons, the notion of "term" just defined reduces to "variable", so the present notion of "formula" reduces to the one in § 16.)

For each choice of a (nonempty) domain D, each term has a table with values in D, and each formula has a truth table. The tables must be entered with all n-place functions with arguments and values in D as value of each n-place meson (for $n = 0$, simply with all members of D).

EXAMPLE 1. We give the tables for $D = \{1, 2\}$ of the term f(f(x)) and of the formula $\forall x[P(f(f(x))) \supset \neg P(f(x))]$. First, we list all possible 1-place functions with values in D (and repeat from § 17 Example 1 the list of 1-place logical functions).[64]

x	$f_1(x)$	$f_2(x)$	$f_3(x)$	$f_4(x)$	$l_1(x)$	$l_2(x)$	$l_3(x)$	$l_4(x)$
1	1	1	2	2	t	t	f	f
2	1	2	1	2	t	f	t	f

	f(x)	x	f(f(x))
1.	$f_1(x)$	1	1
2.	$f_1(x)$	2	1
3.	$f_2(x)$	1	1
4.	$f_2(x)$	2	2
5.	$f_3(x)$	1	1
6.	$f_3(x)$	2	2
7.	$f_4(x)$	1	2
8.	$f_4(x)$	2	2

	P(x)	f(x)	$\forall x[P(f(f(x))) \supset \neg P(f(x))]$
1.	$l_1(x)$	$f_1(x)$	f
2.	$l_1(x)$	$f_2(x)$	f
3.	$l_1(x)$	$f_3(x)$	f
4.	$l_1(x)$	$f_4(x)$	f
5.	$l_2(x)$	$f_1(x)$	f
6.	$l_2(x)$	$f_2(x)$	f
7.	$l_2(x)$	$f_3(x)$	t
8.	$l_2(x)$	$f_4(x)$	t
9.	$l_3(x)$	$f_1(x)$	t
10.	$l_3(x)$	$f_2(x)$	f
11.	$l_3(x)$	$f_3(x)$	t
12.	$l_3(x)$	$f_4(x)$	f
13.	$l_4(x)$	$f_1(x)$	t
14.	$l_4(x)$	$f_2(x)$	t
15.	$l_4(x)$	$f_3(x)$	t
16.	$l_4(x)$	$f_4(x)$	t

Here is the computation of Line 5 of the table for f(f(x)) (left):

$$\begin{array}{c} f(f(x)) \\ f_3(f_3(1)) \\ f_3(2\quad) \\ 1 \end{array}$$

We now compute the entry for Line 7 of the table for the formula (right). This requires us to compute the following supplementary table.

x	$\mathfrak{l}_2(f_3(f_3(\mathrm{x}))) \supset \neg\mathfrak{l}_2(f_3(\mathrm{x}))$
1	t
2	t

Here are the computations for this supplementary table.[106]

$$\begin{array}{llll}
\mathfrak{l}_2(f_3(f_3(1))) & \supset \neg\mathfrak{l}_2(f_3(1)) & \qquad \mathfrak{l}_2(f_3(f_3(2))) & \supset \neg\mathfrak{l}_2(f_3(2)) \\
\mathfrak{l}_2(1\qquad) & \supset \neg\mathfrak{l}_2(2\quad) & \qquad \mathfrak{l}_2(2\qquad) & \supset \neg\mathfrak{l}_2(1\quad) \\
\mathrm{t} & \supset \neg\mathrm{f} & \qquad \mathrm{f} & \supset \neg\mathrm{t} \\
\mathrm{t} & \supset \mathrm{t} & \qquad \mathrm{f} & \supset \mathrm{f} \\
 & \mathrm{t} & & \mathrm{t}
\end{array}$$

Since the supplementary table has a solid column of t's, $\forall \mathrm{x}[\mathrm{P}(\mathrm{f}(\mathrm{f}(\mathrm{x}))) \supset \neg\mathrm{P}(\mathrm{f}(\mathrm{x}))]$ is t in Line 7.

The definitions of *valid* and of *valid consequence* (again to be abbreviated using "$\models$") are as before (also of "$\bar{\bar{D}}$-$\models$").

A little consideration shows that all of the model-theoretic results stated above for the predicate calculus remain good under this extension of the language, where now in THEOREM 15 r can be any term (not just any variable) which is *free for* x *in* A(x) in the extension of the earlier notion (§ 18) obtained by calling an occurrence of a term in a formula *free* if each variable occurrence in it is free in the formula. Thus, briefly, r is free for x in A(x), if no (free) occurrence of a variable in r becomes bound as a result of substituting r for the free x's in A(x). Similarly, in the definitions preceding THEOREM 17, and in THEOREM 18, $\mathrm{r}_1, \ldots, \mathrm{r}_{p_i}$ or $\mathrm{r}_1, \ldots, \mathrm{r}_p$ can be any terms under the conditions stated there for variables.

The proof theory of the predicate calculus with functions differs only by our allowing the r in the $\forall$-schema and the $\exists$-schema to be any term free for x in A(x). The corresponding changes can then be made in $\forall$-elimination and $\exists$-introduction in THEOREM 21 and in its COROLLARIES.

Now that we are allowing 0-place mesons to name individuals (i.e. members of D), we can use these for "Jane" in (a)–(c) beginning § 16, and for "Socrates" and "five" in Examples 19 and 20 § 26. The only difference this makes from using variables (held constant for the valid consequence and deducibility relationships) as we did in Chapter II is that there we might have used the same variables also in bound occurrences. That would make no trouble for our logical analyses, though it seems unnatural since in ordinary languages we normally use proper nouns etc. for such purposes.

[106] We are saving a step here by using the table for f(f(x)) (above left). But of course it is not necessary first to set up tables for terms occurring in our formula.

The substantial gain in using functions comes when functions of $n > 0$ variables are employed in conjunction with the equality predicate (§§ 29, 30).

EXERCISES. 28.1. How many lines does the table have when $D = \{1, 2\}$? Compute the indicated line.
(a) g(f(g(x, f(h))), x) when f(x), g(x, y), h, x are $f_4(x)$, $f_7(x, y)$, 1, 2, respectively (where $f_7(1, 1) = f_7(2, 2) = 1$, $f_7(1, 2) = f_7(2, 1) = 2$).
(b) ∀xP(g(f(g(x, f(h))), x)) ∨ Q, with the above assignment extended by $\mathfrak{l}_2(x)$, f to P(x), Q.

28.2. Which of the following terms are free for x in the formula ∀w(P(x, y) ∨ ∃zQ(y, x) ⊃ R(w))? (a) x. (b) g(z, f(x, y)). (c) f(x, y). (d) g(w, y). (e) g(y, f(h, x)).

28.3. Fix the proof of Theorem 15 (b) to apply here.

★ § 29. Equality.

The *predicate calculus with equality* (or *with identity*) can be described as arising from the predicate calculus (Chapter II or § 28) by giving one of the 2-place ions a special treatment. We write this ion as "—=—" and read it "— equals —".

For our model theory, we evaluate x=y by the logical function $\mathfrak{l}(x, y)$ whose value is t when x and y have the same value, and 0 otherwise (e.g. when $D = \{1, 2\}$, by $\mathfrak{l}_7(x, y)$ Example 3 § 17). So in the predicate calculus with equality, the truth tables are not entered with a value of x=y, as its value is fixed in advance (as are the values of the logical symbols ∼, ⊃, &, ∨, ¬, ∀, ∃). In the predicate calculus with equality, by a *proper ion* we shall mean an ion other than —=—.

EXAMPLE 2. Here is the truth table for $D = \{1, 2\}$ of ∀x[P(f(x)) ∨ ∃y x=f(f(y))] (written in two columns to save space).

∀x[P(f(x)) ∨ ∃y x=f(f(y))]

	P(x)	f(x)			P(x)	f(x)	
1.	$\mathfrak{l}_1(x)$	$f_1(x)$	t	9.	$\mathfrak{l}_3(x)$	$f_1(x)$	f
2.	$\mathfrak{l}_1(x)$	$f_2(x)$	t	10.	$\mathfrak{l}_3(x)$	$f_2(x)$	t
3.	$\mathfrak{l}_1(x)$	$f_3(x)$	t	11.	$\mathfrak{l}_3(x)$	$f_3(x)$	t
4.	$\mathfrak{l}_1(x)$	$f_4(x)$	t	12.	$\mathfrak{l}_3(x)$	$f_4(x)$	t
5.	$\mathfrak{l}_2(x)$	$f_1(x)$	t	13.	$\mathfrak{l}_4(x)$	$f_1(x)$	f
6.	$\mathfrak{l}_2(x)$	$f_2(x)$	t	14.	$\mathfrak{l}_4(x)$	$f_2(x)$	t
7.	$\mathfrak{l}_2(x)$	$f_3(x)$	t	15.	$\mathfrak{l}_4(x)$	$f_3(x)$	t
8.	$\mathfrak{l}_2(x)$	$f_4(x)$	f	16.	$\mathfrak{l}_4(x)$	$f_4(x)$	f

Here is the computation of Line 7.

	∀x[P(f(x)) ∨ ∃y x=f(f(y))]
(i)	∀x[$\mathfrak{l}_2(f_3(x))$ ∨ ∃y x=$f_3(f_3(y))$]
(ii)	t

To get (ii), we need the following supplementary table.

(a)

x	$\mathfrak{l}_2(f_3(x)) \vee \exists y\, x{=}f_3(f_3(y))$
1	$\mathfrak{t}$
2	$\mathfrak{t}$

The two computations for this are as follows (explained below).

$\mathfrak{l}_2(f_3(1)) \vee \exists y\, 1{=}f_3(f_3(y))$	$\mathfrak{l}_2(f_3(2)) \vee \exists y\, 2{=}f_3(f_3(y))$
$\mathfrak{l}_2(2\quad) \vee\ \mathfrak{t}$	$\mathfrak{l}_2(1\quad) \vee\ \mathfrak{t}$
$\mathfrak{f}\qquad \vee\ \mathfrak{t}$	$\mathfrak{t}\qquad \vee\ \mathfrak{t}$
$\mathfrak{t}$	$\mathfrak{t}$

To get the $\mathfrak{t}$ used here in the second line of each column, we need two more supplementary tables:

(b₁)

y	$1{=}f_3(f_3(y))$
1	$\mathfrak{t}$
2	$\mathfrak{f}$

(b₂)

y	$2{=}f_3(f_3(y))$
1	$\mathfrak{f}$
2	$\mathfrak{t}$

Here are the computations for these.

$1{=}f_3(f_3(1))$	$1{=}f_3(f_3(2))$	$2{=}f_3(f_3(1))$	$2{=}f_3(f_3(2))$
$1{=}1$	$1{=}2$	$2{=}1$	$2{=}2$
$\mathfrak{t}$	$\mathfrak{f}$	$\mathfrak{f}$	$\mathfrak{t}$

The last step in each of these four computations is by the rule for evaluating $=$. Since each of (b_1) and (b_2) has a $\mathfrak{t}$, we get the $\mathfrak{t}$ in the second lines of the two computations for (a) by the rule for evaluating $\exists$. Then by the rule for $\forall$ we get (ii).

EXAMPLE 3. Here are the tables for $\exists x \forall y(P(y) \supset x{=}y)$ and $\exists x[P(x) \,\&\, \forall y(P(y) \supset x{=}y)]$ when $D = \{1, 2, 3\}$. First we list the possible values of $P(x)$ (as in § 17 Example 4).

x	$\mathfrak{l}_1(x)$	$\mathfrak{l}_2(x)$	$\mathfrak{l}_3(x)$	$\mathfrak{l}_4(x)$	$\mathfrak{l}_5(x)$	$\mathfrak{l}_6(x)$	$\mathfrak{l}_7(x)$	$\mathfrak{l}_8(x)$
1	$\mathfrak{t}$	$\mathfrak{t}$	$\mathfrak{t}$	$\mathfrak{t}$	$\mathfrak{f}$	$\mathfrak{f}$	$\mathfrak{f}$	$\mathfrak{f}$
2	$\mathfrak{t}$	$\mathfrak{t}$	$\mathfrak{f}$	$\mathfrak{f}$	$\mathfrak{t}$	$\mathfrak{t}$	$\mathfrak{f}$	$\mathfrak{f}$
3	$\mathfrak{t}$	$\mathfrak{f}$	$\mathfrak{t}$	$\mathfrak{f}$	$\mathfrak{t}$	$\mathfrak{f}$	$\mathfrak{t}$	$\mathfrak{f}$

I(x)	(a) $\exists x \forall y(P(y) \supset x=y)$	(b) $\exists x[P(x) \,\&\, \forall y(P(y) \supset x=y)]$
$I_1(x)$	f	f
$I_2(x)$	f	f
$I_3(x)$	f	f
$I_4(x)$	t	t
$I_5(x)$	f	f
$I_6(x)$	t	t
$I_7(x)$	t	t
$I_8(x)$	t	f

We now compute Line 6 for (a) and (b). We shall need the following supplementary tables.

x	(a′) $\forall y(I_6(y) \supset x=y)$	(b′) $I_6(x) \,\&\, \forall y(I_6(y) \supset x=y)$
1	f	f
2	t	t
3	f	f

To compute these we need in turn three supplementary tables:

y	(a_1'') $I_6(y) \supset 1=y$	(a_2'') $I_6(y) \supset 2=y$	(a_3'') $I_6(y) \supset 3=y$
1	t	t	t
2	f	t	f
3	t	t	t

Here are the computations for (a_1'') (and those for (a_2''), (a_3'') are similar):

$I_6(1) \supset 1=1$	$I_6(2) \supset 1=2$	$I_6(3) \supset 1=3$
f ⊃ t	t ⊃ f	f ⊃ f
t	f	t

Using the tables (a_1'')–(a_3''), the evaluation rule for ∀ gives the values shown above in (a′); and those in (b′) are the same thus:

$I_6(1) \,\&\, \forall y(I_6(y) \supset 1=y)$	$I_6(2) \,\&\, \forall y(I_6(y) \supset 2=y)$	$I_6(3) \,\&\, \forall y(I_6(y) \supset 3=y)$
f & f	t & t	f & f
f	t	f

Since (a′) and (b′) each have a t, the entry in Line 6 for (a) and (b) is t.

Now we note that in Table (a) $\exists x \forall y(P(y) \supset x=y)$ is t for exactly those l(x)'s which have at most one t; and in Table (b) $\exists x[P(x) \& \forall y(P(y) \supset x=y)]$ is t for exactly those l(x)'s which have exactly one t. The student should be able to convince himself that this is the case not just for $D = \{1, 2, 3\}$ but for any (nonempty) D. Thus $\exists x \forall y(P(y) \supset x=y)$ expresses "there is at most one x such that P(x)" and $\exists x[P(x) \& \forall y(P(y) \supset x=y)]$ expresses "there is exactly one x such that P(x)" or "there exists a unique x such that P(x)".

The latter notion is frequently enough used so that we adopt "$\exists!xP(x)$" as an abbreviation for $\exists x[P(x) \& \forall y(P(y) \supset x=y)]$. More generally, "$\exists!xA(x)$" abbreviates $\exists x[A(x) \& \forall y(A(y) \supset x=y)]$, where in unabbreviating we pick for y a variable free for x in A(x) and distinct from x and the free variables of A(x). (Each two legitimate unabbreviations of "$\exists!xA(x)$" are congruent, end § 16.) Another useful abbreviation is "$x \neq y$" for $\neg x=y$, or more generally "$r \neq s$" for $\neg r=s$ where r and s are any terms.

The results stated above in the model theory of the predicate calculus remain good in the predicate calculus with equality. (Now in § 19, (1) is to be a list of only distinct proper ions including all occurring in E.) We also have further results, which we collect in the following theorem.

THEOREM 28. (a) $\models x=x$. (b) $\models x=y \supset (x=z \supset y=z)$.

(Open equality axioms for =.)

(c_1^2) $\models x=y \supset (P(x, a_2) \supset P(y, a_2))$. ($c_2^2$) $\models x=y \supset (P(a_1, x) \supset P(a_1, y))$.

(Open equality axioms for a proper ion $P(a_1, a_2)$.)

(d_1^2) $\models x=y \supset f(x, a_2)=f(y, a_2)$. ($d_2^2$) $\models x=y \supset f(a_1, x)=f(a_1, y)$.

(Open equality axioms for a meson $f(a_1, a_2)$.)

Similarly, for each n, there are n "open equality axioms" (c_1^n), . . . , (c_n^n) *for each proper n-place ion* $P(a_1, \ldots, a_n)$, *and n of them* (d_1^n), . . . , (d_n^n) *for each n-place meson* $f(a_1, \ldots, a_n)$.

PROOFS. (c_1^2). By Corollary Theorem $8_{Pd=}$, it will suffice to show that $x=y, P(x, a_2) \models P(y, a_2)$. So consider any D and any assignment which makes $x=y$ and $P(x, a_2)$ both t. Since $x=y$ is t, the value of y in the assignment is the same as that of x; so, since $P(x, a_2)$ is t, so is $P(y, a_2)$. —

To establish the proof theory of the predicate calculus with equality, we start with the axiom schemata and rules of inference we already have (§ 21, or if functions are allowed § 28), and add as further axioms the formulas shown to be valid in Theorem 28, as we anticipated doing when we called them "open equality axioms". (The "closed equality axioms" are the closures of the open equality axioms, § 20.) Then by Theorem 28, THEOREM 12 and COROLLARY continue to hold.

Now the provability and deducibility results already established for the predicate calculus all hold. This is obvious in the case of the direct rules (as in going from the propositional to the predicate calculus, § 21). Moreover, in the proof of the deduction theorem (§§ 10, 22), applications of the new axioms can be handled under the old Case 3. Then we also have the subsidiary deduction rules based on the deduction theorem which are included among the introduction and elimination rules of THEOREMS 13 and 21.

THEOREM 29. (a) $\vdash x=x$. (e) $\vdash x=y \supset y=x$. (f) $\vdash x=y \,\&\, y=z \supset x=z$.

(Reflexive, symmetric and transitive properties of equality.)

For any terms r, s, t, t_1, t_2, *etc.*:

(g_1) $r=s \vdash r=t \sim s=t$. (g_2) $r=s \vdash t=r \sim t=s$.
(h^2_1) $r=s \vdash P(r, t_2) \sim P(s, t_2)$. (h^2_2) $r=s \vdash P(t_1, r) \sim P(t_1, s)$.
(i^2_1) $r=s \vdash f(r, t_2)=f(s, t_2)$. (i^2_2) $r=s \vdash f(t_1, r)=f(t_1, s)$.

Similarly, for each $n > 0$, *we have n such results* $(h^n_1), \ldots, (h^n_n)$ *for each proper n-place ion, and n of them* $(i^n_1), \ldots, (i^n_n)$ *for each n-place meson.*

(Special replacement results.)

PROOFS. (For the format (B_1) which we use, cf. §§ 13, 25.) (e) Assume $x=y$. By (a), $x=x$. Substituting x for z in (b) (by Theorem 21 Corollary 2 (c) with Γ empty), $x=y \supset (x=x \supset y=x)$. Using $\supset$-elim. twice, $y=x$. (h^2_1) Assume $r=s$. I. Assume $P(r, t_2)$. By substituting r, s, t_2 for x, y, a_2 in (c^2_1), $r=s \supset (P(r, t_2) \supset P(s, t_2))$. By $\supset$-elim. twice, $P(s, t_2)$. II. Assume $P(s, t_2)$. By substitution in (e), $r=s \supset s=r$. By $\supset$-elim., $s=r$. By substitution in (c^2_1), $s=r \supset (P(s, t_2) \supset P(r, t_2))$. By $\supset$-elim. twice, $P(r, t_2)$.

THEOREM 30. (Replacement theorem.) (I) *Let* r *and* s *be any terms*, t_r *be a term containing* r *as a specified (consecutive) part, and* t_s *be the result of replacing that part by* s. *Then*:

$$r=s \vdash t_r=t_s.$$

(II) *With* r *and* s *as in* (I), *let* C_r *be a formula containing* r *as a specified (consecutive) part (not the variable in a quantifier),* C_s *be the result of replacing that part by* s, *and* $x_1, \ldots, x_q$ *be the variables of* r *or* s *which belong to quantifiers in* C_r *having the specified part within their scopes. Let* Γ *be a list of (zero or more) formulas not containing any of* $x_1, \ldots, x_q$ *free. Then*:

$$\textit{If } \Gamma \vdash r=s, \quad \textit{then} \quad \Gamma \vdash C_r \sim C_s.$$

PROOF, as illustrated next.

EXAMPLE 4. In the following, each of 1–3 after the first is deducible from the preceding, by (i_1^1), (i_2^2).

1.	f(x, y)	=	g(x)
2.	h(f(x, y))	=	h(g(x))
3.	f(z, h(f(x, y)))	=	f(z, h(g(x)))
4.	f(z, h(f(x, y)))=g(z)	~	f(z, h(g(x)))=g(z)
5.	∀z f(z, h(f(x, y)))=g(z)	~	∀z f(z, h(g(x)))=g(z)
6.	∃y∀z f(z, h(f(x, y)))=g(z)	~	∃y∀z f(z, h(g(x)))=g(z)

Thus f(x, y)=g(x) ⊢ f(z, h(f(x, y)))=f(z, h(g(x))), as (I) asserts.

Each of 1–6 is successively recognizable as being deducible from ∀y f(x, y)=g(x), by ∀-elim., (i_1^1), (i_2^2), (g_1), *71a (since z does not occur free in ∀y f(x, y)=g(x), *72a (since y is not free in ∀y f(x, y)=g(x)). Thus ∀y f(x, y)=g(x) ⊢ ∃y∀z f(z, h(f(x, y)))=g(z) ~ ∃y∀z f(z, h(g(x)))=g(z), as (II) asserts.

COROLLARY 1. *For* C_r, C_s, $x_1, \ldots, x_q$ *as in* (II) *of the theorem:*

$$r=s \vdash^{x_1 \ldots x_q} C_r \sim C_s.$$

COROLLARY 2. (Replacement property of equality.) *Similarly:* (a) *If* Γ ⊢ r=s, *then* Γ, $C_r \vdash C_s$. (b) $r=s, C_r \vdash^{x_1 \ldots x_q} C_s$.

Also (α)–(ζ) of § 5 hold now with "Γ ⊢" replacing "⊧" and "=" replacing "~" and (except in (α)) terms r, s, t, r_0, r_1, r_2, . . . in place of formulas A, B, C, A_0, A_1, A_2, ... Hence the chain method applies to equalities (as well as to equivalences) in the predicate calculus with equality. Its usefulness will be illustrated in §§ 38, 39.

THEOREM 31. *If* ⊢ E, *then there is a proof of* E *in which there occur no proper ions, and no mesons, not occurring in* E.

PROOF. Consider a given proof of E. In this there may occur proper ions, or mesons, not occurring in E. Take the atomic parts (or wholes) of formulas in this proof formed using such ions, and replace them all by ∀xx=x. Then take the terms occurring as parts of formulas in the resulting figure which have such mesons as their outermost symbols and are maximal, i.e. not parts of other such terms, and replace them all by the same variable v not occurring in the given proof. These two operations do not alter the end formula E or any equality axioms for = or proper ions or mesons of E, and they do not spoil any axiom or inference of the predicate calculus (cf. the 2nd paragraph of Footnote 86 § 25). However, open equality axioms for ions replaced are altered to x=y ⊃ (∀xx=x ⊃ ∀xx=x), and for mesons replaced to x=y ⊃ v=v. But each of these is provable, using *1 or (a), and Axiom Schema 1a. Replacing them by their proofs in the figure last obtained, we have a proof of E using only the proper ions and the mesons occurring in E.

COROLLARY. $\vdash$ E *in the predicate calculus with equality, if and only if* $Q_0, \ldots, Q_s \vdash$ E *in the predicate calculus, where* $Q_0, \ldots, Q_s$ *are the closed equality axioms for* $=$ *and the proper ions and the mesons occurring in* E.[107]

PROOF. IF. Using $\forall$-introds. to prove the closed equality axioms $Q_0, \ldots, Q_s$ from the respective open ones in the predicate calculus with equality. ONLY IF. Using $\forall$-elims. to deduce the open equality axioms from the closed in the predicate calculus.

EXERCISES. 29.1. Do the computation for Line 9 in Example 2.

29.2. Write formulas in the predicate calculus with equality to express each of the following.

(a) There are at most two x's such that P(x).
(b) There are exactly two x's such that P(x).
(c) The number of x's such that P(x) is between 2 and 5.

29.3. Establish (b) in Theorem 28, and (f) and (g_1) in Theorem 29. (HINT: for (f), begin with an appropriate substitution into (b).)

29.4. Give a model-theoretic treatment of replacement under equality, paralleling Theorems 19 and 5_{Pd} (the extension of Theorem 5 in § 19).

★ § 30. Equality vs. equivalence, extensionality. We have now (in both model theory and proof theory) the four basic properties of equality: reflexivity ((a) in Theorems 28, 29), symmetry ((e) in Theorem 29), transitivity ((f) in Theorem 29) and the replacement property (Corollary 2 Theorem 30).[108]

The fourth of these properties, replaceability, is sometimes overlooked in listing the fundamental properties of equality. Model-theoretically, it holds on the basis of our interpretation of "x=y" as meaning that x and y

[107] *If* E *does not contain* $=$, *then* $\vdash$ E *in the predicate calculus with equality if and only if* $\vdash$ E *in the predicate calculus*. However, this is not as easy to prove (cf. Exercise 52.3 and end Footnote 275 § 55).

[108] Using Theorem $12_{Pd=}$, the proof-theoretic results imply like results in model theory.

We might have taken these four properties themselves to establish the proof theory. In deriving the last three, we have followed the tradition in proof theory of starting with as few and simple axioms as possible. This proof-theoretic treatment is due essentially to Hilbert and Bernays 1934 pp. 164 ff.

Another (frequently used) method is to employ the axiom (a) and (instead of our other particular equality axioms) the axiom schema $x=y \supset (A(x) \supset A(y))$ where z is any variable, A(z) is any formula, and x and y are any variables (distinct from each other but not necessarily from z) free for z in A(z) (and A(x), A(y) are the results of substituting x, y respectively for the free occurrences of z in A(z)).

are the same object (Exercise 29.4). We built this interpretation into the rule for evaluating $x=y$ in determining truth tables.

Sometimes "equality" is used in a different sense, so that it possesses only the first three properties (reflexivity, symmetry and transitivity) but lacks the replacement property (except in special contexts). Whatever the terminology and symbolism, it is important to recognize that a relation possessing only the first three properties is not "equality" in the above sense. Such a relation we prefer to call only "equivalence" or "an equivalence relation".[109]

To take an example, consider the set D of all fractions $\frac{p}{q}$ where p and q are integers with $q \neq 0$. Thus D has as members $\frac{1}{2}, \frac{-1}{3}, \frac{2}{5}, \frac{5}{10}, \frac{17}{1}, \frac{5}{-15}, \frac{-6}{-12}$, etc. Two such fractions $\frac{p}{q}$ and $\frac{r}{s}$ are equal (i.e. they are the same fraction) exactly if $p = r$ and $q = s$ (i.e. (p, q) and (r, s) are the same ordered pair of integers). However, we are usually not interested in the fractions as fractions but as expressions representing rational numbers. Now the different fractions $\frac{1}{2}, \frac{5}{10}, \frac{-6}{-12}$, etc. all express the same rational number; $\frac{-1}{3}, \frac{5}{-15}$, etc. express another; etc. This is sometimes described by saying that we "define equality" between fractions (assuming multiplication and equality of integers already known) by the formula $\frac{p}{q} = \frac{r}{s} \equiv ps = qr$. (The student can remember the right member as the result of "cross-multiplying" in the left.)

What objects "ratios" (after Euclid) or "rational numbers" actually are probably does not worry the schoolboy or the layman, provided he gets the "right" rationals in the "right" relationships to one another. Our natural concept of rational numbers seems to be that a rational number is something we get by "abstracting" from the fractions forming a set such as $\left\{\frac{1}{2}, \frac{5}{10}, \frac{-6}{-12}, \ldots\right\}$, i.e. by disregarding what is different and keeping what

[109] We are concerned here with equivalence relations between members of the domain D.

We have already been dealing in §§ 2, 4, 5, 15, etc. with an equivalence relation $\sim$ ("material equivalence") between propositions. There we had not only reflexivity, symmetry and transitivity (*91–*21 in Theorem 2, or (β)–(δ) in § 5), but also replaceability in the formulas considered (Theorems 5 and 23). Nevertheless, we do not consider the relation $\sim$ to be an equality relation between propositions, since under our interpretation different (unequal) propositions can be equivalent (namely, they are when they have equal truth values).[67]

is common. The role of a theory such as we describe next is to show rigorously that a system of objects with the desired properties does exist.

One way to construct these objects is to say that the rational $\frac{1}{2}$ is the set $\left\{\frac{1}{2}, \frac{5}{10}, \frac{-6}{-12}, \ldots\right\}$; $-\frac{1}{3}$ is the set $\left\{\frac{-1}{3}, \frac{5}{-15}, \ldots\right\}$; etc. Another way is to pick from each of these classes one particular fraction, say the one in lowest terms with positive denominator, so the fraction $\frac{1}{2}$ is a rational number while $\frac{5}{10}, \frac{-6}{-12}, \ldots$ are expressions for the rational $\frac{1}{2}$. The first method has become fashionable as part of the current practice in classical mathematics of using set-theoretic concepts as the basic vocabulary. We now follow this method in giving a rigorous theory of the rational numbers, supposing such a theory of the integers $\ldots, -3, -2, -1, 0, 1, 2, 3, \ldots$ already established.

While we are giving this theory, in order to remove any possibility of confusion between a fraction $\frac{p}{q}$ (as a fraction) and the rational number which it represents, we shall write the fractions $\frac{p}{q}$ as ordered pairs (p, q). After the theory has been established, we can revert to writing "$\frac{p}{q}$", as the schoolboy and the practicing mathematician do; the equations one ordinarily writes will then read correctly with "=" meaning equality between rationals by interpreting the fractions as names for the rationals.

So far we have been tacitly assuming that the set D of the fractions falls into disjoint (i.e. nonoverlapping) nonempty classes $D_1, D_2, D_3, \ldots$, where say $D_1 = \{(1, 2), (5, 10), (-6, -12), \ldots\}$, $D_2 = \{(-1, 3), (5, -15), \ldots\}$, under the criterion that (p, q) and (r, s) belong to the same class exactly if $ps = qr$ in the theory of integers. We should justify this.

First, however, we observe the following: (A) *Suppose a nonempty set D is partitioned into disjoint nonempty classes. Then the relation "x and y belong to the same one of the classes", or in symbols $x \simeq y$, is an equivalence relation, i.e. $\simeq$ is reflexive ($x \simeq x$), symmetric ($x \simeq y \rightarrow y \simeq x$) and transitive ($x \simeq y$ & $y \simeq z \rightarrow x \simeq z$).* This is so obvious that we can only ask the student to reflect on it a moment.

Now, conversely: (B) *If a binary relation $\simeq$ with D as its domain is an equivalence relation (i.e. if it is reflexive, symmetric and transitive), then D is partitioned into disjoint nonempty classes such that members x and y of D belong to the same one of the classes exactly when $x \simeq y$ holds.*

PROOF. For any member x of D, let x^* be the class of all those members u of D such that $x \simeq u$.

I. Because of the reflexive property of $\simeq$ ($x \simeq x$), x itself belongs to x^*. Thus each of the classes x^* for x in D is nonempty, and each member x of D is in at least one of the classes (the classes x^* exhaust D).

II. Now consider any members x and y of D. CASE 1: $x \simeq y$. We show that, in this case, x^* and y^* are the same class (which by I contains both x and y); i.e. each member u of x^* is a member of y^*, and vice versa. To prove this, first suppose u is a member of x^* (in symbols, $u \in x^*$), i.e. $x \simeq u$. From the case assumption $x \simeq y$ by the symmetric property of $\simeq$, $y \simeq x$; so by the transitive property of $\simeq$, $y \simeq u$, i.e. $u \in y^*$. Similarly, if $v \in y^*$, i.e. $y \simeq v$, then by the case hypothesis and the transitivity, $x \simeq v$, i.e. $v \in x^*$. CASE 2: not $x \simeq y$. In this case, x^* and y^* have no members in common. For, if they had a common member z, then $x \simeq z$ and $y \simeq z$, and by symmetry and transitivity $x \simeq y$, contradicting the case hypothesis.

When we start with a set D and an equivalence relation $\simeq$ on D to discover the disjoint subsets of D by (B), we call those subsets *the equivalence classes of D under the relation* $\simeq$.

In particular, when D is the fractions, we can use (B) to conclude that D is partitioned into the disjoint nonempty subsets which we are taking as the rationals. Thus, we define $(p, q) \simeq (r, s) \equiv ps = qr$, and verify that this is an equivalence relation (Exercise 30.1 (a)). Then (B) applies.

In going from the fractions to the rationals, we substitute for the domain D of the fractions the new domain D^* of its equivalence classes under the equivalence relation just defined. In doing so, the equivalence relation $\simeq$ on D becomes, or gives rise to, the equality relation $=$ on D^*; i.e. two rationals x^* and y^* are the same ($x^* = y^*$) exactly if, for any $(p, q) \in x^*$ and $(r, s) \in y^*$, $(p, q) \simeq (r, s)$.

The reason why we start with the fractions D to build the rationals D^* is that the p, q of the fractions (p, q) provide us with the means of defining and performing the operations we wish to perform on the rationals.

Thus, we are accustomed to taking the sum $\frac{p}{q} + \frac{r}{s}$ by the following manipulation: $\frac{p}{q} + \frac{r}{s} = \frac{ps}{qs} + \frac{qr}{qs} = \frac{ps + qr}{qs}$ (which may be reduced to lowest terms, if we wish). To emphasize that the operation is initially one on fractions as such, we make the definition $(p, q) + (r, s) = (ps + qr, qs)$. But we want an operation on equivalence classes of fractions, since the rationals are those and not the fractions themselves. To confirm that this addition of fractions $(p, q) + (r, s)$ does give an operation on rationals $x^* + y^*$, we must show that the equivalence class to which $(ps + qr, qs)$ belongs depends only on the equivalence classes x^* and y^* to which (p, q) and (r, s) belong. That is, we need to show that

$$(p, q) \simeq (p_1, q_1) \ \& \ (r, s) \simeq (r_1, s_1) \rightarrow (ps + qr, qs) \simeq (p_1 s_1 + q_1 r_1, q_1 s_1)$$

(Exercise 30.1 (b)). For then the addition of equivalence classes x^* and y^* is defined with a uniquely determined equivalence class z^* as result, thus: pick any member (p, q) of x^* and any member (r, s) of y^*; add by the above rule; and take as z^* the equivalence class to which the resulting fraction $(ps+qr, qs)$ belongs.

Similarly, we can define the predicate $x^* < y^*$ between rationals x^* and y^* by first putting $(p, q) < (r, s) \equiv (ps < qr$ if $qs > 0$; $ps > qr$ if $qs < 0)$, and then showing that the truth or falsity of $(p, q) < (r, s)$ depends only on the equivalence classes of (p, q) and (r, s) (Exercise 30.1 (d)).

Not all operations on fractions as pairs of integers (p, q) with $q \neq 0$ depend only on the equivalence classes to which the fractions belong. E.g. $(p, q) \circ (r, s) = (p+r, q+s)$ is a perfectly well defined operation (function) on fractions which does not induce an operation on rationals.

Equivalence (between members of D), unlike equality, does not in general have the replacement property. For example, from
$(p, q) + (r, s) = (t, u)$ and $(p, q) \simeq (p_1, q_1)$ it does not follow that
$(p_1, q_1) + (r, s) = (t, u)$ (though $(p_1, q_1) + (r, s) \simeq (t, u)$ follows).

To summarize what has been elaborated on the example of the rationals, if someone has introduced a collection D of objects, and then says "let us define 'equality' between these objects", he is in our terminology proposing to define an equivalence relation. He is not free to define "equality"; the equality relation between the objects is already fixed simultaneously with the introduction of the set D. What he is proposing to define will become equality in the new domain D^* consisting of the equivalence classes of D, or in a layman's terms of the objects that we can abstract or invent by disregarding the differences which in D subsist between members of an equivalence class.—

EXAMPLE 5. We saw in § 29 that ∃x[P(x) & ∀y(P(y) ⊃ x = y)] expresses "there is a unique x such that P(x)" in the model-theory of the predicate calculus with equality. This cannot be expressed in the model theory of the predicate calculus without equality (Chapter II). To show this, suppose to the contrary we had a formula E of the latter with just one ion P(—) such that E is true when and only when P(x) is evaluated by a logical function $\mathfrak{l}(x)$ which is $\mathfrak{t}$ for exactly one member of D. Then in particular, E is $\mathfrak{t}$ when $D = \{1, 2, 3\}$ and P(x) is evaluated by $\mathfrak{l}_6(x)$ (which has the table $\mathfrak{f}$, $\mathfrak{t}$, $\mathfrak{f}$). Now consider instead a D with four members, say $D = \{1, 2_1, 2_2, 3\}$, and let P(x) be evaluated by $\mathfrak{l}(x)$ where $\mathfrak{l}(1) = \mathfrak{f}$, $\mathfrak{l}_2(2_1) = \mathfrak{l}_2(2_2) = \mathfrak{t}$, $\mathfrak{l}(3) = \mathfrak{f}$. In the new computation, 2_1 and 2_2 behave just as 2 did before, so again E will be $\mathfrak{t}$. This contradicts our supposition, since this $\mathfrak{l}(x)$ is true for two values of x.

As Example 5 illustrates, the rules for evaluation in the model theory of the predicate calculus without equality provide no barrier against splitting any element of any D into a multiplicity of elements which will behave alike in all the computations previously considered. (This process is the reverse of coalescing elements into equivalence classes as new elements.) —

Now we discuss problems of translation from ordinary language into logical symbols when equality may be involved. The basic presupposition for our model-theoretic treatment of equality is that we are dealing with some collection or set as the domain D. This we think of as constituted of "definite well-distinguished objects".[110] So if x is a member of D and y is a member of D, then either x and y are the same member ($x=y$ is true) or x and y are different (distinct) members ($x=y$ is false). In brief, our notion of a domain D already entails the equality predicate over D.

Any application of the predicate calculus, now with equality, is to come after we have a set D as the domain, which we have either constructed or assumed to exist on the basis of supposed facts or for the purpose of the argument. A domain in our sense may constitute quite a sophisticated (or even problematical) abstraction from perceptual experience.

To illustrate, we take the cases of temperature and color. Our sensations of each are fuzzy, so that close temperatures and colors intergrade; it is often hard to say whether two objects under given conditions have the same temperature or color. But physics and psychology have constructed mathematical models, in which temperatures are represented "exactly" by real numbers ($\geq$ an "absolute zero"), and colors by points in some three-dimensional region (or triples of real numbers subject to certain restrictions). In classical mathematics, each two real numbers x and y are either equal or unequal, by the following rule: write each real number as the sum of an integer (positive, zero or negative) and a nonnegative infinite decimal fraction; e.g. $\pi = 3 + .14159\ldots =$ (briefly) $3.14159\ldots$, $2/3 = 0.66666\ldots$; $75/2 = 37.50000 = 37.49999\ldots$, $-3/4 = -1 + .25000 = -1 + .24999\ldots$. Sometimes there will be two ways to do this, one using infinitely repeating 0's and the other infinitely repeating 9's; in these cases, let us now choose always the same way, say the one with repeating 9's. Now $x = y$ if x and y have the same integer-plus-decimal representation; $x \neq y$ otherwise, i.e. if the integers are different, or the decimals differ in at least one digit.[111]

[110] Cantor 1895 p. 481. Cantor created the theory of sets between 1874 and 1897. We deal with this in §§ 32–35 of Chapter IV.

[111] There are other theories of real numbers, equivalent to this but more elegant (for one, cf. IM pp. 30–32). Intuitionists (§ 36) do not accept these theories of real numbers; cf. Heyting 1934, 1956.

If one starts with our notion of a domain D, and forms all possible one-place predicates $P(x)$ over D (or simply all possible one-place logical functions $\mathfrak{l}(x)$ over D),[67] then $x = y$ can be defined thus: $x = y \equiv \{\text{for all } P, P(x) \sim P(y)\}$. For, if x and y are the same member of D, then by our notion of a predicate, for each P, $P(x)$ and $P(y)$ are the same proposition, so $P(x) \sim P(y)$. On the other hand, if x and y are different members of D, then there are predicates P which "separate" them, i.e. such that $P(x)$ is true while $P(y)$ is false or vice versa, so not, for all P, $P(x) \sim P(y)$. This definition expresses the so-called *principle of identity of indiscernibles* of Leibniz 1685+ (reproduced in Lewis 1918 pp. 373–387): if we can discern no property P in which x and y differ, x and y are identical. In second-order predicate calculus (references in § 17), we can hence regard $x=y$ as an abbreviation for $\forall \mathrm{P}(\mathrm{P}(\mathrm{x}) \sim \mathrm{P}(\mathrm{y}))$, instead of introducing it as a primitive predicate. Since conceptually the idea of equality underlies *our* notion of a domain and of predicates over the domain, it seems more elementary and direct to introduce equality as we do than to define it by reference to all predicates. Of course, the Leibniz definition would only be available to us if we were considering second-order logic.

For the Leibniz definition to seem a natural way to *introduce* the idea of equality (i.e. a way which explains its meaning rather than simply states a necessary and sufficient condition for what is already understood), one must start from a rather different position than in our model theory. For, it requires one to entertain the applicability of properties P to objects x, or to consider the truth or falsity of values $P(x)$ of a predicate, before the objects x themselves are clearly distinguished.

In our theory with a presupposed domain D, whether objects are equal or not is dependent on the picture of the world out of which we have constructed the D.

Is the morning star x equal to (the same as) the evening star y? (In symbols, is $x = y$?) It is when the domain D^{**} is that of astronomical bodies, wherein both the morning star and the evening star are Mercury (seen at different times). To shepherds, unversed in astronomy, the star observed in the morning, and the star observed in the evening (on other dates) are two different natural phenomena; the domain D^* is different. To a child, the evening star seen on one date and the evening star on another date may be different phenomena is still another domain D. We can think of D^* as equivalence classes in D, and of D^{**} as equivalence classes in D^* (or larger ones in D), where in each case the abstraction results from recognizing different members of one set as related by an equivalence relation or thinking of them as different manifestations of the same "underlying reality".

The forward implication in Leibniz' definition of equality, namely that $x = y$ implies $\forall P(P(x) \sim P(y))$, expresses in a little different way what we called the "replacement property of equality" (Corollary 2 Theorem 30). Our principle is stated with respect to the class of formulas under consideration, and includes the case of replacement inside quantifiers. The same property is also sometimes described as "extensionality", from the standpoint of the contexts in which $x = y$ justifies the replacement of x by y. A context in which such replacement is justified is called *extensional*; one in which it is not, *nonextensional* or sometimes *intensional*. This jibes with our previous use of "extensional" vs. "intensional", where replacement under "$\sim$" was involved.[92]

Consider the following. "Let n be the number of planets in the solar system. Kepler did not know that $n > 6$. But $n = 9$ (according to current science). Therefore Kepler did not know that $9 > 6$."[112] The context "Kepler did not know that — > 6" is intensional, since the meaning of what is put in place of the blank, rather than just the value of it as a member of the domain D, affects the truth of the result. So the replacement of "n" by "9" on the basis of the equality "$n = 9$" is not justified in "Kepler did not know that $n > 6$".

By this example, we can add the logic of knowing, believing, etc. to the examples already given end § 12 of nonextensional contexts for replacement under material equivalence, i.e. where equality of truth values is insufficient to justify replacement. For, from "$n = 9$" we get "$n > 6 \sim 9 > 6$" by a correct application of replacement (Theorem 30 (II)); but "$n > 6$" cannot be replaced by "$9 > 6$" in "Kepler did not know that $n > 6$".

When a context is not extensional for replacement under equality, something is being taken into account in it that is not recognized in constituting the elements of the domain D. By splitting up those elements to construct a new domain D', the equality in D may become an equivalence relationship on D' while extensionality is restored. Thus if we construe D' as "mental" objects for Kepler rather than "real" objects, so that n and 9 are different mental objects, then "$n = 9$" is not true. So the occurrence of nonextensional contexts for equality can be regarded as just another way of looking at the difference between an equality and an equivalence relationship. Admittedly, in matters of knowledge or belief, it may be rather hard to describe exactly a domain D' which suffices to restore extensionality.

Since a shift in our understanding of what the domain D of objects is

[112] Kepler, famous for his discovery of the laws of planetary motion, died in 1630. The planets Uranus, Neptune and Pluto were discovered in 1791, 1846 and 1930, respectively.

can make a difference between equality and equivalence, it is not surprising that the terminology is often confused. Here is a table of translations.

x=y	x is the same (object) as y.
	x is identical with y.
	x equals y [usually].
	x is y [sometimes].

The first two readings are the least ambiguous. But we normally use the reading "equals" for "=", because it is briefer than the first two, and indeed is the standard way of reading "=" in mathematics. But mathematicians often use "=" and "equals" for what we consider an equivalence relationship. As a new example, textbooks in plane Euclidean geometry often write "$AB = CD$" to mean that the line segments AB and CD are of the same length. One can't then replace "AB" by "CD" in "$AB \perp EF$" (read "AB is perpendicular to EF"). So the translator into our logical symbolism must be on guard to recognize whether "equals" is being used for our =, or for an equivalence relationship. Of course, "length AB = length CD" is an unambiguous translation of "$AB = CD$" in the present example, and this could be written "$AB \simeq CD$". Besides equality or identity in the fundamental sense, mathematicians need many (other) equivalence notions. So terms like "equivalent", "congruent", "similar", "homologous" (and as we have seen, sometimes "equal") are often employed for such equivalence notions.

We come finally to "is" or other forms of the verb "to be". In addition to its use as an auxiliary verb ("he is going", "this was done", "John is loved by Jane") and in expressing the existential quantifier ("there is"), we need to notice three others. (1) When A and B are objects both considered as members of the same domain D, "A is B" (almost always) means "$A = B$" in our sense, i.e. that A and B are the same member of D. Examples: "The number of planets is 9", "Two plus two is four". (2) When A is an object considered as a member of a domain D, and B is a subset of D, "A is B" means that A is a member of B, and is translated by "$A \in B$" or B(A). Examples: "Socrates is a man", "Snow is white", "Jane is lovely". (3) When A and B are both subsets of D, "A is B" means that the subset A is included in (is a subset of) B, and is translated by "$A \subseteq B$" or $\forall x(A(x) \supset B(x))$. Examples: "Men are mortal", "Cats are animals".

So as not to give away the translation in advance, we have used capital "A" and "B" in all three uses. With symbolism chosen as we have usually chosen it, (1)–(3) would be distinguishable by lower case (small) letters for members of sets and capital letters for sets, thus: (1) "a is b" translated "$a = b$". (2) "a is B" translated "$a \in B$" or B(a). (3) "A is B" translated "$A \subseteq B$" or $\forall x(A(x) \supset B(x))$. (In translation from verbal language the problem is then, "Which terms correspond to small letters, and which to capitals?")

This usage cannot be preserved if we consider sets whose members are also sets whose members we may want to name. "Men are numerous" should be translated "$M \in N$" where M is the set of men and N is the set of numerous sets.

It should hardly need to be pointed out that confusion of these meanings of "is" can lead to fallacies.

Under Meanings (1) and (3), "is" is transitive ($a = b \,\&\, b = c \rightarrow a = c$, and $A \subseteq B \,\&\, B \subseteq C \rightarrow A \subseteq C$ (Barbara, Example 23 § 27)); but not under Meaning (2) (fallacious example: "Socrates is a man. Men are numerous. Therefore Socrates is numerous.").[113] Under (1), "is" is both reflexive and symmetric; under (2), "is" is not symmetric, nor (as set theory is usually understood) reflexive; under (3), "is" is reflexive, but not symmetric.

EXERCISES. 30.1. (a) Show that $(p, q) \simeq (r, s) \equiv ps = qr$ is an equivalence relation between pairs of integers (p, q) with $q \neq 0$. (b) Show that $(p, q) + (r, s) = (ps + qr, qs)$ is an operation depending only on the equivalence classes (cf. the text). (c) Treat similarly $(p, q) \cdot (r, s) = (pr, qs)$. (d) Treat similarly $(p, q) < (r, s)$ as defined in the text.

30.2. Let $D = \{0, 1, 2, 3, 4, 5, 6, 7\}$ and $x \simeq y \equiv (x-y$ or $y-x$ is an even natural number). Show that $\simeq$ is an equivalence relation on D, and list the equivalence classes.

30.3. Translate into logical symbolism, and establish validity or invalidity.

(a) Everybody loves Jane. Jane loves someone. Therefore there are two people who love each other.

(b) Everybody loves Jane. Jane loves someone besides herself. Therefore there are two people who love each other.

(c) π is the ratio of the circumference of a circle to its diameter. π is between 3.1415 and 3.1416. Therefore the ratio of the circumference of a circle to its diameter is between 3.1415 and 3.1416.

(d) Samuel Clemens wrote "Huckleberry Finn". Mark Twain wrote "Huckleberry Finn". "Huckleberry Finn" is the work of one author. Therefore Mark Twain is Samuel Clemens.

(e) Jane has at most one husband. Jane is married to Thomas. Thomas is thin. William is not thin. Therefore Jane is not married to William.

(f) Tom is a brother of Dick. Dick is a brother of Harry. No man is his own brother. Therefore Tom and Harry are different people.

(g)* Every person in this room has a brother and a sister both in this room. No person is his own brother or sister. No brother of anyone is a

[113] Example from Suppes 1957 p. 183.

sister of anyone. Therefore there are either no people or at least four people in this room.

(h) The moon seen today is round. The moon seen twelve days ago was crescent shaped. The moon today and the moon twelve days ago are the same object. Therefore something is both round and crescent shaped.

(i) A rose is red. Red is a color. Therefore a rose is a color.

(j) A gnu is an antelope. An antelope is a mammal. Therefore a gnu is a mammal.

(k) Salt and sugar are white. Nothing is both salt and sugar. Therefore nothing is white.

★§ 31. Descriptions. To complete our survey of first-order logic (cf. § 17), we shall discuss briefly the definite article "the" and the indefinite article "a" or "an".

We can name objects by descriptive phrases containing the definite article "the", such as the following: (a) "the present queen of England", (b) "the present king of France", (c) "the sister of x", (d) "the father of x", (e) "the number which added to x gives y", (f) "the number whose square is x", (g) "the least common divisor of x and y", (h) "the third smallest prime number", (i) "the greatest prime number", (j) "the number w such that, for all x, $x + w = x$". In our model theory, by "objects" we mean members of the domain or universe of discourse D about which we are thinking at the moment.

For some of the objects just named, we have other names not using "the": (a_1) "Elizabeth II" (1966), (e_1) $y - x$, (h_1) 5, (j_1) 0. The usefulness of descriptions as a linguistic device is that it gives us a way to make a name, usually for temporary use, when we do not already have one but have the vocabulary to describe or characterize the object.

As our language is commonly understood, a definite description is not used "properly" unless it really does describe a unique object, or, if free variables are present, a unique object for each choice of values of those variables. The general form of a definite description is "the w such that $F(w)$". If $F(w)$ is a predicate of just the one variable w, the description is *proper* exactly if there is just one object w in the domain D such that $F(w)$; this condition translates into our logical symbolism as $\exists!wF(w)$ (cf. § 29). If $F(w)$ is a predicate of other variables $x_1, \ldots, x_n$ as well, write it then also "$F(x_1, \ldots, x_n, w)$", then (as long as we are not confining our attention to only certain values of $x_1, \ldots, x_n$) the description is *proper* exactly if, for each $x_1, \ldots, x_n$ in D, there is a unique w in D such that $F(x_1, \ldots, x_n, w)$; in our symbols, $\forall x_1 \ldots \forall x_n \exists! wF(x_1, \ldots, x_n, w)$. In this case, the description serves to name a function $f(x_1, \ldots, x_n)$ (as well as its values).[105] We include the previous case under this, by allowing $n = 0$ so an individual (or object) is named.

Since 1848, (b) has been an improper description.[114] When D is all persons, (c) is improper, since not every person x has just one sister; only for smaller D's does (c) define a function $f(x)$. But (d) is proper, though proverbially only a wise x knows the value of $f(x)$. When D is the positive real numbers, (e) is improper (e.g. no positive real number can be added to 5 to give 3), but (f) is proper. When D is all the real numbers, (e) is proper, but (f) is not (e.g. 4 has two square roots, 2 and -2). By a theorem of Euclid, (i) is improper.

In ordinary discourse, improper descriptions hardly occur. If someone speaks about "the w such that $F(w)$" when there isn't a unique such w, we usually conclude that he is either confused himself or is attempting to mislead us. We might try to say that any sentence A containing an improper description is simply false. But then we run into the difficulty that $\neg$A and A $\supset$ B would also be false by this criterion, whereas (by our truth tables for $\neg$ and $\supset$) $\neg$A and A $\supset$ B must be true when A is false. Whitehead and Russell 1910 pp. 69–75 (66–71 in the 1925 ed.; also it's in van Heijenoort 1967) handled this difficulty by requiring the part of a sentence which is declared false because it contains an improper description to be indicated; the truth or falsity of larger parts containing it is then to be determined by the usual rules. There is some awkwardness in this; and the problem of the best way to treat descriptions allowing improper ones is still a subject of research.[115] Improper descriptions, and partially defined functions (and in other connections, multiple-valued functions) have some uses in mathematics. However, for the brief treatment we give here, we shall follow common usage in avoiding improper descriptions.

But then we have another difficulty, that what a sentence is (or in our treatment of logic, what a formula is) will not always be determinable immediately from the way it is assembled out of its parts (as in our definitions of "formula" in §§ 1, 16, 28, 29), but will depend on validity or provability (or valid consequence or deducibility) results. That is, if A contains a part "the w such that F(w)", and would be a formula under the definition in § 28 or § 29 if this part were allowed as a term, we shall not know whether to call A a formula until we know whether we have $\exists !$wF(w).

As the price of abjuring improper descriptions, we shall have to live with this difficulty.[116] It will not be very hard to do so, since in ordinary

[114] Or even since 1830. Louis Phillipe's title was "King of the French", not "King of France".

[115] See Scott 1967, where other references are given.

[116] This difficulty is slightly disguised in the present treatment [Theorem 32], since for the predicate calculus with functions and equality we have not yet said anything about limiting the supply of mesons available. But if Γ be closed formulas which express the axioms of some theory for which only the ions and mesons of Γ are considered available, then the introduction of f entails an enlargement of the class of formulas which we make only after obtaining $\forall x_1 \ldots \forall x_n \exists ! wF(x_1, \ldots, x_n, w)$.

discourse and in mathematics we normally want to use definite descriptions only after we already have $\exists!wF(w)$ or $\forall x_1 \ldots \forall x_n \exists!wF(x_1, \ldots, x_n, w)$.

We might introduce descriptions into our symbolism by adding an operator "ιw" (read "the w such that"). This would bind variable occurrences in terms $\iota wF(x_1, \ldots, x_n, w)$ (and in terms and formulas containing such terms), just as $\forall x$ and $\exists x$ do in formulas.

In our treatment here we shall stay within the notation of the predicate calculus with functions and equality as treated above. The descriptions we want can then be handled simply as introductions of new function symbols f.[117] Thus, when we have

(i) $$\forall x_1 \ldots \forall x_n \exists!wF(x_1, \ldots, x_n, w),$$

we may introduce a new function symbol f with the formula

(ii) $$\forall x_1 \ldots \forall x_n F(x_1, \ldots, x_n, f(x_1, \ldots, x_n)).$$

We may even read $f(x_1, \ldots, x_n)$ as "the w such that $F(x_1, \ldots, x_n, w)$" if we wish; but its properties will be given by (ii) with (i). We shall make this explicit in the theorem below. Any finite succession of descriptions, each after the first possibly involving the preceding ones, can thus be provided, without piling up complicated patterns of bound variables in the terms. This is the usual practice in mathematics, except in simple passages.

In this treatment, we lose the advantage that descriptive definitions are self-explanatory, since each time a new function symbol f has been introduced, we have to remember the formula (ii) with which it was introduced. If the function expressed by the symbol is to be used often, we will quickly become used to the symbol f for it, and the notation $f(x_1, \ldots, x_n)$ will be more convenient than the more cumbersome $\iota wF(x_1, \ldots, x_n, w)$. It is in speech and in short written passages that the descriptive operator "the *w* such that" or "ιw" has a clear advantage over introducing new function symbols. However, for our logical analyses of arguments encountered in ordinary language, it is a simple matter to supply (or imagine supplied) function symbols corresponding to the descriptive names used in the language.

[117] We say "function symbol f" rather than "meson f(—, . . . , —)" since we really have in mind a meson consisting of a single new symbol (a "function symbol f") with argument places following if $n > 0$. However, the following treatment can also be read with "meson f(—, . . . , —)" in place of "function symbol f", so that other convenient notations could be used. In fact, the meson f(—, . . . , —) for a given description could even be $\iota wF(—, \ldots, —, w)$, provided terms arising from different such mesons successively introduced cannot be confused (and the bound variables w are chosen so that the substitutions performed for $x_1, \ldots, x_n$ in $\iota wF(x_1, \ldots, x_n, w)$ are free; cf. IM bottom p. 154).

So we are to get the effect of proper (definite) descriptions by a theory for adding new function symbols when it has become known (absolutely or under the assumptions we have made) that the functions are definable descriptively.

We shall give the treatment only in model theory, where it is almost trivial. A proof-theoretic discussion (after Hilbert-Bernays 1934) would be a little too detailed to fit well here.[118]

THEOREM 32. *Let* Γ *be any list of zero or more formulas in the predicate calculus with functions and equality,* $F(x_1, \dots, x_n, w)$ *be a formula containing free only the distinct variables* $x_1, \dots, x_n, w$ $(n \geq 0)$, f *be a function symbol not occurring in any of the formulas* Γ *and* $F(x_1, \dots, x_n, w)$, *and* C *be a formula not containing* f. *Then*: *If* $\Gamma \models \forall x_1 \dots \forall x_n \exists ! w F(x_1, \dots, x_n, w)$ *and* $\Gamma, \forall x_1 \dots \forall x_n F(x_1, \dots, x_n, f(x_1, \dots, x_n)) \models C$, *then* $\Gamma \models C$.

PROOF. Consider any given domain D, and any given assignment in D to just the free variables, ions and mesons in Γ, $\forall x_1 \dots \forall x_n \exists ! w F(x_1, \dots, x_n, w)$, C which makes all of Γ t. We must show that C is also t. By the hypothesis that $\Gamma \models \forall x_1 \dots \forall x_n \exists ! w F(x_1, \dots, x_n, w)$, the formula $\forall x_1 \dots \forall x_n \exists ! w F(x_1, \dots, x_n, w)$ is t. By Example 3 § 29 and the evaluation rule for $\forall$, hence, for each choice of members $\mathfrak{x}_1, \dots, \mathfrak{x}_n$ of D as values of $x_1, \dots, x_n$, there is a unique member w of D such that $F(x_1, \dots, x_n, w)$ is t when w is the value of w. So for this w and $\mathfrak{x}_1, \dots, \mathfrak{x}_n$, we can write $w = f(\mathfrak{x}_1, \dots, \mathfrak{x}_n)$ where f is an n-place function with arguments and values in D. But now, if we augment the given assignment by taking $f(x_1, \dots, x_n)$ as value of $f(x_1, \dots, x_n)$, then $\forall x_1 \dots \forall x_n F(x_1, \dots, x_n, f(x_1, \dots, x_n))$ is t. So by the hypothesis that $\Gamma, \forall x_1 \dots \forall x_n F(x_1, \dots, x_n, f(x_1, \dots, x_n)) \models C$, the formula C is also t. This is under the augmented assignment; but since f does not occur in C, the formula C is t under the given assignment, as was to be proved. —

The effect of this theorem is that, if $\forall x_1 \dots \forall x_n \exists ! w F(x_1, \dots, x_n, w)$ is a valid consequence of our assumptions Γ, then to the symbols we are using we can add $f(x_1, \dots, x_n)$ expressing "the w such that $F(x_1, \dots x_n, w)$ is t when w is the value of w" and to our assumptions the formula $\forall x_1 \dots \forall x_n F(x_1, \dots, x_n, f(x_1, \dots, x_n))$; any formula C not containing f which is a valid consequence of the augmented assumptions will be a valid consequence of the original assumptions. In the process of recognizing C to be a valid consequence, our reasoning may be helped by the additional assumption $\forall x_1 \dots \forall x_n F(x_1, \dots, x_n, f(x_1, \dots, x_n))$. In fact, there can be a very great gain in efficiency here, through the use of the function

[118] Cf. IM § 74. From our model-theoretic treatment and the Gödel completeness theorem (§ 52 below), some of the proof-theoretic results will follow (but this will not give them in the elementary way which is desired in proof theory; cf. § 37).

symbol $f(x_1, \ldots, x_n)$ in conjunction with the replacement theorem (Theorem 30 § 29).

EXAMPLE 6. Let the formulas be constructed as in § 29 with 1, —— (or also $—^{-1}$) as the mesons. In the predicate calculus with equality,

$$(1) \quad \forall x\forall y\forall z\,(xy)z=x(yz),\ \forall x\, x1=x,\ \forall x\, xx^{-1}=1 \vdash xz=yz \supset x=y$$

(Exercise 31.2). Hence, by Theorem 32 (with Theorems $9_{Pd=}$ and $12_{Pd=}$),

$$(2) \quad \forall x\forall y\forall z\,(xy)z=x(yz),\ \forall x\, x1=x,\ \forall x\exists!w\, xw=1 \models xz=yz \supset x=y.$$

EXAMPLE 7. In Example 6, (2) does not hold omitting $\forall x\exists!w\, xw=1$ (call it then (2′)). For, the two assumption formulas of (2′) are both t and the conclusion is f, when D is the rational numbers and 1, xy, 0, 1, 0 are the values of 1, xy, x, y, z. (In effect, $xz=yz \supset x=y$ lets us prove $0=1$ from $0{\cdot}0=1{\cdot}0$ by "dividing by 0".)

COROLLARY. *If* $\Gamma \models \forall x_1 \ldots \forall x_n\exists!wF(x_1, \ldots, x_n, w)$, *and* Γ *is consistent in the sense that for no formula* A *do both* $\Gamma \models$ A *and* $\Gamma \models \neg$A *hold, then so is* $\Gamma, \forall x_1 \ldots \forall x_n F(x_1, \ldots, x_n, f(x_1, \ldots, x_n))$.

Now we observe that Theorem 32 and its Corollary hold even with "$\exists$w" in place of "$\exists$!w" (call the theorem then THEOREM 32a). The only difference in the proof is that, for each $x_1, \ldots, x_n$ as values of $x_1, \ldots, x_n$, the w is not completely determined, but a particular one must be chosen from the class of w's such that $F(x_1, \ldots, x_n, w)$ is t when w is the value of w.[119]

Theorem 32a justifies the use of "proper indefinite descriptions", i.e. expressions of the form "a w such that $F(x_1, \ldots, x_n, w)$" where $\forall x_1 \ldots \forall x_n\exists wF(x_1, \ldots, x_n, w)$ is available ($n \geq 0$).

Extreme care should be used, however, in adopting $f(x_1, \ldots, x_n)$ as a translation of "a w such that $F(x_1, \ldots, x_n, w)$" in any extended argument. For, with each set of values of $x_1, \ldots, x_n$, the term $f(x_1, \ldots, x_n)$ in all its uses under the assumption (ii) has to express the same one of the w's such that $F(x_1, \ldots, x_n, w)$. (This is reflected in the equality axioms for f, § 29.) In ordinary discourse, different uses of "a w such that $F(x_1, \ldots, x_n, w)$" in an extended context are likely to be independent of one another. In fact, uses of "a" or "an" to describe a member of a set $\hat{w}F(x_1, \ldots, x_n, w)$ are usually so transitory that an adequate translation

[119] The principle that, when $\forall x_1 \ldots \forall x_n\exists wF(x_1, \ldots, x_n, w)$ is t, we can, for each $x_1, \ldots, x_n$ in D, choose such a w in D as the value of a fixed function $f(x_1, \ldots, x_n)$, is a form of the axiom of choice (§ 35 below).

For $n = 0$, Theorem 32a is hardly different from $\exists$-elimination (with $\models$ for $\vdash$, and with F(w) in place of A(x)): the individual symbol f in (ii) takes the place of the variable w held constant in "Γ, F(w) $\models$ C".

can be given just using a variable as the name of the object described. Examples: "Socrates is a man" ($\exists x[s=x \,\&\, H(x)]$, or simply $H(s)$ as in Example 19 § 26), "A child needs affection" ($\forall x[C(x) \supset A(x)]$ as in § 27), "A man was here" ($\exists x[M(x) \,\&\, H(x)]$).

EXERCISES. 31.1. Prove Corollary Theorem 32.

31.2*. Establish (1) in Example 6.[120]

31.3. Show that

$$\forall x \dot{\exists}! w F(x, w),\ \forall x F(x, f(x)) \vdash F(x, w) \sim f(x)=w.$$

Thus, in operating with a proper description, $F(x, w)$ and $f(x)=w$ are equivalent.

31.4. Criticize the following arguments.

(a) Henry and Jane are brother and sister ($F(h, j)$). Therefore Jane is the sister of Henry ($j = f(h)$; cf. Exercise 31.3). Henry and Edith are brother and sister ($F(h, e)$). Therefore Edith is the sister of Henry ($e = f(h)$). Therefore (by (e) and (f) § 29) Edith and Jane are the same person ($e = j$).

(b) By the meaning of "the sister of John", John and the sister of John are brother and sister ($F(j, f(j))$). Therefore (by $\exists$-introd.), John has a sister ($\exists w F(j, w)$).

[120] In substance the solution is the proof of T11 in § 39. (A direct proof of (2) with "$\vdash$" is only slightly longer. In more complicated examples the gain by using Theorem 32 or a proof-theoretic version of it can be greater.)

PART II

MATHEMATICAL LOGIC AND THE FOUNDATIONS OF MATHEMATICS

CHAPTER IV

THE FOUNDATIONS OF MATHEMATICS

§ 32. Countable sets. In the present century, work on mathematical logic and work on the foundations of mathematics have been closely connected. Problems and ideas about the foundations of mathematics have contributed much to the development of logic, and logic has been a primary tool in the investigation of those foundations. In this Part II of the book, we shall survey this common area. We shall both acquaint ourselves with new developments, and examine more carefully some notions which underlay the discussion in Part I.

We begin with some points in Cantor's theory of sets, which dates from the discoveries he published in 1874 concerning the comparison of infinite collections.

Suppose we wish to know whether one collection is less numerous or equally numerous or more numerous than another. For finite collections, we can settle this by attempting to "match" or "pair" the members of the sets, or as we shall say hereafter to put them into *one-to-one* (1–1) *correspondence*. If the two sets can be put into 1–1 correspondence, they are "equally numerous" or as we shall say have the *same cardinal number*. However, the idea of such a correspondence is more primitive than the idea of "cardinal number", as the following example illustrates.

In a tribe of aborigines who cannot count beyond twenty, a chief is to be chosen from two candidates A and B by awarding the position to the candidate with the larger herd of cattle. The two herds are run through a gate, with a pair of animals, one from each herd, always passing through together, until one or the other herd or both are exhausted. If A's herd is exhausted before B's, B becomes chief; and vice versa, A wins if he has animals left when all of B's have gone through. If the last two animals walk through together, a different method of selection must be used, or a co-chieftancy established. Though each herd may have more than twenty cattle, so it could not be counted by the tribe, this method of pairing works.

In 1638, Gallileo noted the "paradox" that the squares of the positive integers can be placed in 1–1 correspondence with all the positive integers, contrary to the axiom of Euclid that the whole is greater than any of its *proper* parts, i.e. parts not the whole.[121]

Thus with infinite collections, putting one collection into 1–1 correspondence with a proper part of the other does not exclude the possibility that under a different method of pairing the wholes may correspond 1–1. With the two herds, this could not happen: if B wins the chieftainship on one way of running the herds through the gate, A is certain (though he has not seen a mathematical proof) that he could not have tied or won by sending his animals through in a different order.

We assume familiarity with the sequence of the *natural numbers* (or *nonnegative integers*)[122]

$$0, \quad 1, \quad 2, \quad 3, \quad 4, \quad 5, \quad \ldots.$$

Collections which can be placed in 1–1 correspondence with the natural numbers we call *countably infinite* or *denumerably infinite* or *enumerably infinite*. To show that an infinite set is countable, we need only put its members into an infinite list, or 1–1 correspondence to the natural numbers as written above. A particular such list, or 1–1 correspondence to the natural numbers, we call an *enumeration* of the set. Examples of countably infinite sets are, besides the ***natural numbers*** themselves, the set of the ***positive integers***, the set of the ***squares of the positive integers***, and the set of all the ***integers***, as we see from the following enumerations:

$$1, \quad 2, \quad 3, \quad 4, \quad 5, \quad 6, \quad \ldots$$

$$1, \quad 4, \quad 9, \quad 16, \quad 25, \quad 36, \quad \ldots$$

$$0, \quad 1, \quad -1, \quad 2, \quad -2, \quad 3, \quad \ldots$$

A *countable* set or *denumerable* set or *enumerable* set is a set which is either countably infinite or finite. Here we can describe a *finite* set as a set which can be put into 1–1 correspondence with an initial segment $0, \ldots, n-1$ of the natural number sequence, possibly the empty segment ($n = 0$). This is equivalent to saying that a finite set is a set that

[121] That an infinite collection can be put into 1–1 correspondence with a proper part of itself was noticed a number of times before Gallileo. Steele (in his historical introduction to his 1950 translation of Bolzano 1851, who noted the same) cites Plutarch (46?–125? A.D.) and Proclus (412–485 A.D.). Thomas 1958 cites Adam of Balsham (Parvipontanus) 1132, and Pierre Duhem "Le système du monde" 1954 ed. vol. 7 p. 123 cites Robert Holkot (d. 1349). Julius R. Weinberg supplied these references and also "Centriloquium Theologicum", Conclusio 17 (erroneously attributed to William Ockham (d. 1349?)).

[122] Some authors use "natural numbers" as a synonym for "positive integers" 1, 2, 3, . . ., obliging one to use the more cumbersome name "nonnegative integers" for 0, 1, 2, ... Besides, in this age, we should accept 0 on the same footing with 1, 2, 3, ...

has a natural number n as its "cardinal number", as this use of natural numbers is commonly understood.[123]

Another example of a countably infinite set is the set of the ***rational numbers.*** This is surprising, if one first compares them with the integers in the usual algebraic order. The points on the x-axis with integral abscissas are isolated, while those with rational abscissas are "everywhere dense", i.e. between each two, no matter how close, there are always others. We can nevertheless enumerate them by the following device. We begin with the fact that each rational can be written as a fraction of integers with a positive denominator. We arrange all such fractions in an infinite matrix, thus:

$$
\begin{array}{ccccccccccccc}
\frac{0}{1} & & \frac{1}{1} & \rightarrow & \frac{-1}{1} & & \frac{2}{1} & \rightarrow & \frac{-2}{1} & & \frac{3}{1} & & \dots \\
\downarrow & \nearrow & & \swarrow & & \nearrow & & \swarrow & & & & & \\
\frac{0}{2} & & \frac{1}{2} & & \frac{-1}{2} & & \frac{2}{2} & & \frac{-2}{2} & & \frac{3}{2} & & \dots \\
& \swarrow & & \nearrow & & \swarrow & & & & & & & \\
\frac{0}{3} & & \frac{1}{3} & & \frac{-1}{3} & & \frac{2}{3} & & \frac{-2}{3} & & \frac{3}{3} & & \dots \\
\downarrow & \nearrow & & \swarrow & & & & & & & & & \\
\frac{0}{4} & & \frac{1}{4} & & \frac{-1}{4} & & \frac{2}{4} & & \frac{-2}{4} & & \frac{3}{4} & & \dots \\
& & & & & & \dots & & & & & &
\end{array}
$$

We can enumerate these fractions by following the arrows. Finally, we can go through that enumeration striking out each fraction which, interpreted as a rational number, is equal in value to one that has preceded it. This gives us the following enumeration of the rational numbers:

$$0,\ 1,\ -1,\ \tfrac{1}{2},\ \tfrac{1}{3},\ -\tfrac{1}{2},\ 2,\ -2,\ -\tfrac{1}{3},\ \tfrac{1}{4},\ \dots.$$

Another enumerably infinite set is the set of the ***real algebraic numbers,*** that is, the set of real numbers which are roots of algebraic equations (polynomial equations) in one variable x with integral coefficients, such as the equation

$$4x^5 - 17x^3 + 2x^2 + 5 = 0.$$

The general form of an algebraic equation of *degree* n (≥ 1) is

$$a_0x^n + a_1x^{n-1} + \dots + a_{n-1}x + a_n = 0 \qquad (a_0 \neq 0).$$

[123] This is surely the more intuitive definition of finiteness of a set. But the property of the set of the positive integers noted in Gallileo's "paradox" is characteristic of infinite sets. So, as Peirce 1885 and Dedekind 1888 proposed, a *finite* set can be defined alternatively as a set which cannot be put into 1–1 correspondence with a proper part of itself. Cf. IM pp. 13, 14 (after reading § 34 below).

If we can enumerate the algebraic equations, we can enumerate the real algebraic numbers. For then, in the enumeration of the equations, we can replace each equation by its distinct real roots, which will be finitely many (at most as many as its degree), to obtain an "enumeration with repetitions" of the real algebraic numbers. Then the repetitions can be removed. The algebraic equations with integral coefficients can be enumerated by first noticing that we can without ambiguity simply write the exponents on the line with the rest of the symbols (thus: $4x5-17x3+2x2+5=0$). Then the equations become finite sequences of just fourteen different symbols:

$$0 \quad 1 \quad 2 \quad 3 \quad 4 \quad 5 \quad 6 \quad 7 \quad 8 \quad 9 \quad x \quad + \quad - \quad =$$

The first symbol in an equation is not a 0. Now we can regard the fourteen symbols as the digits in a quattuordecimal number system, i.e. a number system based on 14 in the same way that the decimal system is based on 10. When we do so, each equation becomes an expression for a natural number, indeed a positive integer, in that system. Of course, not all natural numbers when written in the 14-system with the above symbols as the digits will read as algebraic equations. By leaving out the natural numbers which do not (closing up the "gaps"), we get an enumeration of the algebraic equations; that is, the algebraic equations can be listed in the order of magnitude of the positive integers which they become on interpreting their symbols as the digits in a 14-system.

We call the method just used for enumerating the algebraic equations the *method of digits*. We use it now to establish the following general principle.

(A) *If all the members of a set S can be named unambiguously by non-empty finite sequences of (occurrences of) symbols from a given finite list of symbols or alphabet* $s_0, \ldots, s_{p-1}$ *(or even from a countably infinite alphabet* $s_0, s_1, s_2, \ldots$*), then the set S is countable.*

In the case of a finite alphabet $s_0, \ldots, s_{p-1}$, we could proceed as above (where $p = 14$ and $s_0, \ldots, s_{p-1}$ are the fourteen symbols displayed), except for one contingency. After regarding a finite sequence of (occurrences of) $s_0, \ldots, s_{p-1}$ as an expression for a natural number in the p-system with $s_0, \ldots, s_{p-1}$ as the digits, we cannot tell from that number *as a number* how many initial s_0's there were in the sequence. Thus, in the 14-system with the above digits, the same number is expressed by each of: $4x5-17x3+2x2+5=0$, $04x5-17x3+2x2+5=0$, $004x5-17x3+2x2+5=0, \ldots$. This ambiguity did not matter above, since we could and did exclude algebraic equations beginning with a 0. To avoid the ambiguity about the number of initial s_0's when it does matter, we can instead regard the symbols $s_0, \ldots, s_{p-1}$ as the digits for $1, \ldots, p$ in a $p+1$-system having a different symbol for 0.

Notice that it does not matter whether the members of S have only one name each or several names using the alphabet $s_0, \ldots, s_{p-1}$. If members of S may have several names, then when we eliminate the numbers not arising from names we can also eliminate all but the smallest number which we get from the several names of any one member of S.

The case of a countably infinite alphabet $s_0, s_1, s_2, \ldots$ can be reduced to the case of a finite alphabet by replacing each of the infinitely many symbols by a suitable combination of finitely many symbols. For example, we can pick two symbols a and b, and then replace $s_0, s_1, s_2, s_3, \ldots$ by a, ab, abb, abbb, ... Thereby e.g. the name $s_0s_3s_1s_1$ becomes aabbbabab, from which without ambiguity we can recover $s_0s_3s_1s_1$. The method of digits can be applied to the new two-letter alphabet a, b; thus we can interpret aabbbabab as expressing a number in the ternary system with 0, a, b as the digits for 0, 1, 2. In particular cases, some other reduction of $s_0, s_1, s_2, \ldots$ to a finite alphabet may be more convenient to use.

Conversely: (B) *If a set S is countable, its members can be named unambiguously by nonempty finite sequences of (occurrences of) symbols from a given finite alphabet.* For, if S is infinite and $a_0, a_1, a_2, \ldots$ is a particular enumeration of S, then a_0 can be named by 0, a_1 by 1, a_2 by 2, ..., using the 10 symbols 0, 1, ..., 9. Briefly, each member a_i of S can be named by (the numeral for) its *index* i in a given enumeration $a_0, a_1, a_2, \ldots$ of S. Similarly, if S is finite.

Together, (A) and (B) should make it virtually obvious what sets are countable (i.e. finite or countably infinite). Other devices than the method of digits are often convenient for enumerating sets.

Exercises. 32.1. Show that, in the method of digits for a finite alphabet $s_0, \ldots, s_{p-1}$, the names of the members of S are simply enumerated by taking all the one-letter names first, then all the two-letter names, then all the three-letter names, ..., using alphabetic order within each group.

32.2. Use the method of digits to show the rational numbers countable. Give the first six rational numbers in the enumeration you obtain.

32.3. Show that the following sets are countable (use both the idea employed in the text for the rationals, and the method of digits):[124]

(a) The ordered triples (a, b, c) of natural numbers, or of the members of any given countably infinite set.

(b) The finite sequences of such members.

(c) The finite sets of such members.

(d) The finite sequences of finite sequences of such members.

[124] (1, 2, 3), (2, 1, 3), (2,1,1,3) are different finite sequences; but the sets with members shown in curly brackets {1, 2, 3}, {2, 1, 3}, {2, 1, 1, 3} are the same.

32.4. Criticize the following argument: "Every real number x can be named unambiguously by an integer plus an infinite decimal fraction $X + .x_1x_2x_3 \ldots$ (e.g. $\pi = 3 + .14159\ldots$, $-\frac{3}{4} = -1 + .24999\ldots$). Only finitely many symbols $0, \ldots, 9, ., +, -$ are used in the names. Therefore by the method of digits (or (A)), the set of the real numbers is countable."

§ 33. Cantor's diagonal method. The results in § 32 on the application of the notion of 1–1 correspondence to infinite sets would perhaps have found their place in mathematical history as an interesting curiosity, not noticed before (or noticed earlier and forgotten) but leading nowhere especially, if it had turned out that each two infinite sets can be put into 1–1 correspondence. The fruitfulness of the idea of comparing infinite sets under 1–1 correspondences appears in that this is not the case. For, as we show next, there are *uncountable* or *nondenumerable* sets, i.e. infinite sets which cannot be put into 1–1 correspondence with the natural numbers.

We begin with the set of the ***one-place number-theoretic functions.*** These are the functions of one variable a ranging over the natural numbers with values also natural numbers. Examples are a^2, $3a+1$, 5 (a constant function), a (the identity function), $[\sqrt{a}]$ (the greatest integer $\leq \sqrt{a}$), etc.[125]

To prove the uncountability of the set of (all) these functions, suppose given an enumeration $f_0(a), f_1(a), f_2(a), \ldots$ of some one-place number-theoretic functions (not necessarily all). We shall thence construct a one-place number-theoretic function $f(a)$ different from each function in the given enumeration. This will prove that the given enumeration cannot be an enumeration of all the one-place number-theoretic functions. To help us in visualizing the construction of $f(a)$, let us tabulate the sequences of values of the functions $f_0(a), f_1(a), f_2(a), \ldots$ as the rows of an infinite matrix:

	a	0	1	2	$\ldots$
	$f_0(a)$	$f_0(0)$	$f_0(1)$	$f_0(2)$	$\ldots$
(1)	$f_1(a)$	$f_1(0)$	$f_1(1)$	$f_1(2)$	$\ldots$
	$f_2(a)$	$f_2(0)$	$f_2(1)$	$f_2(2)$	$\ldots$
	$\ldots$		$\ldots$		

[125] We have formulas (i.e. finite names) for these five functions. But also functions are to be included whose successive values are determined by laws not corresponding to ordinarily used names, and even functions whose successive values are picked "irregularly" or in a "random" manner.

We define $f(a)$ to be the function whose sequence of values is obtained by taking the successive values along the diagonal (indicated by the arrows) and changing each of them, say by adding 1 to it; thus,

$$f(a) = f_a(a) + 1.$$

This function does not occur in the given enumeration $f_0(a)$, $f_1(a)$, $f_2(a)$, ... For, it differs from $f_0(a)$ in the value taken for 0, from $f_1(a)$ in the value taken for 1, and so on.

(Thus, if $f_0(a) = a^2, f_1(a) = 3a+1, f_2(a) = 5, f_3(a) = a, f_4(a) = [\sqrt{a}], \ldots,$
then $\quad f_0(0) = 0,\ f_1(1) = 4, \quad f_2(2) = 5, f_3(3) = 3, f_4(4) = 2, \ldots$
but $\quad f(0) = 1,\ f(1) = 5, \quad f(2) = 6, f(3) = 4, f(4) = 3, \ldots$.)

To phrase the argument differently, suppose that the function $f(a)$ were in the enumeration $f_0(a)$, $f_1(a)$, $f_2(a)$, ...; i.e. suppose that for some natural number p,

$$f(a) = f_p(a)$$

for every natural number a. Substituting the number p for the variable a in this and the preceding equation,

$$f(p) = f_p(p) = f_p(p) + 1.$$

This is impossible, since the natural number $f_p(p)$ does not equal itself increased by 1.

The method we have just used is called *Cantor's diagonal method*. By restricting the set of the functions to which we apply it, we next obtain some other examples of uncountable sets.

We can take just those one-place number-theoretic functions, each of which has only 0, 1, 2, 3, 4, 5, 6, 7, 8, 9 as values and which does not have all its values 0 from some value on. Then the rows of the matrix (1) can be interpreted as the nonterminating decimal fractions for ***real numbers*** x ***in the interval*** $0 < x \leq 1$. Each such real number has just one such decimal expansion; e.g. $\frac{3}{4} = .74999\ldots$ (we don't use $.75000\ldots = .75$, which is "terminating"), $1/\sqrt{2} = .20711\ldots$, $1 = .99999\ldots$, $\pi - 3 = .14159\ldots$, $2/3 = .66666\ldots$. The alteration performed along the diagonal must not take us out of the class of functions considered. We can, say, change each value $\neq 5$ to 5 and 5 to 6; thus

$$f(a) = \begin{cases} 5 & \text{if } f_a(a) \neq 5, \\ 6 & \text{if } f_a(a) = 5. \end{cases}$$

(If $x_0 = \frac{3}{4}$, $x_1 = 1/\sqrt{2}$, $x_2 = 1$, $x_3 = \pi - 3$, $x_4 = \frac{2}{3}$, ..., we obtain a real number x whose decimal expansion (necessarily nonterminating) starts out .55565....) This application of the diagonal method establishes that the set of the real numbers x in the interval $0 < x \leq 1$ is uncountable.

It follows almost at once that the set of all the ***real numbers*** is uncountable (Exercise 33.1 (a)).

It is interesting historically to note how Cantor's discoveries in 1874 illuminated an earlier discovery of Liouville in 1844. Liouville constructed by a somewhat complicated special method certain *transcendental* (i.e. nonalgebraic) real numbers. Cantor's diagonal method makes the existence of transcendental numbers apparent from only the very general considerations presented above. In fact, from any given enumeration of the algebraic numbers, particular transcendentals can be obtained by the diagonal method.

Finally, let us apply the diagonal method to the set of the one-place number-theoretic functions taking only 0 and 1 as values. Now we have no choice about the alteration performed along the diagonal. We must interchange 0 and 1; thus the "diagonal function" is defined by

$$f(a) = \begin{cases} 1 & \text{if } f_a(a) = 0, \\ 0 & \text{if } f_a(a) = 1. \end{cases}$$

In this application, each function can be interpreted as describing the set of natural numbers at which the function value is 0; we call these functions the *representing functions* of these sets. We show some examples of sets of natural numbers at the left below, opposite the sequences of values of their representing functions.

	0	1	2	3	4	...
$S_0 =$ all numbers $\{0, 1, 2, 3, 4, \ldots\}$	0 ↘	0	0	0	0	...
$S_1 =$ the even numbers $\{0, 2, 4, \ldots\}$	0	1 ↘	0	1	0	...
$S_2 =$ the squares $\{0, 1, 4, \ldots\}$	0	0	1 ↘	1	0	...
$S_3 =$ the primes $\{2, 3, \ldots\}$	1	1	0	0 ↘	1	...
$S_4 =$ the empty set $\{\ \}$	1	1	1	1	1	...
...			...			

For $S_0, S_1, S_2, S_3, S_4, \ldots$ as shown, the diagonal method gives the set $S = \{1, 2, 4, \ldots\}$ whose representing function starts out with the values $1, 0, 0, 1, 0, \ldots$. Thus 0 is not a member of S, though it is of S_0; 1 is a member of S, though it is not a member of S_1; etc. So S is not any of S_0, S_1, ... This application of the diagonal method shows that the set of all the ***sets of natural numbers*** is uncountable (in contrast to Exercise 32.3 (c)).

The sets shown in this section (including the exercise) to be uncountable can be put into 1–1 correspondence with one another (they are "equivalent").[126] Closer examination of the diagonal method in § 34 will show that it gives us infinite sets which are neither countable nor "equivalent" to any of these uncountable sets.

EXERCISE 33.1. Show the following sets to be uncountable.

(a) The real numbers.

(b) The transcendental numbers.

(c) The one-place logical functions (§ 17) when $D = \{0, 1, 2, \ldots\}$.

§ 34. Abstract sets. Starting from these discoveries, Cantor developed a theory of *abstract sets*, in which he gave them a general setting and attempted to discuss sets of the most general sort.[127] We can give here only a brief indication of the theory of abstracts sets.[128]

Cantor (1895 p. 481) defined a *set* thus, "By a 'set' we understand any collection M of definite well-distinguished objects m of our perception or our thought (which are called the 'elements' of M) into a whole." We write "$m \in M$" to say that m is an *element* of M, or synonymously that m is a *member* of M or *belongs* to M; and "$m \notin M$" to say that m does not belong to M.[129] Two sets M_1 and M_2 are the same ($M_1 = M_2$) if they have the same members. A finite set can be described by listing its members (the order being immaterial) in curly brackets; using dots, we can even suggest thus an infinite set. For example, $\{1, 2, 3\}$ is a set with three elements, and $\{0, 1, 2, \ldots\}$ is the set of the natural numbers. We say a set M_1 is a *subset* of a set M, and write $M_1 \subseteq M$ (or $M \supseteq M_1$), if each member of M_1 is a member of M. For example, the three-element set $\{1, 2, 3\}$ has eight subsets

$$\{\ \}, \{1\}, \{2\}, \{3\}, \{1, 2\}, \{1, 3\}, \{2, 3\}, \{1, 2, 3\}.$$

The first $\{\ \}$ is the *empty* (or *vacuous*) set (often written $\varnothing$), the next three are *unit* (i.e. one-element) sets, and the last is the *improper subset* of $\{1, 2, 3\}$.

[126] A proof (to follow § 34) is given in IM pp. 16, 17.

[127] Cantor also developed a theory of *point sets*, much of which has now become familiar in the form of point set topology.

[128] Cantor 1895–7 is quite readable, and has been translated into English (cf. our bibliography). Fraenkel 1961 gives an excellent and very full account. A concise treatment is in Bachmann 1955.

[129] We are repeating a little from § 26 above to make the present chapter largely self-contained.[89,91,92]

The cardinal number $\overline{\overline{M}}$ is a notion which we "abstract" from M and the other sets which can be put into 1–1 correspondence with M. Thus a child gets his concept of "two" by abstracting from two parents, two ears, two apples, two kittens, etc. What "two" itself is may not worry him. Cantor wrote, "The general concept which with the aid of our active intelligence results from a set M, when we abstract from the nature of its various elements and from the order of their being given, we call the 'power' or 'cardinal number' of M." This double abstraction suggests his notation "$\overline{\overline{M}}$" for the cardinal of M. Frege 1884 and Russell 1902 defined $\overline{\overline{M}}$ as the set of all the sets N which can be put into 1–1 correspondence with M; this gives a place to the cardinals themselves as objects of a universe whose only members are sets. To put this in terms that were elaborated in § 30, let us say M is *equivalent* to N, and write $M \sim N$, if M can be put into 1–1 correspondence with N. Then $\sim$ is an equivalence relation (i.e. it is symmetric, reflexive and transitive); and $\overline{\overline{M}}$ is the equivalence class to which M belongs under $\sim$ within a universe of sets (cf. (B) in § 30).

Whatever one's ontology of the cardinals, $\overline{\overline{M}} = \overline{\overline{N}}$ if and only if $M \sim N$.

For any set M, we write "2^M" as a notation for the set of all the subsets of M.

In our present notation, our last result in § 33 is that 2^M is uncountable, or $\overline{\overline{2^M}} \neq \overline{\overline{M}}$, when $M = \{0, 1, 2, \ldots\}$. The reasoning applies similarly to any set M to show that $\overline{\overline{M}} \neq \overline{\overline{2^M}}$. Let us repeat the reasoning for the case $M = \{1, 2, 3\}$. Given any set $M_1 \subseteq 2^M$ which is in 1–1 correspondence with M itself, the diagonal method leads us to a subset of M (i.e. a member of 2^M) not belonging to M_1. Thus, if $M_1 = \{\{2\}, \{2, 3\}, \{1, 2\}\}$, the matrix is as follows:

	1	2	3
{2}	1	0	1
{2, 3}	1	0	0
{1, 2}	0	0	1

Interchanging 0 and 1 on the diagonal, we are led to $\{1, 3\}$, which is a member of 2^M (2^M has the eight members displayed above) but not of M_1. If instead $M_1 = \{\{1, 2, 3\}, \{1\}, \{3\}\}$, we get $\{2\}$, which again is not in M_1; etc. Of course, since 2^M has eight members, i.e. $\overline{\overline{2^M}} = 8$, while $\overline{\overline{M}} = 3$, we can say we already knew that $\overline{\overline{M}} \neq \overline{\overline{2^M}}$ for $M = \{1, 2, 3\}$, as part of what we are taking for granted about the natural number sequence. However, this and the concluding example in § 33 with $M = \{0, 1, 2, \ldots\}$

are two illustrations of the general proof that, *for each set* M, $\overline{\overline{M}} \neq \overline{\overline{2^M}}$.

This result can be given a stronger form. First, we define $\overline{\overline{M}} < \overline{\overline{N}}$ (or $\overline{\overline{N}} > \overline{\overline{M}}$) to hold exactly if M is equivalent to some subset of N but N is equivalent to no subset of M (i.e. if there is a set N_1 such that $M \sim N_1 \subseteq N$ but there is no set M_1 such that $N \sim M_1 \subseteq M$). Here it is necessary to verify that the result is independent of which sets M and N with the respective cardinals $\overline{\overline{M}}$ and $\overline{\overline{N}}$ we use, or in the terminology of § 30 of which members M and N we pick from the two equivalence classes $\overline{\overline{M}}$ and $\overline{\overline{N}}$ (Exercise 34.1). We see almost at once that the order relation $<$ between cardinals is irreflexive (i.e. $\overline{\overline{M}} \not< \overline{\overline{M}}$) and transitive (i.e. if $\overline{\overline{M}} < \overline{\overline{N}}$ and $\overline{\overline{N}} < \overline{\overline{P}}$, then $\overline{\overline{M}} < \overline{\overline{P}}$) (Exercise 34.2).

As we already remarked in § 32 assuming familiarity with the natural number sequence, the natural number n is the cardinal of the initial segment $0, \ldots, n-1$ of the natural numbers. Consequently, Cantor's definition of the order relation $<$ between any two cardinals applies in particular to the natural numbers as cardinals. It can be shown (though not without some trouble) that *this* order relation $<$ between the natural numbers as finite cardinals coincides with the familiar one which we have been taking for granted.[130]

By only slight refinements of the argument given above, we can establish Cantor's theorem: (C) *For each set* M, $\overline{\overline{M}} < \overline{\overline{2^M}}$.

Another easily proved theorem concerns the *union* $\cup M$ of a set M whose members are sets. $\cup M$ has as its members each of the objects which is a member of a member of M. For example, if $M = \{\{2\}, \{2, 3\}, \{2, 5\}\}$, then $\cup M = \{2, 3, 5\}$. The theorem is: (D) *If M is a set of sets containing none of greatest cardinal* (i.e. to each member A of M there is a member A' of M with $\overline{\overline{A'}} > \overline{\overline{A}}$), *then* $\overline{\overline{A}} < \overline{\overline{\cup M}}$ *for each member A of M.*

We adopt the notation $\aleph_0$ (read "aleph null" or "aleph zero") for the cardinal of the set of natural numbers. Then for each natural number n, $n < \aleph_0$. This follows from (D) with $n = \overline{\overline{\{0, 1, \ldots, n-1\}}}$ and the fact that the cardinal order of the natural numbers is also their usual order.

Also we write $\overline{\overline{2^M}}$ as "$2^{\overline{\overline{M}}}$". For M finite, this is consistent with the usual arithmetic; e.g. we saw that $\overline{\overline{2^M}} = 8 = 2^3$ when $\overline{\overline{M}} = 3$.

The foregoing results give the existence of the following ascending series of cardinals

$$0 < 1 < 2 < \ldots < \aleph_0 < 2^{\aleph_0} < 2^{2^{\aleph_0}} < \ldots,$$

[130] A treatment is given in IM pp. 12, 13 with Example 1 § 7 p. 22.

while by (D) there is a still greater cardinal after all of these, and so on indefinitely. Thus, by applying the idea of comparing sets by one-to-one correspondences, Cantor discovered that there is not simple one infinity, but a whole hierarchy of different infinite (or "transfinite") cardinals.

EXERCISES. 34.1. Justify the definition of "$\overline{\overline{M}} < \overline{\overline{N}}$" by showing that, if $M \sim M'$ and $N \sim N'$, then (α) holds if and only if (α') holds:

(α) For some N_1, $M \sim N_1 \subseteq N$; but for no M_1, $N \sim M_1 \subseteq M$.

(α') For some N_1, $M' \sim N_1 \subseteq N'$; but for no M_1, $N' \sim M_1 \subseteq M'$.

34.2. Show that the relation $\overline{\overline{M}} < \overline{\overline{N}}$ is irreflexive and transitive.

34.3*. Prove (C) and (D).

34.4. (a) What is the cardinal of the set in Exercise 33.1 (c)? (b) What is the cardinal of the set of the one-place logical functions when the domain is the sets of natural numbers?

§ 35. The paradoxes. The relation between Cantor's set theory and mathematics was like the course of true love; it never did run smooth.

Cantor's set theory deals with "actual" or "completed" infinities. At the beginning there was great resistance to this by the mathematical public, stemming in part from the famous dictum of Gauss (1831): "I protest . . . against the use of an infinite magnitude as something completed, which is never permissible in mathematics: one has in mind limits which certain ratios approach as closely as desirable while other ratios may increase indefinitely." (Werke VIII p. 216.) Gauss had in mind infinite magnitudes, while Cantor's theory employs infinite collections.

Just when Cantor's ideas were well on the way toward winning acceptance from most mathematicians, in the 1890's contradictions appeared in the upper reaches of his set theory. Nevertheless, since then set theory (suitably adapted) has increased its place in mathematics, while the paradoxes have focussed attention on the foundations of set theory and of mathematics generally.

The Burali-Forti paradox which appeared in 1897 (but was known to Cantor in 1895) arises in Cantor's theory of ordinal numbers, which we have not discussed.[183]

Russell's paradox 1902a concerns the set of all sets which do not contain themselves. Call this set S. *Suppose* **(a)** S contains itself. Then by the definition of S, S does not contain itself. So by reductio ad absurdum (rejecting the supposition (a)), we have *proved*: **(b)** S does not contain itself. But then by the definition of S: **(c)** S does contain itself. Together (b) and (c) constitute a proved contradiction, or paradox.

Russell in 1919 gave the following popularized version: The barber in a certain village shaves all and only those persons in the village who do not shave themselves. Question: Does he shave himself?

We now give in detail a third paradox of the theory of sets, the Cantor paradox (found by him in 1899). Let T be the set of all sets. Now 2^T is a set of sets, and hence $2^T \subseteq T$. By the definition of $<$ for cardinals (§ 34), if $M \subseteq N$ then $\bar{\bar{M}} \not> \bar{\bar{N}}$. (Why?) So $\overline{\overline{2^T}} \not> \bar{\bar{T}}$. But by Cantor's theorem (C), $\overline{\overline{2^T}} > \bar{\bar{T}}$. Thus we have a contradiction.

One may try to dismiss this by saying that the set T of all sets does not constitute a set. Then in Cantor's theorem "For each set M, $\bar{\bar{M}} < \overline{\overline{2^M}}$", what is the range of the variable M (what we called the domain D in Chapter II)?

Of a somewhat different type is the Richard paradox (Jules Richard, 1905), which runs as follows.

By a "phrase" we shall mean any finite sequence each of whose members is either a blank or one of the twenty-six letters of our alphabet (with a blank not first or last). Thus, "abracadabra", "of cabbages and kings", "the square of a", and "xtu rlbp" are phrases. We can enumerate the phrases by the method of digits (§ 32), using a 27-digit or 28-digit number system to represent the natural numbers. Certain phrases, such as our example "the square of a", describe in the English language one-place number-theoretic functions. We now strike out from our enumeration of all the phrases those which do not describe such functions; thereby we obtain an enumeration $P_0, P_1, P_2, \ldots$ of the phrases which do. Say the functions described are $f_0(a), f_1(a), f_2(a), \ldots$, respectively.

Now consider the following phrase: "the function whose value for each natural number a is obtained by adding one to the value for a of the function described by the phrase corresponding to a in our enumeration of the phrases which describe one place numbertheoretic functions". In this phrase, we could replace the last part "in our enumeration of the phrases which describe one place numbertheoretic functions" by a detailed description of the exact construction of the enumeration, and so obtain from the whole another phrase P *fully* describing the same function.

This phrase P describes a number-theoretic function, namely

$$f(a) = f_a(a) + 1.$$

Hence P occurs in the enumeration $P_0, P_1, P_2, \ldots$. This is impossible, since the function described by P differs from that described by P_0 in its value for $a = 0$, from that described by P_1 in its value for $a = 1$, from that described by P_2 in its value for $a = 2$, and so on. Otherwise expressed: Since the phrase P occurs in the enumeration $P_0, P_1, P_2, \ldots$, it is P_p for some p. Then

$$f(a) = f_p(a).$$

A contradiction arises by substituting p for a in this and the preceding equation. (In Richard's original version, real numbers were used instead of one-place number-theoretic functions.)

This paradox is closely connected with the facts that, on the one hand, only a countable infinity of number-theoretic functions are describable in a given language (because the set of all the phrases in the language is only countably infinite, § 32), while, on the other hand, the set of all the number-theoretic functions is uncountable (by Cantor's diagonal method, § 33).

A similar paradox is due to G. G. Berry (in Russell 1906 p. 645). Consider the expression "the least natural number not nameable in fewer than twenty-two syllables". This expression names a definite natural number, say n, since each nonempty set of natural numbers (in this case, the set of natural numbers not nameable in fewer than twenty-two syllables) has a least element. By its definition, n is not nameable in fewer than twenty-two syllables. But our expression naming n has in fact exactly twenty-one syllables!

These modern paradoxes are related to the paradox of "The Liar", which comes from antiquity.[131] The following statement is attributed to the Cretan philosopher Epimenides, sixth century B.C. "All Cretans are liars". Let us suppose that by "liar" Epimenides meant a person who *never* tells the truth.

Suppose his statement is true; then by what it says and by his being a Cretan, it is false, which is a contradiction. So by reductio ad absurdum, the statement is not true, i.e. it is false. This means that at some time some Cretan has told the truth or eventually some Cretan will tell the truth. This, however, should be a matter for the historian to decide; it should not be demonstrable on logical grounds only, as we appear to have demonstrated it.

The direct form of the paradox of "The Liar" was given by Eubulides in the fourth century B.C. We can give it as follows: "The statement I am now making is a lie." One sees directly that the quoted statement cannot be true and that it cannot be false.

In the ancient "dilemma of the crocodile", a crocodile has stolen a child, but offers to return the child to its father, if the father can guess whether or not the crocodile will return the child. If the father guesses that the crocodile will not return the child, the crocodile is in a dilemma.

A missionary, fallen among cannibals, discovers that he is about to become their supper. They offer him the opportunity to make a statement, under the conditions that, if the statement is true, he will be boiled, and if it is false, he will be roasted. What should the missionary say?

[131] For some historical details and references, see Weyl 1949 p. 228 Footnote 2.

Cantor's set theory, as historically it first arose and as we met it in §§ 32–34, is called "naive" set theory. In using Cantor's "definition" of a set (§ 34), Cantor and we were guided only by our imagination in deciding which objects are sets.

Cantor's and the other paradoxes of set theory show the difficulties inherent in the attempts to develop the theory on an intuitive basis starting from Cantor's definition of set. These difficulties pose the problem how to modify set theory so that contradictions do not arise. In fact, the problem goes further, and forces us to ask ourselves wherein we were deceived by methods of constructing and reasoning about objects which had seemed convincing before they were found to eventuate in paradoxes. Complete agreement among mathematicians on the cause of the paradoxes and the cure has not been attained yet (1967), and it seems problematical that it ever will.

In the remainder of this section, we indicate briefly the least radical kind of reformulation of mathematics toward avoiding such paradoxes as the Burali-Forti, Cantor's and Russell's. (Ramsey 1926 classified the known paradoxes into two sorts, one now called "logical" including the three just mentioned, the other "epistemological" or "semantical" including Richard's, Berry's and "The Liar".)

This reformulation of mathematics begins with the observation that the paradoxes of set theory (the logical paradoxes) are associated with using "too large" sets, such as the set T of all sets in Cantor's paradox. Since free use of our conceptions starting with Cantor's definition of set led to the difficulty, Zermelo proposed in 1908 to restrict the sets to those provided by a list of axioms. These axioms are drawn up so that there is no apparent means to derive the known paradoxes from them. On the other hand, the axioms do suffice for the deduction of the usual body of classical mathematics, including abstract set theory short of the paradoxes.

We give now, in our words, the list of seven axioms or principles which appear in Fraenkel 1961 with the pages where they appear.[132] (It is not intended that the student, so far as this course is concerned, should learn these axioms. We give them here to exemplify what sort of axioms are used.) Choosing this particular list of axioms has the advantage that the excellent expositions in Fraenkel 1961 and in Fraenkel and Bar-Hillel 1958 are built around them. Another axiomatic treatment is in Bernays and Fraenkel 1958.

(I) (Axioms of extensionality, p. 14.) Two sets A and B are equal if (and only if) they contain the same members; i.e. $A = B \sim (A \subseteq B$ and $B \subseteq A)$.

(II) (Axiom of subsets, p. 16.) Given a set A and a predicate $P(x)$ meaningful for the members of A (i.e., for each $x \in A$, either $P(x)$ is true

[132] The pages in the 1953 edition are 21, 22, 24, 28, 42, 97, 123.

or $P(x)$ is false), there exists the set $\hat{x}[x \in A \,\&\, P(x)]$ containing exactly those members of A for which $P(x)$ is true. (This axiom is also called the "axiom of selection", the "axiom of segregation" or the "Aussonderungs-axiom".)

(III) (Axiom of pairing, p. 18.) If a and b are different objects, there exists the set $\{a, b\}$ containing exactly a and b.

(IV) (Axiom of union, p. 20.) Given a set S of sets, there exists the set $\cup S$ containing just the members of the members of S.

(V) (Axiom of infinity, p. 32.) There exists at least one infinite set: the set $\{0, 1, 2, \ldots\}$ of the natural numbers. (Fraenkel used $\{1, 2, 3, \ldots\}$.)

(VI) (Axiom of the power set, p. 72.) Given a set A, there exists the set 2^A whose members are all the subsets of A.

(VII) (Axiom of choice, p. 90.) Given a disjointed set S of nonempty sets, there exists a set C which has as its members one and only one element from each member of S. (S is disjointed if no two distinct members of S have an element is common.)

A form of the axiom of choice was first explicitly noted as an assumption in Zermelo's proof of his "well-ordering theorem" 1904, 1908a, from which it follows that each two cardinal numbers $\overline{\overline{A}}$ and $\overline{\overline{B}}$ are comparable (i.e. that either $\overline{\overline{A}} < \overline{\overline{B}}$ or $\overline{\overline{A}} = \overline{\overline{B}}$ or $\overline{\overline{A}} > \overline{\overline{B}}$). The form given here, which is Russell's "multiplicative axiom" 1906a, suffices for the derivation of Zermelo's form (and vice versa).

The axiom of choice has been the subject of much research with a view to minimizing its use, singling out its consequences, or (Gödel 1938, 1939, 1940) defending it as an assumption which can be added to the other axioms of set theory without entailing a contradiction if the other axioms by themselves lead to no contradiction.[133] In 1963–4 Cohen showed that instead the negation of the axiom of choice can be added consistently (detailed exposition in 1966). Another treatment is in Scott 1966.

At one point these axioms (as did the axioms given by Zermelo in 1908a) lack definiteness. This is in (II), where the notion of a predicate $P(x)$ meaningful for elements $x \in A$ is incorporated. This lack of definiteness was first remedied by Fraenkel 1922 and somewhat differently by Skolem 1922–3. What is required is a specification of a class of admissible predicates $P(x)$. In Skolem's method, the rules for constructing the $P(x)$'s are formulated simply in the process of specifying the symbolism of the language in which the axioms are stated.

The specification of the symbolism of a language, which is necessary for the purpose of being exact in logical deduction, must really be presupposed for the rigorous treatment of logic as in Chapters I–III above. The semantical paradoxes (Richard's, Berry's, "The Liar") show that care is necessary in this. That is, the language of a mathematical theory must be subjected

[133] Gödel 1947 provides a very readable exposition.

to rules governing the formation of propositions somewhat akin to the rules listed above governing the existence of sets. We shall deal further with this.

In some systems of axiomatic set theory (particularly Gödel's, 1940) two sorts of collections are considered explicitly, collections called "sets" which can not only possess members but also be members of collections, and others called "classes" which may not be taken as members of collections. Each "set" is a "class". The collection of all "sets" constitutes a "class"; but we are stopped from obtaining the Cantor paradox because this "class" is not a "set".

In the definition of validity in the predicate calculus (§ 17), we simply said the domain D is to be a "nonempty set or collection". That was before we were ready to talk about a distinction between "sets" and "classes" in the present sense. We can get a little more leeway now for model theory by allowing D to be a nonempty "class". For in the notion of validity, we do not need to take D as a member of anything. The difficulties which ensue when "too large" collections are treated as members are not involved in just using a collection as the range of variables. This gives an answer to the question raised above of what the domain D can be for set theory itself (where we asked about the range of M in Cantor's theorem, $\bar{\bar{M}} < \overline{\overline{2^M}}$). (Except when we refer explicitly to this passage, "class" will continue in this book to be a synonym for "set" as in ordinary usage.)

§ 36. Axiomatic thinking vs. intuitive thinking in mathematics. Partly in connection with the broader aspects of the problem posed by the paradoxes (§ 35), we inquire now into the nature of mathematics and the scope of mathematical methods.

The axiomatic-deductive method in mathematics is known to us from Euclid's "Elements" (c. 330–320 B.C.), although there is a tradition that credits Pythagoras (sixth century B.C.) with the introduction of the method. By use of it, the body of geometrical knowledge was systematized. Euclid's axiomatic system may be described roughly thus: "definitions" of certain *primitive terms*, such as "point", "line", "plane" are given, which are intended to suggest to the reader what is meant by those terms; certain propositions concerning the primitive terms, felt to be acceptable as immediately true on the basis of the meanings suggested by the definitions, are taken as *axioms* or *postulates*; then other terms are defined in terms of the primitive ones; and other propositions, called *theorems*, are deduced by logic from the axioms. Axiomatics such as Euclid's, in which meanings are given to the primitive terms from the outset, is called *material axiomatics*.

One of Euclid's postulates seemed less evident than the rest, the fifth postulate or "parallel postulate".[134] He used this in proving the theorem that, through a given point *P* not on a given line *l*, exactly one line can be drawn parallel to *l* (i.e. not meeting *l* in a point). Efforts were made from Euclid's time on to prove this postulate from the others as a theorem. We now know these efforts could not succeed.

For, Lobatchevski in 1829 and Bolyai in 1833 worked out a system of geometry in which, through a given point *P* not on a given line *l*, infinitely many lines parallel to *l* can be drawn. It is apparent that the meanings of Euclid's primitive terms in terms of physical space do not enable one to decide whether Euclid's parallel postulate is true or the contrary postulate of Lobatchevsky and Bolyai. The differences in the resulting geometries may be too small to show up in any measurements we can make in the portion of space accessible to us, just as in some other times people have thought the earth flat from the portion of it they could see.

So whether a proposition of Euclidean geometry is exactly true must be a property of the geometry as a logical system. But if Euclidean geometry is a valid logical structure, so is the Lobatchevskian geometry. For, as Felix Klein pointed out in 1871, the axioms of the plane Lobatchevskian geometry are all true when the primitive terms in them are reinterpreted so that "plane" is taken to mean the interior of a given circle in the Euclidean plane, "point" means a point inside this circle, "line" means a chord of this circle, and distances and angles are computed by formulas due to Cayley 1859. (Another such Euclidean model, applicable to a bounded portion of the non-Euclidean plane, was given in 1868 by Beltrami, who reinterpreted line segments as segments of shortest paths between points, or "geodesics", on a "surface of constant negative curvature".)

In these models, we may observe that something new has been done with the axioms, not to be found in the earlier axiomatic thinking: the meanings of the primitive terms have been varied, holding the deductive structure of the theory fixed. Thus *formal axiomatics* arose, in which the meanings of the primitive terms, instead of being specified in advance, are left unspecified for the deductions of the theorems from the axioms. One is then free to choose the meanings of the primitive terms in any way that makes the axioms true. We have been representing this standpoint in our definition of "valid consequence" in model theory (§§ 7, 20). Especially in modern algebra, it has proved very fruitful to develop the consequences of systems of axioms regarded formally, such as the axioms of abstract

[134] It reads: If two straight lines in a plane meet another straight line in the plane so that the sum of the interior angles on the same side of the latter straight line is less than two right angles, then the two straight lines will meet on that side of the latter straight line.

group theory (cf. § 39). The results deduced from the axioms of group theory, while leaving unspecified the set of elements and the multiplication operation, constitute a body of theory ready-made for diverse applications.

In formal axiomatics, the system of axioms may be investigated for such properties as the independence of one axiom from the others (by seeking an interpretation of the primitive terms which makes that axiom false and the others true), categoricity (i.e. that the elements in any two interpretations can be put into 1–1 correspondence preserving all properties), etc.[135]

In this approach to axiomatics some questions arise. Why do we choose the axioms we do, and why should the resulting systems interest us? The answer is evidently that we may apply the resulting theory to systems of objects provided from outside the axioms by an interpretation of the primitive terms. Sometimes essentially different interpretations are possible (the axioms are then not categorical, but *ambiguous*); e.g. this is the case for the axioms for abstract groups. We should not wish to employ a system of axioms satisfied under *no* interpretation; such a system we call *vacuous*. One of the problems in formal axiomatics is to show axiom systems non-vacuous. However, a system of objects used as an interpretation is often drawn from some other axiomatic theory; then we have a regress, which merely brings us to the question of the significance of that axiomatic theory instead. If at no stage is an application made outside of formal axiomatics, the whole activity must appear to be futile. We therefore conclude that, if we are not to adopt a mathematical nihilism, formally axiomatized mathematics must not be the whole of mathematics. At some place there must be meaning, truth and falsity. At the very least, when we say that in a given formal axiomatic theory a certain proposition is a theorem, we must believe this is true, i.e. that the proposition does follow from the axioms, though whether the proposition itself is true is being left out of account, since in formal axiomatics the deductions are carried out in advance of assigning meanings to the primitive terms (or disregarding any such assignment).

As a further illustration of a mathematical proposition which is not intended to be asserted merely as a formal but meaningless consequence of axioms, consider the theorem (proved in number theory) that, for given integers a, b, c, we can find out whether or not integers x and y exist such that $ax + by + c = 0$, i.e. the theorem that there is a method of deciding whether or not the equation $ax + by + c = 0$ (a, b, c integers) is solvable in integers. Although the theory of the integers may have been established

[135] This kind of treatment of axiomatic systems is well presented in J. W. Young 1911. Several of the topics mentioned briefly now will be elaborated below: independence proofs in § 57; categoricity in § 53; consistency proofs by interpretation in § 52.

axiomatically, this proof is intended to mean that, for any particular *a*, *b*, *c*, we *can discover* whether or not there are solutions. A student who could merely give a proof from axioms of the theorem that one can find out whether there are solutions or not, but could not do a problem in which he found out, would not have acquired what the teacher intended to teach. Nevertheless, he would be doing all that should be asked of him if the theorem (that one can find out) were intended only in the sense of formal axiomatics.

As the least drastic method of meeting the situation posed by the paradoxes, we described axiomatic set theory at the end of § 35. Here the axiomatics is to be understood in the formal sense, unless one is to try to retain an intuitive conception of sets, which it was presumed was exactly what the axioms were intended to supplant. However the present considerations show that the resort to a formal axiomatic theory, though it may offer considerable advantages, leaves open such problems as why the axioms are significant and whether or not they apply to any system of objects not merely similarly postulated as existing for some other formal axiomatic theory.

Hilbert undertook to deal with these problems. He admitted that *classical* mathematics (i.e. the familiar mathematics using classical logic) contains much that goes beyond what is clearly meaningful and justifiable on intuitive grounds, as indeed mathematicians generally were made to realize when in set theory they went too far and encountered paradoxes. But he proposed to save classical mathematics (short of paradoxes) by a program which we can roughly describe as follows. Classical mathematics should be formulated as a formal axiomatic theory, and then the theory should be shown to be consistent, i.e. free from contradiction.

Before this proposal of Hilbert, first made in 1904, but not seriously undertaken by him and his co-workers until after 1920, consistency proofs had been given for formal axiomatic theories by means of a model or interpretation, in which all the axioms are found to be true when the primitive terms are interpreted in terms of another theory. We saw an example of this above, by which the non-Euclidean plane geometry of Lobatchevsky is shown to be consistent if Euclidean geometry is consistent.[135] In each case, a proof of consistency by a model only shows the theory consistent if another is. By René Descartes' method of analytic geometry 1619, the consistency of geometries generally is reduced to that of the theory of real numbers, i.e. to analysis. But how is one to establish the consistency of analysis? Certainly not by using a geometrical model; this would be a vicious circle. Nor, according to Hilbert and Bernays 1934, by appeal to the physical world. For limitations of our measurements in the physical world prevent us from saying that a continuum is actually given by experience; rather it is an idea we obtain by extrapolating or idealizing what is actually given.[136]

So Hilbert's proposal to prove classical mathematics as embodied in a formal system consistent called for a new method in place of the method of giving a model. This method consists in a direct application of the idea of consistency, namely, that there be no contradiction or paradox consisting of two theorems, one of which is the negation of the other. To show that this cannot happen, Hilbert proposed to make the proofs in the axiomatic theory the object of a mathematical investigation, called *metamathematics* or *proof theory*. Of course, such a demonstration of consistency would be relative to the methods used in the metamathematics. Hilbert therefore aimed to use in his metamathematics only methods, which we call "finitary", that are intuitively convincing. Specifically, these methods should avoid using an "actual" or "completed" infinity. Hilbert's new approach avoids the completed infinite in the statement of the problem of proving consistency. For there is only a countable infinity of proofs in a given theory, and the consistency proposition only concerns any pair of proofs, not the set of all proofs as a completed object. (What the theory is supposed to be about may be much less elementary.) So it seemed not unreasonable to hope that the problem of consistency, now that it was stated in finitary terms, might be solved by finitary methods.

In §§ 37–39 we shall take a closer look at how parts of mathematics can be made into formal axiomatic theories and studied in metamathematics.

Brouwer was the champion of intuitive thinking in mathematics, as Hilbert was of axiomatic thinking. Brouwer's and Hilbert's approaches may also be called "genetic" (or "constructive") and "existential", respectively.

According to Weyl 1946, "Brouwer made it clear, as I think beyond any doubt, that there is no evidence supporting the belief in the existential character of the totality of all natural numbers . . .. The sequence of numbers which grows beyond any stage already reached by passing to the next number, is a manifold of possibilities open towards infinity; it remains forever in the status of creation, but is not a closed realm of things existing in themselves."

While Hilbert proposed to shore up the structure of classical mathematics by a consistency proof, Brouwer was ready to abandon those parts of mathematics in which mathematicians had been carried away by words that had outrun clear meanings. Brouwer proposed instead to develop an "intuitionistic" mathematics, which would go only as far as intuition would lead it. For Brouwer, the systems of objects for mathematics should

[136] Hilbert and Bernays 1934 pp. 15–17; quoted on IM pp. 54–55.

be generated by some principles of construction, not brought into existence all at once as sets satisfying a list of axioms.

Since Brouwer took only the "uncompleted" or "potential" infinite as intuitive, he declined to accept logical principles which require for their justification a conception of infinite sets as completed. So, in a paper entitled "The untrustworthiness of the principles of logic" 1908, he challenged the assumption that the laws of classical logic have an absolute validity, independent of the subject matter to which they are applied. Particularly, he criticized the law of the excluded middle, $P \vee \neg P$. Consider a predicate $P(x)$ where x ranges over some set D. Applied to $\exists x P(x)$ as the P, the law says that either there is an x in D such that $P(x)$ or there is no x in D such that $P(x)$; in symbols: $\exists x P(x) \vee \neg\exists x P(x)$. In case D is a finite set (and $P(x)$ is a predicate such that, for each value x in D, we can test whether $P(x)$ holds or not), Brouwer finds $\exists x P(x) \vee \neg\exists x P(x)$ true. For, we can find out whether $\exists x P(x)$ or $\neg\exists x P(x)$ by testing, for each member x of D in turn, whether or not $P(x)$ holds for that x. Since D is finite, this process will (in principle) terminate. But if D is an infinite set, say a countable one, such as the set of the natural numbers, the testing process cannot humanly be completed. If we are lucky, we may part way through the testing find an x such that $P(x)$. But if there is no such x, or if such an x comes too late and doomsday comes too soon, we could search till doomsday and still not have an answer to our question. Accordingly Brouwer finds no ground for taking $\exists x P(x) \vee \neg\exists x P(x)$ to be always true, when D is infinite. Quoting from Weyl 1946, "According to his view and reading of history, classical logic was abstracted from the mathematics of finite sets and their subsets. . . . Forgetful of this limited origin, one afterwards mistook that logic for something above and prior to all mathematics, and finally applied it, without justification, to the mathematics of infinite sets."

Brouwer's path, like Hilbert's as we shall see, was beset with difficulties. An intuitionistic mathematics was developed (beginning in 1918), which in part falls short of classical mathematics in the results obtained, in part takes a different direction. In parts common to classical and intuitionistic mathematics, the intuitionistic (or "constructive") proofs, though often harder, often give more information. To prove an existence statement $\exists x A(x)$, an intuitionist insists that it be shown how to find an x such that $A(x)$. An "indirect proof" by showing that the assumption $\neg\exists x A(x)$ leads to contradiction is not accepted by him as showing that $\exists x A(x)$; it establishes only $\neg\neg\exists x A(x)$.[137]

Now we touch on the controversy between Brouwer and Hilbert.

[137] We gave a little fuller sketch in IM § 13. Excellent introductions are in Heyting 1934, 1955 and 1956. Kleene and Vesley 1965 is intended for readers familiar with IM Chapters I–XII (or as a minimum IV–VIII) or the equivalent.

Brouwer argued that, even if Hilbert should succeed in giving a consistency proof for classical mathematics, that would not make classical mathematics correct. Thus he wrote in 1923, "An incorrect theory which is not stopped by a contradiction is none the less incorrect, just as a criminal policy unchecked by a reprimanding court is none the less criminal." Hilbert retorted in 1928, "To take the law of the excluded middle away from the mathematician would be like denying the astronomer the telescope or the boxer the use of his fists." This controversy between the "formalists", represented by Hilbert, and the "intuitionists", represented by Brouwer, led eventually to the acknowledgement by the intuitionists that Hilbert's program would be unobjectionable if and only if the formalists refrain from taking a consistency proof as justification for attaching a real meaning to those parts of mathematics which the intuitionists reject as having no intuitive basis (Brouwer 1928).

It remains then for the formalist to explain, after having admitted that classical mathematics goes beyond intuitive evidence, how nevertheless its nonintuitionistic parts can be of value. Addressing himself to this problem, Hilbert 1926, 1928 drew a distinction between *real statements* which have an intuitive meaning, and *ideal statements* (involving the completed infinite) which do not. It is a common device of modern mathematics to adjoin "ideal elements" to a previously constituted system in order to achieve theoretical objectives, such as to simplify the theorems, comprehend them under a more unified viewpoint, etc. An example occurs in projective geometry, where a line at infinity is adjoined to the finite part of the plane so that any two (distinct) parallel lines intersect in a point of that line. Thereby the exception for parallel lines to the "incidence relations" between points and lines is removed. Thus, in projective geometry, not only do each two distinct points contain a unique line (which passes through both), but dually each two distinct lines contain a unique point (in which they intersect). Hilbert argued that just this kind of theoretical gain is achieved by adjoining the ideal statements to the real statements in classical mathematics; it is through this procedure that classical mathematics achieves its power and elegance.

In this way, mathematics becomes a theoretical construction in which, Hilbert says, it should not be expected that each separate statement should have a real meaning, any more than that each proposition in a system of theoretical physics should be capable of immediate experimental verification; in the latter case, it is the theory as a whole that is tested against reality.

A concrete example of the theoretical gain obtained by going through ideal statements in the process of proving real statements is provided by *analytic number theory*, in which theorems about integers are proved via the theory of real or complex numbers. Many propositions of elementary number theory have been proved thus, which either we do not know how to prove, or can establish only by much more complicated proofs, if only nonanalytic methods are used.

Closely related to this defense of classical mathematics as a simple and elegant systematizing scheme is the defense provided by its success in applications to the theoretical sciences, especially physics. This led Weyl 1926 to pronounce Hilbert correct when mathematics is merged with physics in the process of theoretical world construction, while he sided with Brouwer in restricting himself to intuitive truths when mathematics is pursued for itself alone.[138]

§ 37. Formal systems, metamathematics. In § 36, we discussed formal axiomatics, stressing that the primitive terms are to be treated as meaningless for the purpose of deductions from the axioms by logic; i.e. either they have been assigned no meanings, or the meanings they have been assigned are to be left out of account. If they had meanings which have to be taken into account, this would amount to saying that the theorems depend not only on the properties of the primitive terms expressed by the axioms, but also on further properties which enter through the use of the meanings. But then those further properties should be stated as additional axioms.

Euclid in his "Elements" failed to state in his "axioms and postulates" all the properties he used. Figures accompany many of his proofs. It had long been taken for granted that the figures are inessential to the proofs, serving only to make it easier to discover them or to follow them or to remember them. But they do in fact sometimes introduce information that is used.

This has been dramatically illustrated by devising "proofs" of "false" theorems, such as that every triangle is isosceles.[139] There is nothing different in these "proofs" from proofs given in Euclid except the use of slightly distorted figures.

The "hidden" assumptions of Euclid have been brought into the light in modern times, and stated as axioms by Pasch 1882, Hilbert 1899 and others. Thus Hilbert's axioms include the following, adapted from Pasch: *If a line in the plane of a triangle not passing through a vertex cuts a side, it cuts one or the other of the remaining two sides.* Our figures do come out this way; but nothing in Euclid's text enables us to prove that they must. Hilbert's "Grundlagen der Geometrie" 1899 gives an elegant treatment of Euclidean geometry from the standpoint of formal axiomatics, with all assumptions explicitly stated.

[138] See Weyl 1949 pp. 50–62.

[139] W. W. Rouse Ball 1892 (pp. 80–81 in the 11th ed. 1939). Reproduced in J. W. Young 1911 pp. 143–145.

Now in formal axiomatics, while the *primitive terms* are to be meaningless, in carrying out the deductions by logic the meanings of the *ordinary words* are used. However, we have seen that theories may differ in their logic as well as in their mathematical assumptions. So, to make it perfectly explicit what the theorems of a theory are to be, we should carry out the step for all the words which is carried out in formal axiomatics for the primitive terms; i.e. we should divest them of meaning for the purpose of deductions, and carry out the deductions entirely by stated rules applying only to the form of the sentences. The logic used in the deductions in formal axiomatics must be represented in part by these rules, but may in part also be provided by logical axioms.

To carry out such a complete formalization would not be practicable if the theory were kept in an ordinary language such as English. For, the word languages have irregularities and ambiguities which would greatly complicate the task.

Indeed mathematics has in the entire modern period profited greatly by using special symbolisms, though it has customarily left parts of its sentences, including the parts involved in logical deduction, in ordinary language. The symbolic equation not only represents a great economy in writing, but presents an opportunity for manipulations (such as transposing: e.g. $x + 5 = 2$ gives $x = 2 - 5$ and $x = -3$), which, though justified by the meanings, are in practice usually carried out speedily without stopping to think through the justifications. This is indeed a semi-formal kind of reasoning, which greatly increases the power of modern mathematics.

The complete formalization which we now desire, for Hilbert's and other purposes, is obtained by combining the symbolization prevalent in modern mathematics with the symbolic treatment of logic available from the work of Boole, Peirce, Frege, Whitehead and Russell, and others. Using these two ingredients, we construct a completely symbolic language for the theory we wish to formalize. We specify both its syntax (by "formation rules") and its logic (by "deductive rules" or "transformation rules").[28] The result we call a *formal system* or *formalism* or *logistic system*. This method of making a theory explicit is sometimes called the *logistic method*.

To discuss a formal system, which includes both defining it (i.e. specifying its formation and transformation rules) and investigating the result, we operate in another theory or language, which we call the *metatheory* or *metalanguage* or *syntax language*. In contrast, the formal system is the *object theory* or *object language*. The study of a formal system, carried out in the metalanguage as part of informal mathematics, we call *metamathematics* or *proof theory*.

For the metalanguage we use ordinary English and operate informally, i.e. on the basis of meanings rather than by formal rules (which would require a metametalanguage for their statement and use). Since in the metamathematics English is being applied to the discussion only of the symbols, sequences of symbols, etc. of the object language, which constitute a relatively tangible subject matter, it should be free in this context from the kind of lack of clarity that was one of the reasons for formalizing.

Since a formal system (usually) results by formalizing portions of existing informal or semiformal mathematics, its symbols, formulas etc. will have meanings or interpretations in terms of that informal or semiformal mathematics. These meanings together we call the (*intended* or *usual* or *standard*) *interpretation* or *interpretations* of the formal system.[140] If we were not aware of this interpretation, the formal system would be devoid of interest for us. But the metamathematics, to accomplish its purpose, must study the formal system as just itself, i.e. as simply a system of meaningless symbols, and may not take into account its interpretation. When we speak about the interpretation, we are not doing metamathematics.

Furthermore, as we saw in § 36, for Hilbert's program the methods used in metamathematics must be ones, called "finitary", which are intuitively convincing.

Now we relate the present terminology to that used in Part I. We do not consider a language, taken as an object of study, to be a "formal system" unless it is a symbolic language (not a part of a word language, like English) and unless axioms and rules of inference are specified for it (not simply model-theoretic concepts, like "valid" and "valid consequence"). We do not refer to a language used in studying an object language as the "metalanguage" or "syntax language" unless in it only finitary methods are used (though some authors do). Thus "object language" and "observer's language" as used in Part I are broader terms than "formal system" and "metalanguage".

In Part I, we used "proof theory" a little loosely compared to Hilbert's intended sense, to which we shall now adhere. For, we had not actually specified a symbolic language there.[141]

[140] If that informal or semiformal mathematics is ambiguous (like abstract group theory), admitting different interpretations, then we can use "interpretation" here in either the singular or the plural, depending on where the focus of our attention is, on the semiformal mathematics as the mathematician works with it while developing it, or on its interpretations in turn.

[141] In our "proof theory" in Part I we did use only finitary methods, though we didn't say so. In "model theory", the methods are not restricted to be finitary. In modern "model theory", we would likewise work with a completely symbolic language. Just as Hilbert is responsible for "proof theory" in its present refinement, Tarski 1933, 1935 etc. is the originator of much of modern model theory. Carnap 1935 did some work similar to Tarski 1933. Sometimes proof theory is called "syntax" and model theory "semantics". For a full bibliography of model theory, see Addison, Henkin and Tarski 1965.

If we supply suitable definitions (referring to a symbolic language) of "prime formula" or "atom" for the propositional calculus, and of "prime predicate expression" or "ion" for the predicate calculus (or with functions § 28, also of "prime function expression" or "meson"), then using them as the basis for the definition of "formula" in § 1 or § 16 or § 29 (or of "term" and "formula" in § 28 or § 29), we get formal systems of propositional calculus and of predicate calculus without or with equality.[142] (In §§ 38, 39 we shall actually do this in a variety of ways.) After this is done, the theorems in Part I constitute metamathematical theorems, except those involving the validity or "valid consequence" notion for the predicate calculus (without or with equality), whose definitions are not finitary. (However, Corollary Theorem 12 as extended to the predicate calculus in §§ 23, 28, 29 is metamathematical when proved using 1-⊧, which is finitary, although the extension of Theorem 12 itself is not.)

Hilbert's best-publicized objective, to save classical mathematics short of the paradoxes by a consistency proof (§ 36), calls for a formal system embracing (elementary) number theory (i.e. the theory of natural numbers, or also of similar "systems" of $\aleph_0$ objects), analysis (i.e. the theory of real numbers, etc.), and presumably quite a bit more. However the problems encountered by metamathematics in treating number theory proved so challenging that Hilbert and his coworkers in the two decades 1920–1940 gave most of their attention to the metamathematics of a system (or several systems) of number theory. We shall describe such a formal system **N** in the next section, and it will serve as an example for much of the work in Chapter V. Some other formal systems will be described in § 39.

Of course, other problems than the consistency problem are of interest to metamathematicians; and metamathematics has proved fruitful in a variety of ways, not all of them foreseen. We shall see some of its other applications below.

Another application is to the description and investigation of "machine languages" or "computer languages" for use with modern high-speed computers. Information needs to be given to computers in exact form, by sequences of symbols on tapes, cards, or otherwise, leaving nothing to the imagination. The formation of the sequences of symbols used for this purpose must be subject to exact syntactical rules, of the sort which first became an object of mathematical study in Hilbert's metamathematics.[143]

§ 38. Formal number theory. We now describe a specific formal system **N**, which is designed to formalize elementary number theory. We

[142] At the same time, as will appear in § 38, we need to be a little more explicit about how parentheses are used.

[143] Inversely, computers may be applied in metamathematics to seeking proofs of theorems, or to checking proposed proofs, etc.[153]

first introduce the *formal symbols*, which structurally play the part of letters of the alphabet in our formal language (although for the interpretation most of them correspond to words in English). These symbols are:

$$\sim, \supset, \&, \vee, \neg, \forall, \exists, =, +, \cdot, ', 0, a, b, c, \ldots, {}_{|}, (,).$$

The commas, and the three dots near the end, are not formal symbols but punctuation marks used in displaying the formal symbols on the page.

The symbols $a, b, c, \ldots$ are "variables"; we need to have a countable infinity of them available potentially. Since the Latin alphabet has only 26 letters, for definiteness we shall assume that the *variables* consist of the 26 Latin script small letters, and also any of these followed by one or more occurrences of the symbol ${}_{|}$, so that a, $a_{|}$, $a_{||}$, $a_{|||}$, b, $b_{|}$, $b_{||}$, $b_{|||}$, etc. are variables.[144] The variables other than the 26 script letters are then not single formal symbols, but finite sequences of formal symbols. The system **N** then has an alphabet of exactly 41 formal symbols.

Notice that here $a, b, c, \ldots, a_{|}, a_{||}, a_{|||}, \ldots$ with the letters in script are the variables in the object language themselves, not names in the metalanguage for variables in the object language, as are "a", "b", "c", ..., "x", "y", "z", "x_1", "x_2", "x_3", ... with the letters in Roman type (following usage begun in § 16). That is, here a is just a, whereas x can be a or b or c or $a_{|}$, etc., in different metamathematical statements about a variable x.[145]

We call a finite sequence of (occurrences of) formal symbols a *formal expression*. Just as formal symbols correspond structurally to letters in ordinary languages, formal expressions correspond structurally to words, though for the interpretation some of them will represent entire sentences. Most formal expressions, like $))a0=$ and aaa will not interest us. But we shall now define two particular classes of significant formal expressions: "terms", which for the interpretation correspond to nouns, and "formulas", which for the interpretation correspond to declarative sentences. Each definition consists of several clauses.

DEFINITION OF "TERM". 1. 0 is a *term*. 2. The variables $a, b, c, \ldots$ are *terms*. 3–5. If r and s are terms, so are (r)', (r)+(s), and (r)·(s). 6. The only *terms* are those given by 1–5.

Here "r" and "s" are not formal symbols, but metamathematical variables used in the syntax language to represent formal expressions, in this

[144] If it is more convenient to use "a_1", "a_2", "a_3", ... than "$a_{|}$", "$a_{||}$", "$a_{|||}$", ..., we can do so in practice by regarding the former as metamathematical abbreviations for the latter.

[145] A point arises here with respect to all the formal symbols including the variables, which was discussed in § 1 Footnote 6 for $\sim, \supset, \&, \vee, \neg$ when they are symbols of the object language (as here). To get names of the formal symbols in the syntax language, we simply use specimens of those symbols as names for themselves ("autonymously").

case any terms already constructed. Thus "(r)+(s)" is not a formal expression, but a metamathematical expression which becomes a formal expression when "r" and "s" are replaced by terms.

Examples of terms: 0, a, b, c, $a_|$, $a_{||}$, (0)′, ((0)′)+(a), (((0)′)+(a))·(b).

DEFINITION OF "FORMULA". 1. If r and s are terms, then (r)=(s) is a *formula*. 2–6. If A and B are *formulas*, so are (A) ∼ (B), (A) ⊃ (B), (A) & (B), (A) V (B) and ¬(A). 7–8. If A is a *formula* and x is a variable, then ∀x(A) and ∃x(A) are *formulas*. 9. The only *formulas* are those given by 1–8.[146]

As was the case with "r" and "s" in the definition of term, "A" and "B" are metamathematical variables representing any formulas, and "x" is a metamathematical variable representing any formal variable. Thus "∀x(A)" becomes a formula when "x" is replaced by any variable, e.g. a, and "A" is replaced by any formula, e.g. (a)=(b), giving ∀a((a)=(b)). Using e.g. b instead, we get a different formula ∀b((a)=(b)). This is why the metamathematical variable "x" is necessary in Clause 7; had we written a instead, we would be allowing only ∀a((a)=(b)) (but not ∀b((a)=(b)), ∀c((a) = (b)), etc.) as a formula.

The definition of "term" here agrees basically with that in § 28 for the case of four mesons 0, (—)′, (—)+(—), (—)·(—) which we form using the *individual symbol* (or 0-*place function symbol*) 0, the 1-*place function symbol* ′, and the two 2-*place function symbols* + and ·. The definition of "formula" then agrees with that in § 28 (extending § 16) for the case of one ion —=— formed using the 2-*place predicate symbol* =.[142]

If a formal expression is given, how can we determine whether or not it is a term, or is a formula? Consider the following example.

(1) (∃c((((c)′)+(a))=(b))) ⊃ (¬((a)=(b))).

We first observe that each of c, a, b are terms. We then proceed outwards, using the parentheses as a guide, verifying successively that (c)′, ((c)′)+(a) are terms, and then that (((c)′)+(a))=(b), ∃c((((c)′+(a))=(b)), (a)=(b),

[146] It is important to remember that the formulas of a formal system are not just any formulas of informal mathematics, or any finite sequences of (occurrences of) the formal symbols of the system; but they are just those finite sequences of the formal symbols which are formed in accordance with the rules defining "formula" (here the nine rules just listed).

To emphasize this, many authors call them "well-formed formulas" or "wffs". (To be consistent, if we did so here, we ought to say also "well-formed terms" or "wfts", and "well-formed proofs" or "wfps".) We find "well-formed formulas" and "wffs" a little aneuphonious (unwell-sounding). So we prefer, after the well-taken point has been well-emphasized, to say simply "formulas" (following Hilbert and Bernays 1934, 1939). For the infrequently needed arbitrary finite sequences of formal symbols, we have the longer name "formal expressions".

$\neg((a)=(b))$, and finally (1), are formulas. In practice in the case of quite long formal expressions, prior to thus testing to see whether or not they are terms or formulas, we may pair parentheses in the following way. First, pair a left parenthesis "(" and right parenthesis ")" which occur, the left parenthesis to the left of the right one, with no other parenthesis between; this pairing may be indicated by attaching subscripts $_1$. Then repeat the process using subscripts $_2$, then $_3$, etc., each time taking into account only parentheses not already subscripted. The result of carrying out this process on (1) is as follows:

$$({}_7\exists c({}_6({}_4({}_2({}_1 c)_1{}')_2+({}_3 a)_3)_4=({}_5 b)_5)_6)_7 \supset ({}_{11}\neg({}_{10}({}_8 a)_8=({}_9 b)_9)_{10})_{11}.$$

Now by following the order of the subscript pairs, we can carry out the verification stage by stage that (1) is a formula. By a *proper pairing* of $2n$ parentheses, n of them left and n of them right, we mean a pairing such that a left parenthesis is always paired with a right parenthesis to the right of it and no two pairs separate each other thus: $({}_i\,({}_j\,)_i\,)_j$. It can be proved that $2n$ such parentheses admit at most one proper pairing. Also it can be proved that in a term or formula there always is a proper pairing of the parentheses, by which indeed we always can find out in what order the parts were put together under the clauses of the definitions of terms and formula.[147]

The reader will have observed that here we call for parentheses to be introduced with each of our operators for building terms or formulas already constructed into larger terms or formulas. This contrasts with §§ 1, 16, where we only asked for parentheses to be supplied as required to avoid ambiguity about the scopes (§ 1 Example 1). But we shall arrange that our practice here will in effect be the same as before. To do so, we provide that, as a kind of abbreviation in our metamathematical writing, we may omit parentheses to any extent such that we can see how to restore them (as in §§ 1, 16 we supplied them), using the following ranking of the present operators:

$$\sim,\ \supset,\ \&,\ \vee,\ \neg,\ \forall x,\ \exists x,\ =,\ +,\ \cdot,\ '.$$

Now (1) can be abbreviated to $\exists c(c'+a=b) \supset \neg a=b$ or even to $\exists c\, c'+a=b \supset \neg a=b$. We shall regard these omissions of parentheses as being only in the exposition of the metamathematics, and not as altering what a term and a formula strictly are. This keeps the fundamental metamathematical definitions of "term" and "formula" simpler than if we incorporated into them explicit rules for supplying parentheses selectively. Of course, here the logistic method requires complete explicitness in these

[147] See IM pp. 23–24, 73–74.

fundamental definitions. Similarly, in our exposition we may change some parentheses to square or curly brackets to assist the eye in pairing them.

We also introduce new metamathematical symbols to permit abbreviations of terms and formulas. For instance, "$a \neq b$" is an abbreviation for $\neg a=b$; and "$a<b$" for $\exists c(c'+a=b)$ or $\exists d(d'+a=b)$, etc. Here "$\neq$" and "$<$" are symbols used for abbreviation, not formal symbols. In the case of the abbreviation "$a<b$", there is ambiguity what variable to use in the quantifier which we supply in unabbreviating. The general rule shall be that "r<s" abbreviates $\exists$x(x$'$+r=s), where x can be any variable not occurring in r or in s. Under this rule, two legitimate unabbreviations of "r<s" will be congruent (§ 16), so it will be immaterial for our usual purposes which legitimate unabbreviation we choose. (For, in the first place, two congruent formulas have the same meaning under the interpretation of the symbolism. And in the metamathematics, either will be provable if the other is, by Theorem 25 with Corollary 2 Theorem 23 § 24, both of which will be available here; and likewise either can replace the other in a deducibility relationship.) Now (1) can be written $a<b \supset a \neq b$. Further useful abbreviations are "$a>b$" for $b<a$, "$a<b<c$" for $a<b$ & $b<c$, etc.; and "1" for 0$'$ (i.e. for (0)$'$), "2" for 1$'$ (i.e. for ((0)$'$)$'$), "3" for 2$'$, etc.

The list of formal symbols and the definitions of "term" and "formula" constitute the *formation rules* of our formal system (analogous to the rules of syntax in ordinary grammar). We shall now give the definitions that establish the deductive structure of the system (the *transformation rules* or *deductive rules*). These begin with a list of axiom schemata, particular axioms and rules of inference (which we may call collectively the "postulates"). When we have given this list, then "(formal) proof (of B_l)", "B is provable" or in symbols "$\vdash$ B", "(formal) deduction (of B_l) from $A_1, \ldots,$ A_m (holding all variables constant)", "B is (formally) deducible from $A_1, \ldots, A_m$ (holding all variables constant)" or in symbols "$A_1, \ldots,$ $A_m \vdash$ B", etc. are to be defined for **N** from its postulate list as before for the propositional and predicate calculi from their smaller postulate lists (§§ 9, 21).

Here we reemphasize from § 9 that a (formal) proof of B is an object of the object language (§§ 1, 37); specifically, it is a suitable finite sequence of formulas, which in turn are suitable finite sequences of formal symbols, where "suitable" means of the respective kinds defined above. We may talk about such proofs, construct them, or "prove" that they exist. Here the "prove" in quotes is the ordinary word in English, and means that there is a proof in the intuitive sense (an *informal* proof) in the observer's language, now called the metalanguage. A *formal* proof is a proof of a formula, which (for the metamathematics) is a meaningless sequence of symbols. An *informal* proof (in metamathematics) is a proof of a meaningful statement about the meaningless formal objects; and this informal proof is an intuitive demonstration of the truth of that statement. Thus a "proof that $\vdash$ B" or "a proof of '$\vdash$ B' " is an informal proof of the fact that a formal proof of B exists. We might try to use a different word than "proof" for informal proofs, but it is not convenient to do so; so we rely on the context to make it clear when we mean a formal proof (in the object language) and when an informal proof (in the metalanguage).

To begin with, **N** shall have all the postulates of the predicate calculus. That is, it shall have the three rules of inference, the $\supset$-rule or modus ponens (§ 9 and Theorem 3), the $\forall$-rule and the $\exists$-rule (§ 21 and Theorem 16); and it shall have Axiom Schemata 1a–10b (§ 9 and Theorem 2), and the $\forall$-schema and the $\exists$-schema (§ 21 and Theorem 15), i.e. all formulas of these forms shall be axioms. Moreover, we now permit as the r for the $\forall$-schema and the $\exists$-schema not merely a variable as in § 21 but more generally as in § 28 any term r such that, when r is substituted for the free occurrences of x in A(x) with result A(r), no occurrence of a variable in any of the *resulting* occurrences of r will be bound. We say such a term r is *free for* x *in* A(x) (generalizing § 18 from variables to terms, as in § 28). For example, taking x to be a, r to be $d'+e$, and A(x) to be $\exists c(c'+a=b)\ \&\ \neg a=0$, the condition is satisfied; but not for the same x and r when A(x) is $\exists d(d'+a=b)\ \&\ \neg a=0$.

Therefore, statements of the form "$\vdash$ B" or "$A_1, \ldots, A_m \vdash$ B" or "$A_1, \ldots, A_m \vdash^{x_1 \ldots x_q}$ B" (direct rules) which hold for the predicate calculus now hold for **N** (cf. § 21). Moreover (as in § 28), in the direct rules of Theorem 21 r can be any term in the present sense, and in its Corollaries $r_1, \ldots, r_p$ can be any list of terms not necessarily distinct, in each case subject to the freedom condition stated there.

In addition to the postulates of the predicate calculus (with the more general r for the $\forall$-schema and the $\exists$-schema), there shall be one axiom schema (13 below) and eight particular axioms (14–21 below). For Axiom Schema 13, x is any variable, A(x) is any formula, and A(0), A(x′) are the results of substituting 0, x′ respectively for the free occurrences of x in A(x). (These substitutions are automatically free.)

13. $A(0)\ \&\ \forall x(A(x) \supset A(x')) \supset A(x)$.

14. $a'=b' \supset a=b$.	15. $\neg a'=0$.
16. $a=b \supset (a=c \supset b=c)$.	17. $a=b \supset a'=b'$.
18. $a+0=a$.	19. $a+b'=(a+b)'$.
20. $a \cdot 0=0$.	21. $a \cdot b'=a \cdot b+a$.

Thus **N** consists of the predicate calculus plus some "nonlogical axioms" (or "mathematical axioms"), namely eight particular axioms 14–21 and $\aleph_0$ axioms by Axiom Schema 13.[148]

The deduction theorem (THEOREM 11, §§ 10, 22) extends to **N**. For (as in § 29), we can handle the new axioms under the old Case 3. Hence in **N** we have the introduction and elimination rules of THEOREMS 13 and 21 (5 subsidiary deduction rules based on the deduction theorem, and 13 direct rules).

As we remarked in § 37, a formal system formalizing a portion of informal mathematics has an "intended" (or "usual" or "standard") interpretation. (When we discuss this, or model theory more generally, we are going outside of metamathematics.) The informal mathematics that we aim to formalize in **N** is elementary number theory. So for the intended interpretation, the variables range over the natural numbers {0, 1, 2, . . .}, i.e. this set is the domain. The logical symbols $\sim$, $\supset$, &, $\vee$, $\neg$, $\forall$, $\exists$ are interpreted as in Chapters I and II (in classical logic). The function symbol $'$ is interpreted as expressing the successor function $+1$, and 0 ("zero"), $+$ ("plus"), $\cdot$ ("times") and $=$ ("equals") have the same meanings as those symbols convey in informal mathematics.[149] All the terms in **N** are names for natural numbers, specified or unspecified, just as the formulas express propositions.[53,105]

The nonlogical axioms play a role under the interpretation like assumption formulas for the valid consequence relationship (II) of § 20 with all their free variables having the generality interpretation. Thus in taking Axiom 14 as a permanent assumption, or axiom, for elementary number theory, we intend to assume that, for every pair of natural numbers a and b, if $a+1 = b+1$, then $a = b$. So Axiom 14 is to be regarded as synonymous with its closure $\forall a \forall b(a'=b' \supset a=b)$.

The deductive rules of **N** are in keeping with this interpretation. For, by uses of $\forall$-introd. in Theorem 21 (with the Γ empty, so Proviso (B) does not inhibit us), the closure of each axiom is provable using the axiom. Inversely, if we had taken the closures of the present nonlogical axioms as the nonlogical axioms, then the present ones would be provable by applications of $\forall$-elim. Thus in setting up **N**, it made no essential difference whether the nonlogical axioms were written down open or closed.

[148] Most of the next remarks (down through (A) and (B)) apply similarly to any formal system consisting of the predicate calculus, or the predicate calculus with equality, plus nonlogical axioms. This includes the predicate calculus with equality itself as based on the predicate calculus without equality (§ 29). For systems based on the predicate calculus with equality, the "nonlogical axioms" are those added to the axioms of the predicate calculus with equality.

[149] We rely on the context to make it clear when 0, $'$, $+$, $\cdot$, $=$ are being used as formal symbols and $\neq$, $<$, $>$, 1, 2, 3, . . . in abbreviating formal expressions, and when they are being used informally.

The open axioms (and provable formulas) are simpler to write, and their use fits with common mathematical practice.[150] We now naturally extend the terminology used with (temporary) assumptions in § 20 to say the free variables in the axioms of **N**, and also in any provable formulas (cf. Exercise 23.4), have the "generality interpretation". (In a deducibility relationship in **N**, the free variables of the assumption formulas which have conditional interpretation stand for the same natural numbers in the conclusion B as in the assumption formulas $A_1, \ldots, A_m$, i.e. for ones satisfying the conditions expressed by $A_1, \ldots, A_m$, while the other free variables of B have the generality interpretation; cf. Exercise 23.5, where $x_1, \ldots, x_q$ may be taken to include all the free variables of B not free in $A_1, \ldots, A_m$.)

Using ∀-introds. and ∀-elims. as just shown: (A) ⊢ B *in* **N**, *if and only if, for some list* $A_1, \ldots, A_m$ *of nonlogical axioms of* **N**, $\forall A_1, \ldots, \forall A_m \vdash B$ *in the predicate calculus.* For the "only if" part, $A_1, \ldots, A_m$ can be the (finitely many) nonlogical axioms actually used in some particular proof of B.[151]

In logic (Chapters II, III), we did not have in mind a particular domain and a particular interpretation of the ions and mesons (except of —=— in § 29). Let us see how the model-theoretic concepts of validity and "valid consequence" (symbolized by "⊨") which we used there apply now. We need the versions at least with functions (§ 28), since in **N** we have the function symbols +, ·, ′, 0 as mesons. We *can* use those of § 29 with both functions and equality, where the predicate symbol = is given the meaning of equality or identity (which it has in the intended interpretation of **N**). For, although we based **N** proof-theoretically on the predicate calculus without equality, the postulates of the latter are all good for the predicate calculus with equality. As the name "nonlogical axioms" suggests, those axioms of **N** are not valid (or with the validity notion of § 29, all but Axioms 16 and 17 are invalid). So Theorem 12 does not simply extend to **N** to say that "If ⊢ E, then ⊨ E". Instead, using (A) with Corollary Theorem 11 (as extended to **N**), and Theorem 12 and Corollary Theorem 8 (in the versions of § 28, or of § 29): (B) *If* ⊢ B *in* **N**, *then, for some list* $A_1, \ldots, A_m$

[150] A closed formula is called a *sentence* by Tarski (in the Eng. tr. of 1933, in Tarski, Mostowski and Robinson 1953, etc.). We are using "sentence" simply for the linguistic objects (declarative sentences) in the informal language which are formalized by formulas, open or closed.

[151] A proof of B in **N** using as nonlogical axioms only $A_1, \ldots, A_m$ is not in general a deduction of B from $A_1, \ldots, A_m$ in the predicate calculus with all variables held constant. For in the latter (§ 21), we are restricted in using the ∀-rule and the ∃-rule below the first use of $A_1, \ldots, A_m$ to variables not occurring free in $A_1, \ldots, A_m$. In a proof in **N**, where $A_1, \ldots, A_m$ count as axioms rather than as assumption formulas with their variables held constant, this restriction is lifted. Hence the ∀'s are necessary in (A).

of nonlogical axioms of N, $\forall A_1, \ldots, \forall A_m \models B$. Thus, if $\vdash B$ in N, then B is t for every nonempty domain D and every assignment in D which gives the value t to the closures of all the nonlogical axioms of N (and hence to any finite list $\forall A_1, \ldots, \forall A_m$ of those closures).

Now what are these domains D and assignments in D? In less technical language, what interpretations make all the nonlogical axioms of N true under the generality interpretation of their free variables? One of them is the intended interpretation of N already described. Whether the nonlogical axioms are also true under some other interpretation than this is a question for investigation, the result of which we shall report in Chapter VI (§ 53).

Now let us consider the significance of each of the nonlogical axioms of N under the intended interpretation.

Axioms 14 and 15 and Axiom Schema 13 formalize the third, fourth and fifth of Peano's list of five axioms for the natural numbers, 1889.[152] Peano's first axiom, that 0 is a natural number, and his second axiom, that if n is a natural number so is $n+1$, are taken care of instead by the formation rules which make 0 a term and whenever r is a term make (r)′ a term, since all the terms in the system are interpreted as expressing natural numbers.

Axioms 16 and 17 are axioms for equality. We do not postulate the reflexive law of equality, $a=a$, because it is deducible from Axioms 16 and 18 with a, b, c general by the predicate calculus (as we shall show below), and so is provable in N. From $a=a$ with Axiom 16 the symmetric and transitive laws can be proved (below). Axiom 17 asserts that the value of the successor function ′ is determined by the value of its variable. In § 29 we called this the "(open) equality axiom for ′". The two equality axioms for + and the two for · are provable, as will be indicated below.

Axioms 18–19 and 20–21 provide "recursive definitions" or "definitions by induction" of the functions + (plus or addition) and · (times or multiplication). Let us see how they "define" these functions. They are not definitions which simply introduce abbreviations for combinations of symbols already available. Instead, the two equations 18 and 19 enable us, for any fixed value of a (e.g. 3), to determine the value of $a+b$ successively for 0, 1, 2, . . . as value of b, thus (using the symbolism informally for the moment):

(A_0^3) $3+0 = 3$ [by Axiom 18],
(A_1^3) $3+1 = 3+0' = (3+0)'$ [by Axiom 19] $= 3'$ [by (A_0^3)] $= 4$,
(A_2^3) $3+2 = 3+1' = (3+1)'$ [by Axiom 19] $= 4'$ [by (A_1^3)] $= 5$,

. . . .

[152] Peano dealt instead with the positive integers. A quite full explanation of these axioms is in IM §§ 6, 7 (and in § 8, which overlaps the present § 36). That explanation is not essential to the present survey. The role of the fifth Peano axiom (Axiom Schema 13) is illustrated in Example 2 below; of the third and fourth, in Exercise 38.5.

With equations (A^a_0) giving the values of $a+b$ for any values a and b of a and b available now, we can proceed similarly to determine the values of $a \cdot b$ (e.g. with 3 as value of a):

(M^3_0) $3 \cdot 0 = 0$ [Ax. 20],
(M^3_1) $3 \cdot 1 = 3 \cdot 0' = 3 \cdot 0 + 3$ [Ax. 21] $= 0+3$ [(M^3_0)] $= 3$ [(A^0_3)],
(M^3_2) $3 \cdot 2 = 3 \cdot 1' = 3 \cdot 1 + 3$ [Ax. 21] $= 3+3$ [(M^3_1)] $= 6$ [(A^3_3)],
. . . . .

EXAMPLE 1. The following is a (formal) proof in **N**. Strictly speaking, the proof is the sequence of 17 formulas. In addition, numbers have been supplied at the left, and explanations at the right.

1. $a=b \supset (a=c \supset b=c)$ — Axiom 16.
2. $0=0 \supset (0=0 \supset 0=0)$ — Axiom Schema 1a.
3. $\{a=b \supset (a=c \supset b=c)\} \supset \{[0=0 \supset (0=0 \supset 0=0)] \supset [a=b \supset (a=c \supset b=c)]\}$ — Axiom Schema 1a.
4. $[0=0 \supset (0=0 \supset 0=0)] \supset [a=b \supset (a=c \supset b=c)]$ — $\supset$-rule, 1, 3.
5. $[0=0 \supset (0=0 \supset 0=0)] \supset \forall c[a=b \supset (a=c \supset b=c)]$ — $\forall$-rule, 4.
6. $[0=0 \supset (0=0 \supset 0=0)] \supset \forall b \forall c[a=b \supset (a=c \supset b=c)]$ — $\forall$-rule, 5.
7. $[0=0 \supset (0=0 \supset 0=0)] \supset \forall a \forall b \forall c[a=b \supset (a=c \supset b=c)]$ — $\forall$-rule, 6.
8. $\forall a \forall b \forall c[a=b \supset (a=c \supset b=c)]$ — $\supset$-rule, 2, 7.
9. $\forall a \forall b \forall c[a=b \supset (a=c \supset b=c)] \supset \forall b \forall c[a+0=b \supset (a+0=c \supset b=c)]$ — $\forall$-schema.
10. $\forall b \forall c[a+0=b \supset (a+0=c \supset b=c)]$ — $\supset$-rule, 8, 9.
11. $\forall b \forall c[a+0=b \supset (a+0=c \supset b=c)] \supset \forall c[a+0=a \supset (a+0=c \supset a=c)]$ — $\forall$-schema.
12. $\forall c[a+0=a \supset (a+0=c \supset a=c)]$ — $\supset$-rule, 10, 11.
13. $\forall c[a+0=a \supset (a+0=c \supset a=c)] \supset [a+0=a \supset (a+0=a \supset a=a)]$ — $\forall$-schema.
14. $a+0=a \supset (a+0=a \supset a=a)$ — $\supset$-rule, 12, 13.
15. $a+0=a$ — Axiom 18.
16. $a+0=a \supset a=a$ — $\supset$-rule, 15, 14.
17. $a=a$ — $\supset$-rule, 15, 16.

Thus $a=a$ (the reflexive law of equality) is a provable formula; or in symbols, $\vdash a=a$. (Here "$\vdash a=a$" is not a formula, but a statement in our metamathematical shorthand that the formula $a=a$ is formally provable.)

Now we give informal proofs in the formats (A) and (B_1) (cf. §§ 13, 25) of the fact that $\vdash a=a$ (i.e. that a formal proof of $a=a$ exists).

(A)

1. $\vdash a=b \supset (a=c \supset b=c)$ — true because the formula concerned is Axiom 16.
2. $\vdash a+0=a \supset (a+0=a \supset a=a)$ — from 1 by substituting $a+0$, a, a for a, b, c (Theorem 21 Corollary 2 (d) for Γ empty).
3. $\vdash a+0=a$ — true because $a+0=a$ is Axiom 18.
4. $\vdash a=a$ — from 3 and 2 by two applications of $\supset$-elim.

(B_1) Substituting $a+0$, a, a for a, b, c in Axiom 16, $a+0=a \supset (a+0=a \supset a=a)$. From this and Axiom 18 by $\supset$-elims., $a=a$.

Repeating in part some points made in Chapters I and II (especially §§ 10, 13, 25), formal proofs even of simple formulas tend to be long, as Example 1 illustrates. We are interested in what formulas there are formal proofs of, i.e. in what formulas are formally provable. We shall be satisfied to know whether formal proofs of these formulas exist, and we shall not ordinarily care about actually seeing such formal proofs when we have shown that they exist. Therefore we find it satisfactory to use informal proofs to show that the formal proofs exist, when it is easier to do so and when the *informal* proofs are by "finitary" methods (§§ 36, 37).[153] If challenged to find the formal proofs we have demonstrated to exist, we are then able to supply them. Continuing in the manner illustrated, our methods for informally proving (in metamathematics) the existence of formal proofs (now in the system **N**) can be brought quite close to the methods of the ordinary mathematician in developing the elementary theory of the natural numbers. However, we must not forget that we are doing something somewhat different than he; i.e. we must not lose sight of the path by which our informal proofs can be converted into formal proofs in **N**.

We shall go a bit further in illustrating the beginnings of the development of number theory in **N**, by giving a few more metamathematical or informal proofs that certain formulas expressing theorems of elementary number theory have formal proofs in **N**.

From the reflexive property of equality $\vdash a=a$ (*100) and Axiom 16, we readily obtain the symmetric and transitive properties $\vdash a=b \supset b=a$

[153] If computing machines are to be applied to seek or check proofs, a further step is appropriate. Some selection of the quicker and easier methods to which our metamathematical investigations have led us, or (for seeking proofs) possibly of other methods tailored to the machines strengths and weaknesses, should be stereotyped as a new formal system in which the computer is to make its searches or perform its checks. Here we are not referring just to formal number theory, but also to the predicate calculus and other systems obtained by adding mathematical axioms to it. Cf. Wang 1960, Davis and Putnam 1960, J. A. Robinson 1963, 1965.

(*101) and $\vdash a=b \,\&\, b=c \supset a=c$ (*102).[154] (If it isn't obvious, refer to Theorem 29 in § 29.)

Having the reflexive, symmetric and transitive properties of equality, we can use chains of equalities analogously to chains of equivalences in § 5, except that we do not yet have the general replacement theorem (Theorem 30, analogous to Theorem 5) to use in justifying links of the chains. We use chains to simplify the presentation in the next example.

EXAMPLE 2. We show that $\vdash a=b \supset a+c=b+c$ (*104). (In the terminology of § 29, this is one of the two open equality axioms for +.) Preparatory to using $\supset$-introd., assume **(a)** $a=b$. Preparatory to applying Axiom Schema 13 ("mathematical induction") with c, $a+c=b+c$ as the x, A(x), we need to deduce (from $a=b$) the two formulas $a+0=b+0$ and $\forall c(a+c=b+c \supset a+c'=b+c')$. I. (BASIS.) $a+0 = a$ [Ax. 18] $= b$ [(a)] $= b+0$ [Ax. 18 with substitution; this gives $b+0=b$ (cf. Example 1 (A) Step 2), whence by the symmetric property of equality $b=b+0$]. So (using the transitive property of equality, implicit in the chain method) $a+0=b+0$. II. (INDUCTION STEP.) Preparatory to $\supset$-introd., assume **(b)** $a+c=b+c$. Then $a+c'=(a+c)'$ [Ax. 19 with subst.] $= (b+c)'$ [using (b) and Ax. 17, with subst. and $\supset$-elim.] $= b+c'$ [Ax. 19 with subst.]. By the planned $\supset$-introd. (discharging the assumption (b)), $a+c=b+c \supset a+c'=b+c'$. Thence by $\forall$-introd. (since our remaining assumption (a) does not contain c free), $\forall c(a+c=b+c \supset a+c'=b+c')$. III. By Ax. Sch. 13 with the results of I and II and &-introd. and $\supset$-elim., $a+c=b+c$. By $\supset$-introd. (discharging (a)), $a=b \supset a+c=b+c$.

Next we can establish (the provability of) the other equality axiom for +, namely $a=b \supset c+a=c+b$ (*105), though this isn't quite so easy (Exercise 38.2). The two for · can be shown to be provable similarly. Then we will have all the axioms of the predicate calculus with equality (§ 29) which go with the symbolism of **N**. So by Theorem 30 in § 29, we will have the replacement property of equality in general; so chains of equalities can be constructed using any replacements based on available equalities. (Before we had to be careful to use replacement steps only when we had special justification, as by Axiom 17 in II of Example 2.) Now (or earlier, after establishing just the equality axioms for +), we can prove the associative law for addition $(a+b)+c=a+(b+c)$ (*117; Exercise 38.3), etc.

The foregoing gives only enough of a beginning to convey an impression of how the development of number theory in **N** would go. A week or two of study would bring us quite far in this development.

By carrying such a development quite far, partly in a straightforward way as illustrated, and partly through investigations of a general nature,

[154] We give parenthetically the numbers with which these results appear in IM Chapter VIII.

it can be said to be established that the system **N** is adequate for the usual elementary number theory such as occurs in standard texts (but not for analytic number theory, end § 36). By this we mean that, first, the predicates and propositions commonly used in informal elementary number theory can be expressed by formulas in **N**, and, second, for those propositions which are commonly proved as theorems in informal number theory, the formulas expressing them are formally provable in **N** (as illustrated by a few simple examples just now).

The first part calls for a little explanation. We have already seen that, although $a<b$ is not provided directly in the formal symbolism, a formula $\exists c(c'+a=b)$ is so provided which under the (intended) interpretation expresses $a<b$. By introducing "$<$" as a symbol of abbreviation, we can thus express inequalities. That a divides b can be expressed by $\exists c(a\cdot c=b)$, which we abbreviate "$a|b$". That a is a prime number can be expressed by $1<a\ \&\ \neg\exists c(1<c\ \&\ c<a\ \&\ c|a)$, which we abbreviate "Pr(a)". Euclid's theorem that there are infinitely many primes can be expressed by $\exists b(\mathrm{Pr}(b)\ \&\ a<b)$ or $\forall a\exists b(\mathrm{Pr}(b)\ \&\ a<b)$. The provability of this formula is established in IM p. 192 *161, where it appears about 60 results later than our Exercise 38.3. (By no means all intervening results are used in its proof.)

Since $0, {}', +, \cdot$ are the only function symbols in **N**, besides which variables are available, no function other than a polynomial can be expressed by a term of **N**. This is undoubtedly a limitation in the symbolism of **N**, but it can be circumvented. For what can be expressed informally using functions can in fact be paraphrased using predicates instead. Thus, say $f(x_1, \ldots, x_n)$ is a number-theoretic function of n variables, and let $F(x_1, \ldots, x_n, y)$ be the predicate of $n+1$ variables which is true for exactly those $n+1$-tuples $(x_1, \ldots, x_n, y)$ such that $f(x_1, \ldots, x_n) = y$. We call $F(x_1, \ldots, x_n, y)$ the *representing predicate of* $f(x_1, \ldots, x_n)$. What can be stated using $f(x_1, \ldots, x_n)$ can be paraphrased using $F(x_1, \ldots, x_n, y)$ instead. For example, consider the function $x!$ (where $0! = 1$ and $(x+1)! = 1 \cdot 2 \cdot \ldots \cdot (x+1)$ with $x+1$ factors). Let $F(x, y)$ express $x! = y$. Take the proposition $(x+1)! = x!(x+1)$, which is an (informal) theorem about the factorial function. This can be paraphrased in terms of the predicate $F(x, y)$ e.g. thus: $\exists u\exists v[F(x+1, u)\ \&\ F(x, v)\ \&\ u=v\cdot(x+1)]$.

Now it turns out that, although only polynomials can be expressed in **N** by means of terms, the representing predicates $F(x_1, \ldots, x_n, y)$ of a vastly greater class of functions can be expressed in **N**.[155]

Furthermore, not only can the propositions using such functions be expressed indirectly by using their representing predicates, but all the

[155] This is a consequence of work of Gödel 1931 and Kleene 1936 (cf. IM §§ 48, 49, 57, or specifically Theorem VII (b) p. 285).

reasoning that could be carried out with the functions can be paralleled too.[156]

Thus it comes about that **N** is adequate for the usual elementary number theory despite its obvious shortage of function symbols.

One might ask: Would it not be better to remedy this deficiency by constructing another system with more function symbols? One can do this, but for foundational questions it is often best to keep the system as simple in its structure as possible. In fact, the theory just cited allows us to use such systems richer in functions than **N** and construe all the results in terms of **N**.

Another question is whether we could still get the same results with even fewer function symbols in the symbolism. Specifically, could we omit $\cdot$ as a formal symbol, and omit Axioms 20 and 21, and then get the effect of $\cdot$ in the same way that in **N** we get the effect of $x!$ etc.? The answer is "No".[157]

So far, in the case of **N**, we have used Hilbert's idea of studying the formal system from outside by "finitary" methods (in metamathematics) mainly in developing short cuts by which we prove metamathematically (informally) that various formulas are formally provable. Hilbert intended of course that metamathematics should concern itself also with general questions about the formal system, such as those as to its consistency and completeness. (We did deal with both of these for the propositional calculus in §§ 11, 12, and with consistency for the predicate calculus in § 23.)

Ackermann in 1924–5 thought he had proved metamathematically the consistency of **N**. But von Neumann in 1927 pointed out that Ackermann's proof is limited to the subsystem of **N** in which the use of Axiom Schema 13 (the induction schema) is restricted to the case of an A(x) which does not contain any free occurrence of x in the scope of a quantifier, i.e. in the B of a part ∀yB or ∃yB; at the same time von Neumann gave another metamathematical consistency proof for that subsystem.

The failure of metamathematicians to establish the consistency of **N** in the next few years was illuminated in 1931 by results of Gödel, which we will present in a general setting in Chapter V (especially §§ 43, 44). These results begin with an answer to the completeness question (in which we ask whether **N** suffices for *all* elementary number-theory, not just for what is usually developed).

[156] The basic step is the elimination of "proper definite descriptions" (§ 31), using a theorem of Hilbert and Bernays 1934 (cf. pp. 422–457 and 460 ff.). There is a treatment in IM pp. 407–419. The theorem gives directly conditions under which the effect of $f(x_1, \ldots, x_n)$ can be secured in a system in which we have $F(x_1, \ldots, x_n, y)$. It is a proof-theoretic version of Theorem 32 § 31.

[157] This follows by combining a result of Presburger 1930 and Theorem IV in § 43 below. (Cf. IM pp. 204, 407.)

In the following chapters, **N** can always refer to the particular formal system described in this section (which is the simplest way to read those chapters), and sometimes (when we so indicate) **N** can also refer more generally to any system with similar properties.

EXERCISES. 38.1. Translate Example 2 (as given in the format (B_1)) into statements using the symbol $\vdash$ (in the format (B_3)), and check it in the latter format (cf. §§ 13, 25).

38.2*. Show that $\vdash a=b \supset c+a=c+b$.

38.3. Assuming the result of Exercise 38.2, and using the general method of Example 2, show that $\vdash (a+b)+c=a+(b+c)$.

38.4. Show that $\vdash 3+0=3$, $\vdash 3+1=4$, $\vdash 3+2=5, \ldots$; i.e. establish statements of formal provability corresponding to (A_0^3), (A_1^3), (A_2^3), ...

38.5. Show that, using besides predicate calculus only Axioms 14 and 15: $\vdash 1\neq 0$, $\vdash 2\neq 0$, $\vdash 2\neq 1$, $\vdash 3\neq 0$, $\vdash 3\neq 1$, $\vdash 3\neq 2, \ldots$.

★§ 39. Some other formal systems. In this section, we give further examples [2]–[50] of formal systems (where [1] is the system **N** of § 38).

We now describe [2] a system **G**, which formalizes the elementary theory of an unspecified "group" (explanation follows).

The *formal symbols* shall be

$$\sim, \supset, \&, \vee, \neg, \forall, \exists, =, \cdot, {}^{-1}, 1, a, b, c, \ldots, {}_{|}, (,).$$

Variables are to be constructed as in § 38 for **N**. (The only difference in the formation rules between **N** and **G** is that $+, \cdot, {}', 0$ are replaced by $\cdot, {}^{-1}, 1$.)

DEFINITION OF "TERM". 1. 1 is a *term*. 2. The variables $a, b, c, \ldots$ are *terms*. 3–4. If r is a *term*, so are (r)·(s) and $(\text{r})^{-1}$. 5. The only *terms* are those given by 1–4.

Given this definition of "term", the definition of "formula" from "term" reads as in § 38.

Parentheses are omitted under the same conventions as there. Furthermore (as we could also have done in § 38), we shall abbreviate r·s (i.e. (r)·(s)) as "rs" (omitting the dot).

As postulates, we take those of the predicate calculus, with the present notions of "term" and "formula" (§§ 21, 28); just as for **N**, the r for the $\forall$-schema and the $\exists$-schema can be any term free for x in A(x). In addition, there shall be the following six particular axioms:

E1. $a=b \supset (a=c \supset b=c)$. E2. $a=b \supset ac=bc$. E3. $a=b \supset ca=cb$.
G1. $(ab)c=a(bc)$. (Associative law.)
G2. $a1=a$. (Right identity.)
G3. $aa^{-1}=1$. (Right inverse of a.)

We do not intend a single interpretation of this formal system **G**, as we did of **N**. The system **G** may be interpreted by any "group" G, as we shall explain and illustrate in a moment. For any choice of a "group" G, the terms are interpreted as naming members (specified or unspecified) of G (or more precisely of G_0, below).

Axioms E1–E3 give what we need to postulate of the properties of equality (identity).[158]

The axioms G1–G3 for groups will be familiar to many students. A *group* G is briefly any "system" of objects which "satisfies" these three axioms (with the variables in the generality interpretation §§ 20, 38, and with = expressing "identity" or "equality" § 29).

That is, a *group* G consists of a nonempty set G_0, a 2-place function $a \cdot b$ with arguments and values in G_0, a 1-place function a^{-1} with arguments and values in G_0, and a member of G_0 (or 0-place function) 1, such that the closures of G1–G3 are all t in our model theory of the predicate calculus with equality § 29 when G_0 is the domain and $a \cdot b$, a^{-1}, 1 are the values of $a \cdot b$, a^{-1}, 1.[159]

To give only a few particular interpretations, G can be (1) the positive rational numbers with $\cdot$, $^{-1}$, 1 having their usual meanings. (The student should have no trouble verifying that G1–G3 are satisfied by each of our interpretations.) Alternatively, G can be (2) the rational numbers except 0, (3) the positive real numbers, (4) the real numbers except 0, or (5) the complex numbers except 0, with $\cdot$, $^{-1}$, 1 in their usual meanings. Still again, G can be (6) all the integers, (7) all the rational numbers, (8) all the real numbers, or (9) all the complex numbers, with $a \cdot b$, a^{-1}, 1 meaning $a+b$, $-a$, 0, respectively.

As an interpretation not by a number system, G can be (10) or (11) all the possible rotations of a square within its plane, or through three-dimensional space, into itself. That is, we are to rotate the square so that after the rotation it occupies the same portion of space as before, but the original corners may be in new positions; e.g. if the original square is $ABCD$, then after the rotation A may be where D formerly was, B where A formerly was, etc. Also, turning the square by $-90°$ or $270°$ or $630°$ about

[158] Alternatively, we could have started from the predicate calculus with equality (§ 29), thus automatically postulating five open equality axioms (or their equivalent),[108] and have added just the three axioms G1-G3.

[159] Textbooks on (informal) group theory are e.g. Marshall Hall 1959 and Rotman 1965.

The symbols "$\cdot$", "$^{-1}$", "1" are often chosen differently. Particularly, "$\cdot$" is often written as "$\circ$" to suggest a variety of interpretations, in some of which $\circ$ will not be the usual multiplication, and "1" as "i" (the "identity").

"Group theory" includes nonelementary parts (e.g. subgroups, isomorphism, representations), not formalized in **G**.

an axis perpendicular to its center is considered the same rotation; so "rotation" really means "result of rotation". Now $a \cdot b$ is interpreted to mean (the result of) the rotation a followed by the rotation b. The further details are left to the student (Exercise 39.1).

Similarly, G can be (12) or (13) all the possible rotations of a circle into itself, or (14) all the possible rotations in 3-dimensional space of a cube into itself, or (15) of a sphere into itself.

Now we demonstrate (informally) the (formal) provability of a number of formulas (writing "T" for "theorem"). Proofs not given are exercises. As was remarked for **N** in § 38, the direct rules of the predicate calculus, and also all of the introduction and elimination rules of THEOREMS 13 and 21, hold good for **G**.

T1. $\vdash a=a$. (Reflexive property of equality.)

PROOF (in the format (B_1)). By substituting $a1$, a, a for a, b, c in Axiom E1, $a1=a \supset (a1=a \supset a=a)$. Thence using G2 and $\supset$-elim. twice, $a=a$. — The next two follow as did *101 and *102 in § 38.

T2. $\vdash a=b \supset b=a$. (Symmetric property of equality.)

T3. $\vdash a=b \;\&\; b=c \supset a=c$. (Transitive property of equality.)

Having now the reflexive, symmetric and transitive properties of equality, we can use chains of equalities (though without unrestricted replacement yet; cf. § 38).

T4. $\vdash a^{-1}a=1$. (Left inverse of a, the same as our right one.)

PROOF (by a chain of equalities). $a^{-1}a = (a^{-1}a)1$ [G2 (and substitution)] $= (a^{-1}a)(a^{-1}(a^{-1})^{-1})$ [G3, E3] $= a^{-1}(a(a^{-1}(a^{-1})^{-1}))$ [G1] $=$ $a^{-1}((aa^{-1})(a^{-1})^{-1})$ [G1, E3] $= a^{-1}(1(a^{-1})^{-1})$ [G3, E2, E3] $=$ $(a^{-1}1)(a^{-1})^{-1}$ [G1] $= a^{-1}(a^{-1})^{-1}$ [G2, E2] $= 1$ [G3].

T5. $\vdash 1a=a$. (Left identity, the same as our right one.)

PROOF. $1a = (aa^{-1})a$ [G3, E2] $= a(a^{-1}a)$ [G1] $= a1$ [T4, E3] $=$ a [G2].

T6. $\vdash ax=a \supset x=1$. (Uniqueness of right identity.)

PROOF. Assume for $\supset$-introd., $ax = a$. Now $x = 1x$ [T5] $=$ $(a^{-1}a)x$ [T4, E2] $= a^{-1}(ax)$ [G1] $= a^{-1}a$ [by the assumption $ax=a$ with E3] $= 1$ [T4].

T7. $\vdash xa=a \supset x=1$. (Uniqueness of left identity.)

T8. $\vdash ax=1 \supset x=a^{-1}$. (Uniqueness of right inverse of a.)

PROOF. Assume $ax = 1$. Now $x = 1x$ [T5] $= (a^{-1}a)x$ [T4, E2] $=$ $a^{-1}(ax)$ [G1] $= a^{-1}1$ [by the assumption $ax=1$ with E3] $= a^{-1}$ [G2].

T9. $\vdash xa=1 \supset x=a^{-1}$ (Uniqueness of left inverse of a.)

T10. $\vdash a=b \supset a^{-1}=b^{-1}$. (Equality axiom for $^{-1}$.)

PROOF. Assume $a = b$. Then $ba^{-1} = aa^{-1}$ [by the assumption $a=b$ with E2] $= 1$ [G3]. So by T8, $a^{-1}=b^{-1}$.

Now we have the equality axioms for all our symbols (in T1, E1, E2,

E3, T10). So hereafter we can replace on the basis of equalities without restriction (Theorem 30 § 29).

Furthermore, having now both the replacement property of equality and the associative law G1, we shall hereafter write (rs)t and r(st) as "rst", so applications of G1 become tacit.

T11. $\vdash ac=bc \supset a=b$. (Right cancellation.)

PROOF. Assume $ac = bc$. Then $a = a1 = acc^{-1} = bcc^{-1} = b1 = b$.

T12. $\vdash ca=cb \supset a=b$. (Left cancellation.)

T13. $\vdash (a^{-1})^{-1}=a$. (Inverse of inverse of a.)

PROOF. $(a^{-1})^{-1} = 1(a^{-1})^{-1} = aa^{-1}(a^{-1})^{-1} = a1 = a$.

T14. $\vdash (ab)^{-1}=b^{-1}a^{-1}$. (Inverse of a product.)

HINT FOR PROOF: Use T8.

Next we describe [3] a simpler formal system **Gp** of group theory. For this, we omit the function symbol $^{-1}$ and the individual symbol 1 from the list of formal symbols, correspondingly simplifying the definition of "term" and thence of "formula". We replace G2 and G3 by the following two axioms.

Gp2. $\exists b\ ab=c$. (Existence of right quotient.)
Gp3. $\exists a\ ab=c$. (Existence of left quotient.)

That is, the postulates are those of the predicate calculus and the six axioms E1–E3, G1, Gp2, Gp3. This corresponds to defining a *group G* as a system satisfying the axioms G1, Gp2, Gp3.

Th1. $\vdash a=a$. (Reflexive property of equality.)

PROOF. Substituting a for c in Gp2, $\exists b\ ab=a$. Preparatory to $\exists$-elim., assume $ab=a$. Substituting into E1, $ab=a \supset (ab=a \supset a=a)$. Using $\supset$-elim. twice, $a=a$. Now we complete the $\exists$-elim.

Th2. $\vdash a=b \supset b=a$. (Symmetric property of equality.)

Th3. $\vdash a=b \,\&\, b=c \supset a=c$. (Transitive property of equality.)

Th4. $\vdash \exists i \forall a\ ai=a$. (Existence of right identity.)

PROOF (using E2, E3, G1 tacitly). Assume for $\exists$-elim. from formulas obtained from Gp2 and Gp3 by changes of bound variables (*74 § 24) and substitutions: **(1)** $\underline{i}b=b$, **(2)** $a\underline{j}=a$, **(3)** $\underline{k}j=i$, **(4)** $\underline{l}a=k$, **(5)** $i\underline{m}=j$, **(6)** $b\underline{n}=m$. (The x of each $\exists$-elim. will be the underlined variable.) Now $i = kj$ [(3)] $= laj$ [(4)] $= la$ [(2)] $= k$ [(4)], so by (3): **(7)** $ij=i$. Also $j = im$ [(5)] $= ibn$ [(6)] $= bn$ [(1)] $= m$ [(6)], so by (5): **(8)** $ij=j$. By (7) and (8), $i=j$, so by (2): **(9)** $ai=a$. Now we complete five $\exists$-elims., discharging successively the assumptions (6)–(2), use $\forall$- and $\exists$-introd. to get $\exists i \forall a\ ai=a$, and finally discharge (1) by the sixth $\exists$-elim.

Now we shall sketch why **Gp** formalizes essentially the same theory as **G**. Since the axioms of **Gp** (unlike those of **G**) are symmetric with respect to ·, without further work we can write down Th5.

Th5. $\vdash \exists j \forall b\, jb = b$. (Existence of left identity.)
Th6. $\vdash \forall a\, ai = a \;\&\; \forall b\, jb = b \supset i = j$. (Equality of right and left identity.)
Th7. $\vdash \forall a\, ai = a \;\&\; \forall a\, ax = a \supset i = x$. (Uniqueness of right identity.)
(HINT: use Th5, Th6.) For the abbreviation "$\exists!$xA(x)", cf. § 29.
Th8. $\vdash \exists i[\forall a\, ai = a \;\&\; \forall x(\forall a\, ax = a \supset i = x)]$, abbreviated
$\vdash \exists! i\, \forall a\, ai = a$. (Unique existence of right identity.)
(HINT: use Th4, Th7.) The formula $\forall a\, ai = a$ expresses the representing predicate $1 = i$ of the right identity 1 as a 0-place function (cf. end § 38), and the provable formula of Th8 says that $\forall a\, ai = a$ is a representing predicate. Now the conditions for the application of the eliminability theorem cited in Footnote 156 § 38 are fulfilled. By that theorem, we can add the individual symbol 1 to the symbolism of **Gp**, and the formula $\forall a\, a1 = a$, or with the same effect G2, to its axioms, to obtain a system $\mathbf{Gp_1}$ with the following properties. In $\mathbf{Gp_1}$, exactly the same formulas not containing 1 are provable as in **Gp**. Each provable formula of $\mathbf{Gp_1}$ containing 1 can be paraphrased, as illustrated at the end of § 38, by a provable formula of **Gp**.

Now we develop similarly the theory of the inverse a^{-1} in $\mathbf{Gp_1}$.
$\text{Th}_1 9.$ $\vdash 1a = a$. $\text{Th}_1 10.$ $\vdash \exists b\, ab = 1$. $\text{Th}_1 11.$ $\vdash \exists c\, ca = 1$.
$\text{Th}_1 12.$ $\vdash ab = 1 \;\&\; ca = 1 \supset b = c$. $\text{Th}_1 13.$ $\vdash ab = 1 \;\&\; ax = 1 \supset b = x$.
$\text{Th}_1 14.$ $\vdash \exists! b\, ab = 1$. (Unique existence of right inverse of a.)
The formula $ab = 1$ expresses the representing predicate $a^{-1} = b$ of a^{-1}. By a second application of eliminability, we can add to $\mathbf{Gp_1}$ the function symbol $^{-1}$ and the axiom G3, obtaining a system $\mathbf{Gp_2}$. But by Exercise 39.2, Gp2 and Gp3 are redundant in $\mathbf{Gp_2}$ (being provable in its "subsystem" **G**), and thus they can be omitted as axioms of $\mathbf{Gp_2}$, doing which gives **G**.

We conclude hence that in **G** exactly the same formulas not containing $^{-1}$ or 1 are provable as in **Gp**, while the provable formulas of **G** containing $^{-1}$ or 1 can all be paraphrased by provable formulas of **Gp**. Thus **G** is in the same relation to **Gp** as the richer systems of number theory alluded to at the end of § 38 are to the system **N**.

Our examples of groups except (11), (13), (14) and (15) satisfy an additional axiom:

G4. $ab = ba$. (Commutative law.)

Such groups are called *commutative* or *Abelian*. Formal systems [4] **AG** and [5] **AGp** for Abelian groups are obtained by adding G4 to the postulates of **G** and **Gp**, respectively.

To see that (11) does not satisfy G4, say the square is in a horizontal plane with edges running north and east. Let a be a clockwise rotation of 90° (as seen from above) about a vertical axis through the center of the square, and b be a rotation of 180° about a horizontal axis running east through the center. Then the results ab and ba of performing these two rotations in different orders are different, as the student may verify. Similarly with (13), (14) and (15).

Our groups are infinite, i.e. have infinitely many elements in the set G_0, except (10), (11) and (14), which are finite. —

As we remarked in § 37, we get specific formal systems of propositional and predicate calculus by supplying suitable definitions of "atom", and of "ion" or of "ion" and meson".

We begin with [6] the *pure propositional calculus* **Pp**. For this, we introduce a new species of formal symbols $\mathscr{A}, \mathscr{B}, \mathscr{C}, \ldots$ (script capitals), called *proposition letters.*[160] The problem of what to do after the twenty-sixth letter can be handled as we did for number theory **N**, by using a formal symbol $_|$ to manufacture additional proposition letters. Most treatments simply assume there is a potentially infinite list of proposition letters (and likewise of variables for number theory). The other formal symbols are $\sim$, $\supset$, &, $\vee$, $\neg$ and parentheses. "Formulas" are defined as in § 1 with the proposition letters now being the atoms (and with parentheses handled as in § 38, or in some other suitably explicit manner). If we have to talk about both these formulas and those of formal number theory **N** in the same context, the two kinds may be distinguished as *proposition letter formulas* and *number-theoretic formulas*. The postulates, of course, are Axiom Schemata 1a–10b and the $\supset$-rule.

Similarly, we obtain [7] the *pure predicate calculus* **Pd** by using as the ions $\mathscr{A}$, $\mathscr{A}(-)$, $\mathscr{A}(-,-)$, . . ., $\mathscr{B}$, $\mathscr{B}(-)$, $\mathscr{B}(-,-)$, ... We call these *predicate letters.*[160] Infinitely many of them are to be available with each number $n \geq 0$ of open places; we may get these either by assuming an infinity of script capital letters as formal symbols, or by using $_|$ to manufacture others after the twenty-six the printer really has. The other formal symbols are $\sim$, $\supset$, &, $\vee$, $\neg$, $\forall$, $\exists$, variables a, b, c, . . . (infinitely many, or twenty-six with $_|$ to form others), the comma and parentheses. Thence "formula", or more specifically "predicate letter formula", is defined as in § 16 with the predicate letters now being the ions (and an explicit treatment of parentheses). The postulates are Axiom Schemata 1a–10b, the $\forall$- and $\exists$-schema (as originally stated in § 21 with just variables as the terms), and the $\supset$-, $\forall$- and $\exists$-rule.

[160] In systems having particular axioms and a postulated substitution rule (in contrast to axiom schemata and a derived substitution rule),[86] $\mathscr{A}, \mathscr{B}, \mathscr{C}, \ldots$ are called *proposition variables*. Cf. § 9 just after Example 4.

In similar treatments of the predicate calculus, what we call "predicate letters" are called "predicate variables" (or some by authors "functional variables");[51] and similarly with functions, our "function letters" becoming "function variables".

By adding to the predicate letters as ions also (—)=(—) (read "— equals —"), and to the postulates the open equality axioms for = and for each predicate letter with $n > 0$ open places, we obtain [8] the *pure predicate calculus with equality* **Pd=**.

Now let us take as mesons (§ 28) the expressions $\mathscr{f}, \mathscr{f}(—), \mathscr{f}(—,—), \ldots, \mathscr{g}, \mathscr{g}(—), \mathscr{g}(—,—), \ldots$, which we call *function letters.*[160] Here $\mathscr{f}, \mathscr{g}, \mathscr{h}, \ldots$ are lower case script letters from the middle of the alphabet, which are to be removed from use as variables. (We may either assume an infinite supply of these letters as well as of the letters for variables, or reserve finitely many for each use and manufacture others using $_|$.) With the rest of the formation rules and the postulates taken as in § 28 or § 29, we thence obtain [9] the *pure predicate calculus with functions* **Pdf** and [10] the *pure predicate calculus with functions and equality* **Pdf=**.

In these pure systems, the atoms or ions (or also mesons) are provided completely generally, without reference to any particular application of logic.

But also we can have formal systems of propositional calculus and of predicate calculus without or with equality by using the formation rules of a more complicated or more special system. Thus using "formula" as for **Pd**, **Pd=**, **Pdf**, **Pdf=**, **N**, **G** or **Gp** with just the postulates of the propositional calculus (Axiom Schemata 1a–10b and the ⊃-rule), we get respective systems [11]–[17] of propositional calculus. Using "formula" (or "term" and "formula") as for **Pd=**, **Pdf=**, **N**, **G** or **Gp** with the postulates of the predicate calculus **Pd** or **Pdf**, we get systems [18]–[22] of predicate calculus. Using "term" and "formula" as for **N**, **G** or **Gp** with the postulates of the predicate calculus with equality **Pdf=**, we get systems [23]–[25] of that. The treatments of propositional calculus, of predicate calculus and of predicate calculus with equality in Part I were framed to apply to any of these systems and others.

The systems of logic [15]–[17], [20]–[22] and [23]–[25] using the formation rules of **N**, **G** or **Gp** are examples of *applied* systems of logic, as their formation rules are chosen with a view to the application of the system to a more or less specific subject matter. Thus, an *applied predicate calculus* (illustrated by the "number-theoretic predicate calculus" [20]) has a list of $s \geq 1$ predicate symbols or possibly other ions $P_1, \ldots, P_s$ (each with a specified number $p_i \geq 0$ of open places) and a list of $t \geq 0$ function symbols or possibly other mesons $f_1, \ldots, f_t$ (each with a specified number $q_i \geq 0$ of open places).[161] Thence "term" (≡ "variable" if $t = 0$) and "formula" are defined as above (§ 16, or for $t > 0$ § 28, but with an

[161] This covers most uses of the term "applied predicate calculus". One can also consider systems with infinitely many predicate or function symbols, and systems with a mixture of these notations and those of the pure systems.

explicit treatment of parentheses § 38). For the number-theoretic predicate calculus, $P_1, \ldots, P_s$ is (—)=(—) simply ($s = 1, p_1 = 2$), and $f_1, \ldots, f_t$ is (—)+(—), (—)·(—), (—)′, 0 ($t = 4$, $q_1 = q_2 = 2$, $q_3 = 1$, $q_4 = 0$). In an applied predicate calculus, the predicate and function symbols or other notations have one or more "intended" or "standard" interpretations. (Of course, these can play no part in the metamathematics.) Applied systems of logic give a formalization of the logic to be used in a particular subject, e.g. number theory, directly in the formalized language of that subject. This corresponds closely to how the mathematician or layman uses logic.

By using in place of Axiom Schema 8 the intuitionistic Axiom Schema 8^1, we obtain intuitionistic systems [26]–[50] corresponding to the classical systems [1]–[25]. (Cf. ends §§ 12, 25.)

EXERCISES. 39.1. For the group of rotations of a square into itself, show that there are 4 or 8 elements according as the square must remain in its plane or not during the motion. What are the interpretations of 1 and $^{-1}$ for this group?

39.2. Using T1–T5, show that $\vdash \exists b\; ab=c$ and $\vdash \exists a\; ab=c$.

39.3. Prove T7, T9, T12, T14.

39.4. The part of the proof of Th4 ending with (9) demonstrates that (1), (2), (3), (4), (5), (6) $\vdash$ (9). Show explicitly the rest of the steps (after (9)) in the format (B_3), and verify the correctness of each ∃-elim. and the ∀-introd.

39.5. Prove Th6–Th8 and $Th_1 9$–$Th_1 14$.

CHAPTER V

COMPUTABILITY AND DECIDABILITY

§ 40. Decision and computation procedures. Consider a given countably infinite class of mathematical or logical questions, each of which calls for a "yes" or "no" answer.

Is there a method or procedure by which we can answer any question of the class in a finite number of steps?

In more detail, we inquire whether for the given class of questions a procedure can be described, or a set of rules or instructions listed, once and for all to serve as follows. If (*after* the procedure has been described) we select *any* question of the class, the procedure will then tell us how to perform successive steps, after a finite number of which we will have the answer to the question we selected. In performing the steps, we have only to follow the instructions mechanically, like robots; no insight or ingenuity or invention is required of us. After any step, if we don't have the answer yet, the instructions together with the existing situation will tell us what to do next.[162] The instructions will enable us to recognize when the steps come to an end, and to read off from the resulting situation the answer to the question, "yes" or "no".

In particular, since no human performer can utilize more that a finite amount of information, the description of the procedure, by a list of rules or instructions, must be finite.

If such a procedure exists, it is called a *decision procedure* or *algorithm* for the given class of questions. The problem of discovering a decision procedure is called the *decision problem* for this class.

For example, there is a decision procedure for the class of questions "Does a divide b?" (or "Is a a factor of b?") where a and b are any positive integers. It consists in performing the ordinary long division of b by a and observing whether or not the remainder is 0.

Likewise, there is an algorithm for determining whether or not an algebraic equation

$$(1) \qquad a_0x^n + a_1x^{n-1} + \ldots + a_{n-1}x + a_n = 0 \qquad (a_0 \neq 0, n > 0)$$

[162] In practice such procedures are often described incompletely, so that some inessential choices may be left to us. For example, if several numbers are to be multiplied together, it may be left to us in what order we multiply them.

with integral coefficients $a_0, a_1, \ldots, a_{n-1}, a_n$ has a rational root. It depends on the theorem that, if such an equation (1) has a rational root p/q (p, q integers), then p must be a factor of a_n and q a factor of a_0. So only finitely many rational numbers p/q are candidates for roots; and we can try them each in turn.

As a third illustration, there is a decision procedure for deciding whether or not, for given integers a, b and c, the equation $ax + by + c = 0$ has integral solutions for x and y. For this procedure, which is based on Euclid's "greatest common divisor algorithm", we refer to textbooks on elementary number theory (e.g. MacDuffee 1954 § 9).

For a given formal system **S**, consider the following three general questions, i.e. (countably infinite) classes of particular questions: "Is a given formal expression a formula?", "Is a given finite sequence of formulas a proof?", "Is a given formula provable?".

As we illustrated in § 38, any particular question of the first class can be answered by seeking a proper pairing of the parentheses in the expression. If one is found, then with the help of it we can attempt to retrace the steps by which the expression, if it is a formula, was built up under the definitions of "term" and "formula". In so doing, we shall find out whether the expression can or cannot be thus constructed. To answer any given question of the second class, we merely consider each formula in the given sequence in order, and examine whether it is an axiom or follows from earlier formula(s) by one of the rules of inference. The objects which must be examined to answer a question of either of these two classes are contained as parts of the finite object to which the question applies.

The third class of questions is fundamentally different. To show directly from the definition that a formula is provable, one must exhibit a proof of it. But the proof, if there is one, need not be made up only of parts of the formula itself. One must therefore look elsewhere than within the given object to answer the equation. The definition of a proof of a given formula sets no bound on the length of the proof. To examine all possible proofs without bound on their length is not a procedure which leads to the answer to the question in finitely many steps in the case the formula is not provable. So this third decision problem, i.e. the decision problem for provability in the system **S**, unlike the first two, is not trivial. If there is a decision procedure, it is one which is not afforded almost immediately by the definition of "provable formula". Also this decision problem for a formal system **S** is of especial interest. So it is often called *the* decision problem for the formal system.

For provability in the propositional calculus, there is the decision procedure found by Post in 1921; i.e., to determine whether or not $\vdash$ E, it suffices to compute the truth table for E, and see whether or not it has all t's (for, by Theorems 14 and 12, $\vdash$ E if and only if $\models$ E).

The recognition of the decision problems for formal systems goes back to Schröder 1895, Löwenheim 1915, and Hilbert 1918.

It would be of especial import to have a decision procedure for the system **N** of elementary number theory (§ 38). For then the solution to many long-unsolved particular problems of elementary number theory could be obtained mechanically. This is assuming that only formulas true under the interpretation are provable in **N**. For example, we could then settle the problem of Fermat's "last theorem". About 1637 Fermat claimed he had a proof that the equation $x^n + y^n = z^n$ has no solution in positive integers x, y, z, n for $n > 2$. No one since has succeeded in proving or disproving this proposition. Fermat's "last theorem" can be expressed in **N** by the formula

$$\text{F:}\quad \neg\exists x\exists y\exists z\exists n[x>0 \,\&\, y>0 \,\&\, z>0 \,\&\, n>2 \,\&\, \text{A}(x, y, z, n)]$$

where $\text{A}(x, y, z, n)$ is obtained by paraphrasing $x^n + y^n = z^n$ to avoid the exponent function (cf. end § 38). Thus to get $\text{A}(x, y, z, n)$, we first find by known methods (Footnotes 155 and 156 § 38) a formula $\text{E}(x, n, u)$ expressing $x^n = u$; then $\text{A}(x, y, z, n)$ can be

$$\exists u\exists v\exists w[\text{E}(x, n, u) \,\&\, \text{E}(y, n, v) \,\&\, \text{E}(z, n, w) \,\&\, u+v=w].$$

If Fermat's "last theorem" is false, this fact can be shown by computation with a suitable quadruple (x, y, z, n) as counterexample. Corresponding to this informal remark, it can be shown that in this case the formula $\neg$F is provable in **N**. We cannot give the details in this brief treatment; it comes under the claim we made in § 38 that **N** is adequate for the usual elementary number theory. If Fermat's "last theorem" is true, then, by the supposition we made here that in **N** only true formulas are provable, $\neg$F is unprovable. So a decision procedure for provability in **N** would enable us to decide by a finite number of mechanical steps whether Fermat's last theorem is true or false, by applying the procedure to determine whether $\neg$F is unprovable or provable, respectively.[163]

[163] This would constitute at least a significant *theoretical* gain over the present situation (1967), in which no sequence of mechanical steps is known to lead after finitely many steps to the answer to the question whether Fermat's "last theorem" is true or false.

Practically, the answer to the question might still be beyond our reach by the means described. For, in applying the given decision procedure for the system **N** to the formula $\neg$F, more space and time than we have available might be required to carry out the finitely many steps that lead to the decision.

For what we do in this book, it is not essential that we concern ourselves with this matter of whether a given decision procedure for a class of questions is or is not practical to use in answering certain questions of the class. (That falls under "computer sciences".)

The efforts expended over centuries in searching for solutions to this and other famous problems of number theory make it implausible that a decision procedure for **N** should exist. It might have seemed equally implausible in 1918 that mathematics could find a way to *prove* that there can be no decision procedure for **N**. But exactly this was done by Church in 1936 on the basis of a thesis of his which we shall propound in § 41.

We begin by observing that, just as we may have a decision procedure or algorithm for a countably infinite class of questions each calling for a "yes" or "no" answer, we may have a *computation procedure* or *algorithm* for a countably infinite class of questions which require as answer the exhibiting of some object.

For example, there is a computation procedure for the class of questions "What is the sum of two natural numbers a and b?". We learned this procedure in elementary school when we learned to add. The long division process constitutes an algorithm for the class of questions "For given positive integers a and b, what are the natural numbers q (the quotient) and r (the remainder) such that $a = bq + r$ and $r < b$?". Thirdly, there is the algorithm which bears Euclid's name for the class of questions, "What is the greatest common divisor of two positive integers a and b?". There is also an algorithm for the class of questions "Given a formula E in the propositional calculus and a list $P_1, \ldots, P_n$ of distinct atoms including all occurring in E, what is the truth table for E entered from these atoms?". The last three of these algorithms for what-questions are incorporated into three of the algorithms mentioned above for yes-or-no-questions.

The problem of finding a computation procedure or algorithm for a class of what-questions is the *computation problem* for the class of questions.

We have chosen here to formulate the idea of decision problems and computation problems only for countably infinite classes of questions (yes-or-no-questions or what-questions, respectively).

For a finite class of questions, the decision or computation problem (analogously formulated) is trivial from the classical standpoint. For (theoretically at least), it could be solved simply by preparing a list of the answers to all the questions of the class.

"What is the shortest distance by major highways between any two of the principal cities of the United States?" Assuming agreement on what are major highways and which are the principal cities, one can find the answer to any question of the class from the tabulations given on some highway maps.

"Given any permissible position in chess, can white win (irrespective of black's responses)?" It is notorious that in this case, although a finite list of the answers for all positions exists theoretically, for practical purposes it is unavailable. If it were, the fun in chess would be lost.

"Is a given one of a certain finite set of propositions $\{A, B, C, D, E\}$ true?" Here A, B, C, D, E are supposed to be five fixed propositions, so there are only five questions in our class. An algorithm would be provided by the right list of five yes's and no's. This would be short and simple. However, if A is Fermat's "last theorem", no one to date can provide the algorithm. Only a classical mathematician considers it established that the algorithm exists. An intuitionist would consider the question whether the algorithm exists to be open until someone has solved the problem of Fermat's "last theorem". (Cf. § 36.)

The results to be developed in the rest of this chapter are independent of such differences of opinion between classical and intuitionistic mathematicians on when an algorithm exists.

Returning to our case of algorithms for countably infinite classes of questions, they would likewise always exist trivially from a classical standpoint, if we allowed the (descriptions of) algorithms to be infinite objects; we could again simply list all the answers. But we said an algorithm has to a method or procedure or set of rules which can be used, and which therefore must be finitely described. A *finite* set of instructions must be provided that will suffice to guide one to the answer of any one of an *infinite* set of questions.

A discussion of algorithms for uncountably infinite classes of questions is outside the scope of this book.

When we have a countably infinite class of questions, the different questions of the class will usually arise by giving different values ("arguments") to one or several variables or "parameters" in the general statement of the (class of) questions. In several examples above, a and b (or a, b and c) play the role of these variables. In our second example of a decision procedure, $a_0, \ldots, a_n$ are such variables but their number n also varies.

Because our class of questions is always (outside of the digression above) to be countably infinite, we can always enumerate the questions, say as $Q_0, Q_1, Q_2, \ldots, Q_a, \ldots$. Then a can be the variable or parameter. In the case of yes-or-no-questions, by putting $P(a) \equiv$ {the answer to Q_a is "yes"}, the infinite class of yes-or-no-questions becomes a one-place number-theoretic predicate $P(a)$.

Suppose instead the questions are what-questions. We now further assume that the objects to be produced in answering the questions come from a countably infinite class, which we can enumerate as y_0, y_1, $y_2, \ldots, y_b, \ldots$. By putting $f(a) = b$ if the answer to Q_a is b, the infinite class of what-questions becomes a one-place number-theoretic function $f(a)$.

So via an enumeration, each countably infinite class of yes-or-no questions we are to consider can be reduced to the form "Is the value of the number-theoretic predicate $P(a)$ for a as argument true?"; of what-questions, to "What is the value b of the number-theoretic function $f(a)$ for a as argument?".

In cases when the class of questions is already given by using a fixed number of variables with suitable ranges, it is often more convenient to render it directly by a number-theoretic predicate or function. For example, "Does a divide b?" can be rendered by a predicate $P(a, b)$, which indeed is commonly written "$a|b$". To standardize in the following treatment, we take our variables $a, b, c, \ldots, x, y, z$ to range over the natural numbers. For the present example, we can extend the "divides" notion to include 0 ($a|0$ is true for every a; $0|b$ is false except for $b = 0$). "What is the sum of a and b?" is obviously rendered by the function $a+b$.

Inversely, starting with any number-theoretic predicate $P(a_1, \ldots, a_n)$, we have the countably infinite class of questions "For given $a_1, \ldots, a_n$, is $P(a_1, \ldots, a_n)$ true?". Starting with any number-theoretic function $f(a_1, \ldots, a_n)$, we have the questions "For given $a_1, \ldots, a_n$, what is the value of $f(a_1, \ldots, a_n)$?".

Thus it comes to the same thing whether we talk about countably infinite classes of questions or about number-theoretic predicates or functions. If there is a decision procedure for a predicate (or for the class of questions arising from it), we call the predicate (or class of questions) *decidable*. Similarly, if there is a computation procedure for a function, we call the function *computable*.

Furthermore, the case of predicates can be reduced to the case of functions by defining the *representing function* $f(a_1, \ldots, a_n)$ of a predicate $P(a_1, \ldots, a_n)$, thus:

$$f(a_1, \ldots, a_n) = \begin{cases} 0 \text{ if } P(a_1, \ldots, a_n) \text{ is true,} \\ 1 \text{ if } P(a_1, \ldots, a_n) \text{ is false.} \end{cases}$$

In the one-variable case, $f(a)$ is what in § 33 we called the "representing function" of the set of a's for which $P(a)$ is true (in symbols, the set $\hat{a}P(a)$). It comes to the same thing to compute the value of $f(a_1, \ldots, a_n)$ and observe whether it is 0 or 1 as to decide whether $P(a_1, \ldots, a_n)$ is true or false. In other words, a decision procedure can be handled as a computation procedure by taking 0 for "yes" and 1 for "no".[164]

To emphasize the idea of an algorithm by contrast, we consider some more cases in which it is not clear that there is one. We have already mentioned that there is an algorithm for "Does $ax + by + c = 0$ have a solution in integers?", or written as a predicate $(Ex)(Ey)[ax + by + c = 0]$. Here we are using "$(Ex)$" and "$(Ey)$" in the same meaning as $\exists x$ and $\exists y$ in Chapter II; but we prefer henceforth to keep the logical symbolism we use outside a given formal system distinct from the symbolism we use inside, as far as is feasable.[165] The existence of this algorithm is not obvious from the start (unlike those for $a|b$ and $a+b$). Some theory due to Euclid is used in setting it up.

Let us generalize this example to consider, instead of the first-degree equation $ax + by + c = 0$, the second-degree equation $ax^2 + bxy + cy^2 + dx + ey + f = 0$ (a, b, c not all 0) or even any algebraic equation (polynomial equation) of any degree $n > 0$ in any number $m > 0$ of variables $x_1, \ldots, x_m$, and make our class of questions "Does any given algebraic equation with integeral coefficients have a solution for its variables in integers?". The decision problem for this class of questions is "Hilbert's tenth problem", included in his famous list of 23 "future" problems of mathematics 1900a. It has not been solved.

To make matters as simple as possible, consider some 2-place predicate $P(a, x)$ for which we do have an algorithm. It will not directly follow that there is an algorithm for $(Ex)P(a, x)$. The only procedure the *definition* of this predicate suggests *directly* is, after choosing a value of a, to start trying $P(a, 0), P(a, 1), P(a, 2), \ldots$ in the hope of finding one which is true. We can't be sure we will by this procedure find out in a finite number of steps whether $(Ex)P(a, x)$ is true or not. For (as we already noted for "$\exists x P(x)$" in § 36), if after any finite number of steps we haven't already found an x which makes $P(a, x)$ true, we won't know whether it's because

[164] Since we are concerned here only with the questions whether the propositions taken as values of the predicates are true or false, the predicates in this theory are treated extensionally, i.e. not distinguished from the corresponding logical functions, which become their representing functions on changing $\mathfrak{t}$, $\mathfrak{f}$ to 0, 1.[92]

Below we emphasize that sometimes the definition of a predicate or function does not directly give an algorithm for it, but an algorithm may be afforded by some theory about the function or predicate. For predicates, if we restore the intensional concept of them, we could then say that an (intensional) predicate P may be undecidable while an equivalent predicate P_1 is decidable.

[165] Beginning with this chapter (and in a few instances before)[23] we use the following informal logical symbolism: $\equiv$ ("equivalent"), $\rightarrow$ ("implies"), & ("and"), $\vee$ ("or"), $\bar{}$ (thus $\bar{A}$; "not"), (x) ("for all x"), (Ex) ("(there) exists (an) x (such that)").

$(Ex)P(a, x)$ is false for that a, or because we haven't gone far enough in our search.

To give this example in terms of number-theoretic functions rather than predicates (i.e. propositional functions), suppose we have a computation procedure for $f(a, x)$. We do not at once know whether there is one for the function

$$f(a) = \begin{cases} 0 \text{ if } (Ex)f(a, x)=0, \\ 1 \text{ otherwise.} \end{cases}$$

If $f(a, x)$ is the representing function of $P(a, x)$, then $f(a)$ is the representing function of $(Ex)P(a, x)$.

The recognition that for some (countably infinite) classes of mathematical questions we have algorithms, and that for others we at least don't know any, goes far back in mathematical history. The Greeks, including Euclid, sought algorithms; and the term "algorithm" is derived from the name of the ninth century Arabian arithmetician al-Khuwarizmi.

To recapitulate, we have seen that computation and decision problems can all be reduced to the computation problems for number-theoretic functions.

So to advance toward our objective (Church's theorem), we have to deal with the question: For what number-theoretic functions are there computation procedures (or algorithms)? Briefly: What is the class of "computable" functions?

What we have to go on in considering this question is a somewhat vague intuitive idea of what constitutes a computation procedure. A computation procedure may consist in essentially a direct application of the definition of the function, or it may consist in something considerably different which mathematical theory shows must lead to the same function values as the original definition requires.

Although our intuitive notion of a computation procedure is vague, it is nevertheless real, as is shown by two circumstances. First, it does not leave mathematicians in any doubt or disagreement that they do have computation procedures for many particular functions, e.g. $a+1$, $a+b$, $a \cdot b$, a^b, $a!$, $\max(a, b)$ (= the maximum of a and b, i.e. the greater if $a \neq b$, the common value if $a = b$), $\min(a, b)$ (= the minimum of a and b), $a \dot{-} 1$ (= $a-1$ if $a \geq 1$, 0 if $a = 0$), $a \dot{-} b$ (= $a-b$ if $a \geq b$, 0 if $a < b$), $[a/b]$ (= the quotient when a is divided by b if $b \neq 0$, 0 if $b = 0$), $[\sqrt{a}]$ (= the greatest natural number whose square is $\leq a$), $[e^a]$ (= the greatest natural number $\leq e^a$), etc. Second, there is likewise no doubt, in other particular cases, that the definition of a function or a given equivalent of its definition does not directly give a computation procedure. For example, we agree that the definition of $f(a)$ displayed above does not give a computation procedure for it.

We are only recapitulating what has not been disputed among mathematicians for over two millenia (if we correctly understand mathematical history).

This intuitive notion of a computation procedure, which is real enough to separate many cases where we know we do have a computation procedure before us from many others where we know we don't have one before us, is vague when we try to extract from it a picture of the totality of all possible computable functions. And we must have such a picture, in exact terms, before we can hope to prove that there is no computation procedure at all for a certain function, or briefly to prove that a certain function is uncomputable. Something more is needed for this.

Most mathematical logicians agree that it was found in 1935 (and published in 1936), as we shall see in the next section. —

Hereafter, we may say that something can be done "effectively", or that an operation or process is "effective", as a brief way of saying that there is an algorithm for it (i.e. a decision or computation procedure).

EXERCISES. 40.1. For the following predicates and functions, or classes of questions, do we have an algorithm or do we lack such (at least without more knowledge)? For (a) and (b), assume $P(a, x)$ decidable.

(a) $f(a) = \begin{cases} \text{the least } x \text{ such that } P(a, x) \text{ if } (Ex)P(a, x), \\ 0 \text{ otherwise.} \end{cases}$

(b) $f(a, b) = \begin{cases} \text{the least } x \leq b \text{ such that } P(a, x) \text{ if } (Ex)[x \leq b \;\&\; P(a, x)], \\ b+1 \text{ otherwise.} \end{cases}$

(c) Is a a prime number?

(d) What is the nth prime number? (Assume Euclid's theorem that there are infinitely many primes, § 38.)

(e) $f(a) = \begin{cases} a+1 \text{ if Fermat's "last theorem" is true,} \\ a \text{ if Fermat's "last theorem" is false.} \end{cases}$

(f) For given formulas $A_1, \ldots, A_m$, B in the propositional calculus, does $A_1, \ldots, A_m \vdash B$ hold in the propositional calculus?

(g) For given formulas $A_1, \ldots, A_m$, B in the predicate calculus, does $A_1, \ldots, A_m \vdash B$ hold in the predicate calculus?

(h) Given formulas $A_1, \ldots, A_m$ in the predicate calculus, is a given finite sequence of formulas a deduction from $A_1, \ldots, A_m$ in the predicate calculus holding are all variables constant?

(i) For a given finite domain D and a given formula E in the predicate calculus, does $\bar{\bar{D}}$-$\models$ E hold? (Cf. § 17).

(j) For a given formula E in the predicate calculus, does $\models$ E hold?

(k) (For calculus students.) Does a given elementary function (say one composed from rational numbers and the variable x using finitely many times addition, multiplication, division, powers, roots, and trigonometric and exponential functions and their inverses) have an indefinite integral which is elementary?

(l) (For algebra students.) Does a given system of m linear equations in n unknowns with integral coefficients have a solution?

40.2. Show that $[\sqrt{a}]$ is computable.

40.3*. (For calculus students.) Show that $[e^a]$ is computable. (Use the result of Hermite in 1873 that e is transcendental; cf. § 33.)

§ 41. Turing machines, Church's thesis. The situation in 1935 was that a certain exactly defined class of computable number-theoretic functions considered by Church and Kleene during 1932–35, called the "λ-definable functions", had been found to have properties strongly suggesting that it might embrace all functions which can be regarded as computable under our vague intuitive notion. This result was somewhat unexpected, since initially it was not clear that the class contained even the particular computable function $a \dot{-} 1$ mentioned above, and a proof in 1932 (publ. 1935) that it did was the present author's first piece of mathematical research. Another class of computable functions, called the "general recursive functions", defined by Gödel in 1934 building on a suggestion of Herbrand, had similar properties. It was proved by Church 1936 and Kleene 1936a that the two classes are the same, i.e. each λ-definable function is general recursive and vice versa.

Under these circumstances Church proposed the thesis (published in 1936) that all functions which intuitively we can regard as computable, or in his words "effectively calculable", are λ-definable, or equivalently general recursive. This is a thesis rather than a theorem, in as much as it proposes to identify a somewhat vague intuitive concept with a concept phrased in exact mathematical terms, and thus is not susceptible of proof. But very strong evidence was adduced by Church, and subsequently by others, in support of the thesis.

A little later but independently, Turing's paper 1936–7 appeared in which another exactly defined class of intuitively computable functions, which we shall call the "Turing computable functions", was introduced, and the same claim was made for this class; this claim we call *Turing's thesis*. It was shortly shown by Turing 1937 that his computable functions are the same as the λ-definable functions, and hence the same as the general recursive functions. So Turing's and Church's theses are equivalent. We shall usually refer to them both as *Church's thesis*, or in connection with that one of its three versions which deals with "Turing machines" as *the Church-Turing thesis*. Post in 1936, independently of Turing, published rather briefly a formulation fundamentally the same as Turing's. In 1943 he published a fourth equivalent, using ideas from unpublished work of his in 1920–22.[166] Still another equivalent formulation is provided by Markov's theory of algorithms 1951c.

Turing's machine concept arises from a direct effort to analyze computation procedures as we know them intuitively into elementary operations. Turing argued that repetitions of his elementary operations would suffice for any possible computation. For this reason, Turing computability suggests the thesis more immediately than the other equivalent notions, and so we choose it for our exposition.

Turing described a kind of theoretical computing machine. This differs in two respects from a human computer working under preassigned instructions or an actual *digital* computing machine (such as a desk calculator, or a high-speed computer with electronic tubes or transistors). These are respects in which we idealize by divesting the human computers and the physical machines of their practical limitations.

First, a "Turing machine" is not liable to errors; i.e. it obeys the intended laws for its action without any deviations.

Second, a "Turing machine" is given a potentially infinite memory. That is, although the amount of information stored at any one time is finite, there is no upper bound on this amount. Information to be stored may include (at one time or another) the statement of the particular question given the machine, the scratch work performed by the machine while arriving at the answer, and the answer. To allow for unlimited storage of such information, we consider as separate the machine proper and a peripheral storage facility, which we will take to be an infinite "tape".

The machine proper, which does the computing and thus determines what function is computed, admits only a fixed finite number of possible "states". It represents the finite list of rules or finite description of a procedure in our intuitive notion of an algorithm, § 40. (Also information, but only up to a fixed amount, can be stored momentarily by the machine's assuming one or another of its "states".)

Now we formulate our notion of a *Turing machine* in detail. We number moments of time for the operation of the machine as 0, 1, 2, ... At any given moment, the machine shall be in one of $k+1$ *states*, which we number 0, 1, . . ., k. *State* 0 we call the *passive state*; the others, *active states*. A *linear tape* ruled in *squares* passes through the machine (when it is set up for operation). The tape is potentially infinite to the right. Each square is either *blank* s_0 or has printed on it one of a given finite list of *symbols* $s_1, \ldots, s_j$; so $s_0, \ldots, s_j$ are the possible *square*

[166] An account (from 1941) of this work of Post in the 1920's was published posthumously in Davis 1965 pp. 338–433.

conditions. However, only a finite number of squares are printed at any moment in the use of the machine. At each moment, beginning with Moment 0, one square of the tape is *scanned* by the machine.

Now consider any moment when the machine is in one of its active states $1, \ldots, k$. Between this moment and the next, the machine performs an *act* consisting of a sequence of three operations (a), (b), (c), each operation coming from the respective category as follows: (a) print one of the symbols $s_1, \ldots, s_j$ on the scanned square (supposed blank at the given moment), or erase the scanned square (supposed printed at the given moment), or erase and print one of $s_1, \ldots, s_j$ on the scanned square (supposed printed at the given moment), or make no change in the scanned square; (b) move the tape so as at the next moment to scan the square next to the left of the scanned square (briefly, move left), or leave the tape unmoved (briefly, stay centered), or move the tape so as at the next moment to scan the square next to the right of the scanned square (briefly, move right); (c) change to another state, or remain in the same state. What act (within these possibilities) is performed, between a given moment at which the machine state is active (one of $1, \ldots, k$) and the next moment, is determined by the machine state and the scanned-square condition (one of $s_0, \ldots, s_j$) at the given moment, with one exception to be explained presently. We call the machine state together with the scanned-square condition at a given moment the *configuration*. In contrast, the machine state together with the specification of the scanned square and the entire printing on the tape we call the (*machine vs. tape*) *situation*.

The exceptional case in which the configuration at the given moment does not determine the act is that the configuration would call for a motion left when the scanned square is already the leftmost square of the tape. Then the (b) and (c) parts of the act are altered to: stay centered, and assume the passive state (briefly, stop). The machine "jams". We could avoid this exception by assuming a 2-way infinite tape.[167]

If at a given moment the machine is in the passive state 0, no act is performed between this and the next moment, i.e. the machine does not print or erase, does not move, and does not change from state 0.

We shall give an illustration of the operation of a Turing machine presently. However, let us first define how such a machine is to be used to compute a number-theoretic function. (Turing used his machines primarily to carry out continuing computations of decimal fractions for real

[167] We begin thus in IM p. 357; but the results are essentially the same.[168,170]

Our discussions of Turing machines (beginning with one in our seminar on the foundations of mathematics at Wisconsin in 1941) follow Turing 1936–7 in the general conception of the behavior of the machines, but not in the detailed formulation and development. Cf. IM top p. 361.

numbers.) For this purpose, we must agree how the argument(s) (i.e. value(s) of the independent variable(s)) are to be represented on the tape, and how the machine is to give us the resulting value of the function. We shall make the supposition that all machines to be considered have among their symbols the tally mark "|"; say it is s_1. We shall represent natural numbers by sequences of tallies, "|" for 0, "||" for 1, "|||" for 2, ... To set up the machine and tape to compute for a given argument *a*, we shall arrange that: at the moment 0 the system consisting of machine and tape is started off so that the leftmost square of the tape is blank, *a* is represented by tallies on the next $a+1$ squares, all squares to the right of these are blank, the machine is scanning the rightmost printed square, and is in its first active state 1. In this situation, we say the machine is *applied to a as argument*. We say the machine *computes a value c for a as argument*, if, starting from this situation at Moment 0, the machine at some later moment assumes the passive state 0 ("stops") with a blank and $c+1$ tallies printed on the tape after the $a+1$ tallies representing the argument *a*, the tape being otherwise blank, and the rightmost printed square being again the scanned square.[168]

A given machine may compute a value for each natural number *a* as argument, or for some *a*'s but for others, or for no *a*'s. If, for each *a*, it computes a value *c* where $c = f(a)$, we say that the machine *computes* the function $f(a)$, and that $f(a)$ is *Turing computable*.

Similarly, for functions of more than one variable.

For example, a machine is applied to 1 as argument, if at Moment 0 it is in the following situation

where the "1" written over the third square shows that this is the scanned square and the machine is in state 1, and all squares to the right of those shown are blank. A machine computes the value 2 for 1 as argument if, having been stated at Moment 0 in the situation above, it will at some later moment x reach the situation

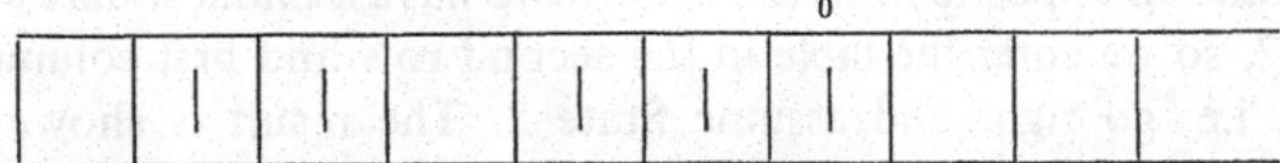

where again all squares to the right of those shown are blank. If in similar fashion, when the machine is started with any $a+1$ tallies on the

[168] In this case, the machine cannot during the computation attempt a motion leftward from the leftmost tape square; for, in that case State 0 would be assumed without the last of two successions of tallies separated by a blank being scanned.

tape, it will eventually stop with these followed by a blank and $a+2$ tallies (the last of them scanned), the machine computes the function $f(a) = a+1$; the above illustrates this for $a = 1$.

Now we shall describe a machine $\mathfrak{S}$ which does compute this function $f(a) = a+1$ (the successor function), and we shall follow it through its acts in computing for the argument $a = 1$. The machine will have only the one symbol "|". To save drawing a picture of the tape each time, we shall use sequences of numerals "0" and "1" where a "0" stands for a blank square and a "1" for a square printed with tally. Thus in the illustration above we would write the initial situation

$$0 \ 1 \ 1^1 \ 0 \ 0 \ 0 \ 0 \ 0 \ 0 \ldots,$$

and the situation at the moment the computation is completed

$$0 \ 1 \ 1 \ 0 \ 1 \ 1 \ 1^0 \ 0 \ 0 \ldots.$$

To describe a machine, we need merely tell for each of its active states $1, \ldots, k$ and each of the $j+1$ scanned-square conditions (blank s_0 or printed with $s_1, \ldots, s_j$) what act it is to perform. For the machine $\mathfrak{S}$ we describe now, there are to be 11 active states, and 2 conditions of the scanned square, namely blank and printed with "|" (or as we write them more conveniently, "0" and "1"). The acts it is to perform for each of the configurations of these 11 active states and 2 scanned-square conditions can be shown by the machine table at the left below (top p. 237). In the table "*P*" means "print", "*E*" means "erase"; "*L*", "*C*", "*R*" mean "left", "center" (i.e. don't move), "right"; and the number at the end of each table entry is the state the machine is to assume at the succeeding moment.

At the right, we follow out the acts of $\mathfrak{S}$ in computing $a+1$ for the argument $a = 1$. At Moment 0, in the sample computation at the right, we see that a printed square (i.e. a 1) is scanned in state 1; so we enter the machine table in the first row and second column finding "*R*2", i.e. go right and assume State 2. The resulting situation is shown in the sample computation opposite Moment 1. Now we have a blank square scanned in State 2, so we enter the table in the second row and first column, finding "*R*3", i.e. go right and assume State 3. The result is shown opposite Moment 2. Now the scanned square is still blank but the state is 3, so by the table (third row, first column, where it reads "*PL*4"), the machine prints a "|", goes left, and assumes State 4, with result shown at Moment 3. Continuing, we find that at Moment 23 the machine has computed the desired value 2 ($= a+1$ for $a = 1$) shown by three tallies. To see that the machine computes $f(a) = a+1$, we must convince ourselves that it will compute the value $a+1$ for every value of a as argument. We have done

Table for Machine $\mathfrak{S}$ which computes $f(a) = a+1$

Machine state	Scanned square condition 0	1
1	*C*0	*R*2
2	*R*3	*R*9
3	*PL*4	*R*3
4	*L*5	*L*4
5	*L*5	*L*6
6	*R*2	*R*7
7	*R*8	*ER*7
8	*R*8	*R*3
9	*PR*9	*L*10
10	*C*0	*ER*11
11	*PC*0	*R*11

Computation by Machine $\mathfrak{S}$ for $a = 1$

Moment	Machine vs. tape situation						
0	0	1	1^{1}	0	0	0	0
1	0	1	1	0^{2}	0	0	0
2	0	1	1	0	0^{3}	0	0
3	0	1	1	0^{4}	1	0	0
4	0	1	1^{5}	0	1	0	0
5	0	1^{6}	1	0	1	0	0
6	0	1	1^{7}	0	1	0	0
7	0	1	0	0^{7}	1	0	0
8	0	1	0	0	1^{8}	0	0
9	0	1	0	0	1	0^{3}	0
10	0	1	0	0	1^{4}	1	0
11	0	1	0	0^{4}	1	1	0
12	0	1	0^{5}	0	1	1	0
13	0	1^{5}	0	0	1	1	0
14	0^{6}	1	0	0	1	1	0
15	0	1^{2}	0	0	1	1	0
16	0	1	0^{9}	0	1	1	0
17	0	1	1	0^{9}	1	1	0
18	0	1	1	1	1^{9}	1	0
19	0	1	1	1^{10}	1	1	0
20	0	1	1	0	1^{11}	1	0
21	0	1	1	0	1	1^{11}	0
22	0	1	1	0	1	1	0^{11}
23	0	1	1	0	1	1	1^{0}
24	0	1	1	0	1	1	1^{0}

this only for $a = 1$; but the reader should not find it too hard to get the "hang" of how this machine operates to see that it does so for every a.

To illustrate computation of a function of 2 variables, a machine to compute $f(a, b) = a+b$, when started (for $a = 3$, $b = 1$) in the situation

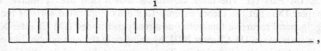

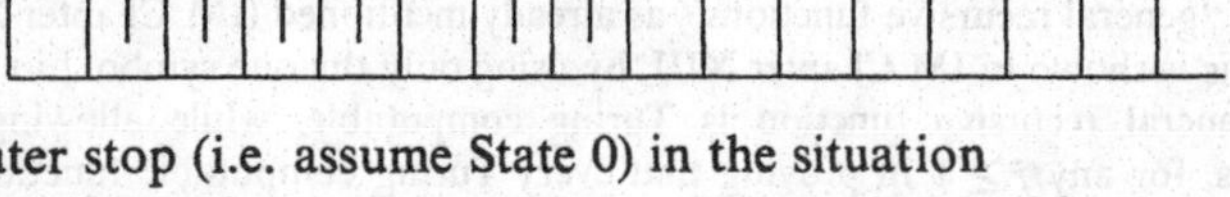

should later stop (i.e. assume State 0) in the situation

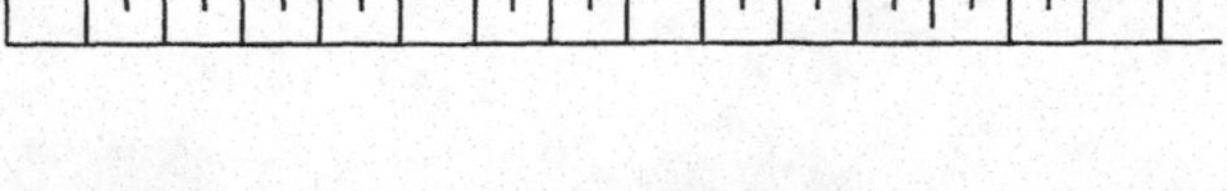

The machine that computes $a+1$ is sufficiently complicated so the reader may well wonder how to find machines to compute complicated effectively calculable functions. Of course we are only interested *here* in the theoretical possibility of finding a machine to compute any given effectively calculable function, and not in whether the machine operates efficiently. The business of finding machines can be systematized by starting from the theory of recursive functions.[169] This theory deals with recursive definitions of functions, such as

$$\begin{cases} a+0 = a, \\ a+b' = (a+b)', \end{cases} \quad \begin{cases} a \cdot 0 = 0, \\ a \cdot b' = a \cdot b + a, \end{cases} \quad \begin{cases} a^0 = 1, \\ a^{b'} = a^b \cdot a. \end{cases}$$

How these definitions define the functions was illustrated in § 38. The functions commonly used in number theory are definable by use of such recursions, and proceeding from the recursive definitions one can in a systematic way find corresponding Turing machines, after first setting up Turing machines for such simple operations as filling in with tallies all but the rightmost of a sequence of blank squares preceded and followed by a tally, copying a sequence of tallies, etc.

In giving the arguments to the machines and receiving the values, we have used only one symbol "|" besides the blank (i.e. two square conditions "0" and "1"), and likewise in all the action of Machine $\mathfrak{S}$. The definition of "Turing machine" allows more symbols. To write a *machine table* for a machine with j symbols $s_1, \ldots, s_j$, there will be $j+1$ columns. In the case $j > 1$, "P" is ambiguous; then in place of "P" or "E" or the absence of either we can write i in decimal notation to indicate that the square condition at the next moment is to be s_i $(0 \leq i \leq j)$.

While we are thus leaving open the possibility of using $j+1 > 2$ square conditions instead of only 2, the fact is that we get no larger class of computable functions thereby.[170]

It may seem a little anomalous that, after we have argued that any intuitive computation should be performable using only such operations as a Turing machine admits, we should find it quite an exercise to see that a simple function like $a+1$ is Turing computable. However we have

[169] E.g. IM Part III. The three definitions shown next are "primitive recursions" (IM Chapter IX); but the theory extends to more complicated sorts of recursions, all defining "general recursive functions" as already mentioned (IM Chapter XI).

[170] This is shown in IM Chapter XIII, by using only the one symbol | in proving that every general recursive function is Turing computable, while allowing j symbols $s_1, \ldots, s_j$ for any $j \geq 1$ in proving that every Turing computable function is general recursive.

The same method shows that we could have allowed a 2-way infinite tape without getting more (or less) computable functions.

started at the very beginning, and we have done it with a Turing machine with only one symbol "|" (besides the blank), though we didn't need to. Use of more symbols would make the Turing machine action better resemble informal computation.

Moreover, our illustration may unduly suggest that the computer is restricted to take an ant's eye view of his work, squinting at one square at a time. However, when more than the one symbol "|" (besides the blank) is allowed, it is possible to interpret what we mean by a symbol liberally. There is no reason why individual tape squares cannot be considered to correspond to whole sheets of paper, each ruled into finitely many squares each of which may receive one of finitely many primary symbols. What can be written on the whole sheet can then be construed as a single symbol for the Turing machine. If the sheets are ruled into 20 columns and 30 lines and 100 primary symbols are allowed, there are 600^{101} square conditions $s_0, \ldots, s_j$ ($j = 600^{101} - 1$), and we are at the opposite extreme. The schoolboy doing arithmetic on $8\frac{1}{2}'' \times 11''$ sheets of ruled paper would never need all this variety. He would usually operate just by changing the condition of the sheet of paper before him (= the scanned square), only occasionally transferring figures from one sheet to the next or preceding sheet in his stack of paper (= moving right or left).

From this point of view, the tape square represents whatever one looks at, at a given moment, which may not be small.

Actually the schoolboy at a given moment would directly perceive only a part of what appears on the sheet of paper; the rest, to the extent it affects his act from that moment to the next, he must remember in the form of a state of mind. Thus, psychologically, it is something between the small squares on the paper and the whole sheet that plays the role of the Turing machine tape square.

Another representation of the Turing machine tape is a stack of IBM cards, each card constituting a single square for the Turing machine.

Returning to the schoolboy, it is necessary for our concept that the paper be ruled in squares (at least imaginary squares), and that the symbols be from a given finite list, so as to make it exact what the whole finite range of possibilities is. That is, there must be only finitely many possible conditions for a sheet of paper or a Turing machine tape square. Likewise, there are to be only finitely many states of mind for the human computer or machine states for a given Turing machine. As Turing wrote, "the number of states of mind which need to be taken into account is finite. . . . If we admitted an infinity of states of mind, some of them will be 'arbitrarily close' and will be confused." The finite numbers of square conditions and of states of mind can of course be very large. We are considering *digital* computation, where the data are discrete. This is in contrast to *analog* computation, such as is performed by a slide rule or a differential analyzer, where the data are positions on a scale or scales, representing real numbers only to within some limit of accuracy. The inputs for our computations are to be natural numbers, or they could be something similar, like words in an ordinary language, or formal expressions in a formal system; likewise, the output. Discreteness must be preserved throughout the whole computation. At each moment, the existing (finite) configuration, comprising both the condition of the paper or the tape square, and the state of mind of the computer or the machine state, must completely determine the next configuration. Only thus will the computation lead to a discrete result, not to something approximate, and to such a result that is fully predetermined (when it exists at all) by the initial situation and the computer's instructions or the machine table.

There is however an intrinsic difficulty that our theory of computation must meet. This arises because we define a function $f(a)$ to be computable exactly if there is a machine which will compute its value for *all* choices of a. Thus we must be prepared to carry out computations for arbitrarily big values of a. The schoolboy computing on sheets of paper is not normally required to work with numbers too big to go on one of his sheets. If he had to do this, he also would then be forced to work in a groping manner as our machine $\mathfrak{S}$ does in computing values of $a+1$.

For further discussion of Church's thesis, or the Church-Turing thesis, we refer to the literature.[171] In the next sections, we deal with consequences of it.

[171] In IM the attempt is made to summarize all the evidence for Church's thesis: § 62 gives a general summary, supplemented by p. 352; and § 70 gives the part of the evidence which pertains to Turing's machines, supplementing the present discussion.

The following argument is given in IM, concluding on p. 352. A great stock of intuitively computable functions (all that have been investigated in this connection) are known to be Turing computable. Likewise, we have a great stock of methods or operations (for obtaining new intuitively computable functions from others) which are paralleled by operations for building new Turing machines from given Turing machines (or the analog in terms of recursiveness). If there were a function which is intuitively computable but not Turing computable, it would have to be "inaccessible" by any process of building up toward it from this stock of functions and operations already mastered. IM § 66 gives a theorem (the RECURSION THEOREM, Kleene 1938) applicable to the building process in general. It is hard to imagine how one could give a description, or a set of instructions, for a computation procedure that a human computer could follow, except by putting the description together out of simpler elements, already known, and then it would come under this theorem.

For an argument against Church's thesis and a reply, see Kalmár 1959 and Mendelson 1963. In reading such discussions, we must not lose sight of our intuitive concept of an algorithm or computation procedure (§ 40). An algorithm in our sense must be fully

EXERCISES. 41.1. (a) Write out the computation by Machine $\mathfrak{S}$ when the argument is 0, i.e. starting from the situation $0\ 1^1\ 0\ 0\ 0 \ldots$.
(b) Explain the operation of Machine $\mathfrak{S}$ so as to make it clear that $\mathfrak{S}$ does compute the function $a+1$.

41.2. Construct the table for a machine $\mathfrak{E}$ which decides whether the argument a is even; i.e. $\mathfrak{E}$ is to compute the function

$$f(a) = \begin{cases} 0 \text{ if } a \text{ is even,} \\ 1 \text{ if } a \text{ is odd.} \end{cases}$$

41.3. Modify the table for the machine $\mathfrak{S}$ to obtain:
(a) a machine $\mathfrak{I}$ which computes the identity function $f(a) = a$ (i.e. $\mathfrak{I}$ just copies a).
(b) a machine $\mathfrak{P}$ which computes the predecessor function $f(a) = a \dot{-} 1$ (end § 40).

41.4*. Show that, if $f(a)$ and $g(a)$ are Turing computable, so is the function $h(a) = f(g(a))$.

41.5. Show that no function $f(a)$ would be Turing computable under the above definition modified to allow only a finite tape in the machine.

41.6*. Show that no more functions $f(a)$ would be Turing computable under the above definition modified so as not to require the scratch work to be erased.

---

and finitely described before any particular question to which it is applied is selected. When the question has been selected, all steps must then be predetermined and performable without any exercise of ingenuity or mathematical invention by the person doing the computing. None of the skeptics about Church's thesis has come forward with a plausible suggestion of what an algorithm (in our sense) which would not be mechanizable in Turing's manner could be like. (Of course, Church's thesis would be disproved, if someone should describe a particular function, authenticated unmistakably as "effectively calculable" by our intuitive concept, but demonstrably not general recursive.)

We have been assuming without close examination the CONVERSE OF CHURCH'S THESIS: *If a function is Turing computable* (*or general recursive, or λ-definable*), *then it is intuitively computable* (*or effectively calculable*). In defending this implication to an intuitionist, or to any other kind of constructivist who considers an algorithm to exist only when it is proved by his standards that it always works, we only ask him to accept the following: if the hypothesis that a function is Turing computable holds by his standards, so does the conclusion. Put thus, it is hard to see how it can be questioned. Only if one allows a nonconstructive interpretation of the hypothesis, and yet insists on a constructive interpretation of the conclusion, is the converse of Church's thesis in doubt. ("Church's thesis" is sometimes understood to include this converse, as indeed Church 1936 maintained both implications in proposing to identify the effectively calculable functions with the general recursive functions or the λ-definable functions.)

§ 42. Church's theorem (via Turing machines).[172] We have seen that the pattern of behavior of a given Turing machine is determined by the table for it; if we know the table, essentially we know the machine.

As we remarked in § 41, the use of "P" for "print" is ambiguous when there is more than just the one symbol "|" or s_1 ($j=1$). We now write the table entries using the other method explained there, which is suitable for any number $j \geq 1$ of symbols $s_1, \ldots, s_j$. Under this method, a table entry "7*R*3" would mean that the square scanned at the given moment t is at the next moment $t+1$ to have Condition s_7 (i.e. it is to have the symbol s_7 printed on it), and (as before) the machine is to shift the tape between Moments t and $t+1$ so that at Moment $t+1$ the scanned square is next to the right of the square scanned at Moment t, and at Moment $t+1$ the machine is to be in its state numbered 3. This table entry could appear only in the table for a machine with at least 7 symbols and at least 3 active states.

Rewriting the table for our machine $\mathfrak{S}$ in this manner, it becomes:

Machine state	Scanned square condition	
	0	1
1	0*C*0	1*R*2
2	0*R*3	1*R*9
3	1*L*4	1*R*3
	. . .	
10	0*C*0	0*R*11
11	1*C*0	1*R*11

The table for a machine can be written in code form. Consider the table for $\mathfrak{S}$, originally given in § 41 and repeated now. Using the latter way of writing it, let us insert semicolons at the end of each row of entries, and commas separating entries within a row, and then string the entire body of the table along as one sequence of symbols:

0*C*0,1*R*2;0*R*3,1*R*9;1*L*4,1*R*3; . . . ;0*C*0,0*R*11;1*C*0,1*R*11

This sequence of symbols is the *code* for the machine $\mathfrak{S}$.

The code for any machine can thus be written on a typewriter with the following 15 symbols:

L C R , ; 0 1 2 3 4 5 6 7 8 9

[172] In this and the next two sections we follow the treatment in Kleene 1958 pp. 145–147 (some of which appeared earlier in 1956a, 1957b).

Such a code does not begin with the symbol L. By reinterpreting these symbols as the digits of a number in the number system based on 15, we get a positive integer which describes the machine table and thence the pattern of behavior of the machine; call this number the *index* of the machine.[173]

Now let $T(i, a, x)$ stand for the following:

> i is the index of a Turing machine (call it "Machine $\mathfrak{M}_i$") which, when applied to a as argument, will at Moment x (but not earlier) have completed the computation of a value (call that value "$\varphi_i(a)$").

This predicate (i.e. propositional function) $T(i, a, x)$ is "decidable". For suppose values of i, a, x are given. Then we can determine whether i, in the 15-system of notation, does describe the table for a machine. If it does not, $T(i, a, x)$ is false. If it does, then we can follow out the operations performed by that machine $\mathfrak{M}_i$ (as in the illustration in § 41), starting it at Moment 0 to compute for a as argument, and continuing to Moment x. Finally, in this case, we can see whether at this moment $\mathfrak{M}_i$ has *just* completed the computation of a value. If so, $T(i, a, x)$ is true; if not, false. (For example, if i is the number shown above in the 15-notation, then $T(i, 1, 23)$ is true, but $T(i, 1, x)$ is false for every $x \neq 23$.)

This should make it clear that $T(i, a, x)$ is decidable in the vague intuitive sense (§ 40). It is then implied by Church's thesis, or the Church-Turing thesis (§ 41), that it is (*Turing*) *decidable* in the strict sense that a Turing machine exists which decides it, i.e. computes its representing function $\tau(i, a, x)$, whose value is 0 when $T(i, a, x)$ is true and 1 when $T(i, a, x)$ is false (§ 40). A full treatment of the subject would call for showing this without appeal to Church's thesis. One would not do this from scratch, but by taking advantage of theory developed for this purpose, and alluded to in § 41. In this way we would establish the Part (A) of the following theorem. (In this chapter we are only giving a survey, in which sometimes we will only be able to describe the considerations on which the full proofs are based.)

[173] We are using the "method of digits" end § 32 (and IM § 1), except not bothering to close up the gaps in the resulting numbers (which there is no need for us to do for our present purposes).

We have chosen this particular method of indexing, in expositions since 1956a, as being easy to explain. If we were to deal now with the subjects below in greater detail, it would be advantageous to select a method to make the work as easy as possible. Some other methods have been more commonly used in the literature. But work done in Smullyan 1961 (not quite in the present connection) establishes that the present system is no harder to use.

THEOREM I. (A) *The predicate $T(i, a, x)$ is decidable.*
(B) *$\varphi_i(a)$ as a partial function of i and a is computable.*

To explain Part (B) of the theorem, first observe that the "quantity" $\varphi_i(a)$ in the definition of $T(i, a, x)$ is not defined for every i and a; indeed it is defined, for a given i and a, exactly if there exists an x such that $T(i, a, x)$, or in symbols exactly if $(Ex)T(i, a, x)$. In Chapter II we would have written this "$\exists x T(i, a, x)$", but here we prefer to save "$\exists$x" for use in formal systems, while employing "(Ex)" informally in discussing those systems.[165]

For a value of i which is the index of a Turing machine that computes a one-place number-theoretic function, φ_i is the function computed. Such values of i are described by $(a)(Ex)T(i, a, x)$, where "(a)" means "for all a".[165]

As a function of i and a both, $\varphi_i(a)$ as we remarked is defined exactly if $(Ex)T(i, a, x)$. So it is a partially defined number-theoretic function of two variables i and a, or briefly a *partial function*.

However, for i and a for which it is defined, we can find its value, thus: given such i and a, from i we find the table for Machine $\mathfrak{M}_i$, then we apply (or imitate) $\mathfrak{M}_i$ by performing its steps, starting at Moment 0 with a as argument, up to the moment x for which $T(i, a, x)$ is true, and finally we read from the resulting situation the value computed. This process is an algorithm or decision procedure in the sense of § 40; here the countably infinite class of questions is "What is the value of $\varphi_i(a)$?" where (i, a) ranges, not over all pairs of natural numbers, but over exactly those pairs for which $(Ex)T(i, a, x)$.

We extend the definition given in § 41 for "total" number-theoretic functions to say now that a partial number-theoretic function of two variables is *computed* by a Turing machine, if that machine computes the value of the function for just those pairs of arguments for which the function is defined, and computes no value for other pairs of arguments. Similarly, for n-place partial functions for any n. The Church-Turing thesis applies to partial functions on the same grounds as to total functions (§ 41).[174]

Via this extension of the thesis, it follows from our having an algorithm for finding the values of $\varphi_i(a)$ that there is a machine $\mathfrak{U}$ which computes $\varphi_i(a)$ as a partial function of i and a. Turing showed directly, without appeal to the thesis, that such a machine $\mathfrak{U}$ exists, though in a little different situation (he dealt with decimal expansions of real numbers); and we can do so in our situation. Doing so constitutes the full proof of Theorem I (B).

A machine $\mathfrak{U}$ which computes $\varphi_i(a)$ as a partial function of i and a we

[174] For a fuller discussion, cf. IM pp. 331–332 and § 68.

call a *universal machine*, since it can be used to compute any computable function $\varphi(a)$. To use it to compute $\varphi(a)$, say that $\varphi(a)$ is computed by Machine $\mathfrak{M}_i$; then apply $\mathfrak{U}$ to compute for i, a as a pair of arguments. Thus i plays the role of a set of *instructions* or *program* for $\mathfrak{U}$ which tells $\mathfrak{U}$ what function of a to compute.

The next theorem we can prove in full detail.

THEOREM II. *The function* $\psi(a)$ *defined by*

$$\psi(a) = \begin{cases} \varphi_a(a) + 1 & \textit{if} \quad (Ex)T(a, a, x), \\ 0 & \textit{otherwise} \end{cases}$$

is not computable.

PROOF. Suppose $\psi(a)$ were computable; say machine $\mathfrak{M}_p$ computes it, so that $\psi(a) = \varphi_p(a)$ for all a. Substituting p for a,

$$\psi(p) = \varphi_p(p).$$

But since $\mathfrak{M}_p$ computes $\psi(a)$, we have, for all a, $(Ex)T(p, a, x)$, and in particular $(Ex)T(p, p, x)$. Using this in the definition of $\psi(a)$,

$$\psi(p) = \varphi_p(p) + 1.$$

The two displayed equations contradict each other.

To state the proof a little differently, we can consider each Turing machine $\mathfrak{M}_p$, and see that it will fail to compute $\psi(a)$ correctly for $a = p$, as follows. To begin with, $\mathfrak{M}_p$ may fail to compute any value for p as argument. But if $\mathfrak{M}_p$ does compute a value for p as argument, that value is $\varphi_p(p)$, by the definition of $\varphi_i(a)$; and in this case $(Ex)T(p, p, x)$, so that the correct value of $\psi(p)$ is $\varphi_p(p) + 1$ by the definition of $\psi(a)$. —

The significance of this result comes from the Church-Turing thesis, by which computability in Turing's sense agrees with the intuitive notion of computability. Accepting the thesis, as most workers in foundations do, the director of a computing laboratory must fail if he undertakes to design a procedure to be followed, or to build a machine, to compute this function $\psi(a)$. This refutes the notion, which news reports on modern developments in high-speed computing tend to foster in the public mind, that machines can do everything. The theorem does not assert that there is any particular value of $\psi(a)$ that we cannot learn. But in whatever model we freeze the design of a computing procedure or Turing machine, we will be short of having a procedure or machine that can compute all values of $\psi(a)$. If the values it computes are correct, there must be some values it cannot compute; in particular, it cannot compute the value $\psi(p)$ where p

is its own index (or for a procedure, the index of a machine which mechanizes the procedure). To improve the procedure or machine must take ingenuity, something that cannot be built into the machine.

It should have been apparent that there must be uncomputable number-theoretic functions, as soon as we met the Church-Turing thesis in § 41, by which the set of different possible machines is countably infinite because each machine is describable by a finite table in a fixed symbolism (cf. § 32). The set of the machines being countable, the set of the functions computable by machines is countable, while the set of all the number-theoretic functions is uncountable (§ 33). But it remains of interest to see how simple examples of uncomputable functions we can give. Our example ($\psi(a)$ in Theorem II) is really quite simple, as it is obtained by completing the definition (by the "0 otherwise") of a suitable computable partial function.

Why can't we compute $\psi(a)$ from its definition? Answering this question gives the next theorem.

THEOREM III. *The predicate* $(Ex)T(a, a, x)$ *is undecidable; i.e. the function*

$$\chi(a) = \begin{cases} 0 & \textit{if} \quad (Ex)T(a, a, x), \\ 1 & \textit{otherwise} \end{cases}$$

is uncomputable.

PROOF. If we could decide $(Ex)T(a, a, x)$, we could compute the function $\psi(a)$ of Theorem II thus: Given a, decide whether $(Ex)T(a, a, x)$ or not. If this decision gives "yes", imitate the behavior of $\mathfrak{M}_a$ for a as argument to compute $\varphi_a(a)$, and add 1 to the result. If this decision gives "no", simply write 0.

Again, as in proving Theorem I, we are giving the idea of the proof, not the full technical details. What needs to be supplied is the hypothetical construction of a machine to compute $\psi(a)$ from one (supposed given) to compute $\chi(a)$. This is pretty easy to give, once one has reached the stage in the development that he would have reached in proving Theorem I.

DISCUSSION. Theorem III is essentially Church's theorem, which appeared with his thesis in his 1936 paper entitled "An unsolvable problem of elementary number theory". The difference is that we have given an example in terms of Turing computability, whereas Church's example was in terms of "λ-definability" (beginning § 41). The problem that is "unsolvable" is to find a decision procedure for the predicate $(Ex)T(a, a, x)$. Of course the problem is solved in another sense, by its being shown that there cannot be the required decision procedure. The problem of trisecting the general angle by a ruler and compass construction is unsolvable in one sense; but in another is solved by its being shown that the required construction cannot exist.

We note the especially simple logical form of the undecidable predicate, namely $(Ex)T(a, a, x)$, where as an almost immediate corollary of Theorem I (A) $T(a, a, x)$ is a decidable predicate. This is what is accomplished in Church's theorem, compared to simply contrasting the nonenumerability of the set of all (extensional) number-theoretic predicates (given by Cantor's diagonal method) with the enumerability of the set of the computable number-theoretic predicates (given by Church's thesis).[164]

EXERCISES. 42.1. Show that there is no algorithm for deciding whether a given Turing machine started in a given situation eventually stops (the "halting problem" for a Turing machine).

42.2. Find an undecidable predicate of the form $(x)P(a, x)$ where $P(a, x)$ is decidable. ("(x)" means "for all x".)

42.3. Show that the function $f(a)$ of Exercise 40.1 (a) (often written "$\varepsilon x P(a, x)$") is not computable when $P(a, x) \equiv T(a, a, x)$.

42.4*. Show that there is no algorithm for telling whether, for given i and j, $\varphi_i(a)$ and $\varphi_j(a)$ are the same partial function (i.e. whether, for each a, both functions have either the same value or no value).

42.5*. Show that the computable partial function $\varphi_i(a)$ of Theorem I (B) cannot be extended to a computable total function; i.e. there is no completely defined number-theoretic function $\varphi(i, a)$ such that: $(i)(a)[(Ex)T(i, a, x) \rightarrow \varphi(i, a) = \varphi_i(a)]$, and φ is Turing computable.[175]

§ 43. Applications to formal number theory: undecidability (Church) and incompleteness (Gödel's theorem). Recall that in the foregoing we have been using "decision procedure" and "decidable" (i.e. "a decision procedure exists") in relation to some given countably infinite class of questions (§ 40). Thus in Theorem I (A), the questions are "Is $T(i, a, x)$ true?" for the various values of i, a, x ($i, a, x = 0, 1, 2, \ldots$); and in Theorem III the questions are "Is $(Ex)T(a, a, x)$ true?" for the various values of a ($a = 0, 1, 2, \ldots$). In Theorem IV below, we shall infer from Church's theorem (Theorem III) that the decision problem for the system **N** of formal number theory of § 38 is "unsolvable" (or is solved in the negative); here the questions are "Is **A** provable in **N**?" where **A** ranges over all the formulas of **N**. In Theorem II, which concerns "computation procedure" and "computable", the questions are "What is the value of $\psi(a)$?" for the various values of a ($a = 0, 1, 2, \ldots$).

It should also be recalled that "decidable" or "computable" have both a vague intuitive sense (§ 40), and an exact sense which we have defined via Turing machines (§ 41). The Church-Turing thesis, and its converse that every Turing computable function is intuitively computable, assert

[175] Using recursiveness, with $\Phi(i, a)$ corresponding to the present $\varphi_i(a)$, the solution is in IM p. 341 following Theorem XXII.

that these two senses are equivalent. We understand our theorems now to be proved using the second sense, which requires more of us when we are establishing decidability or computability, and is the only sense in which they can be proved when we are establishing undecidability or uncomputability. But by the Church-Turing thesis, the latter theorems have their significance in terms of the intuitive sense of those terms.

We are now in a position to achieve our goal (§ 40) of showing there is no decision procedure for **N** (Theorem IV). Continuing the analysis a little further, we shall obtain Gödel's famous incompleteness theorem in a generalized version, as well as his second theorem (in § 44), which will clarify the situation described at the end of § 38.

We defined the predicate $T(i, a, x)$ (§ 42) informally in quite elementary terms, though its definition fully written out (with our explanation of Turing machine tables and the indexing of them) is long. It would therefore be disappointing if $T(i, a, x)$ could not be expressed in the symbolism of **N**. Indeed, if it could not, the first part of our claim of the adequacy of **N** for the usual elementary number theory (end § 38) would be false. For, though $T(i, a, x)$ is not commonly included in number-theoretic texts, there is nothing in its definition to prevent it from being included. In fact, the results cited in § 38 do enable us by established methods to find a formula T(i, a, x) (containing free exactly three distinct variables i, a, x) in the symbolism of **N** which expresses $T(i, a, x)$ under the intended interpretation of the symbols.[176] The like, as we have just argued, would have to be the case for any formal system **N** which we would consider adequate for the usual elementary number theory.

In **N**, the particular natural numbers 0, 1, 2, . . . are expressed by the respective terms 0, (0)′, ((0)′)′, . . . (which in § 38 we abbreviated "0", "1", "2", . . .); these terms we call the *numerals* (*for the respective natural numbers* 0, 1, 2, . . .). For any natural number a, we denote the numeral for a by "**a**". Now, for each $a = 0, 1, 2, \ldots$, the proposition $(Ex)T(a, a, x)$ is expressed under our interpretation of the symbolism of **N** by the formula ∃xT(**a**, **a**, x). Call this formula "C_a". Similarly, in *any* adequate formal system of elementary number theory, given any value of a, we can find effectively (end § 40) a closed formula C_a expressing under the intended interpretation the proposition $(Ex)T(a, a, x)$.

[176] To give somewhat more specific references for this than in Footnote 155 § 38, one can show by IM that $T(i, a, x)$ is primitive recursive (Chapter IX), using the technique of § 69, and then apply Corollary Theorem I p. 242. The predicate written "$(Ey)T_1(x, x, y)$" in Chapter XI pp. 281 ff. plays a role for general recursiveness analogous to that of the $(Ex)T(a, a, x)$ of this book for Turing computability. (In some papers "T_1" is written simply "T"; the notation goes back to Kleene 1936, which slightly antedates Turing 1936–7.)

If, for a given x, the proposition $(Ex)T(a, a, x)$ is true, then we can prove it informally, by the mechanical process of exhibiting the computation steps of the machine $\mathfrak{M}_a$ applied to a up to the moment x at which a value will have just been computed. This will prove $T(a, a, x)$ for that x, and $(Ex)T(a, a, x)$ will follow by informal $\exists$-introduction. Now this informal proof can also be carried out within **N** (i.e. it can be "formalized in **N**"). Thus

(a) $$(Ex)T(a, a, x) \rightarrow \{\vdash \mathrm{C}_a \text{ in } \mathbf{N}\}.$$

Of course to show this would require a detailed investigation of the proof theory of **N**, which we are not giving in this book. But if it were not the case, the deductive apparatus of **N** (i.e. the list of its axiom schemata, axioms, and rules of inference) would be inadequate for the usual elementary number theory, contrary to the second part of our claim in § 38.[177]

Now let us assume about **N** that in it only true formulas are provable. Since under our interpretation C_a expresses $(Ex)T(a, a, x)$, this gives in particular

(b) $$\{\vdash \mathrm{C}_a \text{ in } \mathbf{N}\} \rightarrow (Ex)T(a, a, x).$$

In the next section we shall see why at this stage we simply *assume* (b). Clearly, if (b) were not so, we should reject **N** as a formal system for number theory. We certainly believe (b); indeed a proof of sorts, though not a finitary one (§ 36) and thus not one in metamathematics, is available, as follows. Under the usual interpretation, the axioms of **N** are true; and each of the rules of inference, when applied to one or two true formula(s) as premise(s) produces a true formula as conclusion. Thus all provable formulas are true. Hence, (b). (Cf. § 38 (B).)

THEOREM IV. *There is no decision procedure for provability in the formal system* **N** *of* § 38; *or briefly,* **N** *is undecidable.*

More generally, this applies not just to the formal system **N** *of* § 38, *but to any formal system* **N** *in which, to each a, there can be found effectively a closed formula* C_a *such that* (a) *and* (b) *hold.*

PROOF. Suppose there were a decision procedure for provability in **N**. Then we could, given a, decide as follows whether $(Ex)T(a, a, x)$ or not, which would contradict Theorem III. Given a, find (as we can effectively) the formula C_a, and apply the assumed decision procedure for provability in **N** to settle whether this formula C_a is provable. By (b) and (a), according as C_a is or is not provable, $(Ex)T(a, a, x)$ holds or does not hold. —

[177] If the method of finding T(i, a, x) proposed in Footnote 176 is used, then (a) will follow by IM Corollary Theorem 27 p. 244 and $\exists$-introduction.

Preparatory to the next theorem, we further apply our supposition that in **N** only true formulas are provable to the formulas $\neg C_a$ ($a = 0, 1, 2, \ldots$). Thus we assume (where "$\overline{}$" expresses "not")[165]

(c) $\{\vdash \neg C_a \text{ in } \mathbf{N}\} \rightarrow \overline{(Ex)}T(a, a, x)$.

The same remarks we made about (b) apply to (c).

We now ask whether we can prove $\neg C_a$ when it is true; that is, whether the converse of (c), namely

(*) $\overline{(Ex)}T(a, a, x) \rightarrow \{\vdash \neg C_a \text{ in } \mathbf{N}\}$,

also holds. If it did, we should have a decision procedure for $(Ex)T(a, a, x)$ as follows, again contradicting Theorem III. First, observe that all the proofs in the **N** of § 38 can be written on a typewriter with the 41 formal symbols of **N** and the comma to separate successive formulas in a proof, making 42 symbols in all. So the proofs are enumerable (say using the method of digits, end § 32), and indeed effectively (since there is a decision procedure for being a proof, § 40). So, given a, we could search through an enumeration of the proofs in **N** for one of C_a or of $\neg C_a$. By the classical law of the excluded middle with (a) and (*), we shall find one or the other. By (b) and (c), according to which we find we shall learn that $(Ex)T(a, a, x)$ is true or not true.

Thus (*) does not hold for all a, which gives us the next theorem.

THEOREM V. *In the system* **N** *of* § 38 *there is a closed formula* C_p *such that* (i) $\neg C_p$ *is true,* (ii) *not* $\vdash \neg C_p$ *in* **N**, *and* (iii) *not* $\vdash C_p$ *in* **N**.

More generally, this applies to any formal system **N** *in which, to each a, there can be found effectively a closed formula* C_a *(expressing* $(Ex)T(a, a, x)$*) such that* (a)–(c) (*or* (b)–(d) *below*) *hold.*

FIRST PROOF, COMPLETED. Since (*) does not hold for all a, there is some number p such that $\overline{(Ex)}T(p, p, x)$ (i.e. (i), by what $\neg C_p$ expresses) but not $\vdash \neg C_p$ in **N** (i.e. (ii)). By $\overline{(Ex)}T(p, p, x)$ with (b), then not $\vdash C_p$ in **N** (i.e. (iii)).

REMARKS. This gives us Gödel's famous incompleteness theorem 1931, generalized to apply to all formal systems **N** satisfying very general conditions, and with the "formally undecidable sentence" C_p expressing the value of a preassigned predicate $(Ex)T(a, a, x)$ for an argument p depending on the particular system. This generalized form of Gödel's theorem (with a predicate written "$(Ex)T(a, a, x)$") is due to Kleene 1943.[178]

[178] Kleene used "general recursive functions" (§ 41) instead of Turing machines. The first use of Church's thesis to give a generalized version of Gödel's theorem was by Kleene in 1936.

In a formal system with notation like that of **N**, we say a formula E is *formally decidable* if $\vdash$ E or $\vdash \neg$E; and we say the system is *simply complete*, if every closed formula E is formally decidable. Thus by (iii) and (ii) of Theorem V, **N** is simply incomplete, with C_p as an example of closed formally undecidable formula. This notion of "formal decidability" applies to a particular formula, while decidability in the sense of § 40 or of § 41 applies to an infinite class of questions (or propositions). We restrict the E in the definition of simple completeness to be closed, because e.g. we would not wish to have $\vdash 2|a$ or $\vdash \neg 2|a$ (cf. § 38). For, under the generality interpretation (which applies to free variables of provable formulas, § 38), $2|a$ as a provable formula would say "all natural numbers are even" and $\neg 2|a$ would say "all natural numbers are odd". We likewise confine the definition to systems with notation like that of **N**, excluding systems like the propositional and predicate calculi, because e.g. neither $\vdash$ P $\supset$ Q nor $\vdash \neg$(P $\supset$ Q) in those calculi. Although P $\supset$ Q is a closed formula, its atoms P and Q function in the definition of validity like variables ranging with the generality interpretation over all propositions.

The foregoing proof of Theorem V is indirect, the existence of the *p* being inferred from the absurdity of (*) holding for all *a*. We now give a direct proof.

Second proof of Theorem V. Let $\mathfrak{M}_p$ be a Turing machine which, applied to *a* as argument, searches through an enumeration of the proofs in **N** for one of $\neg C_a$, and if one is found writes 0, but otherwise never computes a value (indeed, never stops). Considerations employed in similar situations above should make it clear that such a machine $\mathfrak{M}_p$ exists. A detailed proof for the particular **N** of § 38 is available on the basis of the developments in Turing-machine theory to which we referred late in § 41 and early in § 42. From what the machine $\mathfrak{M}_p$ does, with the definition of $T(i, a, x)$,

(d) $$(Ex)T(p, a, x) \equiv \{\vdash \neg C_a \text{ in } \mathbf{N}\}.$$

Now (b)–(d) give us the three parts of Gödel's theorem, thus.

(i) Suppose $(Ex)T(p, p, x)$; then by (d)

$$\vdash_{\mathbf{N}} \neg C_p.$$

Thence by (c)

$$\overline{(Ex)}T(p, p, x).$$

This contradicts $(Ex)T(p, p, x)$; so by reductio ad absurdum, $\overline{(Ex)}T(p, p, x)$, i.e. $\neg C_p$ is true.

(ii) By (i) $\neg C_p$ is true, i.e. $\overline{(Ex)}T(p, p, x)$. Thence by (d)

$$\text{not } \vdash_{\mathbf{N}} \neg C_p.$$

(iii) Suppose $\vdash_{\mathbf{N}} C_p$; then by (b)

$$(Ex)T(p, p, x),$$

i.e. C_p is true. This contradicts (i); so not $\vdash_{\mathbf{N}} C_p$.

DISCUSSION. Here we have used the feature of formal systems, essential to the purposes which they are intended to serve (§§ 36, 37), that a proof of a formula can be effectively recognized as being such (and also that C_a can be effectively found from a). Without this feature, we would have a trivial counterexample to Theorem V by taking all the true closed formulas as the axioms of N. With it, by Church's thesis we conclude that, for any such system, an $\mathfrak{M}_p$ exists.[179] Here the computability notion can be applied directly to the linguistic symbolism of the system, or that symbolism can be converted to natural numbers, e.g. by the method of digits with Footnote 173, as we have already done with machine tables. The numbers correlated to the linguistic objects are called *Gödel numbers*, and the correlation is called a *Gödel numbering*, after Gödel, who introduced this device in 1931.[180]

The application of Church's thesis, by which we obtain Theorem V for all systems N, can be avoided for a particular system by actually constructing the $\mathfrak{M}_p$ for it. This *in effect* Gödel did in 1931 in proving his theorem for a particular system before Church's thesis had appeared (1936);[181] and this, as we have implied, can be done by known methods to get the theorem, without leaning on Church's thesis, for the particular system N of § 38.

Substituting p for a in (d) and negating both sides (as we already did in proving (ii)),

$$(\overline{\mathrm{d}}) \qquad \overline{(Ex)T(p, p, x)} \equiv \{\text{not} \vdash \neg C_p \text{ in } \mathbf{N}\}.$$

Remembering that under the interpretation $\neg C_p$ expresses $\overline{(Ex)T(p, p, x)}$, we thus have that $\neg C_p$ *is a formula which, under the interpretation, expresses (a proposition equivalent to) its own unprovability*. This was the point of

[179] "In particular, the nature of the intuitive evidence for the deductive processes which are formalized in the system plays no role.

"Let us imagine an omniscient number theorist, whom we should expect, through his ability to see infinitely many facts at once, to be able to frame much stronger systems than any we could devise. Any correct system which he could reveal to us, telling us how it works without telling us why, would be equally subject to the Gödel incompleteness." (Kleene 1943 p. 65.)

[180] Gödel used another method of numbering in 1931; and Hilbert and Bernays 1939 and IM use still another method. In 1931–2, Gödel used what we are calling the "method of digits". This type of numbering is studied in detail in Smullyan 1961, which gives a fresh approach to formal systems from ideas of Post 1943.

[181] In the period immediately after Gödel's results appeared, there was some uncertainty among logicians whether there might not be an escape from them by using some other particular system quite different in its details from the system Gödel used.

departure in Gödel's original proof, or at least in the heuristic explanation he gave for it; namely, he constructed a formula expressing its own unprovability. This is very close to the paradox of the Liar (§ 35), where we have a sentence expressing its own falsity. Only now, by Gödel's substitution of "unprovability" for "falsity", there is a way out. We wish (and believe in the case of **N**) that all provable formulas are true. Then, if all true formulas were provable, we would have "unprovable ≡ false", so we would have the paradox of the Liar. The way out now is that not all true formulas are provable; in particular, $\neg C_p$ is unprovable though true.[182]

The first part of Hilbert's program (§§ 36, 37) called for formalizing number theory, analysis, and a suitable part of set theory in a formal system **S**. Gödel's theorem shows this cannot be done completely even for number theory. For $\neg C_p$ expresses a number-theoretic proposition which by the theorem is true, yet unprovable, provided of course **S** satisfies the hypotheses for the **N** of the theorem. However these hypotheses are connected with the purposes for devising formal systems, so we have no prospect of escaping them. These hypotheses are simply consequences of the structural feature of formal systems discussed above and of the supposition that **N** is adequate and correct for some elementary number theory.

We do not consider that Theorem V means we must give up our emphasis on formal systems. The reasons which make a formal system the only accurate way of saying explicitly what assumptions go into proofs are still cogent. Rather, Theorem V indicates that, contrary to Hilbert's program, the path of mathematical conquest (even within the already fixed territory of arithmetic) shall not consist solely in discovering new proofs from given axioms by given rules of inference, but also in adducing new axioms or rules. There remains the question whether mathematicians can agree on the correctness of the new axioms or rules.

In Theorem V, no sooner are we aware that $\neg C_p$ is unprovable than we also know that $\neg C_p$ is true, so we can extend **N** (call it "$\mathbf{N}_0$") by adding $\neg C_p$ as a new axiom. But then Gödel's theorem will apply to the extended system $\mathbf{N}_1$, and we shall have in this system a true but unprovable formula $\neg C_{p_1}$. The process can be repeated to obtain successively stronger formal systems $\mathbf{N}_0, \mathbf{N}_1, \mathbf{N}_2, \ldots$. We can combine these to form another formal system, if these systems are formed systematically enough so that after they are combined it will be decidable which formulas are axioms, and hence

[182] Nagel and Newman 1956, 1958 give a popular exposition of Gödel's theorem along the lines of Gödel's original proof 1931. They give a misleading impression that the generalized Gödel theorem is implied by Gödel's 1931 reasoning and thus without the Church-Turing thesis.

In Popper 1954, Theaetetus explains Gödel's theorem to Socrates.

which finite sequences of formulas are proofs (otherwise we could not consider the result a formal system). Then, starting from this system, we can again use the extension process based on Gödel's theorem. Proceeding in this way does not provide an escape from the consequences of Gödel's theorem.[183]

§ 44. Applications to formal number theory: consistency proofs (Gödel's second theorem). In establishing (i) in Theorem V we have proved that $\neg C_p$ is true, though by (ii) $\neg C_p$ cannot be proved in **N**. It is illuminating to consider wherein our intuitive proof of (the truth of) $\neg C_p$ transcends the resources of **N**.

In our second proof of (i) in Theorem V we used (c), which we regarded as an assumption. Using (a), this assumption (c) can be derived from the assumption that **N** is *simply consistent*, i.e. that for no formula E do both $\vdash$ E and $\vdash \neg$E hold in **N** (Exercise 44.1 (A)).

Alternatively, we can recast the second proof of (i) to use simple consistency directly instead of (c), thus.

(i) Suppose $(Ex)T(p, p, x)$; thence by (a) and (d)

$$\vdash_{\mathbf{N}} C_p \text{ and } \vdash_{\mathbf{N}} \neg C_p,$$

contradicting the simple consistency. So by reductio ad absurdum, $\overline{(Ex)}T(p, p, x)$, i.e. $\neg C_p$ is true.

In either version of the second proof of (i), the part written out in detail is entirely elementary; indeed, it is only informal use of the predicate calculus. Also the reasoning by which (a) and (d) can be established for the **N** of § 38, or a similar particular system, is elementary, on the level of informal number theory, though it is quite long when executed in full detail. There remains (c) or the simple consistency as the sole component not known to be elementary.

In summary, our informal proof of (i) is entirely elementary except for

[183] Classes of such systems are called "ordinal logics" because the systems are indexed by finite and infinite (or "transfinite") "ordinal numbers". Ordinal logics have been studied by Turing 1939, Feferman 1958 abstracts, 1962, and Kreisel 1958c.

The familiar natural numbers serve both as finite cardinals (e.g. "There are 10 houses on this street.") and as finite ordinals (e.g. "He lives at 10 Downing Street."). Cantor gave a theory in which *ordinal numbers* can be defined as equivalence classes of "well-ordered" sets under the relation of admitting a 1–1 correspondence preserving the ordering. A *well-ordered* set is a "linearly ordered" set each nonempty subset of which possesses a least element. A *linearly ordered* set is a set S together with an "order" relation $<$ which is irreflexive, transitive, and total (for each a and b in S, $a < b \vee a = b \vee a > b$). Under an obvious definition of $<$ for ordinal numbers, the ordinal numbers themselves are well-ordered.

Ordinal logics use the theory of ordinal numbers in a constructive or computable version, due to Church and Kleene 1936, Church 1938 and Kleene 1938.

our having to use in it (c) or the simple consistency of **N**. Thus it is indicated that (c) or that simple consistency is the respect in which it transcends **N**.

Preparatory to formulating this more carefully, we note for reference that the following can be shown in an entirely elementary way:

(1) $\{\mathbf{N} \text{ is simply consistent}\} \rightarrow \overline{(Ex)}T(p, p, x).$

The proposition that **N** is simply consistent can be expressed in the symbolism of **N**. Let $\mathfrak{M}_r$ be a Turing machine which, applied to any number a, searches through an enumeration of the proofs in **N** for a proof of a formula of the form E & ¬E, and if it finds one writes 0, but otherwise never computes a value. The consistency of **N** is equivalent to $\overline{(Ex)}T(r, a, x)$, for any a, and in particular to $\overline{(Ex)}T(r, r, x)$, which is expressed in **N** by $\neg C_r$. Call this formula "Consis".[184]

Now the informal implication (1) can be translated into the symbolism of **N** as the formula

$$\text{Consis} \supset \neg C_p,$$

since $\neg C_p$ expresses $\overline{(Ex)}T(p, p, x)$.

THEOREM VI. (Gödel's second theorem.) *In the system* **N** *of* § 38, *not* ⊢ Consis; *i.e. the formula* Consis *expressing in* **N** *the consistency of* **N** *is unprovable in* **N**.

More generally, this applies to any simply consistent formal system **N**, *with any choice of a formula* Consis *expressing in* **N** *the consistency of* **N**, *such that there can be found effectively, to each a, a closed formula* C_a, *and a number p, for which* (a) *and* (d), *and* (2) *below, hold.*[185]

PROOF (COMPLETED). Because (1) is established by elementary intuitive reasoning, our belief in the adequacy for elementary number theory of the system **N** of § 38 gives us reason to believe that the formula Consis ⊃ $\neg C_p$ expressing (1) can be proved in **N**, i.e. that the informal proof of (1) in elementary number theory can be formalized in **N**. A person having some familiarity with the development of number theory in **N**

[184] Alternatively, we can correlate Gödel numbers to the objects in the symbolism of **N** (end § 42), and take as Consis a formula which *directly* translates the consistency property into a statement about Gödel numbers. Cf. IM p. 210.

[185] So much has been put into the hypotheses of the generalized version of this theorem (the second paragraph) that it only amounts to an application of modus ponens to (2) with the fact that simple consistency, (a) and (d) imply Theorem V (ii). The significance of this second paragraph is that the hypotheses are ones that would be satisfied by any system of number theory, and any choice of Consis for it, that we would normally consider. A fuller treatment of the domain of applicability of Gödel's second theorem was first given by Feferman 1960.

(further than we went in § 38) would find it hard to doubt this. Hilbert and Bernays did verify it to be the case.[186] That is,

(2) $$\vdash_{\mathbf{N}} \text{Consis} \supset \neg C_p.$$

Now suppose $\vdash_{\mathbf{N}}$ Consis. By $\supset$-elimination (modus ponens) from (2), then $\vdash_{\mathbf{N}} \neg C_q$, contradicting (ii) of Theorem V.

Discussion. The second part of Hilbert's program for foundations called for showing by finitary metamathematical reasoning that a formal system chosen as a formalization of classical mathematics is consistent. As the mathematics formalized in **N** § 38 is not all finitary, the hope must have been that the part of the methods formalized in **N** obtained by excluding the nonfinitary ones would suffice for the consistency proof. Gödel's second theorem shows that not even all the methods formalizable in **N**, i.e. a metalanguage isomorphic to **N** itself, can prove the consistency of **N**, if **N** is consistent (as we have been assuming).

Some mathematicians judged that this ended forever the hope of getting a guarantee for classical mathematics by a metamathematical consistency proof.

Others thought it possible that methods could be found which could be considered as finitary even though not formalizable in **N**. Then the set of the finitary methods **F** and the set of the methods formalized in **N** could be pictured as overlapping circles; the law of the excluded middle for countably infinite sets (§ 36) would be in **N** but not in **F**, while some new finitary method would be in **F** but not in **N**.

Progress toward proving the consistency of **N** had been stalled since the consistency proof for a subsystem of **N** by Ackermann in 1924–5 (end § 38). Interesting new proofs, but still for essentially the same subsystem, were given by von Neumann in 1927, Herbrand in 1931–2, and Gentzen in 1934–5. After the necessity of using some nonelementary method was revealed by Gödel's second theorem in 1931, it was not too long before Gentzen in 1936 gave a consistency proof for **N**. This employed, as its method not formalizable in **N**, induction over a certain segment of the finite and transfinite ordinal numbers, which Cantor had obtained by an extension of the counting process or ordinal use of numbers beyond the natural numbers, in association with "well-ordered" sets (analogously to his introduction of transfinite cardinal numbers in association with unordered sets, § 34).[183] The induction was over the ordinals less than the ordinal called "ε_0" by Cantor.[187] A different consistency proof for **N** was

[186] Hilbert and Bernays 1939 pp. 283 ff., especially pp. 300–324. Their work is done in a system essentially equivalent to the **N** of § 38, though using a somewhat different choice of Consis and $\neg C_p$ than the ones we are basing on Turing machines.

[187] For a little more of an indication, cf. IM pp. 476–478.

given by Ackermann in 1940, also using transfinite induction over the ordinals $< \varepsilon_0$. Schütte 1960 gives such a proof in a new, quite perspicuous form. Denton and Dreben 1968 greatly simplify the Ackermann proof by using ideas from Herbrand 1930.

It is a rather subjective matter whether this should make us feel safer about **N** than we already felt on the basis of its axioms being true, and its rules of inference preserving truth, under an interpretation ("truth definition") that as classical mathematicians we presumably accept (§ 43 following (b)). By a simple reduction of classical logic to intuitionistic logic given by Gödel 1932–3, Gentzen 1936 and Bernays, the consistency proof by a truth definition can even be managed intuitionistically.[188] Tarski, asked whether he felt more secure about classical mathematics from Gentzen's consistency proof, replied, "Yes, by an epsilon". (Calculus students will recognize epsilon "ε" as commonly used for a small positive number.)

It is clear that the consistency proofs by induction over the ordinals less than the ordinal ε_0 work very hard to accomplish something, but less clear what.

Kreisel (1951–2, 1958) finds the significance of the consistency proofs by transfinite induction up to ε_0 in by-products. Suppose that, for a given natural number i, the formula $\forall a \exists x T(\boldsymbol{i}, a, x)$ is provable, where $\boldsymbol{i}$ is the numeral for i. Then, assuming that in **N** only true formulas are provable, $(a)(Ex)T(i, a, x)$ holds, and thus $\varphi_i(a)$ is a computable total function (cf. § 42). It is not hard to see that the computable (total) functions which can thus be proved to exist in **N** constitute a proper subclass (i.e. not all) of the computable functions (Exercise 44.2). Indeed Kleene 1936 gave a proof of Gödel's incompleteness theorem from this idea.[178] Kreisel however extracts from Ackermann's consistency proof 1940 a different characterization (not directly from **N**) of this subclass of the computable functions. The possibility thus appears that some true formula $\forall a \exists x T(\boldsymbol{i}, a, x)$ might be shown to be unprovable in **N** because for that i the computable function $\varphi_i(a)$ is not in this subclass.

Gentzen had already claimed in the first (1936) version of his consistency proof to have established a property of provable formulas of classical formal number theory **N** that can be regarded as an intuitive interpretation. But the property was complicated, and received little attention after his proof appeared in another version 1938a easier to follow and not adducing the property.

To see that there is a problem of interpretation, we recall from § 36 that Hilbert 1926, 1928 made a distinction between "real" statements having a clear intuitive meaning, and other statements called "ideal". In classical mathematics, the "ideal" statements are adjoined to the "real". One

[188] Cf. IM § 81. A related reduction was given by Kolmogorov 1924–5.

might have supposed that the "real" statements would include all the propositions of elementary number theory.

However the picture has not turned out to be this simple. For, in elementary number theory there are statements proved classically that are not true on the basis of their meanings for an intuitionist. Kleene 1943 argued this as follows.

The intuitionist understands an existential statement $(Ey)P(y)$ to mean that one can actually find a y such that $P(y)$. From this standpoint, what can $(a)(Ey)P(a, y)$ mean? Only that there is an effective procedure by which, given any a, one can find a y such that $P(a, y)$. By the Church-Turing thesis, this must mean that the y is a computable function of a. Thus we are led to the thesis that intuitionistically $(a)(Ey)P(a, y)$ only if there is a computable function $g(a)$ such that $(a)P(a, g(a))$.

By the classical law of the excluded middle (which intuitionists decline to affirm), for each a, $(Ex)T(a, a, x) \vee \overline{(Ex)}T(a, a, x)$.
Thence $[(Ex)T(a, a, x) \;\&\; 0=0] \vee [\overline{(Ex)}T(a, a, x) \;\&\; 1=1]$.
Thence $(Ey)[(Ex)T(a, a, x) \;\&\; y=0] \vee (Ey)[\overline{(Ex)}T(a, a, x) \;\&\; y=1]$.
Thence $(Ey)\{[(Ex)T(a, a, x) \;\&\; y=0] \vee [\overline{(Ex)}T(a, a, x) \;\&\; y=1]\}$.
This is for each a, so we have proved classically

$$(\alpha) \qquad (a)(Ey)\{[(Ex)T(a, a, x) \;\&\; y=0] \vee [\overline{(Ex)}T(a, a, x) \;\&\; y=1]\}.$$

We have presented this proof informally, but it is a simple matter to formalize it in the system **N** of § 38, so that we have

$$(\beta) \qquad \vdash_{\mathbf{N}} \forall a \exists y\{[\exists x T(a, a, x) \;\&\; y=0] \vee [\neg\exists x T(a, a, x) \;\&\; y=1]\}.$$

Let us abbreviate (α) as $(a)(Ey)P(a, y)$. By the above thesis, (α) holds intuitionistically only if $(a)P(a, g(a))$ for some computable function $g(a)$. But from what $P(a, y)$ is in this example, the only $g(a)$ for which $(a)P(a, g(a))$ holds is the representing function of $(Ex)T(a, a, x)$; and by Theorem III § 42, this $g(a)$ is not computable.

Summarizing: (α) holds in classical informal number theory, and translates into a formula which (as (β) states) is provable in **N**, but (α) cannot be affirmed as true intuitionistically.

Specker 1949 gave similar examples in which the propositions (α) which hold classically but not intuitionistically (if the above thesis of Kleene 1943 is accepted) are instances of well-known theorems of analysis, e.g. the theorem that a bounded monotone sequence of rationals is convergent. For definiteness, take a monotone nondecreasing sequence. Using variables n, b, m, n_1, n_2 ranging over the natural numbers, and writing $f(0)$, $f(1)$,

$f(2), \ldots$ for any sequence of rationals, the theorem can be stated thus:

If

(A) $(n) f(n) \leq f(n+1)$ (f is monotone nondecreasing),

and

(B) $(Eb)(n) f(n) \leq b$ (f is bounded),

then

(C) $(m)(En)(n_1)(n_2)\{n_1, n_2 \geq n \rightarrow |f(n_1)-f(n_2)| < 1/2^m\}$ (f converges).

It should be reasonably obvious that the notion of Turing computability can be applied to functions $f(n)$ with rational numbers as values, either directly or by talking about the natural numbers which index the rationals in some fixed enumeration of the rationals § 32. Specker gave a particular sequence $f(0), f(1), f(2), \ldots,$ for which the function f is Turing computable,[189] and such that (A) and (B) hold but

(C′) $(m)(n_1)(n_2)\{n_1, n_2 \geq g(m) \rightarrow |f(n_1)-f(n_2)| < 1/2^m\}$

holds for no computable function g. Thus, for this particular f, (A)–(C) are expressible in the elementary theory of natural and rational numbers (which is in the common part of the classical and intuitionistic languages of mathematics), the hypotheses (A) and (B) are true both intuitionistically and classically, but the conclusion (C) is true only classically.

Kreisel's proposal uses Ackermann's consistency proof for **N** to correlate to (α) on the basis of (β) a considerably more complicated statement, which is both meaningful and true for a "finitist". A "finitist", if not an intuitionist, is at least of similar ilk.[190]

Formal systems can be set up to study the foundations of intuitionistic mathematics, as Heyting did in 1930, 1930a, though the intuitionists have always maintained on philosophical grounds (from before Gödel's theorem was known) that such systems cannot be complete. (Cf. ends §§ 12, 25, 39.) Kleene, David Nelson and others have since 1941 been applying computable (or general recursive) functions to elucidate the differences between intuitionistic and classical formal systems.[191]

Exercises. 44.1. Using (a), show that: (A) the simple consistency of **N** implies (c), and (B) conversely (using weak ¬-elimination § 11).

44.2. Show that $\vdash_{\mathbf{N}} \forall a \exists x T(i, a, x)$ does not hold for all computable total functions $\varphi_i(a)$. (Assume that only true formulas are provable in **N**.)

44.3*. Assuming **N** simply consistent, show that there is a formal system

[189] In fact, via indices in a standard enumeration of the rationals, f is primitive recursive IM Chapter IX.

[190] Introductions to these ideas of Kreisel are 1953, 1958. Kreisel's current thinking on a broad range of foundational problems is in 1965.

[191] Cf. IM § 82, and Kleene and Vesley 1965.

M of number theory which is simply consistent but in which not all provable formulas are even classically true under the usual interpretation.[192]

44.4*. Let **S** be a system satisfying the hypotheses of Gödel's second theorem (except not assumed to be simply consistent). Let **T** be **S** minus some of its postulates. What would have been an appropriate response by a logician in 1930 to the following report by one of his colleagues? In 1932?
(a) "In **S**, I have proved **S** consistent."
(b) "In **T**, I have proved **S** consistent."

★§ 45. Application to the predicate calculus (Church, Turing).

THEOREM VII. *There is no decision procedure for provability in the (pure) predicate calculus; briefly, the predicate calculus* **Pd** *is undecidable.* (Church 1936a, Turing 1936–7.)

PROOF. Theorem IV followed from Theorem III because the theory of the predicate $(Ex)T(a, a, x)$ can be formalized in **N** to the following extent: to each natural number a, a formula C_a can be found effectively such that C_a is provable in **N** if and only if $(Ex)T(a, a, x)$ ((a) and (b) in § 43).

Theorem VII will follow from Theorem III because the theory of $(Ex)T(a, a, x)$ can similarly be formalized in simply the pure predicate calculus **Pd** (end § 39). Our argument that the theory of $(Ex)T(a, a, x)$ can be formalized in **Pd** will be in three parts. (α) Only *finitely* many function symbols $f_1, \ldots, f_k$ and ("nonlogical") axioms have to be added to the predicate calculus with the predicate symbol $=$ to obtain a formal system $\mathbf{S}_k$ in which the theory of $(Ex)T(a, a, x)$ can be formalized to the extent required. (β) The system $\mathbf{S}_k$ can be transformed into an essentially equivalent system **S** with the symbolism of the pure predicate calculus and still only finitely many nonlogical axioms. (γ) The closed nonlogical axioms of **S** can be regarded as assumption formulas for deductions in **Pd**, to which the deduction theorem can be applied.

(α) To outline the first part, we start with the fact that the representing function (§ 40)

$$\tau(i, a, x) = \begin{cases} 0 & \text{if } T(i, a, x), \\ 1 & \text{otherwise} \end{cases}$$

of the predicate $T(i, a, x)$ can be defined by the last of a finite list of (primitive) recursive definitions, beginning with those of $+$ and $\cdot$ (§§ 38, 41). The reader must take our word for this part of the detailed proof.[193]

Now we describe the system $\mathbf{S}_k$. We start with the predicate calculus, using the symbolism of the system **N** of § 38 (i.e. we start with the system

[192] Solution in § 47 below.

[193] A fair amount of technique such as is provided in IM Chapters IX, X and XIII is helpful in constructing this list of recursive definitions.

[20] of § 39). To this we add function symbols to express the functions after + and · which are defined by the rest of the recursive definitions culminating in the definition of τ. Say that altogether the (individual and) function symbols expressing 0, ′, +, ·, . . . , τ are $f_1, \ldots, f_k$ (so "f_1" is a name for 0, "f_2" for ′, etc.). As the axioms $A_1^k, \ldots, A_{m_k}^k$ we add to the axioms of the predicate calculus, we take the last six particular axioms 16–21 of **N**, also the pairs of equations of the additional recursive definitions, and finally the open equality axioms for the function symbols $f_3, \ldots, f_k$ (§ 29). For f_3 (i.e. +) these are the two formulas which were proved in **N** in § 38 Example 2 and Exercise 38.2 (but now they are to be axioms).

In $\mathbf{S}_k$ the equations expressing the values of the functions +, ·, . . . , τ will all be provable, as is illustrated for + by Exercise 38.4. Thus, when i, a, x are natural numbers such that $T(i, a, x)$ holds, $f_k(\boldsymbol{i}, \boldsymbol{a}, \boldsymbol{x})=0$ will be provable in $\mathbf{S}_k$. So when $(Ex)T(a, a, x)$ holds, we can in $\mathbf{S}_k$ prove the equation $f_k(\boldsymbol{a}, \boldsymbol{a}, \boldsymbol{x})=0$ for a suitable x, and thence by $\exists$-introduction prove $\exists x f_k(\boldsymbol{a}, \boldsymbol{a}, x)=0$; call this formula "$D_a^k$". Thus

$$(1) \qquad (Ex)T(a, a, x) \rightarrow \{\vdash D_a^k \text{ in } \mathbf{S}_k\}.$$

We get an easy proof of the converse of (1) by applying the theory of valid consequence in the predicate calculus with functions (§§ 20, 28). For, similarly to (B) in § 38, if $\vdash D_a^k$ in $\mathbf{S}_k$, then $\forall A_1^k, \ldots, \forall A_{m_k}^k \models D_a^k$. Now take as the domain D the natural numbers {0, 1, 2, . . .}, and consider the assignment in D in which $=, f_1, f_2, f_3, f_4, \ldots, f_k$ are interpreted by the equality predicate =, the natural number 0, and the functions ′, +, ·, . . . , τ respectively. Under this assignment, $\forall A_1^k, \ldots, \forall A_{m_k}^k$ are all t, so (by $\forall A_1^k, \ldots, \forall A_{m_k}^k \models D_a^k$) D_a^k is also t. But under this assignment, D_a^k being t means that $(Ex)T(a, a, x)$. Thus

$$(2) \qquad \{\vdash D_a^k \text{ in } \mathbf{S}_k\} \rightarrow (Ex)T(a, a, x).$$

This proof of (2) is not metamathematical (cf. § 37). A metamathematical proof can be given.[194] Combining (1) and (2),

$$(3_k) \qquad \{\vdash D_a^k \text{ in } \mathbf{S}_k\} \equiv (Ex)T(a, a, x).$$

(β) There is a method by which, starting with $\mathbf{S}_k$, we can replace successively the function symbols $f_k, \ldots, f_4, f_3, f_2, f_1$ expressing τ, . . . , ·, +, ′, 0

[194] The metamathematical proof of (2), and the details missing in our outline of the argument, appear e.g. in Kleene IM Part IV, with certain results of Parts II and III (cf. p. 434 Remark 2), though with a different $T(i, a, x)$ than the present one. (There the treatment is based on general recursiveness instead of on Turing computability; cf. § 41 above.)

Thus for Theorem VII we don't need a nonelementary assumption, like the assumption of (b) for Theorem IV.

by respective predicate symbols $F_k, \ldots, F_4, F_3, F_2, F_1$ expressing the representing predicates of $\tau, \ldots, \cdot, +, ', 0$ (cf. end § 38). We shall give the idea by discussing the step in which f_4 (i.e. $\cdot$) is replaced, assuming that the function symbols $f_k, \ldots, f_5$ have already been replaced by predicate symbols $F_k, \ldots, F_5$. (By giving just this step, we can illustrate the process, while keeping the notation simple.)

So we suppose that, as a result of the previous replacement steps, we have a system S_4 in which the only function symbols are f_4, f_3, f_2, f_1 (i.e. $\cdot, +, ', 0$) and which has only finitely many nonlogical axioms $A_1^4, \ldots, A_{m_4}^4$. Furthermore in S_4, to each natural number a, there is a formula D_a^4 such that

$$(3_4) \qquad \{\vdash D_a^4 \text{ in } S_4\} \equiv (Ex)T(a, a, x).$$

We seek a system S_3 with only the function symbols f_3, f_2, f_1, with only finitely many nonlogical axioms $A_1^3, \ldots, A_{m_3}^3$, and in which to each a there is a formula D_a^3 such that

$$(3_3) \qquad \{\vdash D_a^3 \text{ in } S_3\} \equiv (Ex)T(a, a, x).$$

To obtain S_3 from S_4 we alter the symbolism by replacing the 2-place function symbol $\cdot$ by a 3-place predicate symbol which we shall also write $\cdot$ (i.e. we use $\cdot$ for both f_4 and F_4). Thus in the formation rules, we omit the clause which says that whenever r and s are terms (r)$\cdot$(s) is a term, but add a clause which says that whenever r, s and t are terms $\cdot$(r, s, t) is a formula. For the interpretation, the formula $\cdot$(r, s, t) of S_3 has the same meaning as (r)$\cdot$(s)=t had in S_4. That is, $\cdot(a, b, c)$ expresses the representing predicate $a \cdot b = c$ of the product function $a \cdot b$ (cf. end § 38).

To get the axioms of S_3 from those of S_4, we replace 20 and 21 by

(A) $\cdot(a, 0, 0)$, (B) $\exists c(\cdot(a, b, c) \;\&\; \cdot(a, b', c+a))$.

We replace the two equality axioms for $\cdot$ in S_4 by

(C) $a=b \supset (\cdot(a, c, d) \supset \cdot(b, c, d))$, (D) $a=b \supset (\cdot(c, a, d) \supset \cdot(c, b, d))$.

We add the two axioms

(E) $a=b \supset (\cdot(c, d, a) \supset \cdot(c, d, b))$,
(F) $\exists c(\cdot(a, b, c) \;\&\; \forall d(\cdot(a, b, d) \supset c=d))$.

Here (C)–(E) are the open equality axioms for the predicate symbol $\cdot$, which express that $\cdot(a, b, c)$ is well-defined as a predicate; and (F) (abbreviated "$\exists ! c \cdot (a, b, c)$" in § 29) expresses that, for given a and b, $\cdot(a, b, c)$ is true for one and only one c, i.e. that $\cdot(a, b, c)$ is the representing predicate of a function. Finally, we replace each other axiom of S_4 which has the function symbol $\cdot$ in it by the result of paraphrasing it to use the predicate symbol ·. Already (A) and (B) are such paraphrases of 20 and 21, and the method of paraphrasing is also illustrated at the end of § 38 for the function symbol !.

Now in the system S_3 which we have just constructed from S_4, the same situation exists relative to the function · that exists in **N** of § 38 relative to functions like !. By the general theory cited in Footnote 156 § 38, we can in S_3 carry out the reasoning about the function · via the paraphrases.[195] So in this way, from (3_4) it can be inferred that (3_3) holds when D^3_a is the result of paraphrasing D^4_a to replace the function symbol · by the predicate symbol ·. (In fact, D^4_a will not contain ·, so D^3_a is D^4_a. But D^{k-1}_a will differ from D^k_a, D^1_a from D^2_a, and D^0_a from D^1_a.)

The successive replacements (illustrated by the step from S_4 to S_3) lead to a system S_0 with no function symbols, but instead only predicate symbols, and with only finitely many axioms $A^0_1, \ldots, A^0_{m_0}$. In S_0, to each a there is a formula D^0_a such that

$$(3_0) \qquad \{\vdash D^0_a \text{ in } S_0\} \equiv (Ex)T(a, a, x).$$

From S_0 we now pass to a system **S** having the notation of the pure predicate calculus **Pd**, by changing the predicate symbols $=, F_1, \ldots, F_k$ of S_0 to respective predicate letters $P_0, \ldots, P_k$ (used with the same respective numbers of arguments). Let $A_1, \ldots, A_m$ ($m = m_0$) and D_a result by this change in notation from $A^0_1, \ldots, A^0_{m_0}$ and D^0_a; so $A_1, \ldots, A_m$ are the nonlogical axioms of **S**. A proof of D^0_a in S_0 will clearly become a proof of D_a in **S** when the notation is thus changed (it will make no difference to any step in the proof whether we are using the predicate symbols or the respective predicate letters). Thus[196]

$$(4) \qquad \{\vdash D^0_a \text{ in } S_0\} \rightarrow \{\vdash D_a \text{ in } S\}.$$

The converse is not quite so simple, since a given proof of D_a in **S** might use some predicate letters other than $P_0, \ldots, P_k$. (In **S** we have available infinitely many predicate letters with each number of arguments, but in S_0 only the $k+1$ predicate symbols $=, F_1, \ldots, F_k$.) However a proof of D_a in **S** will remain one when the atoms in it formed using other predicate letters are changed to $\forall x P_1(x)$. The resulting proof of D_a in **S** will become one of D^0_a in S_0 when $P_0, \ldots, P_k$ are changed to $=, F_1, \ldots, F_k$. Thus[196]

$$(5) \qquad \{\vdash D_a \text{ in } S\} \rightarrow \{\vdash D^0_a \text{ in } S_0\}.$$

[195] The application of the results cited in Footnote 156 § 38 to replace a function symbol by a predicate symbol is discussed in general in Hilbert and Bernays 1934 pp. 460–467 and IM pp. 417–419; and for the case of multiplication · (but not for exactly the system S_4) in IM Example 11 p. 419.

[196] The implications (4) and (5) constitute simple applications of the proof-theoretic substitution rule cited in the second paragraph of Footnote 86 § 25.

Combining (4) and (5) with (3_0),

$$(6) \qquad \{\vdash \mathrm{D}_a \text{ in } \mathbf{S}\} \equiv (Ex)T(a, a, x).$$

(γ) Now $\{\vdash \mathrm{D}_a \text{ in } \mathbf{S}\} \equiv \{\forall \mathrm{A}_1, \ldots, \forall \mathrm{A}_m \vdash \mathrm{D}_a \text{ in } \mathbf{Pd}\}$ (similarly to (A) § 38 p. 208) $\equiv \{\vdash \forall \mathrm{A}_1 \supset (\forall \mathrm{A}_2 \supset \ldots (\forall \mathrm{A}_m \supset \mathrm{D}_a) \ldots) \text{ in } \mathbf{Pd}\}$ (by Corollaries Theorems 11_{Pd} and 10_{Pd}). Using this in (6),

$$(7) \quad \{\vdash \forall \mathrm{A}_1 \supset (\forall \mathrm{A}_2 \supset \ldots (\forall \mathrm{A}_m \supset \mathrm{D}_a) \ldots) \text{ in } \mathbf{Pd}\} \equiv (Ex)T(a, a, x).$$

Now we complete the proof, as we planned, by applying Theorem III. Thus, suppose there were a decision procedure for provability in **Pd**. Then, given a, we could decide as follows whether $(Ex)T(a, a, x)$ is true or false, contradicting Theorem III. From a, find the formula $\forall \mathrm{A}_1 \supset (\forall \mathrm{A}_2 \supset \ldots (\forall \mathrm{A}_m \supset \mathrm{D}_a) \ldots)$. Apply the supposed decision procedure for provability in **Pd** to the question whether this formula is provable. By (7), according as the answer is "yes" or "no", $(Ex)T(a, a, x)$ is true or false. —

An interesting new proof of Theorem VII (presupposing ideas given here in Chapter VI) is in Büchi 1962.

The undecidability results made possible by the Church-Turing thesis (§ 41) arose first (like Theorem III) directly in connection with the new notions of λ-definability, general recursiveness and Turing computability, and next (like Theorems IV and VII) for decision problems for formal systems.

Church on May 19, 1936, shortly after enunciating his thesis and obtaining his first results (Theorems III, IV and VII, particularly), wrote the author, "What I would really like to see done would be my results or yours used to prove the unsolvability of some mathematical problems of this order not on their face specially related to logic." This hope was fulfilled, beginning in 1947 when Post and A. A. Markov (the younger) independently of each other showed the "word problem for semi-groups" to be unsolvable.[197] This led to the unsolvability of the "word problem for semi-groups with cancellation" in Turing 1950 (with Boone 1958), and thence to the unsolvability of the "word problem for groups" in a 143-page article by Novikov 1955.[198] Simpler proofs of Novikov's result have since been given from other directions by Boone in 1957 of 1954–7, and in 1959, and by Britton in 1958 with 1956–8, and in 1963, and (as a corollary of another theorem) by Higman in 1961.[199] In 1958 Markov established the unsolvability of the "homeomorphism problem for four-dimensional manifolds" in topology. This is reworked and extended in Boone, Haken and

[197] A treatment is given in IM § 71.

[198] For the problem, see Dehn 1912 p. 117.

[199] Britton's 1963 proof has been worked into a textbook: Rotman 1965 Chapter 12.

Poénaru **1967**. Other major publications in this area include Rabin 1958, Clapham **1964**, Shepherdson **1965**, Boone **1966**, **1966a** and **1967** (with full bibliography). In the papers cited in bold face, the inquiry is refined to consider not just solvability vs. unsolvability but also "degrees of unsolvability" (§ 46).

Unsolvability results have begun to appear in real-variable analysis: Scarpellini 1963; Richardson 1966 abstract (which treats a variant of the integrability problem of Exercise 40.1 (k)).

Undecidability results have also been obtained in connection with grammatical problems for languages for use with computing machines and finite automata: Rabin and Scott 1959; Bar-Hillel, Perles and Shamir 1961.

★§ 46. Degrees of unsolvability (Post), hierarchies (Kleene, Mostowski). We can summarize our proofs of Theorems IV and VII as follows. First, in Theorem III, we established the undecidability of the predicate $(Ex)T(a, a, x)$. Then we "reduced" the decision problem (P) for this predicate to the decision problem (Q) for provability in **N** or in the predicate calculus **Pd**.

To say this in more detail, take the case of **Pd** (Theorem VII). We showed that, if we had a way to answer any question of the class (Q) "Is a given formula E provable in **Pd**?", then we could answer any question of the class (P) "Does a given natural number a have the property that $(Ex)T(a, a, x)$?". But in Theorem III we had shown from the Turing machine concept with the Church-Turing thesis that there can be no algorithm to answer all questions of the class (P). Therefore there can be none to answer all questions of the class (Q).

We reduced (P) to (Q), because (P) is the class of questions we could show directly to be undecidable. However, it is of some interest to see that, inversely, (Q) can be reduced to (P). This is included in the following proposition (and similarly with **N** in place of **Pd**).

(A) *To each predicate of the form* $(Ex_1)\ldots(Ex_m)R(a_1, \ldots, a_n, x_1, \ldots, x_m)$ *where* $R(a_1, \ldots, a_n, x_1, \ldots, x_m)$ *is a given decidable predicate* $(m, n > 0)$, *there is a computable function* $\theta(a_1, \ldots, a_n)$ *such that*

$$(Ex_1)\ldots(Ex_m)R(a_1, \ldots, a_n, x_1, \ldots, x_m) \equiv (Ex)T(\theta(a_1, \ldots, a_n), \theta(a_1, \ldots, a_n), x).$$

Similarly for $n = 0$, *with* $\theta(a_1, \ldots, a_n)$ *becoming simply a natural number* h.

(Proof follows.) Thus, any question of the class "Is $(Ex_1)\ldots(Ex_m)R(a_1, \ldots, a_n, x_1, \ldots, x_m)$ true?" is reduced to answering a corresponding question of the class "Is $(Ex)T(b, b, x)$ true?", namely the question with $b = \theta(a_1, \ldots, a_n)$ (where, since θ is computable, from any $a_1, \ldots, a_n$ we can effectively find the corresponding b).[200]

To prove (A), take any fixed $a_1, \ldots, a_n$, and consider the following intuitive computation procedure. First, suppose all m-tuples of natural numbers $(x_1, \ldots, x_m)$ have been enumerated, i.e. brought into an infinite list (§ 32). Now, using a Turing machine (supposed given) which decides the predicate $R(a_1, \ldots, a_n, x_1, \ldots, x_m)$, test the m-tuples $(x_1, \ldots, x_m)$ in the order listed, searching for one which makes $R(a_1, \ldots, a_n, x_1, \ldots, x_m)$ true, and write 0 if such an m-tuple is found, but continue the search ad infinitum otherwise.

Now we claim, and the reader will have to take our word for the details, that a Turing machine can be constructed which, applied to *any* number b as argument, carries out the above process and writes 0 when an m-tuple $(x_1, \ldots, x_m)$ is found for which $R(a_1, \ldots, a_n, x_1, \ldots, x_m)$ is true, but otherwise never computes a value. What machine we construct depends on the numbers $a_1, \ldots, a_n$ we started with. We further claim that these various machines can be constructed so that their indices can be given by a Turing computable function $\theta(a_1, \ldots, a_n)$. Thus, by what the machines do and the definition of the predicate $T(i, a, x)$ in § 42,

$$(Ex_1) \ldots (Ex_m)R(a_1, \ldots, a_n, x_1, \ldots, x_m) \equiv (Ex)T(\theta(a_1, \ldots, a_n), b, x)$$

for all $a_1, \ldots, a_n, b$. Taking $b = \theta(a_1, \ldots, a_n)$ in this gives the equivalence in (A).

To apply (A) to see that the decision problem (Q) for provability in the predicate calculus **Pd** (or in **N**) can be reduced to that (P) for $(Ex)T(a, a, x)$, we may first give Gödel numbers to the formulas in the predicate calculus in an effective way (cf. § 43 Discussion). Now "A is provable" can be expressed as $(Ex)\text{Pr}(a, x)$ where a is the Gödel number of A, and $\text{Pr}(a, x)$ is the decidable predicate which says that x is the Gödel number of a proof of the formula with the Gödel number a. Now (A) applies with Pr as the R (with $n = m = 1$).

Summarizing, by (A) with the proofs of Theorems IV and VII, each of our three examples of undecidable predicates (or unsolvable decision problems) is reducible to each of the others.

In Cantor's set theory, when we began comparing infinite sets under one-to-one correspondences, we discovered that not all infinite sets are

[200] Post in 1944 gave the first result of this sort, using a different predicate than $(Ex)T(b, b, x)$. (The earlier sections of Post 1944 are less technical than most papers written in this area.) The result for a predicate in the theory of general recursive functions analogous to the present $(Ex)T(b, b, x)$ appeared in IM p. 343. Slightly stretching Post's terminology, $(Ex)T(b, b, x)$ can be called a *complete* predicate for the class of predicates $(Ex_1) \ldots (Ex_m)R(a_1, \ldots, a_n, x_1, \ldots, x_m)$.

equally numerous; indeed, we found a hierarchy of increasing cardinal numbers. Analogously, the question arises now whether all undecidable number-theoretic predicates are equally undecidable, in the sense that the decision problem for any one of them is reducible to the decision problem for any other. As in set theory, the answer is negative; indeed, some predicates are "more undecidable", or (have decision problems) of "higher degree of unsolvability", than others.

So far we have only considered some particular pairs of undecidable predicates which are equally undecidable (the decision problem for either one being reducible in a simple way to that for the other). Now we need a more general notion of when one predicate (or its decision problem) is reducible to another. This requires an extension of Church's thesis or the Church-Turing thesis to decidability or computability relative to a given predicate $Q(b)$.

The concept that we use here is due to Turing 1939, and its use to define "degree (of unsolvability)" to Post 1948 abstract (with some anticipation in 1944).[201] Consider a "Q-machine", which is like an ordinary or "absolute" Turing machine, except that it has access to a second (actually) infinite tape (a "Q-tape") on which are printed the answers to all the questions (for $b = 0, 1, 2, \ldots$) whether $Q(b)$ is true or false. (Turing instead spoke of a machine having access to an "oracle" that would answer any such question the machine asks it about $Q(b)$.) If a suitable Q-machine can answer all the questions (for $a = 0, 1, 2, \ldots$) whether $P(a)$ is true or false, we say (the decision problem for) $P(a)$ is (*Turing*) *reducible* to (the decision problem for) $Q(b)$. If $P(a)$ is reducible to $Q(b)$ and vice versa, we say $P(a)$ and $Q(b)$ (or their decision problems) are of the same *degree* (*of unsolvability*). If $P(a)$ is reducible to $Q(b)$ but not vice versa, we say $P(a)$ is of *lower degree* than $Q(b)$ or $Q(b)$ is of *higher degree* than $P(a)$. Hence, if $P(a)$ is reducible to $Q(b)$, $P(a)$ is of the same or lower degree than $Q(b)$.

Here lower, equal or higher degree between predicates $P(a)$ and $Q(b)$ corresponds to less, equally or more numerous between sets M and N. To put the theory in better form, we should define degrees themselves and not just use the word "degree" to express the above relationships between predicates $P(a)$ and $Q(b)$. Here we can use the same method as in defining "cardinal number" by the Frege-Russell method in § 34. We begin by verifying (Exercise 46.1) that the above "equal-degree" relation "$P(a)$ is reducible to $Q(b)$ and vice versa" is an equivalence relation, i.e. it is reflexive, symmetric and transitive (like the "equally-numerous" relation "M can be put into 1-1 correspondence with N", § 34). So we can define

[201] In IM pp. 314–315 the treatment is based equivalently on the theory of general recursive functions (relativized by Kleene 1943), instead of on Turing computability.

the *degree* of any number-theoretic predicate $P(a)$ to be the equivalence class to which $P(a)$ belongs under this equivalence relation (§ 30). Then to justify the above definition of when one predicate $P(a)$ is of lower degree than another $Q(b)$, which now becomes "the degree of $P(a)$ is $<$ the degree of $Q(b)$", we need to verify (Exercise 46.2) that the result is independent of what particular predicates $P(a)$ and $Q(b)$ are used from their respective equivalence classes.

The relation $<$ between degrees is irreflexive and transitive (Exercise 46.3).

To keep the notation simple, we have spoken just now only of one-place number-theoretic predicates $P(a)$ and $Q(b)$. But the reducibility notion applies likewise to number-theoretic predicates or functions of any numbers ≥ 1 of arguments; and the equivalence classes can be taken in this larger totality.

All the questions about particular values of a decidable predicate $P(a)$ can be answered by a suitable absolute Turing machine. So a suitable Q-machine can answer all those questions, without looking at its Q-tape. Hence each decidable predicate $P(a)$ is of degree $\leq$ the degree of each predicate $Q(b)$. The degree of any decidable predicate we write **0** (since each two decidable predicates are reducible each to the other, the decidable predicates do constitute a degree); this we have just seen is the lowest degree of unsolvability ("solvability").

By Theorem III, the predicate $(Ex)T(a, a, x)$ cannot be decided by an absolute Turing machine. So $(Ex)T(a, a, x)$ cannot be decided by a Q-machine when $Q(b)$ is a decidable predicate. For, the values of $Q(b)$ on the Q-tape would give the Q-machine only unnecessary help; a suitable absolute machine could manufacture those values for itself.

Hence $(Ex)T(a, a, x)$ is of higher degree than the decidable predicates. We write $\mathbf{0}'$ $(= \mathbf{1})$ for the degree of $(Ex)T(a, a, x)$. Thus $\mathbf{0}' > \mathbf{0}$.

Now it is easy to see how to proceed to higher degrees than $\mathbf{0}'$. Though we have not spelled out exactly how the Q-tape of a Q-machine is used, it should be fairly obvious that we can do so, and then construct a predicate $T^Q(i, a, x)$ (relative to a given predicate $Q(b)$) which plays the role for Q-machines which $T(i, a, x)$ plays for absolute machines. Then (similarly to Theorem I (A)) $T^Q(i, a, x)$ is decidable by a Q-machine, and $\varphi_i^Q(a)$ as a partial function of i and a is computable by a Q-machine.

By using a Q-machine instead of an absolute machine in the proof of (A), we can establish:

(B) *For any predicate Q,* (A) *holds when $R(a_1, \ldots, a_n, x_1, \ldots, x_m)$ is allowed to be any predicate decidable by a Q-machine, and $T(a, a, x)$ is replaced by $T^Q(a, a, x)$.*

Similarly, paralleling with Q-machines the proofs of Theorems II and III, $(Ex)T^Q(a, a, x)$ is not decidable by a Q-machine. But since $Q(a) \equiv (Ex)(Q(a) \,\&\, x=x)$, by (B) Q is reducible to $(Ex)T^Q(a, a, x)$. Thus $(Ex)T^Q(a, a, x)$ is of higher degree than Q. We can state this result in the following form.

(C) *If* $Q(b)$ *is a predicate of degree* $\mathbf{d}$, *then* $(Ex)T^Q(a, a, x)$ *is of degree* $\mathbf{d}' > \mathbf{d}$.

Here by using the notation $\mathbf{d}'$ we include that the degree of $(Ex)T^Q(a, a, x)$ depends only on the degree $\mathbf{d}$ of Q; and furthermore that, when Q *is* decidable, the degree of $(Ex)T^Q(a, a, x)$ is the same as the degree of $(Ex)T(a, a, x)$, which we originally designated as $\mathbf{0}'$. These additional facts are easily proved using (B) and (A) (Exercise 46.4).

Using (C) repeatedly, we get a succession of increasing degrees

$$\mathbf{0} < \mathbf{0}' < \mathbf{0}'' < \ldots < \mathbf{0}^{(n)} < \ldots.$$

It is easy to see that:

(D) *If* $P_0(a), P_1(a), P_2(a), \ldots$ *are predicates of increasing degrees, and* $P(n, a) \equiv P_n(a)$, *then* $P(n, a)$ *as a 2-place predicate is of higher degree than any of* $P_0(a), P_1(a), P_2(a), \ldots$.

(C) and (D) play a role in generating increasing degrees like (C) and (D) in § 34 in generating increasing cardinal numbers. This method of generating number-theoretic predicates of increasing degrees is due to Davis 1950, and independently to Kleene and to Post.

Hierarchies of number-theoretic predicates were first discovered from another point of view by Kleene in 1943 (1940 abstract), and independently (with a somewhat different method) by Mostowski in 1947. Using Kleene's approach, we first establish:

(E) *To each decidable predicate* $R(a, x)$, *there is a number* f *such that*

$$(Ex)R(a, x) \equiv (Ex)T(f, a, x).$$

Similarly, to each decidable predicate $R(a_1, \ldots, a_n, x)$, *there is a number* f *such that*

$$(Ex)R(a_1, \ldots, a_n, x) \equiv (Ex)T(f, a_1, \ldots, a_n, x)$$

where $T(i, a_1, \ldots, a_n, x)$ *plays the same role for the Turing machine computation of n-place functions as* $T(i, a, x)$ *for 1-place functions.* (Enumeration theorem.)[202]

[202] $(Ex)T(0, a, x), (Ex)T(1, a, x), (Ex)T(2, a, x), \ldots$ is an enumeration with repetitions of all predicates of the form $(Ex)R(a, x)$ with $R(a, x)$ decidable (the "enumerating predicate" $(Ex)T(i, a, x)$ being of the same form except for the necessary additional variable i).

To prove (E) for the case of one variable a, we need only let $\mathfrak{M}_f$ be a Turing machine which, applied to a, searches for an x such that $R(a, x)$ and writes 0 if one is found, but computes no value otherwise.

(F$_1$) *The predicate* $(x)\bar{T}(a, a, x)$ *is not expressible in the form* $(Ex)R(a, x)$ *with* $R(a, x)$ *decidable. The predicate* $(Ex)T(a, a, x)$ *is not expressible in the form* $(x)R(a, x)$ *with* $R(a, x)$ *decidable. A fortiori,* $(x)\bar{T}(a, a, x)$ *and* $(Ex)T(a, a, x)$ *are not decidable.*

For, suppose to the contrary that $(x)\bar{T}(a, a, x) \equiv (Ex)R(a, x)$ for some decidable predicate $R(a, x)$. Then
$\overline{(x)}\bar{T}(a, a, x) \equiv \overline{(Ex)}R(a, x) \equiv \overline{(Ex)}T(f, a, x)$ (using (E)) $\equiv (x)\bar{T}(f, a, x)$.
Substituting f for a, $\overline{(x)}\bar{T}(f, f, x) \equiv (x)\bar{T}(f, f, x)$, which is absurd
($\vdash \neg(\neg P \sim P)$ in the propositional calculus).

Similarly, suppose that $(Ex)T(a, a, x) \equiv (x)R(a, x)$ for some decidable $R(a, x)$. Then $(Ex)T(a, a, x) \equiv (x)\bar{\bar{R}}(a, x) \equiv \overline{(Ex)}\bar{R}(a, x) \equiv \overline{(Ex)}T(f, a, x)$ (using (E) for $\bar{R}(a, x)$ as its $R(a, x)$). Substituting f for a,
$(Ex)T(f, f, x) \equiv \overline{(Ex)}T(f, f, x)$, which is absurd.

To infer that $(Ex)T(a, a, x)$ is not decidable, suppose that
$(Ex)T(a, a, x) \equiv R(a)$ with $R(a)$ decidable. Then
$(Ex)T(a, a, x) \equiv (x)(R(a) \,\&\, x{=}x)$, which contradicts the second part of (F$_1$), since $R(a) \,\&\, x{=}x$ is decidable. Similarly or by $(x)\bar{T}(a, a, x) \equiv \overline{(Ex)}T(a, a, x)$, $(x)\bar{T}(a, a, x)$ is not decidable.

(F$_2$) $(Ex)(y)\bar{T}(a, a, x, y)$ *is not expressible in the form* $(x)(Ey)R(a, x, y)$ *with* $R(a, x, y)$ *decidable.* $(x)(Ey)T(a, a, x, y)$ *is not expressible in the form* $(Ex)(y)R(a, x, y)$ *with* $R(a, x, y)$ *decidable. A fortiori,* $(Ex)(y)\bar{T}(a, a, x, y)$ *and* $(x)(Ey)T(a, a, x, y)$ *are not expressible using one or zero quantifiers applied to a decidable predicate.*

This is proved similarly to (F$_1$). Likewise we have propositions (F$_3$), (F$_4$), (F$_5$), . . . concerning the predicate forms with 3, 4, 5, . . . quantifiers (alternating between existential and universal) applied to decidable predicates.[203]

(F$_1$), (F$_2$), (F$_3$), . . . are summarized in the hierarchy theorem of Kleene 1943 (1940 abstract):

[203] Since $R(a, x)$ is decidable, $R(a, x) \equiv \bar{\bar{R}}(a, x)$ holds intuitionistically. Also $\overline{(Ex)}T(f, a, x) \equiv (x)\bar{T}(f, a, x)$ and $(x)\bar{\bar{R}}(a, x) \equiv \overline{(Ex)}\bar{R}(a, x)$ are (informal) applications of *82a, which holds intuitionistically. Thus our proof of (F$_1$) is good using only intuitionistic logic. Classical logic is required for (F$_2$), (F$_3$), (F$_4$), ...

(F) *Consider the predicate forms*

$$R(a) \quad \begin{matrix} (Ex)R(a, x) & (x)(Ey)R(a, x, y) & (Ex)(y)(Ez)R(a, x, y, z) & \ldots \\ (x)R(a, x) & (Ex)(y)R(a, x, y) & (x)(Ey)(z)R(a, x, y, z) & \ldots \end{matrix}$$

where in each form R is a decidable predicate. To each form after the first, there is a predicate expressible in that form but not in the form dual to it (i.e. arising from it by interchanging existential and universal quantifiers) nor in any of the forms with fewer quantifiers.

Using a theorem of Post 1948 abstract (IM Theorem XI p. 293), it can be shown that the degrees of a decidable predicate and of the predicates $(Ex)T(a, a, x), (x)(Ey)T(a, a, x, y), (Ex)(y)(Ez)T(a, a, x, y, z), \ldots$ are indeed $\mathbf{0}, \mathbf{0}', \mathbf{0}'', \mathbf{0}''', \ldots$, the same as those of the predicates obtained by starting with a decidable predicate and using (C) repeatedly.

Consider a formal system **N** (like that of § 38) in which, to each a, a formula can be found which expresses $(x)\bar{T}(a, a, x)$; indeed since $(x)\bar{T}(a, a, x) \equiv \overline{(Ex)}T(a, a, x)$, this formula can be $\neg C_a$ for the C_a of § 43. Say the Gödel number of $\neg C_a$ is $\alpha(a)$ where α is a computable function. Using the predicate $\mathrm{Pr}(a, x)$ for **N** mentioned early in this section, $\{\vdash \neg C_a \text{ in } \mathbf{N}\} \equiv (Ex)\mathrm{Pr}(\alpha(a), x)$. Here $\mathrm{Pr}(\alpha(a), x)$ is a decidable predicate of a and x; write it simply $R(a, x)$. Thus

(e) $$\{\vdash \neg C_a \text{ in } \mathbf{N}\} \equiv (Ex)R(a, x).$$

Now it is immediate from the first part of (F_1) that **N** cannot be both correct and complete, so that $\neg C_a$ is provable in **N** if and only if $\neg C_a$ is true, i.e. so that

(⁎⁎) $$\{\vdash \neg C_a \text{ in } \mathbf{N}\} \equiv (x)\bar{T}(a, a, x).$$

For, combining (⁎⁎) with (e) would give $(x)\bar{T}(a, a, x) \equiv (Ex)R(a, x)$, contradicting (F_1). Thus (⁎⁎) cannot hold for all a. Assuming that **N** is correct,

(c) $$\{\vdash \neg C_a \text{ in } \mathbf{N}\} \rightarrow (x)\bar{T}(a, a, x),$$

which is one of the implications in (⁎⁎). So the other

(*) $$(x)\bar{T}(a, a, x) \rightarrow \{\vdash \neg C_a \text{ in } \mathbf{N}\}$$

cannot hold for all a. Since $(x)\bar{T}(a, a, x) \equiv \overline{(Ex)}T(a, a, x)$, this (c) and (*) are the same as in § 43, and by the correctness of **N** we again have (b). Thus (continuing as in the first proof of Theorem V), Gödel's incompleteness theorem, for the system **N** of § 38 or any system **N** that is correct and meets the very general structural condition expressed by (e) holding for some decidable R, is implicit in the first part of (F_1).

This uses classical logic to infer the existence of a value p of a for which (*) is false. (Hierarchy theory is largely classical.) If (instead of assuming (⁎⁎) for reductio ad absurdum) we apply (E) (as we did in proving (F_1)), we get a number f such that

$$\text{(f)} \qquad (Ex)R(f, x) \equiv (Ex)T(f, f, x).$$

Now Theorem V (i)–(iii) with f as the p follow intuitionistically from (b), (c), (e), (f).

Church's theorem (Theorem III) is the third part of (F_1).

In brief, Church's and Gödel's theorems correspond to the two forms $R(a)$ and $(Ex)R(a, x)$ in (F). From this point of view (emphasized in Kleene 1943), the hierarchy theorem (F) can be regarded as a generalization of Church's and Gödel's theorems. Under the other construction using (C), the hierarchy results by the iteration of Church's theorem in a relativized version.[204]

Post in 1944 raised the question whether any predicate of the form $(Ex)R(a, x)$ with R decidable has a degree (strictly) between $\mathbf{0}$ and $\mathbf{0}'$. In 1954, Kleene and Post showed that there exist predicates having degrees between $\mathbf{0}$ and $\mathbf{0}'$; but their method did not show whether any of those predicates are of the form $(Ex)R(a, x)$ with R decidable. In 1956 Friedberg (U.S.A.) then only 20 years old and Mučnik (U.S.S.R.) of similar age, independently of each other, refined the Kleene-Post construction to show that there is a predicate of the form $(Ex)R(a, x)$ with R decidable whose degree is between $\mathbf{0}$ and $\mathbf{0}'$, solving Post's 1944 problem.[205]

There are incomparable degrees (i.e. degrees $\mathbf{a}$ and $\mathbf{b}$ such that neither $\mathbf{a} < \mathbf{b}$ nor $\mathbf{a} = \mathbf{b}$ nor $\mathbf{a} > \mathbf{b}$). The totality of all the degrees possessed by number-theoretic predicates, including degrees jumped over or bypassed in the hierarchies described above, has a very complicated structure.[206]

EXERCISES. 46.1. Show that the relation "$P(a)$ is reducible to $Q(b)$ and vice versa" is reflexive, symmetric and transitive.

46.2. Show that, if $P_1(a)$ and $Q_1(b)$ have the same respective degrees as $P(a)$ and $Q(a)$, then $\{P(a)$ is reducible to $Q(b)$ but not vice versa$\} \equiv \{P_1(a)$ is reducible to $Q_1(b)$ but not vice versa$\}$.

46.3. Prove that for any degrees $\mathbf{a}, \mathbf{b}, \mathbf{c}$: (a) Not $\mathbf{a} < \mathbf{a}$. (b) If $\mathbf{a} < \mathbf{b}$ and $\mathbf{b} < \mathbf{c}$, then $\mathbf{a} < \mathbf{c}$.

[204] For a little more of an indication of hierarchy theory, cf. Kleene 1958 § 2. Mostowski 1954 and Kleene 1955b give fuller but more technical expositions of results to that time.

[205] Friedberg 1956 article concerning, 1957 (abstract 1956); Mučnik 1956, 1958.

[206] The first results to this effect were in Kleene and Post 1954. Much work has been done since by Spector, Lacombe, Shoenfield, Sacks and others. Cf. Sacks 1963.

46.4. Show that: (a) If Q_1 and Q_2 are of the same degree, so are $(Ex)T^{Q_1}(a, a, x)$ and $(Ex)T^{Q_2}(a, a, x)$. (b) If Q is of degree **0**, then $(Ex)T^{Q}(a, a, x)$ is of the same degree as $(Ex)T(a, a, x)$.

46.5. Prove a statement like (E) with universal instead of existential quantifiers.

46.6. Show that for any decidable R,
$(x_1)\ldots(x_m)R(a_1, \ldots, a_n, x_1, \ldots, x_m)$ is of degree $\leq \mathbf{0}'$.

46.7. Establish (D) and (F_2).

46.8*. Show that each of the predicates expressible in the symbolism of **N** § 38 under the usual interpretation (which were called "arithmetical" by Gödel 1931) is expressible in one of the forms of (F).[207]

46.9*. Assume the fact that every decidable predicate is expressible in the symbolism of **N** § 38.[208] Thence establish:

(a) The converse of Exercise 46.8.

(b) For any fixed effective Gödel numbering of the formulas of **N** (end § 43), the predicate "*a* is the Gödel number of a true closed formula of **N**" is not expressible in the symbolism of **N**.

★§ 47. Undecidability and incompleteness using only simple consistency (Rosser). For Theorems IV–VI, we made two non-elementary assumptions (b) and (c) about the system **N**, because they make the approach to the theorems easy, and because we can hardly doubt that they hold for the **N** of § 38 or any system we would want to use instead.

However, to formulate Theorems IV–VI as elementary metamathematical theorems, (b) and (c), or something to the same effect, should be included in the hypotheses. This we have done in the second paragraph of each theorem.

Theorem VI followed from Theorem V by noting that (c) can be replaced by the simple consistency and (a).

There remains (b), which we used in proving Theorems IV and V (iii). Consider any **N** (like that of § 38) in which C_a is of the form $\exists x T(\boldsymbol{a}, \boldsymbol{a}, x)$ where $T(a, a, x) \rightarrow \{\vdash T(\boldsymbol{a}, \boldsymbol{a}, \boldsymbol{x}) \text{ in } \mathbf{N}\}$, whence (a), and

$$\text{(g)} \qquad \bar{T}(a, a, x) \rightarrow \{\vdash \neg T(\boldsymbol{a}, \boldsymbol{a}, \boldsymbol{x}) \text{ in } \mathbf{N}\}.$$

We can then replace (b) in proving Theorem V (iii) by the hypothesis that **N** is "ω-consistent" in the following sense, due to Gödel 1931. A system **S** whose notation includes the numerals (§ 43) is *ω-consistent* if in it, for no variable x and formula A(x), do all of ⊢ A(0), ⊢ A(1), ⊢ A(2), . . . and

[207] Solution in IM p. 285 Theorem VII (d), using general recursiveness instead of Turing computability.

[208] Included in IM p. 285 Theorem VII (b). Using this to generalize Exercise 46.8 to allow symbols for any decidable predicates, we obtain the proposition which led Kleene in 1940 abstract to consider exactly the list of predicate forms in (F).

$\vdash \neg\forall x A(x)$ hold (otherwise, **S** is *ω-inconsistent*). (An ω-consistent system **S** is simply consistent, as follows by writing any formula E as A(x) where x is a variable not occurring in E, so that A(0), A(1), A(2), . . . are all E and $\neg\forall x A(x)$ is equivalent to $\neg E$ by *75.) To prove (iii) from the ω-consistency (with (a), (d) and (g)), suppose $\vdash C_p$, i.e. $\vdash \exists x T(\boldsymbol{p}, \boldsymbol{p}, x)$, whence $\vdash \neg\neg\exists x T(\boldsymbol{p}, \boldsymbol{p}, x)$, whence (using *82a) $\vdash \neg\forall x \neg T(\boldsymbol{p}, \boldsymbol{p}, x)$. But by (i) (already proved in § 44 from (a), (d) and the simple consistency), $\overline{(Ex)}T(p, p, x)$, whence $(x)\bar{T}(p, p, x)$, whence by (g) $(x)\{\vdash \neg T(\boldsymbol{p}, \boldsymbol{p}, \boldsymbol{x})\}$, i.e. $\vdash \neg T(\boldsymbol{p}, \boldsymbol{p}, 0)$, $\vdash \neg T(\boldsymbol{p}, \boldsymbol{p}, 1)$, $\vdash \neg T(\boldsymbol{p}, \boldsymbol{p}, 2)$, ... These with $\vdash \neg\forall x \neg T(\boldsymbol{p}, \boldsymbol{p}, x)$ above contradict the hypothesis of ω-consistency.[209]

Clearly, an ω-inconsistent system violates the interpretation of x as ranging over the natural numbers, expressed by the numerals 0, 1, 2, ...

It is a consequence of Theorem V (ii) that there are simply consistent but ω-inconsistent systems of number-theory. (In substance, this is Exercise 44.3, which we do now.) For, let **N** be the number-theoretic system of § 38, whose consistency we assume (on the basis of the interpretation or of Gentzen's proof). Let **M** come from **N** by adding C_p as an axiom. Then, by arguments just given (using (g) and Theorem V (i) for **N**), **M** is ω-inconsistent. But **M** is simply consistent; for, if $\vdash$ E and $\vdash \neg$E in **M**, then $C_p \vdash E$ and $C_p \vdash \neg E$ in **N**, whence by $\neg$-introd., $\vdash \neg C_p$ in **N**, contradicting Theorem V (ii).

Thus not only does Gödel's second theorem show that simple consistency is hard to prove for number theory, but also his first theorem shows that simple consistency is not the only consistency property that we will ordinarily wish our formal systems for number theory to have. The proofs by

[209] Using (g), ω-consistency implies

$$\text{(b}')\qquad \{\vdash C_a \text{ in } \mathbf{N}\} \rightarrow \overline{\overline{(Ex)}}T(a, a, x),$$

and hence classically (b). For, assume $\vdash C_a$ and $\overline{(Ex)}T(a, a, x)$. Thence $(x)\bar{T}(a, a, x)$ whence by (g): (A) $(x)\{\vdash \neg T(\boldsymbol{a}, \boldsymbol{a}, \boldsymbol{x})\}$. But also (as for $a = p$ in the text): (B) $\vdash \neg\forall x \neg T(\boldsymbol{a}, \boldsymbol{a}, x)$. But ω-consistency says (A) and (B) can't both hold; so by reductio ad absurdum $\overline{\overline{(Ex)}}T(a, a, x)$.

The proof of Theorem IV in § 43 can be reworked to proceed intuitionistically from (b′) instead of (b), thus. Suppose there were a decision procedure for provability in N. We could use it to construct a machine $\mathfrak{M}_p$ which, applied to a, attempts (successfully if $(Ex)T(a, a, x)$) to compute $\varphi_a(a)+1$ if $\vdash C_a$ (CASE 1), and writes 0 if not $\vdash C_a$ (CASE 2). Then as for Theorem II we can deduce a contradiction from $(Ex)T(p, p, x)$ (after using (a) to infer $\vdash C_p$), and also from $\overline{(Ex)}T(p, p, x)$ (after using (b′) to infer not $\vdash C_p$). So by informal use of weak $\neg$-elim. § 11: (C) $(Ex)T(p, p, x) \rightarrow 0\neq 0$, (D) $\overline{(Ex)}T(p, p, x) \rightarrow 0\neq 0$. From (C) by contraposition twice (*13, *12 in § 24), (E) $\overline{\overline{(Ex)}}T(p, p, x) \rightarrow 0\neq 0$. Now we get a contradiction by cases (as above, for $a = p$). In Case 1, by (b′) $\overline{\overline{(Ex)}}T(p, p, x)$, whence by (E) $0\neq 0$, contradicting $0=0$. In Case 2, by (a) and contraposition $\overline{(Ex)}T(p, p, x)$, whence by (D) $0\neq 0$

Gentzen etc. for **N** § 38 do actually establish ω-consistency. At the same time, the problem of interpretation, mentioned at the end of § 44, is emphasized, because we have no reason to stop at securing only simple and ω-consistency.

Meanwhile, in 1936 Rosser found another method of proof of Theorem IV, and of Theorem V with a different number q in place of p, in which all the results are obtained with simple consistency as the only nonelementary hypothesis. We shall obtain these results now as corollaries of a more general theorem (Theorem VIII).

We call a non-empty set or class S of natural numbers *computably enumerable* or *recursively enumerable* if there is a computable function φ such that $\varphi(0), \varphi(1), \varphi(2), \ldots$ is an enumeration (possibly with repetitions) of S.[210] Two sets S_0 and S_1 are *disjoint* if they have no common elements, or in symbols if $S_0 \cap S_1 = \varnothing$ (cf. § 26).

THEOREM VIII. *There are two (non-empty) disjoint recursively enumerable sets C^0 and C^1 with the following property. Given any two disjoint recursively enumerable sets D^0 and D^1 including C^0 and C^1 respectively, i.e. any two recursively enumerable sets D^0 and D^1 such that*

$$(1)\ D^0 \cap D^1 = \varnothing,$$

$$(2^0)\ C^0 \subseteq D^0, \qquad (2^1)\ C^1 \subseteq D^1,$$

a number f can be found which belongs to neither D^0 nor D^1, i.e. such that

$$(3^0)\ f \notin D^0, \qquad (3^1)\ f \notin D^1.$$

(A symmetric form of Gödel's theorem, Kleene 1950.)[211]

PROOF, using an idea of D. Lacombe (reported by Rabin 1958a Footnote 6). With $\varphi_i(a)$ as in § 42 and "$\hat{a}$" meaning "the a's such that" (§ 26), let

$$C^0 = \hat{a}\{\varphi_a(a) \text{ is defined and } \varphi_a(a) = 0\},$$

$$C^1 = \hat{a}\{\varphi_a(a) \text{ is defined and } \varphi_a(a) \neq 0\}.$$

Clearly C^0 and C^1 are disjoint (and non-empty).

[210] Such sets were first considered in Kleene 1936, using the general recursive functions (afterwards proved equivalent to the Turing computable functions; cf. § 41). The term "recursively enumerable" has become the standard one in the literature (Rosser 1936, Post 1944, Smullyan 1959, etc.). Cf. IM pp. 306–307. Also the empty set is often taken to be "recursively enumerable" (Post 1944).

[211] The sense in which this is a form of Gödel's incompleteness theorem is elaborated in IM § 61, which uses the original proof (Kleene 1950).

The sets C^0 and C^1 are disjoint recursively enumerable sets which are "recursively inseparable" in the following sense: for no general recursive set D, $C^0 \subseteq D$ & $C^1 \subseteq \bar{D}$ (where $\bar{D}$ = the compliment of D).

Suppose given recursively enumerable sets D^0 and D^1 satisfying (1), (2^0) and (2^1). Let a Turing machine $\mathfrak{M}_f$ be constructed (incorporating machines which compute functions φ^0 and φ^1 enumerating D^0 and D^1 respectively) which carries out the following operation. Applied to a, $\mathfrak{M}_f$ searches through the enumerations φ^0 and φ^1 of D^0 and D^1 respectively, looking for a. ($\mathfrak{M}_f$ works alternatively, some on φ^0 and some on φ^1, so that if the search is not stopped, it will eventually reach arbitrarily far out in each enumeration.) If $\mathfrak{M}_f$ finds a in the enumeration φ^0 of D^0, it thereupon writes 1 and stops. If $\mathfrak{M}_f$ finds a in the enumeration φ^1 of D^1, it thereupon writes 0 and stops. If neither of these events occurs, $\mathfrak{M}_f$ continues the search ad infinitum, and so computes no value.

To establish (3^0), assume that $f \in D^0$. Then by (1), $f \notin D^1$. So f is in the enumeration φ^0 of D^0, but not in the enumeration φ^1 of D^1. So $\mathfrak{M}_f$ applied to f will find f in the enumeration φ^0 (it cannot be prevented from finding f there by first finding f in the enumeration φ^1). So $\varphi_f(f)$ (the value computed by $\mathfrak{M}_f$ applied to f as argument) is defined, and $\varphi_f(f) = 1$ by the way $\mathfrak{M}_f$ operates. Hence by the definition of C^1, $f \in C^1$. Hence by (2^1) $f \in D^1$, contradicting $f \notin D^1$ (above). By reductio ad absurdum, $f \notin D^0$; i.e. (3^0) holds.

The proof of (3^1) is symmetric to that just given for (3^0). —

A formal system **S**′ is an *extension* of a formal system **S**, and **S** is a *subsystem* of **S**′, if each formula of **S** is a formula of **S**′ and each provable formula of **S** is a provable formula of **S**′. (Any formal system **S** is the "improper" extension of itself.)

Corollary 1. *Assume the system* **N** *of* § 38 *to be simply consistent. To any simply consistent extension* **N**′ *of* **N** *(including* **N** *itself), there is a closed formula* C_f^0 *of* **N** *such that*: (i) $\neg\mathrm{C}_f^0$ *is true*, (ii) *not* $\vdash \neg\mathrm{C}_f^0$ *in* **N**′, (iii) *not* $\vdash \mathrm{C}_f^0$ *in* **N**′.

More generally, this applies to any simply consistent formal system **N** *in which, to each a, there can be found effectively formulas* C_a^0 *and* C_a^1 *(expressing* $a \in C^0$ *and* $a \in C^1$ *respectively) such that* (a^0), (a^1), (b^0), (b^1) *below hold.*

Proof. The same methods which enable us to find in **N** § 38 a formula C_a to express $(Ex)T(a, a, x)$ (beginning § 43) enable us now to find formulas C_a^0 and C_a^1 expressing $a \in C^0$ and $a \in C^1$ respectively. Indeed, let

$$U(i, a, x) \equiv \{\text{Machine } \mathfrak{M}_i\text{, applied to } a\text{, at Moment } x \text{ scans a square next to the right of a blank square}\}$$

(cf. § 41). A formula U(i, a, x) expressing $U(i, a, x)$ can be found, as before we found T(i, a, x) to express $T(i, a, x)$. Let

C_a^0 be $\exists$x[T(***a***, ***a***, x) & U(***a***, ***a***, x)], C_a^1 be $\exists$x[T(***a***, ***a***, x) & $\neg$U(***a***, ***a***, x)].

The same considerations which before gave (a) now give

(a^0) $a \in C^0 \rightarrow \{\vdash C^0_a \text{ in N}\}$, $\qquad$ (a^1) $a \in C^1 \rightarrow \{\vdash C^1_a \text{ in N}\}$,

as can be proved in detail. Also, corresponding to the obvious disjointness of C^0 and C^1, it can be shown that

(b^0) $a \in C^0 \rightarrow \{\vdash \neg C^1_a \text{ in N}\}$, $\qquad$ (b^1) $a \in C^1 \rightarrow \{\vdash \neg C^0_a \text{ in N}\}$.

Now let N′ be any simply consistent extension of N. Let

$$D^0 = \hat{a}\{\vdash C^0_a \text{ in N}'\}, \qquad D^1 = \hat{a}\{\vdash \neg C^0_a \text{ in N}'\}.$$

Then (1) of the theorem is satisfied, by the simple consistency of N′. Also (2^0) holds; for, if $a \in C^0$, then by (a^0) $\vdash C^0_a$ in N, and hence (since N′ is an extension of N) $\vdash C^0_a$ in N′, i.e. $a \in D^0$. Similarly, using (b^1), (2^1) holds.

Because of the nature of any formal system, here N′, Turing machines $\mathfrak{M}^0$ and $\mathfrak{M}^1$ can be found which compute $\varphi^0(n)$ and $\varphi^1(n)$, where $\varphi^0(n)$ is the a of the nth proof (in some enumeration of the proofs in N′) which is a proof of C^0_a for some a, and $\varphi^1(n)$ likewise for $\neg C^0_a$. Thus D^0 and D^1 are recursively enumerable.

So all the hypotheses of the theorem are satisfied. Hence there is a number f satisfying (3^0) and (3^1). Then by the definitions of D^0 and D^1, (iii) and (ii) of the corollary hold. By (3^0) and (2^0), $f \notin C^0$; so (i) also holds.[212] —

We call a formal system **S** *essentially undecidable* (after Tarski 1949 abstract), if **S** is simply consistent, and every simply consistent extension of **S** (including **S** itself) is undecidable.

COROLLARY 2. *Assuming the system* **N** *of* § 38 *to be simply consistent, it is essentially undecidable.*

More generally, any formal system **N** *satisfying the second paragraph of Corollary* 1 *is essentially undecidable.*

PROOF. Assume N simply consistent, and let N′ be any simply consistent extension of N. Now we take

$$D^0 = \hat{a}\{\vdash C^0_a \text{ in N}'\}, \qquad D^1 = \hat{a}\{\text{not } \vdash C^0_a \text{ in N}'\}.$$

Now (1) is immediate. As before (Corollary 1), (2^0) holds by (a^0). Also (2^1) holds; for, if $a \in C^1$, then by (b^1) $\vdash \neg C^0_a$ in N and hence in N′, so by

[212] Symmetrically, (i)–(iii) hold with C^0_f replaced by C^1_f for a different choice of f. At least for the N of § 38, by using (A) in § 46 to write $a \in C^0$ or $a \in C^1$ in the form $(Ex)T(\theta(a), \theta(a), x)$, we have (i)–(iii) as they read in Theorem V but for p replaced by $q = \theta(f)$ with either of the present f's.

the simple consistency of **N**′ not $\vdash C^0_a$ in **N**′, i.e. $a \in D^1$. As before, D^0 is recursively enumerable.

Now assume that there is a decision procedure for provability in **N**′, i.e. there is a machine that decides, for any formula A, whether or not $\vdash$ A in **N**′. Using this machine, we could find a machine to compute a function φ^1 which enumerates the present D^1, so D^1 would also be recursively enumerable. So by the theorem there would be a number f such that $f \notin D^0$ and $f \notin D^1$. With the present D^0 and D^1, this is absurd.

Applications to the decision problems for axiomatic theories (Tarski). Rosser's motivation in 1936 was presumably simply to strengthen Theorems IV and V by using the simple consistency of **N** in place of the ω-consistency. A fertile area of application of Corollary 2 has been cultivated since 1949 by Tarski and his coworkers.[213] This is to showing the undecidability of various axiomatic theories formalized in a logical calculus, which may be either the predicate calculus or the predicate calculus with equality. (Tarski uses the latter.) That is, the axioms of the theory are stated in the symbolism of the calculus, with predicate, (individual) and function symbols; and the calculus provides the logic. Examples of axiomatic theories thus formalized are the formal system **N** for number theory in § 38 (whose undecidability we already know) and the systems **G**, **Gp**, **AG**, **AGp** for groups and Abelian groups in § 39. As we remarked for **N** and **G**, in such systems the deduction theorem and our other introduction and elimination rules hold (Theorems 13, 21).

By the undecidability of a formalized theory or formal system **S** we mean, as above, that there is no decision procedure for answering all of the questions whether a given formula A is provable in the system **S** (formalizing the theory). As we saw in the proofs of Theorems IV and VII, and further in § 46, when we know the decision problem for one class of questions (P) to be unsolvable, we can infer the like for another class (Q) by reducing the questions in the first class to questions in the second (or briefly, by reducing the first decision problem to the second). In particular for formal systems, if $\mathbf{S}_2$ is undecidable, we can infer $\mathbf{S}_1$ to be undecidable if we can find effectively, to each formula B in $\mathbf{S}_2$, a formula B′ in $\mathbf{S}_1$ such that $\vdash$ B in $\mathbf{S}_2$ if and only if $\vdash$ B′ in $\mathbf{S}_1$.

A formal system **S** based on the predicate calculus (without or with equality) as the logic with additional or "nonlogical" axioms is said to be *finitely axiomatizable* (after Tarski 1949 abstract), if the number of those

[213] Theorem VIII itself answered a question concerning analogies between the hierarchies described in § 46 (Kleene 1943, Mostowşki 1947, etc.) and the hierarchies studied in "descriptive set theory" (Borel 1898, Lusin 1930, etc.); cf. Kleene 1950, Addison 1960. Other applications are e.g. in Kleene 1956, Rabin 1958a, Kleene and Vesley 1965 pp. 112, 183.

nonlogical axioms is finite, or all but a finite set of them can be omitted without changing the class of the provable formulas; such a finite set of nonlogical axioms of **S** (or all, if **S** has only finitely many nonlogical axioms) we may call a *finite axiomatization of* **S**.

In the case of a system like **N** which is given with infinitely many nonlogical axioms (the 8 particular axioms 14–21 and the $\aleph_0$-axioms by the Axiom Schema 13), it may not be obvious a priori whether the system is finitely axiomatizable or not. (The author first heard this question asked about the **N** of § 38 during 1949; its solution, in the negative, was published by Ryll-Nardzewski in 1952.)[214]

The application of the reduction method to show a system S_1 undecidable depends on having available a system S_2 already known to be undecidable whose "theory" can be developed in S_1 (via some translation of formulas B in S_2 into formulas B′ in S_1). The simpler S_2 is, the better the chance we can do this in a given S_1.

Tarski 1949 abstract recognized that we will be in a particularly favorable position for doing this, if we have first of all *a system* **S** *which is both essentially undecidable and finitely axiomatizable*, besides being simple. For then *each system* S_1 (*with all the symbols which* **S** *has*) *is undecidable which has a simply consistent common extension* S_3 *with* **S**. To prove this, consider the system S_2 which has as its (nonlogical) axioms those of S_1 and those in a finite axiomatization of **S** (and just the symbols of S_1). Then S_2 is a subsystem of S_3, so S_2 is also (simply) consistent. Also S_2 is an extension of **S**; so by the essential undecidability of **S**, S_2 is undecidable. Now consider the axioms of S_2 which are not axioms of S_1. They are finitely many, since they come from a finite axiomatization of **S**; say they are $A_1, \ldots, A_m$. So $\vdash$ B in S_2 if and only if $\forall A_1, \ldots, \forall A_m \vdash B$ in S_1, which by the deduction theorem etc. (Corollaries Theorems 10 and 11) is the case if and only if $\vdash \forall A_1 \supset (\ldots(\forall A_m \supset B)\ldots)$ in S_1. Thus we have reduced the decision problem for the undecidable system S_2 to that for S_1. So S_1 is undecidable, as was to be shown.

To make things simple, we began with the case S_1 lacks no symbols of **S**. More generally, S_1 may lack predicate or function symbols of **S**, provided they can be "defined" or "interpreted" in a consistent common extension S_3 with **S**. For example, if S_1 has the notation of **N** while **S** has the predicate symbol $<$, the consistent common extension S_3 could have provable in it

[214] By Corollary Theorem 31 § 29, the theory of equality for a finite list of predicate and function symbols is finitely axiomatizable in the predicate calculus.

If some finite set of formulas of a system **S** can be used as the nonlogical axioms instead of the original ones (i.e. without changing the class of the provable formulas), then some finite subset of the original nonlogical axioms can be (so **S** is finitely axiomatizable, under the above definition). Why?

the formula $a < b \sim \exists c\, c' + a = b$. If $\mathbf{S}$ has the symbol !, $\mathbf{S}_3$ could have provable $a! = b \sim \mathrm{F}(a, b)$ where $\mathrm{F}(a, b)$ expresses the representing predicate $a! = b$ of the function $a!$ (cf. end § 38). The additional details which this requires in the above argument that $\mathbf{S}_1$ is undecidable are outside the scope of this book.[215]

Since Rosser 1936, it has been known that **N** is essentially undecidable (though the name "essentially undecidable" was first used by Tarski 1949 abstract). However **N** is not finitely axiomatizable, as we have mentioned (Ryll-Nardzewski 1952). The systems of axiomatic set theory of von Neumann 1925, Bernays 1937–54 and Gödel 1940 (if consistent) are essentially undecidable, since they include **N** (if the symbolism of **N** is suitably identified within them), and they have only finitely many axioms (unlike the system of Zermelo and Fraenkel § 35, where the axiom of subsets (II) gives rise to $\aleph_0$ axioms when formalized in the predicate calculus). Tarski's reduction method for decision problems for elementary axiomatic theories however calls for an essentially undecidable and finitely axiomatizable system which is much more elementary in its interpretation than axiomatic set theory.

Such a system was first given by Mostowski and Tarski 1949 abstract, using basically Rosser's 1936 result (included in our Corollary 2 Theorem VIII); their system deals with the arithmetic of the integers . . . , −2, −1, 0, 1, 2, . . . instead of the natural numbers 0, 1, 2, ...

In 1950 abstract R. M. Robinson showed that a certain subsystem of the **N** of § 38 is essentially undecidable and finitely axiomatizable. Using the predicate calculus as the underlying logic (he used the predicate calculus with equality), it is the system having the following 13 non-logical axioms: 14–21, the four equality axioms for $+$ and $\cdot$, and the formula $a = 0 \vee a > 0$. (This is a subsystem of **N**, since its five axioms which aren't axioms of **N** are provable in **N**.) To see by Corollary 2 above that Robinson's system is essentially undecidable, we need only verify that it satisfies the conditions in the second paragraph of Corollary 1. It is quite clear that, by applying methods used in the proof of Theorem VII from Theorem III (where a system $\mathbf{S}_1$ with finitely many axioms sufficed for the relevant theory of $(Ex)T(a, a, x)$), we could find *a system* with only finitely many axioms that would do. This is the fundamental discovery with which we are concerned now. To make this result more neat, by showing that a *subsystem of* **N**, and indeed exactly Robinson's system, will do, requires

[215] They involve the material cited in Footnote 156 § 38. See Tarski, Mostowski and Robinson 1953 or IM pp. 437–439. In the case $\mathbf{S}_1$ lacks function symbols of S, the logic should be the predicate calculus with equality, or $\mathbf{S}_1$ should have the symbol = and the equality axioms for the symbols of $\mathbf{S}_1$ should be provable in $\mathbf{S}_3$.

additional detailed work of sorts which we have been omitting in this book.[216]

Using Tarski's method with a simple essentially undecidable and finitely axiomatizable system, Tarski and his coworkers have shown the undecidability of a variety of formalized theories in the arithmetic of integers and reals, rings, groups, fields, lattices and projective geometries. In particular, by these means Tarski showed that **Gp** and hence **G** are undecidable.[217] However **Gp** and **G** are not essentially undecidable; for, their respective extensions **AGp** and **AG** are decidable, as Szmielew 1948, 1955 showed.

EXERCISE 47.1*.[218] For each of the following, say whether a Turing machine $\mathfrak{M}$ can be found which performs the described operation. If "yes", give the idea for its construction (as is done for various machines in §§ 42–47), not the full details (as in § 41). If "no", show why not. (We say a machine $\mathfrak{M}_i$ *enumerates* a set C of natural numbers, if $\mathfrak{M}_i$ computes a total function φ_i such that $\varphi_i(0)$, $\varphi_i(1)$, $\varphi_i(2), \ldots$ is an enumeration of C allowing repetitions.)

(a) Applied to i and n, when $\mathfrak{M}_i$ enumerates an infinite set C, $\mathfrak{M}$ computes the nth number $\varphi(n)$ in an enumeration $\varphi(0), \varphi(1), \varphi(2), \ldots$ of C without repetitions.

(b) Applied to i and n, when $\mathfrak{M}_i$ enumerates a nonempty set C, $\mathfrak{M}$ computes the nth number $\varphi(n)$ in an enumeration $\varphi(0), \varphi(1), \varphi(2), \ldots$ of C which is without repetitions if C is infinite.

(c) Applied to f and n, when $\hat{a}(Ex)T(f, a, x)$ is nonempty, $\mathfrak{M}$ computes the nth number $\varphi(n)$ in an enumeration $\varphi(0), \varphi(1), \varphi(2), \ldots$ of $\hat{a}(Ex)T(f, a, x)$ allowing repetitions.

(d) Similarly, with $\hat{a}(x)\bar{T}(f, a, x)$ in place of $\hat{a}(Ex)T(f, a, x)$.

(e) Applied to i and j, when $\mathfrak{M}_i$ and $\mathfrak{M}_j$ enumerate sets D^0 and D^1 for Theorem VIII (1), (2^0), (2^1), $\mathfrak{M}$ computes a number f for (3^0), (3^1).

[216] First, we may establish that $T(i, a, x)$ and $U(i, a, x)$ are primitive recursive, and hence (by IM Corollary 27 p. 244 with Lemma 18b) numeralwise expressed in Robinson's system by formulas T(i, a, x) and U(i, a, x).[176] Then we take for C^0_a and C^1_a the formulas $\exists x[T(a, a, x) \& \forall y(y<x \supset \neg T(a, a, y)) \& U(a, a, x)]$ and $\exists x[T(a, a, x) \& \forall y(y<x \supset \neg T(a, a, y)) \& \neg U(a, a, x)]$. (Here, under the usual interpretation, $\forall y(y<x \supset \neg T(a, a, y))$ is redundant, but its presence facilitates proving (b^0) and (b^1) in the weak system of Robinson.) Now (a^0) and (a^1) follow at once. Also it is not hard to establish (b^0) and (b^1), using the remark in IM p. 198 after the proof of *169. If we do not want to accept the simple consistency of Robinson's system on the basis of its interpretation or the nonelementary consistency proof of Gentzen for the **N** of § 38 (end § 44), an elementary consistency proof is available in IM Theorem 53 (a) p. 470.

[217] See Tarski, Mostowski and Robinson 1953 Chapter 3.

The undecidability of **Gp** also follows from the more recently proved result of Novikov 1955 that the "word problem for groups" is unsolvable (end § 45).

[218] Some of the solutions are in IM pp. 306, 307, 346.

(f) Applied to i, when $\mathfrak{M}_i$ enumerates a nonempty set C of members of $\hat{a}(x)\bar{T}(a, a, x)$, $\mathfrak{M}$ computes another member (not in C) of $\hat{a}(x)\bar{T}(a, a, x)$.
(g) Applied to a, when a is the Gödel number of a formula A of **N** § 38 (in a fixed effective Gödel numbering of those formulas), $\mathfrak{M}$ decides whether A $\vdash_{\mathbf{N}}$ 1=0.
(h) Applied to a, when a is the Gödel number of a formula A of **N** (as in (g)), $\mathfrak{M}$ decides whether it is decidable whether A $\vdash_{\mathbf{N}}$ B where B ranges over all formulas of **N**.

CHAPTER VI

THE PREDICATE CALCULUS (ADDITIONAL TOPICS)

§ 48. Gödel's completeness theorem: introduction. We shall continue our series of theorems about the propositional and predicate calculi, begun in Chapters I–III.[219]

In the propositional calculus, theoretically every question of provability or deducibility is answerable by using truth tables. Of course, practical difficulties arise, if we ask the questions about too many or too complicated formulas. In the predicate calculus, we cannot actually complete the construction of the truth tables when variables are present, except in finite domains.

This difference between the propositional and predicate calculi has now been underlined by Theorem VII in § 45. By its proof with (A) in § 46, there is a formula of the pure predicate calculus **Pd** (§ 39) whose unprovability is equivalent to the truth of Fermat's "last theorem" (§ 40). In effect, mathematicians have labored unsuccessfully for over 300 years to settle the question whether this one particular formula of the predicate calculus is unprovable or provable. Of course the predicate calculus as we know it hadn't been formulated 300 years ago.[220] But this example illustrates the futility of approaching all questions of provability and unprovability in the predicate calculus by using just the ideas of Chapter II. The predicate calculus is such a rich system that a host of particular problems that are commonly considered in mathematics, using the predicate calculus as a tool in successive short arguments, can be clothed entirely in the pure predicate calculus. Indeed, as the proof of Theorem VII illustrates, this is the case for all problems whether a given statement holds in a formal axiomatic theory whose axioms are finite in number and expressible in the symbolism of the predicate calculus with predicate, (individual) and function symbols.

[219] Many of the results of this chapter belong to classical nonfinitary model theory, and thus not to metamathematics (§ 37).

[220] Although characteristic features of the predicate calculus go back to Frege 1879, the first explicit formulation of it as a separate formal system is perhaps in Hilbert and Ackermann 1928, according to Church 1956 pp. 288 ff.

Nevertheless, there is more that can be learned by the study of the predicate calculus, pure or applied (§ 39), as a logical system.

We mentioned in § 23 that Gödel's completeness theorem 1930 will extend Theorem 14 § 12 to the predicate calculus: *For each formula* F *of the predicate calculus* (§ 16), *if* F *is valid* (§ 17), *then* F *is provable* (§ 21), or briefly *if* ⊨ F *then* ⊢ F. The theorem will include some further information, and have some other versions. For the present, we address ourselves to the problem of how to prove the italicized statement just given.

We now adopt the term "parameters" (after Beth 1953 and Craig 1957a) as a name for the symbols or syntactical entities in a formula or formulas to which values are assigned in entering truth tables (and similarly in terms).[221] A *proposition(al) parameter* is an atom in the propositional calculus (§ 1), and in the predicate calculus a 0-place ion (§ 16). A *predicate parameter* is an *n*-place ion for $n \geq 0$. *An individual parameter* is a free variable or 0-place meson (§ 28). A *function parameter* is an *n*-place meson for $n \geq 0$.

A parameter *of* a formula or list of formulas is one which actually occurs in the formula or in some of the formulas. But sometimes we enter tables with values of other parameters; cf. beginning § 4.

In this chapter, for definiteness we will usually speak of formulas as constructed using individual, function, proposition(al) and predicate symbols, as in an applied predicate calculus § 39. But the treatment will also hold good using instead any 0-place mesons, *n*-place mesons, 0-place ions and *n*-place ions allowable under formation rules as in §§ 16, 28.

Returning to our problem, a formula F in the predicate calculus is *not* valid, exactly if F is *falsifiable* in the following sense: there is some (nonempty) domain *D* and some assignment in *D* to the parameters of F for which F takes the value f. In this case, we call such an assignment a *falsifying assignment for* F *in D*, and we say F is *falsifiable in* that *D* or is $\bar{\bar{D}}$*-falsifiable*. The *D* and falsifying assignment together may be called a *counterexample* to F.[222] (Replacing f by t, we get the notions *satisfiable*, *satisfying assignment*, $\bar{\bar{D}}$*-satisfiable* and *example*.)

We now ask whether it is not possible to search for counterexamples to formulas F in such a systematic manner that the following will be the case,

[221] Thus in § 2 (1) the parameters are P, Q, R; in § 17 Example 1 they are P(x), Q, y or P(—), Q, y or simply P, Q, y; in § 28 Example 1, f, x for the term P and f for the formula; and in § 29 Example 2, P, f (but not =). (However in this chapter until § 52, we shall not use § 29, which gives a fixed preassigned value to =.)

[222] If F is a formula of the propositional calculus simply, we don't need to mention any domain *D*, and a falsifying assignment or counterexample can be simply an assignment of t's and f's to the atoms of F which makes F f.

for any given formula F of the predicate calculus: (I) If any counterexample to F exists (i.e. if not ⊨ F), then the search will lead to one. (II) If no counterexample to F exists (i.e. if ⊨ F), then as we pursue the search that fact will eventually manifest itself by the closing to us of all the avenues along which we are searching, whereupon we will be in a position to prove F (i.e. then ⊢ F).

This idea was used independently by Beth 1955, Hintikka 1955, 1955a, Schütte 1956 and Kanger 1957 to give proofs of Gödel's 1930 completeness theorem in which the connection between model theory and proof theory comes in very naturally. The treatment below is quite close to Beth 1955, which gave the present writer the idea for it.[223]

So we consider how we can search *systematically* for counterexamples to formulas F of the predicate calculus.

Example 1. Let F be $\exists x(P \supset Q(x)) \supset (P \supset \forall x Q(x))$. We seek a (non-empty) domain *D* and an assignment in *D* to P, Q(x) which makes **(1)** $\exists x(P \supset Q(x)) \supset (P \supset \forall x Q(x))$ 𝔣. By the truth table for ⊃ in § 2, a *D* and assignment which do this must make **(2)** $\exists x(P \supset Q(x))$ 𝔱 and **(3)** $P \supset \forall x Q(x)$ 𝔣; and doing both the latter is also sufficient for doing the former. For the same reason, to make $P \supset \forall x Q(x)$ 𝔣, it is both necessary and sufficient to make **(4)** P 𝔱 and **(5)** $\forall x Q(x)$ 𝔣.

By the evaluation rule for ∃ in § 17, to make $\exists x(P \supset Q(x))$ 𝔱 (cf. (2)) it is necessary and sufficient to pick the domain *D* so that it contains an element, which we may call a_0, such that **(6)** $P \supset Q(a_0)$ is 𝔱.

To make $P \supset Q(a_0)$ 𝔱, we have two alternatives. It is necessary and sufficient either **(i)** to make **(7)** P 𝔣 or **(ii)** to make **(8)** $Q(a_0)$ 𝔱. We don't need to do both (i.e. it isn't necessary, though of course it would be sufficient). So here our search for a counterexample to F splits, and we can follow either of two paths or avenues.

Consider the first path (i), along which we seek to make P 𝔣. But at (4) we already had to make the same formula P 𝔱. These two requirements are incompatible. Thus, along this path we cannot have a counterexample. This avenue is closed to us; it constitutes a "blind alley" or "dead end".

So, *if* we can get a counterexample at all, it has to be by following the second path (ii). Continuing along this path, to make $\forall x Q(x)$ 𝔣 (cf. (5)), it is necessary and sufficient that the domain *D* contain an element a_1 such that **(9)** $Q(a_1)$ is 𝔣. We have no right to assume this element is the same as

[223] In some respects it more resembles Kanger 1957, as the author learned after working it out.

Some of the other proofs of Gödel's completeness theorem are in Hilbert and Ackermann 1938 (= 2nd ed. of 1928), Hilbert and Bernays 1939, Mostowski **1948b**, Henkin 1949, 1963, Rasiowa and Sikorski **1950**, Rieger **1951**, A. Robinson **1951**, 1963, Beth **1951**, Kleene IM (1952b), 1958. Those listed in bold face apply topology or algebra.

the element a_0 introduced at (6), so we are using another letter a_1 for it. But now, along this path (ii), we indeed have been led to a counterexample. For, our successive analytical steps have shown that it is sufficient for our purpose to pick a domain D containing at least elements named a_0 and a_1, and an assignment to the parameters a_0, a_1, P, Q(x) such that (4) P and (8) Q(a_0) are both t while (9) Q(a_1) is f. This we can do as follows. We pick D to be a domain of exactly two elements; say $D = \{0, 1\}$. To a_0 and a_1 we assign the respective values 0 and 1. To P we give the value t. We evaluate Q(x) by the logical function I(x) such that I(0) is t (so Q(a_0) is t) and I(1) is f (so Q(a_1) is f). (This is the logical function I_2(x) of § 17 Example 1, allowing for the difference in notation, the elements being named "1" and "2" there, "0" and "1" here.) Thus F is falsifiable; so not ⊨ F.

This analysis has taken considerable space to give verbally. We now adopt a symbolic representation of such analyses. We choose our method of representation with the aim of putting before our minds an absolutely clear picture of the structure of our searches for counterexamples, including both the situations obtaining initially and after successive steps, and the over-all structure. These symbolic representations can be somewhat cumbersome. But our purpose is to reason about them, rather than to use them extensively in practice; so it does not much matter if they are cumbersome.

As a search for a counterexample proceeds, initially and after each step, along whichever path we are pursuing (if we have had a choice or successive choices), we have two (finite) lists of formulas: a list Δ of (zero or more) formulas which we are aiming to make t, and a list Λ of (zero or more) formulas which we are aiming to make f. The steps of analysis up to the one just completed (inclusive) have shown that making simultaneously all of Δ t and all of Λ f will suffice to make the original formula F f. (Initially, Δ is empty and Λ is simply F.)

Also at any stage in the search as a whole, our analysis has shown that, to make F f, it is necessary that we make all of Δ t and all of Λ f in the lists Δ and Λ last reached, along at least one of the paths (if we have had choices).

So the situation, initially or after any step, is represented by an ordered pair $\{\Delta; \Lambda\}$. For reasons partly historical, we elect to write "$\Delta \to \Lambda$" instead of "$\{\Delta; \Lambda\}$". Here $\to$ is a new formal symbol (which may be read "give(s)"). The formal expression $\Delta \to \Lambda$ (for any two finite sequences Δ and Λ of zero or more formulas each) we call a *sequent*; and we call Δ its *antecedent* and Λ its *succedent*.

It remains to represent the structure of our searches as a whole. This we do by arranging the sequents in the order in which we are led to them. For reason partly historical, we write the initial sequent $\to$ F at bottom of the figure. Each time we perform a step, we draw a line above and write the one or (in case we have a choice) two sequents to which the step leads.

Thus we represent the analysis or search in Example 1 by the following "tree" (left).

$$\dfrac{\dfrac{\dfrac{\overset{\times}{P \to \forall xQ(x), P} \qquad \dfrac{\overset{\checkmark}{Q(a_0), P \to Q(a_1)}}{Q(a_0), P \to \forall xQ(x)}\,{\to}\forall}{P, P \supset Q(a_0) \to \forall xQ(x)}\,{\supset}{\to}}{\dfrac{P, \exists x(P \supset Q(x)) \to \forall xQ(x)}{\exists x(P \supset Q(x)) \to P \supset \forall xQ(x)}\,{\to}{\supset}}\,{\exists}{\to}}{\to \exists x(P \supset Q(x)) \supset (P \supset \forall xQ(x)).}\,{\to}{\supset}$$

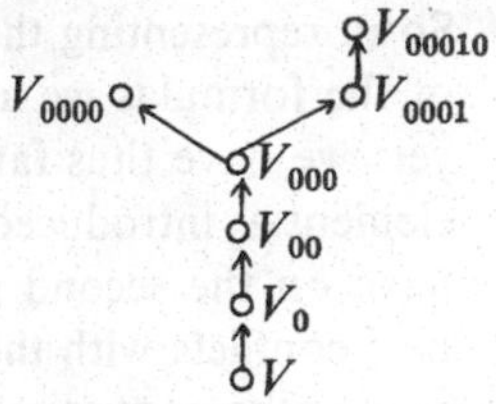

$$\begin{array}{ccc} & & V_{00010} \\ & & \uparrow \\ V_{0000} & & V_{0001} \\ & \nwarrow \quad \nearrow & \\ & V_{000} & \\ & \uparrow & \\ & V_{00} & \\ & \uparrow & \\ & V_0 & \\ & \uparrow & \\ & V & \end{array}$$

We have placed a cross "×" at the top of one path or branch to show that this path is (terminated and) *closed* to us in our search for a counterexample, a check "√" at the top of the other to indicate that the search has terminated successfully there. The symbol "→⊃" indicates that the step in question is an analysis of an implication in the succedent (which we are aiming to make f), "∃→" that it is an analysis of an existence formula in the antecedent (which we are aiming to make t), etc.

The tree of sequents (left) can be considered as the result of placing sequents on the vertices of a purely geometric tree (shown without the sequents at the right). The geometric tree in this example consists of 7 vertices in a certain arrangement (a "partial ordering") shown by the arrows. Thus, the sequent $Q(a_0), P \to \forall xQ(x)$ is placed at the vertex V_{0001}. A *path* or "avenue" in our search for a counterexample is represented by a succession of vertices, starting at the bottom and following arrows in the tree. There are two paths in this example, $VV_0V_{00}V_{000}V_{0000}$ and $VV_0V_{00}V_{000}V_{0001}V_{00010}$; they run together as far as the vertex V_{000}, after which they diverge.[224]

[224] An alternative representation of our search is obtained by simply omitting the verbal explanations given above, but writing the nine numbered formulas in two columns, marked true (t) and false (f). After (6), where we have a choice, the columns split into respective left and right subcolumns (i) and (ii). This gives what Beth 1955, 1959 calls a *semantic tableau:*

t		f	
(2) $\exists x(P \supset Q(x))$ (1)		(1) $\exists x(P \supset Q(x)) \supset (P \supset \forall xQ(x))$	
(4) P (3)		(3) $P \supset \forall xQ(x)$ (1)	
(6) $P \supset Q(a_0)$ (2)		(5) $\forall xQ(x)$ (3)	
(i)	(ii)	(i)	(ii)
	(8) $Q(a_0)$ (6)	(7) P (6)	(9) $Q(a_1)$ (5)

If we intended to use the search method extensively in practice, the tableaux would be

EXAMPLE 2. Let F be $\exists x(P \supset Q(x)) \supset (P \supset \exists xQ(x))$. The analysis is the same as in Example 1, except with $\exists xQ(x)$ in place of $\forall xQ(x)$, down through (8). Now to make (5) $\exists xQ(x)$ $\mathfrak{f}$, it is necessary that **(9)** $Q(a_0)$ be $\mathfrak{f}$. This is sufficient if the domain D contains only a_0, but not otherwise. So in representing the new situation we don't omit $\exists xQ(x)$ from the list of the formulas we are aiming to make $\mathfrak{f}$. (We haven't finally picked D yet; we have thus far only committed ourselves to its having at least the element a_0 introduced at (6).) However, if we look at the situation we now have on the second path (ii), we see that the requirement that (9) $Q(a_0)$ be $\mathfrak{f}$ conflicts with the requirement that (8) $Q(a_0)$ be $\mathfrak{t}$. So in this example *both* paths or avenues open to us in searching systematically for a counterexample to F are closed, or the tree itself is *closed*. This completes an informal proof (in a classical observer's language) that there is no counterexample to this F, i.e. a proof that F is valid. Now it would be surprising if we could not utilize the method of this informal proof that $\models$ F to construct a formal proof of the formula F and thus to show that $\vdash$ F. If our formal system of predicate calculus in Chapter II were not adequate to do this, we would certainly look for ways to augment it. However, we postpone this part of the problem to § 51. The sequent tree in this Example 2 is as follows. (Several formulas are in boldface for later reference.)

$$\dfrac{\dfrac{\dfrac{\dfrac{\overset{\times}{P \to \exists xQ(x), \mathbf{P}} \qquad \dfrac{\overset{\times}{\mathbf{Q(a_0)}, P \to \exists xQ(x), Q(a_0)}}{\mathbf{Q(a_0)}, P \to \exists xQ(x)}\ \to\exists}{\mathbf{P, P \supset Q(a_0)} \to \exists xQ(x)}\ \supset\to}{\mathbf{P, \exists x(P \supset Q(x))} \to \exists xQ(x)}\ \exists\to}{\mathbf{\exists x(P \supset Q(x))} \to P \supset \exists xQ(x)}\ \to\supset}{\to \exists x(P \supset Q(x)) \supset (P \supset \exists xQ(x)).}\ \to\supset$$

V_{00010}, V_{0000}, V_{0001}, V_{000}, V_{00}, V_0, V

Before continuing, we observe that we can codify the steps of analysis we are using in our searches. In terms of the representation by trees of

---

more efficient than the sequent trees, since they do not require us to recopy formulas whose status is not being changed. But they do not show the individual situations as simply (e.g. that when (8) has just been introduced, Δ is (4), (8) and Λ is (5)). Hence we prefer to use the sequent trees in the proof of Gödel's completeness theorem and other theoretical uses.[268] (The tableau here does show explicitly that, when (8) has just been introduced, it suffices to make (2), (4), (6), (8) $\mathfrak{t}$ and (1), (3), (5) $\mathfrak{f}$. This kind of information taken with the structure of such lists, is enough for recognizing the counterexamples. But, altogether, we believe what we shall be doing is more easily visualized in terms of the sequent trees.)

sequents, each step is performable by one of the following 14 rules. Thus, we have several times used the principle that, in order to make an implication $A \supset B$ $\mathfrak{f}$, it is necessary and sufficient to make A t and B $\mathfrak{f}$. This is now codified by the rule at the upper left, called "$\to\supset$" or "the $\supset$-succedent rule". The Γ and Θ tag along to indicate lists of zero or more formulas not changed in the step, those in Γ to be made t and those in Θ to be made $\mathfrak{f}$.

$$\frac{A, \Gamma \to \Theta, B}{\Gamma \to \Theta, A \supset B.} \to\supset \qquad \frac{\Gamma \to \Theta, A \quad B, \Gamma \to \Theta}{A \supset B, \Gamma \to \Theta.} \supset\to$$

$$\frac{\Gamma \to \Theta, A \quad \Gamma \to \Theta, B}{\Gamma \to \Theta, A \,\&\, B.} \to\& \qquad \frac{A, B, \Gamma \to \Theta}{A \,\&\, B, \Gamma \to \Theta.} \&\to$$

$$\frac{\Gamma \to \Theta, A, B}{\Gamma \to \Theta, A \vee B.} \to\vee \qquad \frac{A, \Gamma \to \Theta \quad B, \Gamma \to \Theta}{A \vee B, \Gamma \to \Theta.} \vee\to$$

$$\frac{A, \Gamma \to \Theta}{\Gamma \to \Theta, \neg A.} \to\neg \qquad \frac{\Gamma \to \Theta, A}{\neg A, \Gamma \to \Theta.} \neg\to$$

$$\frac{A, \Gamma \to \Theta, B \quad B, \Gamma \to \Theta, A}{\Gamma \to \Theta, A \sim B.} \to\sim \qquad \frac{A, B, \Gamma \to \Theta \quad \Gamma \to \Theta, A, B}{A \sim B, \Gamma \to \Theta.} \sim\to$$

$$\frac{\Gamma \to \Theta, A(b)}{\Gamma \to \Theta, \forall x A(x)} \to\forall \qquad \frac{A(r), \forall x A(x), \Gamma \to \Theta}{\forall x A(x), \Gamma \to \Theta.} \forall\to$$

where b does not occur free in $\Gamma \to \Theta, \forall x A(x)$.

$$\frac{\Gamma \to \Theta, \exists x A(x), A(r)}{\Gamma \to \Theta, \exists x A(x).} \to\exists \qquad \frac{A(b), \Gamma \to \Theta}{\exists x A(x), \Gamma \to \Theta} \exists\to$$

where b does not occur free in $\exists x A(x), \Gamma \to \Theta$.

In these rules, A and B are (allowed to be) any formulas; x is any variable; A(x) is any formula; b is any variable free for x in A(x) (and unless b is x not occurring free in A(x)); r is any variable not necessarily distinct from the other variables present, or (if formation rules as in § 28 are used) any term, free for x in A(x); A(b) and A(r) are the results of substituting b and r respectively for the free occurrences of x in A(x); and Γ and Θ are any (finite) lists of (zero or more) formulas.

In using the rules $\to\forall$ and $\exists\to$, we must obey the *restriction on variables* (stated with the rules); briefly, b shall not occur free in the sequent below the line. (When the A(x) does not contain the x free, then A(b) is A(x) no matter what variable b is. We agree in such a case to choose for the analysis a variable b not occurring free in the lower sequent, so the restriction will be met.)

The order of listing the formulas within any antecedent and within any succedent is to be immaterial in applying the rules. Thus in Examples 1 and 2, the rule $\exists\rightarrow$ (bottom right) applies from V_{00} to V_{000} (with a_0 as the b), even though $\exists x(P \supset Q(x))$ is not written first in the antecedent.

Let us extend our evaluation process from formulas to sequents, thus: $\Delta \rightarrow \Lambda$ shall take the value $\mathfrak{f}$ when all of Δ are $\mathfrak{t}$ and all of Λ are $\mathfrak{f}$; otherwise, the value $\mathfrak{t}$. We say a sequent $\Delta \rightarrow \Lambda$ is *falsifiable*, if for some (non-empty) domain D and some assignment in D to (at least) all its parameters, it takes the value $\mathfrak{f}$. We say $\Delta \rightarrow \Lambda$ is *valid*, or in symbols $\models \Delta \rightarrow \Lambda$, in contrary case that, for every (non-empty) domain and assignment, $\Delta \rightarrow \Lambda$ is $\mathfrak{t}$.

Each of the 14 rules has been picked (by reasoning illustrated in Examples 1 and 2, (a) for necessity and (b) for sufficiency) to have the property stated in:

LEMMA 6. *For each of the* 14 *rules* $\rightarrow\supset, \ldots, \exists\rightarrow$ *listed above, with the stated stipulations*: *The sequent written below the line is falsifiable*, (b) *if and* (a) *only if the sequent, or at least one of the two sequents, written above the line is falsifiable.* Equivalently: *The sequent written below the line is valid*, (a) *if and* (b) *only if the sequent, or each of the two sequents, written above the line is valid.*

It is of course quicker to use the rules than to think through the principles embodied in them at each step.

Proceeding upward, we can close a path (signifying that we have lost hope of finding a counterexample along it) when we have reached a sequent which we recognize cannot be made $\mathfrak{f}$. We codify this by saying we shall close a path as soon as we reach a sequent of the form

$$(\times) \qquad\qquad \mathrm{C}, \Gamma \rightarrow \Theta, \mathrm{C}.$$

Here we could allow C to be any formula. However it will be useful later (in §§ 55, 56) to know that mistaken attempts at counterexamples can always be rejected using an atom (prime formula) as the C. So we stipulate here that C be such. As with the rules, Γ and Θ are any lists of formulas; and the order of formulas within the antecedent and within the succedent is immaterial.

LEMMA 7. *No sequent of the form* $(\times)$ *is falsifiable.* Equivalently: *Each sequent of the form* $(\times)$ *is valid.*

Example 1 illustrates the situation envisaged in (I) of our proposed plan of attack on the completeness problem (tenth paragraph of this section). Example 2 illustrates (II), provided we can convert the closed sequent tree into a formal proof of F. We give three more examples.

EXAMPLE 3.

$$\begin{array}{cll}
\vdots & & \uparrow \\
P(a_0, a_1), P(a_1, a_2), P(a_2, a_3), \forall x\exists yP(x, y) \to & & V_{0000000} \\
\hline
P(a_0, a_1), P(a_1, a_2), \exists yP(a_2, y), \forall x\exists yP(x, y) \to & \exists\to & V_{000000} \\
\hline
P(a_0, a_1), P(a_1, a_2), \forall x\exists yP(x, y) \to & \forall\to & V_{00000} \\
\hline
P(a_0, a_1), \exists yP(a_1, y), \forall x\exists yP(x, y) \to & \exists\to & V_{0000} \\
\hline
P(a_0, a_1), \forall x\exists yP(x, y) \to & \forall\to & V_{000} \\
\hline
\exists yP(a_0, y), \forall x\exists yP(x, y) \to & \exists\to & V_{00} \\
\hline
\forall x\exists yP(x, y) \to & \forall\to & V_0 \\
\hline
\to \neg\forall x\exists yP(x, y). & \to\neg & V
\end{array}$$

Here there is a single path (no branching). This path can be pursued upward ad infinitum: along it we never reach either a sequent of the form (×) at which we can close it (as at V_{0000} in Example 1, and at both V_{0000} and V_{00010} in Example 2), or a sequent from which no further step upward can be made by our rules (as at V_{00010} in Example 1). Does this tree lead us to a counterexample?

Indeed it does, under either of two approaches. At V_{000}, we know that to falsify $\neg\forall x\exists yP(x, y)$ it is (necessary and) sufficient to make $P(a_0, a_1)$ and $\forall x\exists yP(x, y)$ both $\mathfrak{t}$. There are two possibilities: a_0 and a_1 are the same element of the domain D, or different elements. Under the first approach, we try making $D = \{0\}$ with 0 as the value of both a_0 and a_1. We evaluate P by the logical function $\mathfrak{l}$ for which $\mathfrak{l}(0, 0)$ is $\mathfrak{t}$, so $P(a_0, a_1)$ is $\mathfrak{t}$. Evidently this also makes $\forall x\exists yP(x, y)$ $\mathfrak{t}$; or we can argue that for a D with only the element a_0 we could have omitted $\forall x\exists yP(x, y)$ at V_{00} and hence at V_{000}. Thus we have a counterexample, indicated by the tree just up to V_{000}.

Now let us instead try having a_0 and a_1 different elements of D, and (following this approach consistently) likewise having each of a_2, a_3, a_4, . . . in turn a new element. Let $D = \{0, 1, 2, \ldots\}$, and assign to a_0, a_1, a_2, . . . the values 0, 1, 2, ... To P let us assign the logical function $\mathfrak{l}$ such that $\mathfrak{l}(x, y)$ is $\mathfrak{t}$ when x and y are consecutive natural numbers, and always $\mathfrak{f}$ (or always $\mathfrak{t}$) otherwise. This makes all the atoms in the antecedents $\mathfrak{t}$. The student should have no trouble in seeing that consequently it makes also the molecules there $\mathfrak{t}$, and hence makes $\neg\forall x\exists yP(x, y)$ in the succedent at the bottom $\mathfrak{f}$. So we have another counterexample, corresponding to the whole infinite path.

Let us consider the two approaches in general.[225] When we try for a counterexample, at each step upward by $\rightarrow\forall$ or $\exists\rightarrow$ we have a choice between the two for the variable b introduced in that step if any variables were previously introduced (as at (9) or V_{00010} in Example 1, and at V_{000} in Example 3, with a_1 as the b). Suppose there is a counterexample using the first approach, i.e. with b standing for the same element of D as some other variable a already introduced (as a_0 in Example 3). Then there is *also* a counterexample using the second approach. We can get this from the first counterexample by splitting the element represented by both a and b into two elements (enlarging D by one element) and treating these two elements (one to be used as value of a and the other of b) like the one element they replace in constructing the logical functions (and functions with values in D if mesons are present) for the falsifying assignment.[226] (As we have seen, in Example 3 at V_{000} both approaches work; in Example 1 at V_{00010} only the second works.) So we can never miss finding a counterexample (if a counterexample exists at all) by confining ourselves to the second approach. The treatment in §§ 49, 50 is set up on this basis. In it, we will provide for terminating a path in a sequent tree (even though the rules could be applied further) when a stage is reached from which a counterexample can be read off *under the second approach*. For our primary purpose of proving Gödel's completeness theorem, the additional possibilities for finding counterexamples afforded by the first approach are an unnecessary distraction. Of course, the first approach will often lead to a counterexample more quickly or to a simpler counterexample (as in Example 3).

In Example 3, using the second approach, we saw how an *infinite* counterexample (i.e. one with an infinite D) can be read off a suitably constructed infinite path. In Example 3, there was also a finite counterexample, using the first approach. Our last two examples will illustrate two points: (a) there may be only infinite counterexamples (Example 4); (b) without some over-all plan governing our searches, they may not lead to a counterexample or a closed tree, even though there be one or the other (Examples 4 and 5).

[225] The reasoning in this paragraph is to make our procedure in § 49 understandable in advance. The fact that by it we do succeed in proving Gödel's completeness theorem would be sufficient justification for it.

[226] We showed such a splitting in § 30 Example 5, where it was a question of finding an example for a formula E rather than a counterexample. Until § 52, our evaluation rules are those of the predicate calculus § 17 or the predicate calculus with functions § 28, not those of the predicate calculus with equality § 29. Meanwhile, if = should occur in any of our formulas it is to be treated like any other predicate parameter.

EXAMPLE 4. Let G be
$\forall x\neg P(x, x) \,\&\, \forall x\forall y\forall z(P(x, y) \,\&\, P(y, z) \supset P(x, z))$.

$$\begin{array}{lr}
\quad\vdots & \\
\dfrac{\dfrac{\dfrac{G, \exists yP(a_0, y), \forall x\exists yP(x, y) \to}{G, \forall x\exists yP(x, y) \to}\ \forall\to}{G \,\&\, \forall x\exists yP(x, y) \to}\ \&\to}{\to \neg(G \,\&\, \forall x\exists yP(x, y)).}\ \to\neg
&
\begin{array}{c} \uparrow \\ V_{000} \\ \uparrow \\ V_{00} \\ \uparrow \\ V_0 \\ \uparrow \\ V \end{array}
\end{array}$$

The tree shown is to be continued upward ad infinitum from V_{00} with the same sequents as in Example 3 from V_0 except that G is at the front of each antecedent.

The one path in this tree does not indicate to us a counterexample. For, in building it upward as proposed, we spend forever analyzing the conditions for the truth of $\forall x\exists yP(x, y)$ (as in Example 3) and never get to those for G.

Now we show that there is a counterexample to $\neg(G \,\&\, \forall x\exists yP(x, y))$ with $D = \{0, 1, 2, \ldots\}$ but no finite counterexample.

To do this it will suffice to show that $G \,\&\, \forall x\exists yP(x, y)$ is $\mathfrak{t}$ for a suitable assignment in $D = \{0, 1, 2, \ldots\}$, but is always $\mathfrak{f}$ in any finite (non-empty) domain.

It is easy to see what is going on, if in $G \,\&\, \forall x\exists yP(x, y)$ we not only unabbreviate G, but take the outer conjunctions apart, omit the universal quantifiers, and change P(—, —) to —$<$—:

$$\neg x<x, \quad x<y \,\&\, y<z \supset x<z, \quad \exists y\, x<y.$$

These order axioms are all true (with the free variables in the generality interpretation, §§ 20, 38) when $D = \{0, 1, 2, \ldots\}$ and $<$ stands for the usual order relation between natural numbers. Thus $G \,\&\, \forall x\exists yP(x, y)$ is $\mathfrak{t}$ (and $\neg(G \,\&\, \forall x\exists yP(x, y)$ is $\mathfrak{f}$), when $D = \{0, 1, 2, \ldots\}$ and to P is assigned as value the logical function $\mathfrak{l}$ such that $\mathfrak{l}(x, y)$ is $\mathfrak{t}$ when $x<y$ and $\mathfrak{f}$ otherwise.

But these order axioms cannot be satisfied in any finite (non-empty) domain, as we now show. Consider e.g. a domain D with just three elements. Say one of the elements is a_0. By $\exists y\, x<y$, there is some y such that $a_0<y$; by $\neg x<x$ this y cannot be a_0; say it is a_1 (different from a_0); thus $a_0<a_1$. Again using $\exists y\, x<y$, there is a y such that $a_1<y$; by $\neg x<x$ this cannot be a_1; and also it cannot be a_0, since then $a_0<a_1$ and $a_1<a_0$ with $x<y \,\&\, y<z \supset x<z$ would give $a_0<a_0$, contradicting $\neg x<x$; so this y is the remaining element a_2; thus $a_1<a_2$. Now by $\exists y\, x<y$, there must

be a y such that $a_2 < y$; but in the manner already illustrated, this y cannot be a_0 or a_1 or a_2. So it is impossible to have all three formulas written with $<$ true simultaneously, if D has just three elements. If we construct the truth table for $G \& \forall x \exists y P(x, y)$ with $D = \{0, 1, 2\}$ in the manner of § 17 (it will have $2^9 = 512$ lines), we shall get a solid column of f's. The like will be the case for every finite (non-empty) D (i.e. for $\bar{\bar{D}} = 1, 2, 3, \ldots$).

The fact that there are formulas which are valid (not falsifiable) in every non-empty finite domain, but are not valid (are falsifiable) in $D = \{0, 1, 2, \ldots\}$, was first noticed by Löwenheim 1915; the present example $\neg(G \& \forall x \exists y P(x, y))$ is from Hilbert and Bernays 1934 pp. 123–124.

We shall not take the space now to illustrate how our search procedure, when suitably managed (differently than above), leads to a counterexample in Example 4 (Exercise 49.3).

Example 5. Take the tree as in Example 4, but with $Q \& \neg Q$ as the G. There is no counterexample (since at V_{00} we can't make G t). But as we are misconducting the search, this doesn't manifest itself by the path becoming closed.

From Example 4 (and Example 3 if we confine ourselves to the second approach) we see that we must interpret (I) broadly enough to include counterexamples that are only developed by pursuing the search along some path ad infinitum. But our aim is to show for (II) that, in a properly managed search, if no counterexample exists (F is valid), that fact will always manifest itself by the closing of all paths after a *finite* amount of searching (as in Example 2). This is why, when F is valid, we shall always be able to find a proof of F (§ 51).

We could have anticipated from Theorem VII of Church that, if under (II) we are always to have a finite closed tree, then under (I) we cannot always effectively learn the existence of a counterexample in finitely many steps, whatever procedure we adopt. For, if we could, then we would have an algorithm or decision procedure (§ 40) for answering all questions whether a formula F in the predicate calculus is valid. By Gödel's completeness theorem (which we are aiming to prove) with Theorem 12_{Pd} (§ 23), this would be an algorithm for provability in the predicate calculus, contradicting Theorem VII.[227]

This situation is the reverse of what we might naively have expected back in § 17, where we found some finite counterexamples, but required general reasoning to establish validity.

[227] Also, we can easily adapt the proof of Theorem VII to show directly (without using Gödel's completeness theorem) that there is no algorithm for validity.

In Example 3 (with the second approach) and Example 4, although the counterexamples are infinite, the logical functions or predicates can be described effectively;[164] i.e. there are algorithms for them. This will not always happen under (I).[228]

EXERCISES. 48.1. By a systematic search (presented as a sequent tree), either find (and describe) a counterexample, or show that there is none. (a) $P \vee Q \supset P \& Q$. (b) $(P \supset \neg P) \supset \neg P$.
(c) $P \vee \forall x Q(x) \supset \forall x(P \vee Q(x))$. (d) $\exists x P(x) \& \exists x Q(x) \supset \exists x(P(x) \& Q(x))$.

48.2. Show that (under the definition preceding Lemma 6):
(a) $\models A_1, \ldots, A_m \rightarrow B_1, \ldots, B_n$, if and only if, for each (non-empty) domain D and each assignment in D to the parameters of $A_1, \ldots, A_m \rightarrow B_1, \ldots, B_n$, either $m > 0$ and one of $A_1, \ldots, A_m$ is $\mathfrak{f}$, or $n > 0$ and one of $B_1, \ldots, B_n$ is $\mathfrak{t}$ (briefly: either some of $A_1, \ldots, A_m$ are $\mathfrak{f}$, or some of $B_1, \ldots, B_n$ are $\mathfrak{t}$; cf. § 26). (b) Hence: $\models A_1, \ldots, A_m \rightarrow B$ if and only if $A_1, \ldots, A_m \models B$ (§ 20).

48.3*. Using insight (as in Example 4) rather than the systematic search procedure, find a counterexample in $D = \{0, 1, 2, \ldots\}$ to

$$\neg\{\forall x \neg P(x, x) \& \forall x \forall y \forall z(P(x, y) \& P(y, z) \supset P(x, z)) \\ \& \forall x \exists y P(x, y) \& \forall y \exists x P(x, y)\}.$$

Show that there is no finite counterexample.

§ 49. Gödel's completeness theorem: the basic discovery. Examples 4 and 5 have shown that we must adopt some over-all plan to guide our systematic searches for counterexamples to formulas F, if we are always to obtain either a counterexample or a closed tree.

The forms of the individual analytical steps upward have been described adequately by the list of 14 rules. Prior to adopting a search plan, we shall note some features of the steps (or the rules for them).

In each step upward by one of the rules, we recopy the list of formulas we are aiming to make $\mathfrak{t}$ and the list we are aiming to make $\mathfrak{f}$, with a revision or two alternative revisions. The revision comes from analyzing the conditions for the truth $\mathfrak{t}$ or the falsity $\mathfrak{f}$ of one of the formulas (the *principal formula* in the step) with respect to its outermost propositional connective or quantifier (the *principal operator*). On the basis of this analysis, we introduce into our lists one or two other formulas (the *side formula*(*s*)). Thereupon the principal formula becomes redundant and is omitted, except in the two rules $\forall\rightarrow$ and $\rightarrow\exists$. The remaining formulas

[228] The logical functions or predicates can always be of degree ≤ 1 § 46, by IM Theorems 38 and 40 pp. 398, 401 with Theorem XI (Post's theorem) IM p. 293. Further information about these predicates is given in Mostowski 1954 pp. 284–285.

(*extra formulas*) are recopied unchanged. For example:

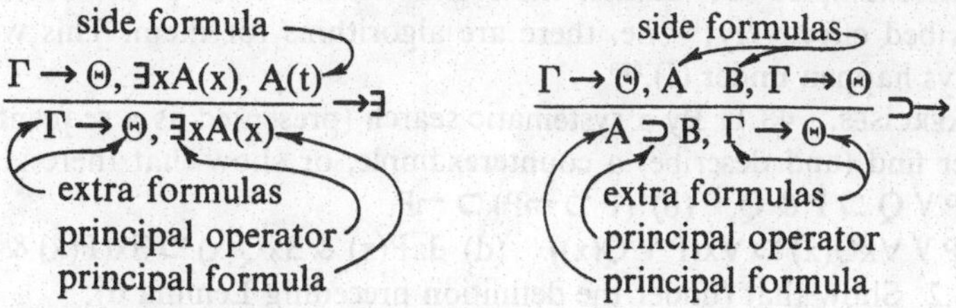

Hence in an obvious way, when we make a step upward by one of the rules, each formula occurrence (as one of the formulas listed as the antecedent or as the succedent) in the sequent or in either sequent above the line "originates from" or *belongs to* or is an (*immediate*) *ancestor of* a particular formula occurrence in the sequent below the line (its *immediate descendant*). Here we assume that, in any case of doubt, an analysis has been adopted fixing which formula occurrence is the principal formula, which are the side formula or respective side formulas, and which extra formula occurrence above comes from each one below.[229]

We can trace these ancestral relationships upward (or descendant relationships downward) through a succession of steps. Also, besides the ancestors of a formula occurrence in sequents above its own (*proper ancestors*), it is convenient to consider it as its own ancestor in its own sequent (its *improper ancestor*); and likewise with descendants. To illustrate this, in Example 2 all 6 ancestors of the formula (occurrence) in the antecedent at V_0 are printed with letters in boldface.

Like relationships apply to formula parts of formulas, to operator occurrences in formulas, and to occurrences of predicate parameters. In Example 2, the first ⊃ in the bottom sequent has four ancestors, namely itself and the three ⊃'s in boldface formulas; the first Q has 6 ancestors, itself and the 5 boldface Q's.

In designating these relationships for formula parts (or wholes) of formulas, we use *image* (or *ancestral image* and *descendant image*) instead of "ancestor" or "descendant" when we wish to restrict the relationship to parts (or wholes) which are the same except possibly for substitutions of terms for variables (reading up) or inversely (reading down). Thus in Example 2 reading down, the boldface Q(a_0) at V_{00010} has as (descendant) images Q(a_0), Q(a_0), Q(a_0), Q(x), Q(x), Q(x), and as descendants Q(a_0),

[229] There could be doubt only when the Δ or the Λ of one of the sequents Δ → Λ contains several occurrences of the same formula, or with the rules ∀→ and →∃ when the Δ of the Λ of the sequent below the line contains congruent formulas ∀xA(x) and ∀yA(y), or ∃xA(x) and ∃yA(y).

$Q(a_0)$, $P \supset Q(a_0)$, $\exists x(P \supset Q(x))$, $\exists x(P \supset Q(x))$, $\exists x(P \supset Q(x)) \supset (P \supset \exists xQ(x))$. Reading up, $P \supset Q(x)$ at V has as (ancestral) images $P \supset Q(x)$, $P \supset Q(x)$, $P \supset Q(x)$, $P \supset Q(a_0)$. For operator or predicate parameter occurrences, "ancestral image" and "descendant image" are synonymous with "ancestor" and "descendant", respectively.

A given composite formula occurrence in a sequent can serve as the principal formula for only one of the 14 (= 2·7) rules, according as the occurrence is in the succedent or the antecedent (2 choices) and according to the kind of the operator which it has outermost (7 choices). So we can classify the composite formula occurrences in sequents by the rules applicable to them as principal formulas.

In treating the problem of falsifying a formula F via our search procedure, the problem quickly generalized itself to that of simultaneously making m formulas $A_1, \ldots, A_m$ t and n formulas $B_1, \ldots, B_n$ f, or equivalently (by the definition preceding Lemma 6) of making a sequent $A_1, \ldots, A_m \rightarrow B_1, \ldots, B_n$ f $(m, n \geq 0)$. In other words, $A_1, \ldots, A_m$ are to be satisfied and $B_1, \ldots, B_n$ to be falsified, simultaneously; or $A_1, \ldots, A_m \rightarrow B_1, \ldots, B_n$ is to be *falsified*.

Now we are ready to adopt a plan to guide our systematic searches for counterexamples.

First (CASE (A)), we undertake to falsify a sequent $E_1, \ldots, E_k \rightarrow F_1, \ldots, F_l$ where $E_1, \ldots, E_k$, $F_1, \ldots, F_l$ are formulas in the sense of § 16, except that they may contain individual symbols (but not other function symbols).[230] By taking $\rightarrow F$ as the sequent, this includes the case of falsifying a single formula F. The sequent $E_1, \ldots, E_k \rightarrow F_1, \ldots, F_l$ may contain free variables, but those variables shall not also occur bound in it.[231]

Let $u_0, \ldots, u_p$ be a list (possibly empty) of the free variables and individual symbols occurring in $E_1, \ldots, E_k \rightarrow F_1, \ldots, F_l$. Let a_0, a_1, a_2, ... be variables not occurring in $E_1, \ldots, E_k \rightarrow F_1, \ldots, F_l$. The terms $u_0, \ldots, u_p, a_0, a_1, a_2, \ldots$, as far as we *activate* them (i.e. make them available), will be used as the b's and r's for our applications of the predicate rules $\rightarrow\forall$, $\forall\rightarrow$, $\rightarrow\exists$, $\exists\rightarrow$. Because the variables among $u_0, \ldots, u_p$ do not occur bound in $E_1, \ldots, E_k \rightarrow F_1, \ldots, F_l$, and a_0, a_1, a_2, ... are "new" variables not occurring in $E_1, \ldots, E_k \rightarrow F_1, \ldots, F_l$, the substitutions with results A(b) and A(r) performed in using these rules

[230] We use E's and F's to allow "$A_1, \ldots, A_m \rightarrow B_1, \ldots, B_n$" to denote various sequents in a tree having the fixed sequent $E_1, \ldots, E_k \rightarrow F_1, \ldots, F_l$ at the bottom.

[231] The reader who is willing to agree that a search plan can be formulated (in this Case (A)) which makes Lemma 8 true may skip the details by passing to Lemma 9.

will be free.[232] Those of $u_0, \ldots, u_p, a_0, a_1, a_2, \ldots$ which we activate are intended to name distinct members of the domain D for the counterexample to $E_1, \ldots, E_k \rightarrow F_1, \ldots, F_l$ we are attempting to construct.[233] As the search for a counterexample to $E_1, \ldots, E_k \rightarrow F_1, \ldots, F_l$ is pursued along any particular path in our sequent tree, we keep track step by step of which of $u_0, \ldots, u_p, a_0, a_1, a_2, \ldots$ have thus far been activated. To represent this concretely, we may employ a barrier $|$. We put the barrier into the list initially just to the right of u_p thus,

$$u_0, \ldots, u_p \mid a_0, a_1, a_2, \ldots,$$

to indicate that $u_0, \ldots, u_p$ are activated initially, if $u_0, \ldots, u_p$ is not empty. But if $u_0, \ldots, u_p$ is empty, we place the barrier initially thus,

$$a_0 \mid a_1, a_2, \ldots,$$

to indicate that in that case a_0 is activated initially. In each step by $\rightarrow\forall$ or $\exists\rightarrow$, we shall use a new variable as the b. Suppose that thus far $u_0, \ldots, u_p, a_0, \ldots, a_{i-1}$ or briefly $t_0, \ldots, t_q$ $(q = p+i+1)$ have been activated, so the list looks so:

$$u_0, \ldots, u_p, a_0, \ldots, a_{i-1} \mid a_i, a_{i+1}, a_{i+2}, \ldots.$$

As the b for the $\rightarrow\forall$ or $\exists\rightarrow$, we then use a_i, at the same time moving the barrier over to indicate that this variable is added to the part of the list which has been activated:

$$u_0, \ldots, u_p, a_0, \ldots, a_{i-1}, a_i \mid a_{i+1}, a_{i+2}, \ldots.$$

The steps along any particular path of the tree will be grouped in *rounds*.

Say that at the beginning of a certain round, call it Round d, we have before us a sequent $\Delta \rightarrow \Lambda$ or more explicitly $A_1, \ldots, A_m \rightarrow B_1, \ldots, B_n$. (For $d = 0$, this is $E_1, \ldots, E_k \rightarrow F_1, \ldots, F_l$. For $d > 0$, it is the sequent reached at the end of Round $d-1$.)

In carrying out Round d, we take each of the formula occurrences $A_1, \ldots, A_m, B_1, \ldots, B_n$ in turn (or more precisely, after the first step, a proper ancestral image of that formula occurrence), and perform if possible one step or a finite number of steps with it as principal formula (unless part way through the round we close the path, as explained below).

[232] The exclusion of variables occurring free in $E_1, \ldots, E_k \rightarrow F_1, \ldots, F_l$ from also occurring bound is necessary. Cf. Exercise 49.2.

[233] For a sequent $E_1, \ldots, E_k \rightarrow F_1, \ldots, F_l$ containing no variables free or bound, and no individual symbols, i.e. a sequent of the propositional calculus simply, we can alternatively forget about the list $u_0, \ldots, u_p, a_0, a_1, a_2, \ldots$.[222] The following discussion, omitting what is then irrelevant, gives a new treatment of the completeness problem for the propositional calculus.

No step can be performed in case the one of $A_1, \ldots, A_m, B_1, \ldots, B_n$ in question is an atom.

In the case of a formula occurrence of one of the kinds $\rightarrow\supset, \ldots, \sim\rightarrow$, one step is performed, which is completely determined by the formula and the corresponding rule.

In the case of a formula occurrence of one of the kinds $\rightarrow\forall$, $\exists\rightarrow$, one step is performed by the respective rule, with the next variable a_i after those previously activated as the b (as explained above).

In the case of a formula occurrence of one of the kinds $\forall\rightarrow$, $\rightarrow\exists$, steps are performed by the respective rule using as the r each one in turn of the terms $t_0, \ldots, t_q$ already activated which has not previously served as the r for that rule with the same principal formula (or more precisely, with a descendant image of it as the principal formula). This calls for from 0 to $q+1$ steps.

A particular path will be *terminated and closed* exactly when a sequent is first reached of the form (×) with C prime, disregarding as usual the order of formulas within antecedent and within succedent.

A path will be *terminated without being closed* exactly when a sequent is reached from which no step as prescribed is possible. This will happen at a given sequent (with $t_0, \ldots, t_q$ the terms thus far activated) exactly when the sequent contains only atoms, and $\forall\rightarrow$- and $\rightarrow\exists$-formulas descendant images of which have already served as the principal formula with each of $t_0, \ldots, t_q$ as the r. (This can only happen at the end of a round, in which case there is no next round.)

We can summarize our search plan thus. We provide for a non-empty domain D by activating $u_0, \ldots, u_p$ or a_0 initially. In each round, we go through the double list of formula occurrences $A_1, \ldots, A_m, B_1, \ldots, B_n$ reached at the end of the preceding round (or given initially), analyzing each molecule in turn, with respect to its outermost operator only (with the terms $t_0, \ldots, t_q$ thus far activated as the r's for $\forall\rightarrow$ and $\rightarrow\exists$), before starting over at the beginning of the list in the next round.

LEMMA 8. *In a sequent tree constructed upward from* $E_1, \ldots, E_k \rightarrow F_1, \ldots, F_l$ *using the* 14 *rules and the described search plan, consider any unclosed path, terminated or infinite. The list* $t_0, \ldots, t_q$ *or* $t_0, t_1, t_2, \ldots$ *of the terms which are eventually "activated" along the path is not empty, and includes all the individual parameters of* $E_1, \ldots, E_k \rightarrow F_1, \ldots, F_l$ *and all the terms used as* b's *and* r's *for* $\rightarrow\forall$, $\forall\rightarrow$, $\rightarrow\exists$, $\exists\rightarrow$ *along the path, and hence every individual parameter of any sequent along the path.*[234] *Each molecule occurring* (*in the antecedent or*

[234] Actually (in this Case (A)), for an infinite path it is easily seen that $\aleph_0$ terms $t_0, t_1, t_2, \ldots$ must be activated, or we would run out of steps to perform at a finite stage. (This remark isn't necessary for our purposes, and it won't always be so in Case (B) § 50.)

the succedent) in any sequent along the path is used as principal formula (in the antecedent or succedent respectively) somewhere along the path, namely, just once, except in the case of an ∀→- *or* →∃-*molecule, which is so used with each of* $t_0, \dots, t_q$ *or* $t_0, t_1, t_2, \dots$ *as the* r.

PROOF. For a molecule occurrence not of the kind ∀→ or →∃, an ancestral image of it will be used as principal formula during the next round after the one in which the formula first appears (or in the case of one of $E_1, \dots, E_k, F_1, \dots, F_l$, in the first round). An ∀→- or →∃-formula occurrence, once it has appeared, will never disappear. An ancestral image of it will be called up for use as principal formula, with every activated term not yet used as the r, in every subsequent round (and in the first round, if it is one of $E_1, \dots, E_k, F_1, \dots, F_l$), either ad infinitum or until the path is terminated. The path can be terminated only when all activated terms have served thus as the r. —

We picked the particular search plan described above to get Lemma 8 quickly, toward proving Gödel's completeness theorem. In practice, the search for a counterexample to $E_1, \dots, E_k \rightarrow F_1, \dots, F_l$ can often be conducted more efficiently (without sacrificing Lemma 8) by allowing some variation from the above plan. Examples 1 and 2 take a little longer to complete under the above plan than in § 48 (Exercise 49.1 (a) and (b)).[235]

LEMMA 9. *In a sequent tree constructed upward from* $E_1, \dots, E_k \rightarrow F_1, \dots, F_l$ *using the 14 rules and the described search plan, corresponding to any unclosed path, terminated or infinite, there is a counterexample to* $E_1, \dots, E_k \rightarrow F_1, \dots, F_l$ *with the domain* $D = \{0, \dots, q\}$ *or* $D = \{0, 1, 2, \dots\}$ *according as only* $t_0, \dots, t_q$ *or* $t_0, t_1, t_2, \dots$ *are activated along the path.*

PROOF. Let the sequents along such a path be $\Delta_0 \rightarrow \Lambda_0, \dots, \Delta_t \rightarrow \Lambda_t$ (terminated unclosed path) or $\Delta_0 \rightarrow \Lambda_0, \Delta_1 \rightarrow \Lambda_1, \Delta_2 \rightarrow \Lambda_2, \dots$ (infinite path), where $\Delta_0 \rightarrow \Lambda_0$ is $E_1, \dots, E_k \rightarrow F_1, \dots, F_l$. Let $\cup\Delta$ be the set of all the formulas which occur in any of the antecedents $\Delta_0, \dots, \Delta_t$ or Δ_0,

[235] In our search plan, when $u_0, \dots, u_p$ is empty, we activate a_0 by fiat at the beginning to be sure of having it. But it is more economical to wait then for a_0 to be activated by the first →∀ or ∃→, when one of these can be reached before we need to perform an ∀→ or →∃; along any unclosed path in which we can't do this, a_0 should be activated by fiat (before termination). Steps may be performed in any convenient order for a while, provided that a routine like the one in our plan is instituted eventually along each infinite path. In applying →∀ or ∃→ to a sequent in which one of the previously activated variables does not occur, that variable may be used as the b instead of a newly activated variable. In applying ∀→ or →∃ when the A(x) does not contain the x free, it is sufficient to use only one r. Also cf. the "first approach" in Example 3 § 48.

$\Delta_1, \Delta_2, \ldots$, and similarly let $\cup\Lambda$ be the union of all the Λ's. We shall show that, with the domain D as described, we can pick an assignment in D to make all the formulas in $\cup\Delta$ (including $E_1, \ldots, E_k$) t and all those in $\cup\Lambda$ (including $F_1, \ldots, F_l$) f.[236]

Because the path is not closed, $\cup\Delta$ *and* $\cup\Lambda$ *contain no atom* C *in common.* For, suppose they did; say C first enters the antecedents in Δ_a and first enters the succedents in Λ_b. Once an atom has entered an antecedent or succedent, it rides up into all higher sequents in antecedent or succedent respectively in the extra formulas Γ or Θ of the steps. So C would be in both Δ_c and Λ_c where $c = \max(a, b)$ (the greater of a and b, or their common value). So the path would have been terminated and closed (by use of $(\times)$) at $\Delta_c \rightarrow \Lambda_c$ (if not before).

It will follow that, *with the domain D as described in the lemma, we can pick an assignment to (at least) all the parameters of* $\cup\Delta$ *and* $\cup\Lambda$ *which makes all atoms in* $\cup\Delta$ t *and all in* $\cup\Lambda$ f. As the parameters, we take **(a)** the variables and individual symbols $t_0, \ldots, t_q$ or $t_0, t_1, t_2, \ldots$, **(b)** the proposition symbols occurring in $\cup\Delta$ or $\cup\Lambda$, and **(c)** the other predicate symbols so occurring. We take $D = \{0, \ldots, q\}$ or $D = \{0, 1, 2, \ldots\}$ respectively, and we assign to $t_0, \ldots, t_q$ the values $0, \ldots, q$, or to $t_0, t_1, t_2, \ldots$ the values $0, 1, 2, \ldots$. To each of (b) we assign t if that proposition atom occurs in $\cup\Delta$, f it occurs in $\cup\Lambda$ (we have seen that it can't occur in both), and an arbitrary value say f if it occurs in neither. To each of (c) we assign the logical function I such that $I(x_1, \ldots, x_n)$ is t if the predicate atom $P(t_{x_1}, \ldots, t_{x_n})$ occurs in $\cup\Delta$, f if $P(t_{x_1}, \ldots, t_{x_n})$ occurs in $\cup\Lambda$ (it can't occur in both), and say f if it occurs in neither.[64] Each predicate atom in $\cup\Delta$ or $\cup\Lambda$ is of the form $P(t_{x_1}, \ldots, t_{x_n})$ for some $x_1, \ldots, x_n$ belonging to D, using the second sentence of Lemma 8. Hence, all atoms in $\cup\Delta$ will now be t and all in $\cup\Lambda$ will be f, as required.

Finally, we show that *the D and assignment picked to make all atoms in* $\cup\Delta$ t *and all in* $\cup\Lambda$ f *also makes all molecules in* $\cup\Delta$ t *and all in* $\cup\Lambda$ f. Suppose it did not. Then, from among all the molecules in $\cup\Delta$ which are not t or in $\cup\Lambda$ which are not f, let us choose one G containing the smallest number (≥ 1) of occurrences of operators. By the last sentence of Lemma 8, G plays the role of principal formula (for the antecedent or succedent rule appropriate to its outermost operator, according as G is picked from $\cup\Delta$ or from $\cup\Lambda$). For G not an $\forall\rightarrow$- or $\rightarrow\exists$-formula, consider the side formula(s) H, or H and I, which are in the premise belonging to the path in question. By the choice of the 14 rules except $\forall\rightarrow$ and $\rightarrow\exists$, the formula(s)

[236] For a terminated unclosed branch, it is sufficient (by the rationale of our searches, or specifically by Lemma 6 (b)) to make just the formulas in the top sequent have the right values (as illustrated in Example 1). However, often the best way to see that we can do this is by the reasoning we give here.

H, or H and I, having the desired value(s) (t in the antecedent, f in the succedent) is sufficient for G to have the desired value. (This property of the rules gave us Lemma 6 (b).) But H, or H and I, having fewer occurrences of operators than G, do have the desired value(s), since G was supposed to have the *minimum* number of operator occurrences for formulas *not* having the desired values. So such a G cannot exist, unless it is an ∀→- or →∃-formula, i.e. ∀xA(x) in the antecedent or ∃xA(x) in the succedent. Then by end Lemma 8, A(r) occurs as side formula for each of $t_0, \ldots, t_q$ or of $t_0, t_1, t_2, \ldots$ as the r. But under our assignment these variables name all the members of the domain *D*. So G must get the desired value, in consequence of all the side formulas $A(t_0), \ldots, A(t_q)$ or $A(t_0), A(t_1), A(t_2), \ldots$ getting the desired values (since they each have one less operator occurrence), contradicting the choice of G. —

In the following lemma dealing with geometric trees, we mean by a *partial path* a succession of vertices connected by following the arrows, starting with the initial vertex *V*. (By a *path*, we have meant such a succession of vertices continued either to termination or ad infinitum.)

LEMMA 10. (König's lemma 1926.)[237] *In a geometric tree with finitely many arrows leading from each vertex, if there are arbitrarily long finite partial paths, then there is an infinite path.*

PROOF AND ILLUSTRATION. In the application we are to make, the number of arrows leading from a vertex can be 0, 1 or 2. Very simple geometric trees of this kind are shown in Examples 1–4; here is a more complicated such tree (drawn horizontally to save space):

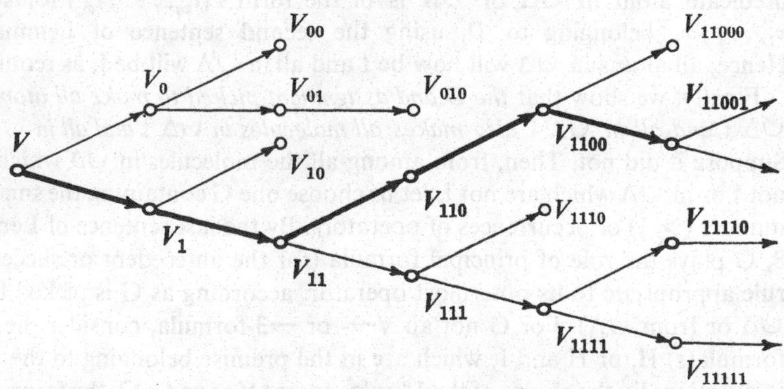

To give the proof in general, suppose we are dealing with a tree as described in which there are arbitrarily long finite partial paths. We wish

[237] In other forms this is in Brouwer 1924, and implicitly in Brouwer 1923a and Skolem 1922–3 p. 222. Cf. Kleene and Vesley 1965 p. 59.

to trace an infinite path. Here is the rule for doing so. Suppose we have traced the desired path as far as vertex V_X (either the initial vertex V, or a later vertex such as V_{1100} in the illustration), and that V_X *belongs to arbitrarily long finite partial paths* (as does V by hypothesis). We want to pick a next vertex that will also have the italicized property. There must be a next vertex after V_X, or V_X could not have the property; say the next vertices are $V_{X0}, \ldots, V_{Xn}$. One at least of these next vertices must have the property; for if the partial paths through $V_{X0}, \ldots, V_{Xn}$ were no longer than $b_0, \ldots, b_n$ vertices respectively, then all partial paths through V_X would be no longer than $\max(b_0, \ldots, b_n)$ vertices. So we can indeed pick a next vertex after V_X also having the property. Altogether, starting with V, we are thus able to pick a next vertex, always with the property, ad infinitum.

In the illustration above, if the boldface vertices are the ones which belong to arbitrarily long finite partial paths, then one infinite path (boldface arrows) starts out $VV_1V_{11}V_{110}V_{1100}V_{11001} \cdots$.

Example 6. For trees in which infinitely many arrows may lead from a vertex, the lemma need not hold. Consider the tree in which $\aleph_0$ arrows each lead from V to a next vertex V_x ($x = 0, 1, 2, \ldots$); and from V_x successive arrows lead to x more vertices $V_{x0}, V_{x00}, \ldots$, thus:

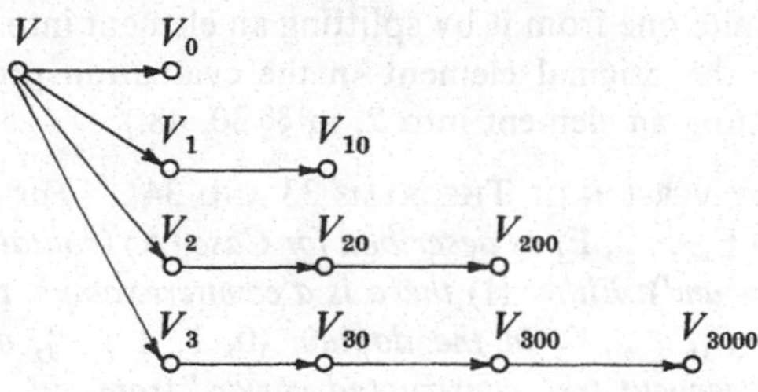

In this tree, there are arbitrarily long finite partial paths, but no infinite path. —

Now, given a sequent $E_1, \ldots, E_k \rightarrow F_1, \ldots, F_l$ as specified for Case (A), let us start constructing the sequent tree upward from it, using the 14 rules and the described search plan. The succession of steps along any path is governed by the search plan. Let us divide our work between the different paths, when there is branching, so that different paths are constructed up to corresponding levels in step. Thus, after having constructed up to their 10th vertices all partial paths which don't terminate sooner, we take each of these which doesn't terminate at its 10th vertex and add the one or two 11th vertices issuing from it, before we put a 12th vertex on any path.

CASE 1: for some b, each path terminates and closes with not more than b vertices. Then the tree itself is closed, and finite (it has at most $1 + 2 + 2^2 + \ldots + 2^{b-1} = 2^b - 1$ vertices). So there is no counterexample, i.e. $\models E_1, \ldots, E_k \rightarrow F_1, \ldots, F_l$. (To review the argument, already illustrated by Example 2, at each partial stage in the tree construction, our sole hope of getting a counterexample is that *one* of the sequents that stand at the tops of branches can be falsified; indeed, we know this by repeated applications of Lemma 6 (a). But with the whole tree closed, each sequent at the top is not falsifiable, by Lemma 7.)

CASE 2: for some b, some path terminates unclosed at its bth vertex. Then by Lemma 9, there is a counterexample to $E_1, \ldots, E_k \rightarrow F_1, \ldots, F_l$ with $D = \{0, \ldots, q\}$. (Along a terminated path, only finitely many terms $t_0, \ldots, t_q$ can be activated in this Case (A).)

CASE 3: otherwise. Then the tree has arbitrarily long finite partial paths. So by König's lemma (Lemma 10), there is an infinite path. Hence by Lemma 9, there is a counterexample with $D = \{0, \ldots, q\}$ or $D = \{0, 1, 2, \ldots\}$.[234]

We have now reached our goal for this section. To simplify the statement (italicized next below), we can take $D = \{0, 1, 2, \ldots\}$ in both Cases 2 and 3. For, if there is a finite counterexample, we can manufacture a countably infinite one from it by splitting an element into $\aleph_0$ elements, all behaving like the original element in the evaluation process. (We used this idea, splitting an element into 2, in §§ 30, 48.)

PRELIMINARY VERSION OF THEOREMS 33 AND 34°. *For a sequent* $E_1, \ldots, E_k \rightarrow F_1, \ldots, F_l$ *as described for Case* (A) (*containing no variables both free and bound*): *Either* (I) *there is a counterexample to* $E_1, \ldots, E_k \rightarrow F_1, \ldots, F_l$ *in the domain* $\{0, 1, 2, \ldots\}$, *or* (II) *there is a* (*finite*) *closed sequent tree constructed upward from* $E_1, \ldots, E_k \rightarrow F_1, \ldots, F_l$ *by the* 14 *rules, and hence there is no counterexample to* $E_1, \ldots, E_k \rightarrow F_1, \ldots, F_l$, *i.e.* $\models E_1, \ldots, E_k \rightarrow F_1, \ldots, F_l$.

It follows as a side result that $E_1, \ldots, E_k \rightarrow F_1, \ldots, F_l$ cannot have uncountably infinite counterexamples and no others:

PRELIMINARY VERSION OF THEOREM 35. *For a sequent* $E_1, \ldots, E_k \rightarrow F_1, \ldots, F_l$ *as described for Case* (A): *If there is any counterexample to* $E_1, \ldots, E_k \rightarrow F_1, \ldots, F_l$, *then there is one in the domain* $\{0, 1, 2, \ldots\}$.

EXERCISES. 49.1. Construct the sequent tree upward (using the described search plan), and conclude (I) (give the counterexample) or (II):

(a) $\rightarrow \exists x(P \supset Q(x)) \supset (P \supset \forall x Q(x))$ (Example 1).

(b) $\rightarrow \exists x(P \supset Q(x)) \supset (P \supset \exists x Q(x))$ (Example 2).

(c) $\rightarrow \neg(Q \ \& \ \neg Q \ \& \ \forall x \exists y P(x, y))$ (Example 5).
(d) $\rightarrow P \ \& \ \exists x(Q(x) \supset Q(x))$. (e) $\exists xP \rightarrow \forall xP$.
(f) $\forall x \exists y P(x, y) \rightarrow \forall x \exists y P(x, y)$. (g) $P \supset Q \rightarrow P \supset Q$.

49.2. Apply the search plan to $\forall a \forall c(P(c) \ \& \ Q(a)) \rightarrow Q(b)$ and $\forall a \forall b(P(b) \ \& \ Q(a)) \rightarrow Q(b)$. Why does it fail for the latter?

49.3*. Consider making a systematic search for a counterexample to $\neg\{\forall x \neg P(x, x) \ \& \ \forall x \forall y \forall z(P(x, y) \ \& \ P(y, z) \supset P(x, z)) \ \& \ \forall x \exists y P(x, y)\}$ (Example 4 § 48), by using the 14 rules upward with the following search plan (different from the plan in the text): in Round 0, perform all steps possible with only a_0 activated; in Round 1, activate a_1 by $\exists\rightarrow$ and then perform all new steps possible with only a_0, a_1 activated; in Round 2, activate a_2 by $\exists\rightarrow$ and then perform all new steps possible with only a_0, a_1, a_2 activated; etc.
(a) Show that Lemma 8 is satisfied by this search plan.
(b) Show that, if the steps are continued to the end of Round d suspending the rule for closing paths, there will be $3(d+1)^3$ "branches" or partial paths. (Thus, activating only a_0, a_1, a_2 as in the informal discussion in Example 4, there will be 81 paths, if the possibilities for closing some of those paths before completing all the disections are disregarded.)
(c) Using as a guide the counterexample which we found by insight in Example 4, pick at each branching the sequent along a path which goes with that counterexample. Write the sequent which will be reached at the end of Round 1 along the resulting infinite path.

§ 50. Gödel's completeness theorem with a Gentzen-type formal system, the Löwenheim-Skolem theorem. Now we would have the first case of Gödel's 1930 completeness theorem (*If* $\models$ F, *then* $\vdash$ F), if under (II) just above (for $k = 0$, $l = 1$) we could infer that $\vdash$ F in the sense of § 21. (We shall do so in § 51.)

However, we already have the basic discovery. This is that, whenever no counterexample to F exists (i.e. when $\models$ F), that fact can be confirmed by a finite mechanical verification process.[238] This process (in the present treatment) consists in verifying that a certain finite figure meets the conditions for being a closed sequent tree constructed upward from $\rightarrow$ F using the 14 rules.

From the standpoint from which formal systems and proof theory were invented (§ 37), this fundamentally is all we were after. We could take the mechanical verification process as itself constituting a proof of F.

In fact, these verifications come under essentially the traditional form

[238] Using a different mechanical verification process than the present one, this basic discovery is quite clearly contained in Skolem 1922–3 pp. 220–222 (which was not known to Gödel in 1930). Cf. Footnote 252 below.

for axiomatic-deductive systems, if we simply read downward in checking the correctness of the trees, instead of upward. At the top of each branch, if the tree is closed, we have a sequent fitting (×) in § 48, which we can now consider to be an axiom schema. (Now "×" instead of meaning "closed" can mean "axiom".) And each step downward is by one of 14 rules (previously read upward), which we can now regard as one- or two-premise rules of inference. (Now "$\supset\rightarrow$" can be called "$\supset$-introduction in the antecedent", etc.)

Now any tree with axioms by (×) at the tops of branches, and each step downward by one of the 14 rules, constitutes a *proof* (*of* its bottom sequent or *endsequent*) in a formal system *G*4 of a new type, called a *Gentzen-type* (*sequent*) *system* or *sequent calculus.*[239] Such systems were introduced by Gentzen 1934–5 (and 1932), in part following Hertz 1929. In contrast, we call the formal system for the predicate calculus of § 21 a *Hilbert-type system H*. More precisely, "*G*4" and "*H*" denote ambiguously several systems, according to how the notions of term and formula (specifically, of prime formula) are regulated (cf. §§ 37, 39).

By Lemmas 6 (a) and 7, *G*4 has the consistency property analogous to Theorem 12_{Pd} for *H*:

THEOREM 33. *Each sequent* $A_1, \ldots, A_m \rightarrow B_1, \ldots, B_n$ *provable in G*4 *is valid*; *in symbols, if* $\vdash A_1, \ldots, A_m \rightarrow B_1, \ldots, B_n$, *then* $\models A_1, \ldots, A_m \rightarrow B_1, \ldots, B_n$.

This is only a restatement of the feature of our searches that closure of the tree means that all possibilities for finding a counterexample have been closed out. Without this consistency property, we would hardly want to use *G*4 as a formal system.

Lemma 6 (b) expresses a novel property of (the rules in) *G*4, not possessed in *H* by the rule of modus ponens.[240]

The fact that *G*4 operates with sequents, so $\rightarrow$ F (rather than F) is provable in *G*4 when the search procedure closes, is not a serious drawback. Anyone who objects to this could easily modify *G*4 to use formulas as did Schütte 1950 (though the sequents are convenient), or he could supplement *G*4 by a rule $\rightarrow$ F / F.

Also *G*4 has the new feature that its proofs are finite trees (they are "proofs in tree form") rather than finite (linear) sequences of formulas ("proofs in sequence form"). We could rewrite the trees in sequence form.

[239] Gentzen-type systems *G*1, *G*2, *G*3, *G*3a are discussed in IM §§ 77–80, which we shall cite more specifically in § 54. The present system *G*4 is very close to *G*3, to the system *L* of Beth 1959 p. 282, and to LC in Kanger 1957. The first 8 rules coincide with the propositional rules of Ketonen 1944.

[240] *G*3 has the feature, but not *G*1, *G*2, *G*3a.

But the trees show the logical structure better, and thus help us in our reasoning about that structure. The linear arrangement of proofs has been traditional, no doubt because oral language is necessarily linear, and written language more conveniently so for ordinary purposes. Proofs (and deductions) in H can be written in tree form.[241]

Now we take up some other cases of Gödel's completeness theorem.

We begin (CASE (B)) by extending the treatment of Case (A) to deal with a countable infinity of formulas $\ldots, E_2, E_1, E_0$ to be made $\mathfrak{t}$ and a countable infinity $F_0, F_1, F_2, \ldots$ to be made $\mathfrak{f}$, simultaneously, in which altogether there occur only a finite list $u_0, \ldots, u_p$ (possibly empty) of free variables and individual symbols (but no other function symbols). Such free variables shall not occur bound in any of the formulas. Alternatively, one of the two lists of formulas may be finite or even empty.[242] For those alternatives, slight alterations of our notation are to be imagined.[243]

To deal with this case, we generalize our notion of sequents and related notions to allow $\aleph_0$ formulas in the antecedent and succedent (or in one of them). We now work with the $\aleph_0$-sequent $\ldots, E_2, E_1, E_0 \rightarrow F_0, F_1, F_2, \ldots$ instead of with the sequent $E_1, \ldots, E_k \rightarrow F_1, \ldots, F_l$.[244]

However, during Round d along any particular path in the tree only the first $d+1$ formulas of the original lists will have been activated. Say that

$$\ldots, E_{d+2}, E_{d+1}, E_d \mid A_1, \ldots, A_m \rightarrow B_1, \ldots, B_n \mid F_d, F_{d+1}, F_{d+2}, \ldots$$

[241] Cf. IM pp. 106–107. In IM "branch" is used where we have been saying "path", with both Hilbert-type and Gentzen-type systems. (Contrary to botany, a "branch" starts from, or reaches to, the bottom, not just from the lowest "branching".) The term in Hilbert and Bernays 1934 is "proof thread (Beweisfaden)".

[242] One of the interesting cases is that $\ldots, E_2, E_1, E_0$ are the closures of the axioms of a formal system like N of § 38, and instead of $F_0, F_1, F_2, \ldots$ we have a single formula F. We shall review this application at the end of § 52. However, it is not required for the present treatment that the formulas $\ldots, E_2, E_1, E_0$ (or $F_0, F_1, F_2, \ldots$) be given effectively, as the axioms of a formal system should be (cf. §§ 37, 43). (For the result mentioned in Footnote 228 with infinitely many formulas, the lists must be effective, so that E_i and F_i can be found effectively from i.)

[243] In this book we do not attempt to consider languages for the predicate calculus with uncountably many symbols, and hence uncountably many formulas. Such languages are highly non-constructive. However, Malcev 1936, Henkin 1950, A. Robinson 1951, etc. have studied them, and found model-theoretic applications. Propositional calculus with uncountably many symbols is allowed in Gödel 1931–2b.

[244] The reader who skipped the details of the search plan in Case (A) should observe one new point (after which he can pass to Cases (C) and (D)): If the tree is closed, and thus finite, then only finitely many of $\ldots, E_2, E_1, E_0$ and $F_0, F_1, F_2, \ldots$ (say at most $E_d, \ldots, E_0$ and $F_0, \ldots, F_d$) can have taken into account in constructing it by the 14 rules and recognizing it as closed by the axiom schema ($\times$).

is the $\aleph_0$-sequent reached at the end of Round $d-1$ (or when $d = 0$, the initial $\aleph_0$-sequent, with $A_1, \ldots, A_m$ and $B_1, \ldots, B_n$ both empty). Then we commence Round d by first moving out the two barriers in the $\aleph_0$-sequent before us to activate E_d and F_d, thus,

$$\ldots, E_{d+2}, E_{d+1} \mid E_d, A_1, \ldots, A_m \rightarrow B_1, \ldots, B_n, F_d \mid F_{d+1}, F_{d+2}, \ldots.$$

Then each of $E_d, A_1, \ldots, A_m, B_1, \ldots, B_n, F_d$ in turn is used to the extent possible as principal formula (as $A_1, \ldots, A_m, B_1, \ldots, B_n$ were in Case (A)). The criterion for closing a path is that the part of the $\aleph_0$-sequent between the barriers (which is a sequent) fit the axiom schema ($\times$), i.e. have an atom common to its antecedent and succedent. After any round with the path not closed, there is always a next round in which at least the barriers are moved one position out. If the formulas E_d and F_d newly activated are atoms, it can happen that there are no steps in that round (i.e. no sequent(s) are added to the tree); namely, this will happen if the sequent previously between the barriers qualified for termination under Case (A). If, in this case, *all* the formulas outside the barriers are atoms, then termination without closure will take place by having an infinity of rounds happen "instantaneously" (the barriers being simply moved out, one position after another). This is the only way termination without closure can take place. Closure, if it takes place, occurs in some finite round.

No change is required in LEMMAS 8 and 9 other than replacing "sequent" by "$\aleph_0$-sequent" and "$E_1, \ldots, E_k \rightarrow F_1, \ldots, F_l$" by "$\ldots, E_2, E_1, E_0 \rightarrow F_0, F_1, F_2, \ldots$" (and LEMMA 10 does not depend on the case). But in conclusion, if we obtain a closed $\aleph_0$-sequent tree, then from it we can obtain a closed sequent tree as follows. Since the $\aleph_0$-sequent tree is closed, it is finite (as argued in Case 1 end § 49). Consider its finitely many top vertices, and let d be the greatest round during which any of them become closed. Then nowhere in the tree do the two barriers stand further out than just right of (an ancestral image of) E_{d+1} in antecedent and just left of F_{d+1} in succedent. The pairs of atoms C in top $\aleph_0$-sequents which occasion their closure are within the barriers. So are the principal and side formulas in each step. Now let us move the barriers out in every $\aleph_0$-sequent up to but not across E_{d+1} and F_{d+1}, and then throw away all formulas still outside the barriers. The result is a closed sequent tree with $E_d, \ldots, E_0 \rightarrow F_0, \ldots, F_d$ at the bottom.

Finally, we take the cases not already treated of finitely many formulas $E_1, \ldots, E_k, F_1, \ldots, F_l$ (CASE (C)), or $\aleph_0$ formulas $\ldots, E_2, E_1, E_0, F_0, F_1, F_2, \ldots$ (CASE (D)), which in the aggregate contain **(i)** finitely or infinitely many free variables (which shall not also occur in any of the formulas bound) and individual symbols and **(ii)** finitely or infinitely many other function symbols.[245] We write out the treatment supposing there are at least one each of the symbols (i) and (ii). Otherwise, slight modifications are to be imagined.[246]

To deal with these cases, we prepare $\aleph_0$ separate lists of $\aleph_0$ terms each, which we show with the initial position of the barrier in each:

$$u_0 \mid u_1, u_2, u_3, \ldots,$$

$$\mid a_0, u_{01}, u_{02}, u_{03}, \ldots \text{ (where } a_0 \text{ is } u_{00}),$$

$$\mid a_1, u_{11}, u_{12}, u_{13}, \ldots \text{ (where } a_1 \text{ is } u_{10}),$$

$$\mid a_2, u_{21}, u_{22}, u_{23}, \ldots \text{ (where } a_2 \text{ is } u_{20}),$$

$$\cdot \quad \cdot \quad \cdot \quad \cdot$$

The first list $u_0, u_1, u_2, \ldots$ is an enumeration of all the terms (§§ 28, 38, 39) constructible using the symbols (i) and (ii). We can always get an enumeration of the terms thus constructible, e.g. by the method of digits ((A) in § 32). For example, if the formulas contain no free variables but exactly one individual symbol e and two other function symbols f(—) and g(—, —), the enumeration could start out thus:

$$e, f(e), g(e, e), f(f(e)), f(g(e, e)), g(e, f(e)), g(f(e), e), \ldots.$$

Each subsequent list $u_{i0}, u_{i1}, u_{i2}, \ldots$ is an enumeration of the *additional* terms which are constructible when the variable a_i is also made available. Thus $u_{i0}, u_{i1}, u_{i2}, \ldots$ is an enumeration of the terms constructible allowing the use of the same symbols as for $u_0, u_1, u_2, \ldots$ and the variables $a_0, \ldots, a_i$ and *actually using* a_i (so we don't include any terms from earlier lists). For example, if the symbols for $u_0, u_1, u_2, \ldots$ are as above, $u_{10}, u_{11}, u_{12}, \ldots$ could start out

$$a_1, f(a_1), g(a_0, a_1), g(a_1, a_0), g(a_1, a_1), f(f(a_1)), f(g(a_0, a_1)), \ldots.$$

When a step by →∀ or ∃→ is performed, we use as the b the first one a_i of $a_0, a_1, a_2, \ldots$ not yet activated, and we move the barrier over one place in the list $u_{i0}, u_{i1}, u_{i2}, \ldots$ which it begins to record this. In ∀→- and →∃-steps, all the terms $t_0, \ldots, t_q$ thus far activated in any of the separate lists are available as the r. The rounds are managed as under Case (A) or (B), except that, after the end of any round at the commencement of the next, we move the barrier one place to the right in each of the lists of terms in which the barrier is not at the extreme left. Thus,

[245] The reader who has been skipping the details of the search plan may pass to the remarks on Lemmas 8 and 9.

[246] If there are none of (ii), the list $a_i, u_{i1}, u_{i2}, u_{i3}, \ldots$ reduces to a_i $(i = 0, 1, 2, \ldots)$; then there must be $\aleph_0$ of the symbols (i), or we would be back in Case (A) or (B). If there are some of (ii) but none of (i), the list $u_0, u_1, u_2, \ldots$ disappears, and a_0 should be activated initially.

along any path in the tree which does not close, once any term in a list is activated, we eventually activate every term in the list; they all go into the consolidated list $t_0, t_1, t_2, \ldots$ of all the terms eventually activated. This will happen even in a terminated but unclosed path, by infinitely many rounds being performed "instantaneously" by just moving barriers over in the lists of terms and for Case (D) also in the lists of formulas.

The second sentence of LEMMA 8 now reads: *The list* $t_0, t_1, t_2, \ldots$ *of the terms which are activated along the path is an enumeration of all the terms constructible using the individual and other function parameters of* $E_1, \ldots, E_k \rightarrow F_1, \ldots, F_l$ *or of* $\ldots, E_2, E_1, E_0 \rightarrow F_0, F_1, F_2, \ldots$ *and those of the new variables* $a_0, a_1, a_2, \ldots$ *which are introduced in using* $\rightarrow\forall$ *and* $\exists\rightarrow$; *every* r *for* $\forall\rightarrow$ *or* $\rightarrow\exists$ *is taken from the list; and hence every term occurring free* (§ 28) *in any sequent along the path is in the list.*

For LEMMA 9, we shall arrange matters so that each of the activated terms $t_0, t_1, t_2, \ldots$ stands for a different member of D. Thereby we will be free to make all predicate atoms $P(t_{x_1}, \ldots, t_{x_n})$ in $\cup\Delta$ t and all in $\cup\Lambda$ f. (We only aim to describe *some* counterexample, as quickly as we can.)

So we take $D = \{0, 1, 2, \ldots\}$, and we shall make the assignment to the parameters in the terms so that the values of $t_0, t_1, t_2, \ldots$ are precisely $0, 1, 2, \ldots$.

To accomplish this, first consider each one e of the free variables or individual symbols; it occurs just once in the list $t_0, t_1, t_2, \ldots$, say as t_i. We give e the value i. Now consider each one of the other function symbols; a 1-place function symbol f(—) will serve for illustration. Each of the terms $f(t_0)$, $f(t_1)$, $f(t_2)$, ... occurs just once in the list $t_0, t_1, t_2, \ldots$; say they occur as $t_{i_0}, t_{i_1}, t_{i_2}, \ldots$. Then we evaluate f(—) by the function f such that $f(0) = i_0, f(1) = i_1, f(2) = i_2, \ldots$.

Now consider the process of evaluating a term (§ 28). For a given assignment, we evaluate its term parts successively from inside out, starting with assigned values of variables and individual symbols, and applying assigned values of functions. But under the assignment just described, at each successive stage (including the last one) the term part t in question is evaluated by the number i such that t is t_i.

The choice of truth values and logical functions to evaluate the proposition and predicate symbols can now read as before.

For the proof that all molecules in $\cup\Delta$ must then be t and all in $\cup\Lambda$ f, we review only the case of an $\forall\rightarrow$-formula (that of an $\rightarrow\exists$-formula is similar). Say e.g. G is the $\forall\rightarrow$-formula $\forall x(P(f(x)) \& Q)$ in $\cup\Delta$ and gets the wrong value f, while all 1-operator (and 0-operator) formulas in $\cup\Delta$ and $\cup\Lambda$ get the right values. But then by Lemma 8 all of $P(f(t_0)) \& Q$, $P(f(t_1)) \& Q$, $P(f(t_2)) \& Q, \ldots$ will be in $\cup\Delta$ as side formulas to G, and so have the right values, namely all t. Then $\forall x(P(f(x)) \,\&\, Q)$ is t (contradicting our assumption); for, consider any x. When x has the value x, the value of f(x) is the value of $f(t_x)$, so the value of $P(f(x)) \,\&\, Q$ is that of $P(f(t_x)) \,\&\, Q$, which is t. This is for any x; so the supplementary table for $P(f(x)) \,\&\, Q$ has a solid column of t's; so $\forall x(P(f(x)) \,\&\, Q)$ is t.

Taking the main conclusion we reached at the end of § 49, restating it using the idea of a proof in *G*4 (beginning this section), and adding the new cases (B)–(D), we have the following theorem.

THEOREM 34°. (Gödel's completeness theorem with a Gentzen-type formal system *G*4.) *Let* $\left\{\begin{matrix} E_1, \ldots, E_k, F_1, \ldots, F_l \\ \ldots, E_2, E_1, E_0, F_0, F_1, F_2, \ldots \end{matrix}\right\}$ *be formulas of the predicate calculus containing bound no variable which occurs in any of them free. Either* (I) $\left\{\begin{matrix} E_1, \ldots, E_k \rightarrow F_1, \ldots, F_l \\ \ldots, E_2, E_1, E_0 \rightarrow F_0, F_1, F_2, \ldots \end{matrix}\right\}$ *is falsifiable in the domain of the natural numbers* {0, 1, 2, . . .} ("$\aleph_0$-falsifiable"), *or* (II) $\left\{\begin{matrix} E_1, \ldots, E_k \rightarrow F_1, \ldots, F_l \\ \textit{for some } d,\ E_d, \ldots, E_0 \rightarrow F_0, \ldots, F_d \end{matrix}\right\}$ *is provable in G*4.

In Cases (A) and (C) (the upper version in Theorem 34), when (II) holds, $E_1, \ldots, E_k \rightarrow F_1, \ldots, F_l$ is not falsifiable. (This part of our conclusion in § 49 has meanwhile been stated separately as Theorem 33.) Hence, if $E_1, \ldots, E_k \rightarrow F_1, \ldots, F_l$ is falsifiable, it is falsifiable in {0, 1, 2, . . .}. Using this for $l = 0$, we have the (a) part of the next theorem. For, the sequent $E_1, \ldots, E_k \rightarrow$ is falsifiable in a given domain exactly if the formulas $E_1, \ldots, E_k$ are simultaneously satisfiable in that domain.

In Theorem 35, we do not need to keep the condition that the formulas contain bound no variable which occurs free in any of them. For, otherwise we could first replace the given formulas by formulas congruent to them (§ 16) and satisfying the condition, without changing their truth tables for any domain and assignment. Then, after applying the theorem with the condition met, we could change back to the original formulas.

THEOREM 35. (The Löwenheim-Skolem theorem.)[247] *In the predicate calculus:*

(a) (After Löwenheim 1915, Skolem 1920.) *If* E *is satisfiable, it is* $\aleph_0$*-satisfiable. If* $E_1, \ldots, E_k$ *are simultaneously satisfiable, they are simultaneously* $\aleph_0$*-satisfiable.*

[247] These versions (except "compactness" in (b)), for Cases (A) and (B), can be established by the reasoning in Skolem 1920 plus the trivial argument (end § 49) that, if we come out with a finite domain, an element can be split into $\aleph_0$ elements. More details of the history are in Footnote 252 below.

(b) (After Skolem 1920.) *If* $E_0, E_1, E_2, \ldots$ *are simultaneously satisfiable* [*or even if, for each d,* $E_0, \ldots, E_d$ *are simultaneously satisfiable* ("compactness", Gödel 1930)], *then* $E_0, E_1, E_2, \ldots$ *are simultaneously* $\aleph_0$-*satisfiable.*

PROOF of (b), supposing $E_0, E_1, E_2, \ldots$ contain bound no variable which occurs free in any of them. We shall apply Theorem 33, and the lower version of Theorem 34 omitting $F_0, F_1, F_2, \ldots$.

Suppose that, for each d, $E_0, \ldots, E_d$ are simultaneously satisfiable. This means that, for each d, there is a respective domain D_d and assignment in D_d which makes the formulas $E_0, \ldots, E_d$ all t, and thus makes the sequent $E_d, \ldots, E_0 \rightarrow$ f, so that not $\models E_d, \ldots, E_0 \rightarrow$, and hence by Theorem 33 not $\vdash E_d, \ldots, E_0 \rightarrow$ in *G4*. Since this is for every d, (II) of Theorem 34 is excluded. So by the remaining alternative (I), $\ldots, E_2, E_1, E_0 \rightarrow$ is falsifiable in $\{0, 1, 2, \ldots\}$, i.e. $E_0, E_1, E_2, \ldots$ are simultaneously satisfiable in $\{0, 1, 2, \ldots\}$, or briefly they are simultaneously $\aleph_0$-satisfiable.

EXERCISES. 50.1. (a) Establish the following for $m, n > 0$:

$\{\models A_1, \ldots, A_m \rightarrow B_1, \ldots, B_n\} \rightleftarrows \{\models \rightarrow A_1 \& \ldots \& A_m \supset B_1 \vee \ldots \vee B_n\}$,
$\{\vdash_{G4} A_1, \ldots, A_m \rightarrow B_1, \ldots, B_n\} \leftrightarrows \{\vdash_{G4} \rightarrow A_1 \& \ldots \& A_m \supset B_1 \vee \ldots \vee B_n\}$,
assuming for the fourth implication that no variable occurs both free and bound in $A_1, \ldots, A_m \rightarrow B_1, \ldots, B_n$. State and prove like propositions (b) for $m=0$ & $n>0$ and (c) for $m>0$ & $n=0$.

50.2. Establish the following additional versions of the Löwenheim-Skolem theorem. *If* $E_1, \ldots, E_k \rightarrow F_1, \ldots, F_l$ *or* $\ldots, E_2, E_1, E_0 \rightarrow F_0, F_1, F_2, \ldots$ *is falsifiable, it is* $\aleph_0$-*falsifiable. If* $\aleph_0$-$\models$ F, *then* $\models$ F. *If* $E_1, \ldots, E_k$ $\aleph_0$-$\models$ F, *then* $E_1, \ldots, E_k \models$ F.

§ 51. Gödel's completeness theorem (with a Hilbert-type formal system). To obtain Gödel's completeness theorem in the form "*If* $\models$ F, *then* $\vdash$ F", it remains for us to establish "*If* $\vdash \rightarrow$ F *in G4, then* $\vdash$ F *in H*". We get this in Corollary Theorem 36, by a lengthy but straightforward exercise in the proof theory of *H*.

THEOREM 36. *If* $\vdash A_1, \ldots, A_m \rightarrow B_1, \ldots, B_n$ *in the predicate calculus G4, then* $A_1, \ldots, A_m, \neg B_1, \ldots, \neg B_n \vdash P \& \neg P$ *in the predicate calculus H.*

LEMMA 11. (a) *In H, for* $n > 0$,

$$\{A_1, \ldots, A_m, \neg B_1, \ldots, \neg B_n \vdash P \& \neg P\} \rightleftarrows^{\circ} \{A_1, \ldots, A_m, \neg B_1, \ldots, \neg B_{n-1} \vdash B_n\}.$$

(b) *In H, for* $m > 0$,

$$\{A_1, \ldots, A_m, \neg B_1, \ldots, \neg B_n \vdash P \& \neg P\} \rightleftarrows \{A_2, \ldots, A_m, \neg B_1, \ldots, \neg B_n \vdash \neg A_1\}.$$

PROOF OF LEMMA. The four implications follow respectively from (1) ¬-introd. and (double) ¬-elim., (2) weak ¬-elim., (3) ¬-introd., (4) weak ¬-elim. (cf. Theorem 13 § 11). We do (1) in detail using the format (A) § 13 (employing Theorem 9 tacitly).

1. $A_1, \ldots, A_m, \neg B_1, \ldots, \neg B_n \vdash P \& \neg P$ — by hypothesis.
2. $A_1, \ldots, A_m, \neg B_1, \ldots, \neg B_n \vdash P$ — &-elim., 1.
3. $A_1, \ldots, A_m, \neg B_1, \ldots, \neg B_n \vdash \neg P$ — &-elim., 1.
4. $A_1, \ldots, A_m, \neg B_1, \ldots, \neg B_{n-1} \vdash \neg\neg B_n$ — ¬-introd. 2, 3.
5. $A_1, \ldots, A_m, \neg B_1, \ldots, \neg B_{n-1} \vdash B_n$ — ¬-elim., 4.

PROOF OF THEOREM. Suppose given a proof of $A_1, \ldots, A_m \rightarrow B_1, \ldots, B_n$ in *G*4. Such a proof is in tree form (§ 50). We shall show that, starting at the tops of branches and working *down* step by step, we can establish, for each sequent $\Delta \rightarrow \Lambda$ in its turn, that $\Delta, \neg\Lambda \vdash P \& \neg P$, where if Λ is $L_1, \ldots, L_s$ then $\neg\Lambda$ is $\neg L_1, \ldots, \neg L_s$ (and if Λ is empty, so is $\neg\Lambda$).

We have 15 cases to consider, according as the sequent $\Delta \rightarrow \Lambda$ we are considering is at the top of a branch (in which case it is an axiom by (×)) or results by a downward step (an "inference") by one of the 14 rules. We give 5 cases, leaving the others to the reader (Exercise 51.1).

CASE (×): $\Delta \rightarrow \Lambda$ is an axiom, i.e. is of the form $C, \Gamma \rightarrow \Theta, C$. So "$\Delta, \neg\Lambda \vdash P \& \neg P$" is "$C, \Gamma, \neg\Theta, \neg C \vdash P \& \neg P$", which is true by weak ¬-elim., or using Lemma 11 (a) and Theorem 9 (i).

CASE →⊃: $\Delta \supset \Lambda$ (of the form $\Gamma \rightarrow \Theta, A \supset B$) comes from $\Delta_1 \rightarrow \Lambda_1$ (of the form $A, \Gamma \rightarrow \Theta, B$) by →⊃. We are establishing the desired property of the sequents in the tree by working downward step by step. So we do not have to deal with this sequent $\Delta \rightarrow \Lambda$ (by establishing "$\Delta, \neg\Lambda \vdash P \& \neg P$", i.e. "$\Gamma, \neg\Theta, \neg(A \supset B) \vdash P \& \neg P$") until we have already handled $\Delta_1 \rightarrow \Lambda_1$ (by establishing "$\Delta_1, \neg\Lambda_1 \vdash P \& \neg P$", i.e. "$A, \Gamma, \neg\Theta, \neg B \vdash P \& \neg P$"). By Lemma 11 (a), our problem reduces to inferring "$\Gamma, \neg\Theta \vdash A \supset B$" from "$A, \Gamma, \neg\Theta \vdash B$". This is immediate by the deduction theorem (Theorem 11, or ⊃-introd. in Theorem 13).

CASE ⊃→. From (α) $\Gamma, \neg\Theta, \neg A \vdash P \& \neg P$ and (β) $B, \Gamma, \neg\Theta \vdash P \& \neg P$, we need to infer $A \supset B, \Gamma, \neg\Theta \vdash P \& \neg P$. Use V-elim. and *59.

CASE →∀. Using Lemma 11 (a), we aim to infer $\Gamma, \neg\Theta \vdash \forall x A(x)$ from (α) $\Gamma, \neg\Theta \vdash A(b)$. Since $\Gamma, \neg\Theta$ do not contain b free (by the restriction on variables for the rule →∀), we can apply ∀-introd. (Theorem 21 § 23) to (α) to infer (β) $\Gamma, \neg\Theta \vdash \forall b A(b)$. By the stipulations following the rules in § 48, the hypotheses of Lemma 5 § 24 are satisfied, so by *73 $\vdash \forall x A(x) \sim \forall b A(b)$. This with (β) gives $\Gamma, \neg\Theta \vdash \forall x A(x)$.

CASE →∃. We must show that, if (α) $\Gamma, \neg\Theta, \neg\exists x A(x), \neg A(r) \vdash P \& \neg P$, then $\Gamma, \neg\Theta, \neg\exists x A(x) \vdash P \& \neg P$. From the ∃-schema $A(r) \supset \exists x A(x)$ by contraposition (*12 § 24) and ⊃-elim., $\neg\exists x A(x) \vdash \neg A(r)$. Using this with (α) and Theorem 9, $\Gamma, \neg\Theta, \neg\exists x A(x) \vdash P \& \neg P$.

COROLLARY. (a) *If* $\vdash A_1, \ldots, A_m \rightarrow B$, *then* $A_1, \ldots, A_m \vdash B$.

(b) *If* $\vdash A_1, \ldots, A_m \rightarrow B_1, \ldots, B_n$, then $\vdash A_1 \,\&\ldots\&\, A_m \supset B_1 \vee \ldots \vee B_n$ $(m, n > 0)$.[248]

(c) *If* $\vdash \rightarrow B_1, \ldots, B_n$, *then* $\vdash B_1 \vee \ldots \vee B_n$ $(n > 0)$.

(d) *If* $\vdash A_1, \ldots, A_m \rightarrow$, *then* $\vdash \neg(A_1 \,\&\ldots\&\, A_m)$ $(m > 0)$.

PROOF. From the theorem using Lemma 11, &→ and →V, etc. —

Using Theorem 36 and Corollary in (II) of Theorem 34, we obtain statements about the Hilbert-type system *H*. We give some other convenient versions of Gödel's completeness theorem with a Hilbert-type system in Theorem 37 below.

Generalizing the notion "$E_1, \ldots, E_k \models F$" (§ 20), we say that F is a *valid consequence of* $E_0, E_1, E_2, \ldots$ (*holding all variables constant*), or in symbols $E_0, E_1, E_2, \ldots \models F$, if, for each domain *D*, F is t for all assignments which make all of $E_0, E_1, E_2, \ldots$ t. Also, replacing "for each domain *D*" by "for the domain $D = \{0, 1, 2, \ldots\}$", we get the like with "$\aleph_0$-$\models$" in place of "$\models$" (and similarly with "$\bar{\bar{D}}$-$\models$" for any other particular nonempty *D*).

Generalizing the notion "$E_1, \ldots, E_k \vdash F$", we say F is *deducible from* $E_0, E_1, E_2, \ldots$ (*holding all variables constant*), or in symbols $E_0, E_1, E_2, \ldots \vdash F$, if there is a deduction of F from $E_0, E_1, E_2, \ldots$ (holding all variables constant) defined as before (§ 21), except with the infinite list of formulas $E_0, E_1, E_2, \ldots$ available for use as assumption formulas instead of only a finite list. But only finitely many of $E_0, E_1, E_2, \ldots$ can be used in a given deduction; so the notion "$E_0, E_1, E_2, \ldots \vdash F$" as just defined is immediately equivalent to "for some *d*, $E_0, \ldots, E_d \vdash F$".

Clearly, "for some *d*, $E_0, \ldots, E_d \models F$" implies "$E_0, E_1, E_2, \ldots \models F$". The converse is not immediate (as it was with "$\vdash$"); but it is given by Gödel's completeness theorem.

THEOREM 37°. (Gödel's completeness theorem 1930.) *In the predicate calculus H*:

(a) *If* $\models F$ [*or even if* $\aleph_0$-$\models F$], *then* $\vdash F$. *If* $E_1, \ldots, E_k \models F$ [*or even if* $E_1, \ldots, E_k\ \aleph_0$-$\models F$], *then* $E_1, \ldots, E_k \vdash F$.

(b) *If* $E_0, E_1, E_2, \ldots \models F$ [*or even if* $E_0, E_1, E_2, \ldots\ \aleph_0$-$\models F$], *then, for some d*, $E_0, \ldots, E_d \vdash F$, *and hence* $E_0, E_1, E_2, \ldots \vdash F$.

(c) *Either* $E_1, \ldots, E_k$ *are simultaneously* $\aleph_0$-*satisfiable, or* $\vdash \neg(E_1 \,\&\ldots\&\, E_k)$ $(k > 0)$.

(d) *Either* $E_0, E_1, E_2, \ldots$ *are simultaneously* $\aleph_0$-*satisfiable, or, for some d*, $\vdash \neg(E_0 \,\&\ldots\&\, E_d)$.

PROOFS. As with Theorem 35, it will suffice to prove each part under

[248] If we interpret "$A_1 \,\&\ldots\&\, A_m$" for $m = 0$ to be $\neg(P \,\&\, \neg P)$ (or $P \supset P$), and "$B_1 \vee \ldots \vee B_n$" for $n = 0$ to be $P \,\&\, \neg P$ (or $\neg(P \supset P)$), then (b) holds without the restriction to $m, n > 0$.

the supposition that no formula contains bound any variable occurring free in any of the formulas. For, otherwise we could first replace the given formulas by ones congruent to them (§ 16) and meeting this condition, without affecting the holding of the model-theoretic hypothesis, and at the end use Theorem 25 and Corollary 2 Theorem 23 (§ 24) to pass back again to the original formulas in the Hilbert-type proof-theoretic conclusion.

(a) Assume $\aleph_0$-$\models$ F. Then $\rightarrow$ F is not falsifiable in {0, 1, 2, . . .}. So by (II) of Theorem 34 (the upper version with $k = 0$, $l = 1$), $\vdash_{G4} \rightarrow$ F. So by Theorem 36 Corollary (a), $\vdash_H$ F. Similarly with $k > 0$.

(b) Assume $E_0, E_1, E_2, \ldots$ $\aleph_0$-$\models$ F. Then it is impossible in {0, 1, 2, . . .} to make all of $E_0, E_1, E_2, \ldots$ t and F f, simultaneously; i.e. $\ldots, E_0, E_1, E_0 \rightarrow$ F is not $\aleph_0$-falsifiable. So by (II) of Theorem 34 (the lower version), for some d, $\vdash E_d, \ldots, E_0 \rightarrow$ F. Now apply Theorem 36 Corollary (a).

(d) The first alternative is equivalent to: (I) $\ldots, E_2, E_1, E_0 \rightarrow$ is $\aleph_0$-falsifiable. So by Theorem 34, if that alternative doesn't apply, then: (II) for some d, $\vdash E_d, \ldots, E_0 \rightarrow$, whence by Theorem 36 Corollary (d), $\vdash \neg(E_0 \;\&\; \ldots \;\&\; E_d)$. —

The parts of Theorem 37 are not independent (Exercise 51.2). Moreover, the Löwenheim-Skolem theorem (Theorem 35) can be inferred from Theorems 12_{Pd} and 37 (with the bracketed versions of the hypotheses), as easily as we inferred it from Theorems 33 and 34 (Exercise 51.3).

The significance of Gödel's completeness theorem and the Löwenheim-Skolem theorem will be discussed at the end of § 52 and in § 53.

Exercises. 51.1. Treat the following cases in the proof of Theorem 36: $\rightarrow$&, &$\rightarrow$, $\rightarrow\neg$, $\rightarrow\sim$ (cf. Exercise 5.3), $\forall\rightarrow$, $\exists\rightarrow$.

51.2°. Infer all parts of Theorem 37, and the results of using Theorem 36 in Theorem 34, from each of:

(a) *If* $E_0, E_1, E_2, \ldots$ $\aleph_0$-$\models$ F, *then* $E_0, E_1, E_2, \ldots \vdash$ F. (This part of Theorem 37 (b) is our most compact form of Gödel's completeness theorem.)

(b) Theorem 37 (d).

51.3. Infer Theorem 35 from Theorems 12_{Pd} and 37 (c), (d).

§ 52. Gödel's completeness theorem, and the Löwenheim-Skolem theorem, in the predicate calculus with equality. In §§ 48–51, = might have been one of the predicate symbols, but we did not give it a special status in the evaluation process (or in the deductive apparatus), as is done in § 29. To have done so would have prevented us from simply making all atoms in $\cup\Delta$ t, and all in $\cup\Lambda$ f, in the proof of Lemma 9. Thus the "$\models$", "$\vdash$" etc. in Theorems 34–37 are those of §§ 17, 21, 28, not of § 29.[249] In the

[249] The proof of Theorem 33 is good under either meaning of $\models$. We aren't defining another formal system *G4* for the case of the predicate calculus with equality, as we did with *H* in § 29.

Löwenheim-Skolem theorem (Theorem 35), if we start with a domain D and satisfying assignment for $E_1, \ldots, E_k$ or $E_0, E_1, E_2, \ldots$ in which $=$ does have the usual meaning of equality or identity, we cannot be sure that $=$ has that meaning in the new satisfying assignment in $\{0, 1, 2, \ldots\}$. This can be remedied for Theorems 35 and 37, while the domain may become finite, by an easy device due to Kalmár 1928–9 and Gödel 1930. The closed equality axioms mentioned in Lemma 12 are described in § 29 following Theorem 28.

LEMMA 12. *If* $E_0, E_1, E_2, \ldots$ *include the closed equality axioms for* $=$ *and all the proper predicate symbols and function symbols in* $E_0, E_1, E_2, \ldots$, *and under the evaluation rules of the predicate calculus without equality* $E_0, E_1, E_2, \ldots$ *are all* t *for some non-empty domain D and assignment, then* $E_0, E_1, E_2, \ldots$ *are all* t *for a domain* D^* *(with* $0 < \overline{\overline{D^*}} \leq \overline{\overline{D}}$*) and an assignment in which* $=$ *has the meaning of equality or identity. Similarly with* $E_1, \ldots, E_k$.

PROOF. Consider a given domain D and assignment which make all of $E_0, E_1, E_2, \ldots$ t.

From equality axioms which are included in $E_0, E_1, E_2, \ldots$, each of the following formulas is deducible in the predicate calculus (§ 21 or § 28), as we see by adapting the proofs for Theorem 29 § 29 (using $\forall$-elims. and $\supset$- and $\forall$-introds.):

(i) $\forall x(x=x)$. (ii) $\forall x \forall y(x=y \supset y=x)$. (iii) $\forall(x=y \;\&\; y=z \supset x=z)$

where $\forall$ is closure § 20 (i.e. in (iii), $\forall x \forall y \forall z$).

(iv) For each proper predicate symbol $P(a_1, \ldots, a_n)$ with $n > 0$ occurring in $E_0, E_1, E_2, \ldots$, and each i ($i = 1, \ldots, n$),

$$\forall[x=y \supset (P(a_1, \ldots, a_{i-1}, x, a_{i+1}, \ldots, a_n) \sim P(a_1, \ldots, a_{i-1}, y, a_{i+1}, \ldots, a_n))].$$

(v) For each function symbol $f(a_1, \ldots, a_n)$ with $n > 0$ occurring in $E_0, E_1, E_2, \ldots$, and each i ($i = 1, \ldots, n$),

$$\forall[x=y \supset f(a_1, \ldots, a_{i-1}, x, a_{i+1}, \ldots, a_n)=f(a_1, \ldots, a_{i-1}, y, a_{i+1}, \ldots, a_n)].$$

Consequently (by Theorem 12_{Pd}, etc.), for the given domain D and assignment, all the formulas (i)–(v) are t.

Let $x \simeq y$ be the logical function or predicate or binary relation assigned to the predicate parameter $=$ in the given assignment; thus for each x, y in D, $x \simeq y$ holds exactly when the formula x=y is t for x and y as the values of x and y.

From the form of (i)–(iii) (and their being t), and the rules for evaluating $\forall$, $\supset$ and &, it follows that the relation $x \simeq y$ is reflexive, symmetric and transitive; i.e. **(a)** for each x in D, $x \simeq x$, **(b)** for each x, y in D,

$x \simeq y \rightarrow y \simeq x$, and (c) for each x, y, z in D, $x \simeq y \mathbin{\&} y \simeq z \rightarrow x \simeq z$ (Exercise 52.1 (a)). Such a relation we called an "equivalence relation" in § 30. By (B) there, $\simeq$ separates D into disjoint (i.e. non-overlapping) nonempty classes (called "equivalence classes") such that any elements x and y of D belong to the same equivalence class if and only if $x \simeq y$. The equivalence class x^* to which x belongs is the class of all elements u of D such that $x \simeq u$ (by (a), x belongs to this). Our new domain D^* will be the set of all these equivalence classes. Then clearly $0 < \bar{\bar{D}}^* \leq \bar{\bar{D}}$.

By the above definitions, when x and y are the values of x and y:

$$\{\text{x=y is t}\} \equiv x \simeq y \equiv$$
$$\{x \text{ and } y \text{ belong to the same equivalence class}\} \equiv x^* = y^*.$$

Hence the truth value of x=y for given values x and y of x and y in D is not changed by changing the value x of x to another member u of D in the same equivalence class; and likewise with the value y of y. Similarly, from the form of (iv) (and its being t), etc., the truth value of $\text{P}(\text{a}_1, \ldots, \text{a}_n)$ for given values $a_1, \ldots, a_n$ of $\text{a}_1, \ldots, \text{a}_n$ in D is not changed by changing the value a_i of a_i to another value c_i in the same equivalence class, i.e. such that x=y is t when x, y are evaluated by a_i, c_i $(i = 1, \ldots, n)$. Likewise from (v), the equivalence class to which the value in D of $\text{f}(\text{a}_1, \ldots, \text{a}_n)$ belongs for given values $a_1, \ldots, a_n$ of $\text{a}_1, \ldots, \text{a}_n$ is unaltered by changing the value a_i of a_i to another value c_i in the same equivalence class $(i = 1, \ldots, n)$.

So in every step in the evaluation of a formula, we can disregard the differences between different members of D which are in the same equivalence class without altering the truth value obtained; in effect we can coalesce the elements of D belonging to each equivalence class into one element which behaves like any one of them in the evaluation. If we construe the element obtained by this coalescing of elements in an equivalence class to be the equivalence class itself, we are then doing the evaluation in the domain D^*, as we proposed. Since this coalescing does not affect the outcome of any evaluation, each of $\text{E}_0, \text{E}_1, \text{E}_2, \ldots$ is t as evaluated in D^*.

The assignment in D^* which accomplishes this, and which we have described as resulting from the given assignment in D by coalescing the elements of equivalence classes, can be described more explicitly as follows.

To x=y is assigned the value t exactly when the values x^* and y^* in D^* of x and y are the same member of D^*. This is what we were aiming to accomplish. A proposition symbol P is evaluated in D^* by the same truth value as in D. An individual parameter e, if evaluated in D by e, is evaluated in D^* by e^* (the equivalence class to which e belongs). An n-place predicate symbol P with $n > 0$, if evaluated in D by P, is evaluated in D^* by P^* where $P^*(a_1^*, \ldots, a_n^*) \equiv \{P(a_1, \ldots, a_n)$, for any choices $a_1, \ldots, a_n$ of members of $a_1^*, \ldots, a_n^*$, respectively}, that value being independent of the choices (as we inferred from (iv) being t in D). We define similarly the value f^* in D^* of an n-place function symbol f with $n > 0$ from the value f of f in D (Exercise 52.1 (b)). —

Let "... $(\leq \aleph_0)$-⊨ ..." stand for "... $\bar{\bar{D}}$-⊨ ... for each domain D with $0 < \bar{\bar{D}} \leq \aleph_0$", and "$(\leq \aleph_0)$-satisfiable" for "satisfiable in some domain D with $0 \leq \bar{\bar{D}} \leq \aleph_0$".

THEOREMS $35_=$ AND $37^{\circ}_=$. *Theorems* 35 *and* 37 *hold good reading "the predicate calculus with equality", "$(\leq \aleph_0)$-⊨" and "$(\leq \aleph_0)$-satisfiable" for "the predicate calculus", "$\aleph_0$-⊨" and "$\aleph_0$-satisfiable", respectively.*

PROOF. We shall prove Theorem $37_=$ (d). Theorem $37_=$ (a)–(c) and Theorem $35_=$ can be proved similarly, or inferred thence similarly to Exercises 51.2 (b) and 51.3 (Exercise 52.2).

For Theorem $37_=$ (d), we treat the case that $E_0, E_1, E_2, \ldots$ contain $\aleph_0$ proper predicate symbols and function symbols, giving rise (with =) to $\aleph_0$ closed equality axioms $Q_0, Q_1, Q_2, \ldots$.

Theorem 37 (d), applied to the list of formulas $E_0, Q_0, E_1, Q_1, E_2, Q_2, \ldots$ as its $E_0, E_1, E_2, \ldots$, gives us two alternatives.

CASE (I): $E_0, Q_0, E_1, Q_1, E_2, Q_2, \ldots$ are simultaneously satisfiable in the domain $D = \{0, 1, 2, \ldots\}$, under the evaluation rules of the predicate calculus. Then by Lemma 12, they are also simultaneously satisfiable in a domain D^* (with $0 < \bar{\bar{D}}{}^* \leq \bar{\bar{D}} = \aleph_0$) with = meaning equality, i.e. under the evaluation rules of the predicate calculus with equality; a fortiori, $E_0, E_1, E_2, \ldots$ are so satisfiable.

CASE (II): for some d, say $d = 2c$ as the case of d odd is similar, $\vdash \neg(E_0 \,\&\, Q_0 \,\&\, \ldots \,\&\, E_c \,\&\, Q_c)$ in the predicate calculus. Thence by propositional calculus, for that c, $Q_0, \ldots, Q_c \vdash \neg(E_0 \,\&\, \ldots \,\&\, E_c)$ in the predicate calculus. But then by § 29, $\vdash \neg(E_0 \,\&\, \ldots \,\&\, E_c)$ in the predicate calculus with equality. —

We conclude this section with some remarks on how Gödel's completeness theorem 1930 and the theorem on consistency with respect to validity establish various equivalences between model-theoretic notions and proof-theoretic ones. These remarks apply both to the predicate calculus and to the predicate calculus with equality. (For the predicate calculus, we could have put them at the end of § 51.)

By Theorem 37 (a) and (b) with Theorem 12_{Pd} and the first paragraph of § 23 (or the like for the predicate calculus with equality):

$$\{\models F\} \equiv \{\vdash F\},$$
$$\{E_1, \ldots, E_k \models F\} \equiv \{E_1, \ldots, E_k \vdash F\},$$
$$\{E_0, E_1, E_2, \ldots \models F\} \equiv \{E_0, E_1, E_2, \ldots \vdash F\}.$$

In the second and third of these equivalences, since we are using the symbols "⊨" and "⊢" without superscripts, the conditional interpretation applies to any free variables that occur in $E_1, \ldots, E_k$ or $E_0, E_1, E_2, \ldots$ (§§ 20, 21). Before applying these equivalences, closures should be taken with respect to any variables intended to have the generality interpretation, by the operator $\forall'$ of § 20 (by $\forall$, if all are to have the generality interpretation).

The second equivalence is of interest when $A_1, \ldots, A_k$ are formulas expressing the axioms of some formal axiomatic theory in the symbolism of the (first-order) predicate calculus (without or with equality), and $E_1, \ldots, E_k$ are their respective closures. For example, $A_1, \ldots, A_k$ can be the 6 nonlogical axioms E1–E3, G1–G3 of the formal system **G** of group theory § 39. (It is natural to use free variables with the conditional interpretation in temporary assumptions for the sake of an argument, but not in the axioms of a mathematical theory. Instead, one would more naturally use individual symbols, like the symbol 1 of the system **G**. Hence we take "full" closures here, by $\forall$, not just $\forall'$.) "$E_1, \ldots, E_k \models F$" expresses the model-theoretic notion of what it means for F to hold in such a theory; it says that F is true in all mathematical systems which make $A_1, \ldots, A_k$ simultaneously true (under the generality interpretation of their free variables). Any such system we now call a "model" of $A_1, \ldots, A_k$ (or of $E_1, \ldots, E_k$); precisely, a *model* of $A_1, \ldots, A_k$ is a (nonempty) domain *D* and an assignment in *D* to the parameters of $E_1, \ldots, E_k$ which make $E_1, \ldots, E_k$ simultaneously t (under the evaluation rules of §§ 17, 28, or of § 29, according as we are working in the predicate calculus without or with equality).[250] "$E_1, \ldots, E_k \vdash F$" expresses the proof-theoretic notion of F holding in the same theory (§§ 21, 28, or § 29).

The third equivalence has the same significance when the formulas A_0, $A_1, A_2, \ldots$ express in the predicate calculus (without or with equality) the axioms of a formal axiomatic theory with $\aleph_0$ axioms, and $E_0, E_1, E_2, \ldots$ are their respective closures.[242] For example, $A_0, A_1, A_2, \ldots$ can be the nonlogical axioms of **N** § 38, namely Axioms 14–21 and the $\aleph_0$ axioms by Axiom Schema 13. (A finite number of axioms will not suffice for **N**; cf. end § 47.) A *model* of $A_0, A_1, A_2, \ldots$ (or of $E_0, E_1, E_2, \ldots$) is a nonempty

[250] Often one speaks loosely of a mathematical system satisfying some axioms, or of some axioms being true under an interpretation, when it is to be understood that free variables occurring in the axioms have the generality interpretation. The precise meaning is then what we have rendered here in the definition of "model".

In technical discussions, when we say that formulas are satisfiable, in general or in a given domain, we mean precisely that the formulas themselves are satisfiable (but not necessarily their closures, if the formulas are open); and similarly with "satisfying assignment", "falsifiable", "counterexample", etc.

domain D and an assignment in D to the parameters of $E_0, E_1, E_2, \ldots$ which make $E_0, E_1, E_2, \ldots$ simultaneously t.[251]

Gödel's completeness theorem also gives the equivalence of Hilbert's proof-theoretic version of the consistency problem for axiomatic theories based on the predicate calculus without or with equality ("Is the theory deducible from the axioms free from contradiction?") to the older model-theoretic version ("Are the axioms true of some system of objects?"), § 36.

It was clear in a rough way back in the days of formal axiomatics that the existence of a mathematical system satisfying the axioms, or precisely what we now call a "model", implies that no contradiction can arise in the theory deducible from the axioms, if the theory in which the model is constructed is free from contradiction. The argument ran: Suppose a contradiction were deducible from the axioms; then in the theory providing the model, by corresponding inferences about the objects constituting the model, a contradiction would be deducible from the corresponding theorems. This was necessarily rough then, because "the theory deducible from the axioms" only became exactly defined with the formalization of the whole language and logic in modern proof theory. We now have the principle that a model establishes consistency simply by contraposing the chain of implications

$$\{E_1, \ldots, E_k \vdash P \,\&\, \neg P\} \rightarrow \{E_1, \ldots, E_k \models P \,\&\, \neg P\} \rightarrow \{E_1, \ldots, E_k \text{ are not simultaneously satisfiable}\},$$

and similarly with $E_0, E_1, E_2, \ldots$ as the closures of the axioms.

But the converse of that principle was by no means clear. Consider its contrapositive. Why, when a system of axioms is vacuous (i.e. not true of any system, § 36), or as we are now expressing it, when their closures are not simultaneously satisfiable, must a contradiction necessarily follow from the axioms in a finite number of elementary logical steps? By Gödel's completeness theorem, however, we now know that this is so. Take the case of $\aleph_0$ axioms $A_0, A_1, A_2, \ldots$. If their closures $E_0, E_1, E_2, \ldots$ are not simultaneously satisfiable, then by Theorem 37 (d) (or Theorem $37_=$ (d)), for some d, $\neg(E_0 \,\&\, \ldots \,\&\, E_d)$ is provable in the predicate calculus (or the predicate calculus with equality), so that formula and the contradictory formula $E_0 \,\&\, \ldots \,\&\, E_d$ are provable in the formal system based on the predicate calculus (or the predicate calculus with equality) with $A_0, A_1,$

[251] In the illustrations by G and N, since the equality axioms for = and their function symbols are all provable, by Corollary Theorem 31 § 29 it makes no essential difference in the proof-theoretic notions whether the logic is the predicate calculus without or with equality (Axioms E1–E3, or Axioms 16, 17, are redundant in the latter case). But (for systems with = as a symbol) the model-theoretic notions of the predicate calculus with equality are usually the ones that primarily interest us.

$A_2, \ldots$ as the nonlogical axioms. Another way, stated for the predicate calculus without equality (but adaptible to the predicate calculus with equality, by letting $E_0, E_1, E_2, \ldots$ include the closed equality axioms and using Lemma 12):

$$\{E_0, E_1, E_2, \ldots \text{ are not simultaneously satisfiable}\}$$
$$\rightarrow \{\ldots, E_2, E_1, E_0 \rightarrow \text{ is not falsifiable}\}$$
$$\rightarrow \{\text{for some } d, \vdash E_d, \ldots, E_0 \rightarrow\} \text{ [Theorem 34]}$$
$$\rightarrow \{\text{for some } d, E_d, \ldots, E_0 \vdash P \,\&\, \neg P\} \text{ [Theorem 36].}$$

EXERCISES. 52.1. Supply details in the proof of Lemma 12 as follows: (a) Prove the transitive property of $\simeq$.
(b) Write out the definition of f^*.

52.2. Prove Theorems $37_=$ (b) and $35_=$ (b).

52.3. Prove (via model theory) the proposition of Footnote 107 § 29.

52.4*. Show that, in the predicate calculus without equality, Axioms 14–18 of **N** § 38 possess no finite model (though they have a countably infinite one).

§ 53. Skolem's paradox and nonstandard models of arithmetic. Gödel's completeness theorem of 1930 is not as famous as his incompleteness theorem of 1931. But a little consideration will show that it is an equally remarkable theorem, when we include with it the older Löwenheim-Skolem theorem, which comes out of it immediately (Exercise 51.3), and of which Skolem's 1922–3 proof comes close to the 1930 theorem.[252]

[252] According to modern scholarship (van Heijenoort 1967), there are serious gaps in Löwenheim's attempted proof 1915 of the theorem which bears his name. What Löwenheim stated is our Theorem $35_=$ (a) (except without all our possibilities for the definition of "formula").

The first correct proof of this proposition is in Skolem 1920, where it is generalized to $\aleph_0$ formulas (giving essentially our Theorem $35_=$ (b) without the "compactness" part). In this proof, Skolem used the set-theoretic axiom of choice (§ 35), and his normal form 1920 for formulas of the pure predicate calculus (cf. Hilbert and Bernays 1934 pp. 158 ff., IM p. 435).

The work of establishing the reducibility to Skolem normal form can be spared, however. By using instead any prenex form (Theorem 27 § 25), we get a very simple proof in the following manner (Skolem 1929 p. 24, Kleene 1958 p. 139). Say e.g. a prenex form of E is $\forall w \exists x \forall y \exists z A(w, x, y, z)$ where $A(w, x, y, z)$ is a quantifier-free formula containing just the distinct variables v, w, x, y, z and two function symbols κ, $\lambda(-)$. Using the axiom of choice, then E is satisfied in a non-empty domain D by a certain assignment, if and only if $\forall w \forall y A(w, \beta(w), y, \delta(w, y))$ (where $\beta(-)$, $\delta(-, -)$ are new function symbols) is satisfied in D by that assignment plus some assignment to β, δ. But a domain and satisfying assignment for the latter formula remain such, when the domain is cut down from D to the countable domain D^* consisting of just the values of v and κ and the other members of D obtainable thence by repeated applications of the functions evaluating λ, β and δ, and the functions and predicates in the assignment are

The equivalences between model-theoretic notions and proof-theoretic notions which we have just recounted (end § 52) are remarkable in the following respect. The model-theoretic notions (validity, satisfiability, etc.) are highly transcendental; thus, for a domain of $\aleph_0$ elements, the set of the logical functions (which is involved in those notions) has the uncountable cardinal $2^{\aleph_0}$ (Exercise 34.4 (b)). The proof-theoretic notions, in contrast, are quite concrete; they are finitary. Still, these are results which were sought for the predicate calculus, whether or not they were fully expected. They confirm that the predicate calculus (without or with equality) fully accomplishes (for first-order theories) what has been conceived to be the role of logic.

But Gödel's completeness theorem (or the Löwenheim-Skolem theorem)

"restricted" to D^*.

These axiom-of-choice proofs establish that, given any model of a formula or list of formulas, there is a finite or countably infinite *submodel* of the given model (in the sense just illustrated). No special attention is required to =, whose meaning of identity, if used in the given model, carries through to the new model.

The device (used in § 48–50) of picking truth values of atoms toward building up a satisfying (or falsifying) assignment, independently of the values the atoms have in a given such assignment, avoids the use of the axiom of choice, but sacrifices the submodel result, and (at least as ordinarily carried out) required a restoration of the usual meaning of = (as in § 52) in the case of the predicate calculus with equality. In 1922–3 pp. 220–222 and 1929 pp. 24–29, Skolem used this device to give proofs of the Löwenheim-Skolem theorem (in versions like our Theorem 35 (a) and (b)) which at least implicitly (but in retrospect quite obviously) establish everything in Gödel's 1930 completeness theorem (including compactness), with two exceptions. First, only a form of the basic discovery (beginning § 50) is established; it is not shown further that, when validity is confirmed by the finite mechanical process in Skolem's treatment, one has provability in Hilbert's sense. Skolem could hardly have done this in 1922–3, since Hilbert's proof-theoretical formulation of the completeness problem only became clearly defined in Hilbert and Ackermann 1928 p. 68. So we can say that Skolem in 1922–3 discovered the completeness of intuitive logic (of first order), and independently Gödel in 1930 (not knowing Skolem 1922–3 or 1929) discovered the completeness of formal logic. Secondly, Skolem does not give the supplementary argument to restore the meaning of =. (There is no explicit discussion in his 1920, 1922–3 or 1929 of the impact of equality on the theorems in question, and it is not always clear what he intends regarding =.)

According to Dreben and van Heijenoort (in the introductory note to the translation of Skolem 1928 in van Heijenoort 1967), one can get Gödel's 1930 theorem (without equality) by combining arguments in Skolem 1920 and 1922–3, or 1929, with ones in Herbrand 1930 (plus now-familiar results on prenex form, say our Theorems 27 with 12_{Pd} and 19).

The treatments of Gödel's completeness theorem cited above do not deal with functions directly, as is done in Cases (C) and (D) § 50 (exceptions: the argument obtained by combining Skolem and Herbrand, the related treatment in Kleene 1958 and 1961 outlined in Footnote 280 below). Of course, since Hilbert and Bernays 1934, the results could be extended to functions via a replacement of functions by their representing predicates (cf. §§ 38, 45; IM p. 424).

has given us more than was sought. What was not sought, and presumably was unexpected when discovered (in 1915, 1920 and 1930), is singled out in Theorems 35 and $35_=$. These extra dividends make the theorem as much an incompleteness theorem for systems of axioms as it is a completeness theorem for logic.

Since Gödel's completeness theorem and the Löwenheim-Skolem theorem involve the non-constructive notions of validity and satisfiability, they do not belong to metamathematics. So their proofs cannot be wholly finitary. Proofs like those above hold this non-constructiveness to a minimum. In fact, the only non-constructive steps in the proofs of Theorem 34 and of Theorem 37 with "$\aleph_0$-⊧" are applications of the classical law of the excluded middle to propositions about countably infinite collections.[253]

It may not at first seem remarkable that, if formulas $E_1, \ldots, E_k$ or $E_0, E_1, E_2, \ldots$ are simultaneously satisfiable in some domain D, they are also simultaneously satisfiable in $\{0, 1, 2, \ldots\}$ or a finite domain, as the Löwenheim-Skolem theorem asserts. But suppose $E_1, \ldots, E_k$ or $E_0, E_1, E_2, \ldots$ are the closures of the axioms of a system of axiomatic set theory. Take e.g. the axioms of Gödel 1940. In them Gödel used two sorts of variables, one for "classes" (end § 35) and one for "sets". But all sets are classes, and no other objects are considered. So the axioms can be restated using only class variables, and three predicate symbols: $x=y$, $x \in y$ and $\mathfrak{M}(x)$ ("x is a set").[254] Mathematicians generally suppose that the axioms $A_1, \ldots, A_{17}$ are all true of a nonempty system of objects (the classes including the sets, for Gödel's set theory). That is, they suppose there is a non-empty domain D and an assignment in D to the predicate symbols $\in$ and $\mathfrak{M}$ (with $=$ meaning equality) for which the closures $E_1, \ldots, E_{17}$ of the axioms are all t. (The D and assignment are a model of the axioms.) But then by the Löwenheim-Skolem theorem (Theorem $35_=$ (a)), there is a domain D^* with $0 < \overline{\overline{D}}{}^* \leq \aleph_0$ in which $E_1, \ldots, E_{17}$ are also simultaneously satisfiable. Inspection of the axioms rules out the possibility that $\overline{\overline{D}}{}^* < \aleph_0$. Thus, if there is any model of the axioms at all, there is a countably infinite model (with $=$ meaning equality). There are only $\aleph_0$ "sets" in the new model. Yet in the axiomatic set theory based on the axioms, Cantor's theorem holds ((C) § 34). By this theorem, the set of the subsets

[253] Utilizing this observation, Hilbert and Bernays 1939 pp. 234–253 formalize their proof of Gödel's completeness theorem in (essentially) the number-theoretic system N, and thus establish the following metamathematical completeness theorem for the pure predicate calculus (cf. IM p. 395).

The addition to the predicate calculus of an unprovable predicate letter formula for use as an axiom schema would cause the number-theoretic formal system based on the predicate calculus and the nonlogical axioms of N (§ 38) *to become* ω*-inconsistent* (§ 47).

[254] Cf. IM p. 420 Example 13. Axiom A1 and the predicate symbol $\mathfrak{Cls}$ (x) ("x is a class") can then be omitted.

of the natural numbers (which is a set in the theory) is uncountable. This is Skolem's "paradox" 1922–3. Skolem used his own axioms for set theory 1922–3, with $\aleph_0$ axioms. The argument is the same with $\aleph_0$ axioms, using Theorem $35_=$ (b), or Theorem 35 (b) for set theories (having only sets as objects) in which $=$ is defined from $\in$ as in § 26.

Skolem's is not a paradox in the sense of an outright contradiction, but rather a kind of anomaly. For, there is an explanation: the "enumerating set" of ordered pairs, by which D^* is put into 1–1 correspondence with the natural numbers, is not itself a set admitted in the axiomatic set theory.

The "paradox" entails that any axiomatization of set theory by countably many axioms in the restricted predicate calculus with equality must fail fully to capture the notions of "set", "set of subsets of a given set", "1–1 correspondence", "countable", etc. These concepts, if we give them a prior status, elude characterization by any such set of axioms. But the paradoxes of set theory (§ 35) make it hard to give them a prior status, independent of any system of axioms. This led Skolem to the view that the concepts of set theory have only a relative status ("relativity of set theory"). Thus a set which is uncountable in one axiomatization can be countable in another, and there is no absolute notion of countability.

Of course, another conceivable explanation of Skolem's "paradox" (for a given system of axiomatic set theory) is that there is no model. In this case, by Gödel's completeness theorem (as at the end of § 52), $\neg(E_1 \& \ldots \& E_k)$, or $\neg(E_0 \& \ldots \& E_d)$ for some d, would be provable in the predicate calculus. Thus there would be a "real" paradox, i.e. a contradiction, in the axiomatic set theory, though (in any of the systems of axiomatic set theory commonly employed) it has not yet been discovered.

We conclude this section with two applications of "compactness", viz. that if each finite subset of $E_0, E_1, E_2, \ldots$ are simultaneously satisfiable, so are all $\aleph_0$ of the formulas (in Theorems 35 and $35_=$).

The first application (Theorem 38) deals with the question whether the natural number sequence $0, 1, 2, \ldots$ can be completely described by a list of axioms $A_0, A_1, A_2, \ldots$ written in the symbolism of the (first-order) predicate calculus with equality.

In particular, are the $\aleph_0$ non-logical axioms $A_0, A_1, A_2, \ldots$ of the number-theoretic system **N** § 38 true only under the intended interpretation described there? The mathematical system $S_0 = (D_0, 0_0, '_0, +_0, \cdot_0)$, consisting of the natural numbers $D_0 = \{0, 1, 2, \ldots\}$ as the domain, and the usual zero, successor, plus and times as the values of 0, ′, + and ·, constitutes a model for them (with $=$ meaning equality); i.e. this domain and assignment makes their closures $E_0, E_1, E_2, \ldots$ simultaneously t in the predicate calculus with equality. The question is whether they have any other model than S_0.

In a trivial way, they obviously do. For, the axioms do not say what the individual members of the domain D shall themselves be. The axioms leave us free to choose those members as we please, provided we can then choose from among them an individual to be called 0, and functions to be called $'$, $+$ and $\cdot$, having the formulated properties. So we can satisfy E_0, E_1, $E_2, \ldots$ by taking as D any countably infinite set, and using for the values of 0, $'$, $+$ and $\cdot$ the member of D, and the 1-, 2- and 2-place functions over D, which are determined from the usual ones via some enumeration of D. Thus the three following systems S_0 (the usual natural numbers), S_1 (the non-positive integers with unusual definitions of $'$ and $\cdot$) and S_2 (the positive integers with unusual definitions of 0, $+$ and $\cdot$) are all models of $A_0, A_1, A_2, \ldots$:

$S_0 = (D_0, 0_0, '_0, +_0, \cdot_0) = (\{0, 1, 2, \ldots\}, 0, x+1, x+y, xy)$,
$S_1 = (D_1, 0_1, '_1, +_1, \cdot_1) = (\{0, -1, -2, \ldots\}, 0, x-1, x+y, -xy)$,
$S_2 = (D_2, 0_2, '_2, +_2, \cdot_2) = (\{1, 2, 3, \ldots\}, 1, x+1, x+y-1, (x-1)(y-1)+1)$.

If we have a prior notion of $\ldots, -2, -1, 0, 1, 2, \ldots$ under which the sets $\{0, 1, 2, \ldots\}$, $\{0, -1, -2, \ldots\}$, $\{1, 2, 3, \ldots\}$ are different, the systems S_0, S_1, S_2 are different. But they are "inessentially" different, in that they have a common structure or form; they are "isomorphic".

To define this idea carefully, we say a given 1-1 correspondence of D^* to D is an *isomorphism* of $S^* = (D^*, 0^*, '^*, +^*, \cdot^*)$ to $S = (D, 0, ', +, \cdot)$, if it "preserves" the notions of the system, i.e. (writing $x^* \Leftrightarrow x$ to say that, in that 1–1 correspondence, x^* as member of D^* corresponds to x as member of D), if $0^* \Leftrightarrow 0$, $x^* \Leftrightarrow x \rightarrow x^{*'*} \Leftrightarrow x'$ and $x^* \Leftrightarrow x \,\&\, y^* \Leftrightarrow y \rightarrow x^* +^* y^* \Leftrightarrow x+y \,\&\, x^* \cdot^* y^* \Leftrightarrow x \cdot y$. We say S^* is *isomorphic to* S, if there is an isomorphism of S^* to S. This defines "isomorphism" and "isomorphic" for systems of the type of $(D, 0, ', +, \cdot)$ where D is a nonempty set and 0, $'$, $+$, $\cdot$ are 0-, 1-, 2- and 2-place functions over D with values in D. The definitions are similar for other types of systems; e.g. for $(D, <)$ where D is a nonempty set and $<$ a 2-place predicate over D, an isomorphism of $(D^*, <^*)$ to $(D, <)$ requires that $x^* \Leftrightarrow x \,\&\, y^* \Leftrightarrow y \rightarrow (x^* <^* y^* \equiv x < y)$.

If we consider systems of the type of $(D, 0, ', +, \cdot)$ without regarding the members of D as known other that through the relationships of the system, then the various isomorphic systems just described become merged into a single "abstract system", of which they are different "representations"; cf. IM § 8. (Such an abstract system can be regarded as an equivalence class of more concrete systems under the equivalence relation "isomorphic to".)

The following system S_3 (the usual integers) is not isomorphic to S_0:
$S_3 = (D_3, 0_3, '_3, +_3, \cdot_3) = (\{\ldots, -2, -1, 0, 1, 2, \ldots\}, 0, x+1, x+y, xy)$.
For, suppose we attempt to establish an isomorphism of S_3 to S_0. Say e.g. that, in the proposed 1–1 correspondence of $\{\ldots, -2, -1, 0, 1, 2, \ldots\}$ to $\{0, 1, 2, \ldots\}$, $2 \Leftrightarrow 0$. Then the requirement for an isomorphism that

$x^*\Leftrightarrow x \to x^{*\prime *}\Leftrightarrow x'$ forces us to put $3\Leftrightarrow 1, 4\Leftrightarrow 2, 5\Leftrightarrow 3, \ldots$, and we have no elements of S_0 left over that can correspond to the remaining elements $\ldots, -2, -1, 0, 1,$ of S_3. The system S_3 is not a model for the axioms of **N**, since Axiom 15 $\neg a'=0$ does not hold for it (under the generality interpretation of a); its closure is f for S_3 as the domain and assignment, since when a has the value -1, $a'=0$ is t and $\neg a'=0$ is f.

Now we rephrase our question about **N** to ask the following. Do the non-logical axioms of **N** have only models isomorphic to the intended model $(D_0, 0_0, '_0, +_0, \cdot_0)$? In the terminology of § 36, is this system of axioms categorical? Are these axioms true of no other abstract mathematical system than the natural numbers (considered abstractly)?

More generally, is there some countable set of formulas $A_0, A_1, A_2, \ldots$ in the symbolism of the predicate calculus with equality (including at least the symbols 0 and ′) which constitutes a categorical set of axioms for the natural numbers?

These questions are answered in the negative by the following theorem.

THEOREM 38. (Skolem's theorem 1934 on nonstandard models of arithmetic.) *Let* $A_0, A_1, A_2, \ldots$ *be formulas in the predicate calculus with the symbols* 0, ′, = *and possibly other (individual,) function and predicate symbols. Suppose* $A_0, A_1, A_2, \ldots$ *are all true (under the generality interpretation of their free variables) of the system S of the natural numbers* $D = \{0, 1, 2, \ldots\}$ *with* 0, ′, = *in their usual meanings (and any appropriate meanings of the other (individual,) function and predicate symbols); i.e.* $S = (D, 0, ', \ldots)$ *is a model of* $A_0, A_1, A_2, \ldots$ *in the predicate calculus with equality. Then there is a model* $S^* = (D^*, 0^*, '^*, \ldots)$ *of* $A_0, A_1, A_2, \ldots$ *in the predicate calculus with equality such that* $\overline{\overline{D}}{}^* = \aleph_0$ *and* $(D^*, 0^*, '^*)$ *is not isomorphic to* $(D, 0, ')$ *(a fortiori, S* is not isomorphic to S).*

PROOF. We can suppose that $A_0, A_1, A_2, \ldots$ include Axioms 14 and 15 of **N**, as otherwise we could add them (since they are both true of S). Let $E_0, E_1, E_2, \ldots$ be the closures of $A_0, A_1, A_2, \ldots$. Let i be some individual symbol not in $A_0, A_1, A_2, \ldots$. Consider the list of formulas

$$E_0, 0\neq i, E_1, 1\neq i, E_2, 2\neq i, \ldots$$

where (as in § 38) 0, 1, 2, ... are the terms 0, 0′, 0″, ... (called "numerals" in § 43) and r≠s is ¬r=s. For each d, the first $d+1$ of the displayed formulas are simultaneously satisfiable in the domain D of the natural numbers; indeed, we can use the given model S of $A_0, A_1, A_2, \ldots$ augmented by d as value of i, since $\boldsymbol{d}\neq$i (where $\boldsymbol{d}$ is 0′ . . . ′ with d accents, as in § 43) is not among the first $d+1$ formulas. So by compactness (in Theorem $35_=$ (b), using the bracketed version of the hypothesis), all the $\aleph_0$ displayed formulas are simultaneously satisfiable in some domain D^* with $0 < \overline{\overline{D}}{}^* \leq \aleph_0$; say they are satisfied in D^* by $0^*, '^*, \ldots$ as values of

0, ′, ... However, since Axioms 14 and 15 are among $A_0, A_1, A_2, \ldots$, we easily exclude $\bar{\bar{D}}^* < \aleph_0$ (Exercise 53.1); thus $\bar{\bar{D}}^* = \aleph_0$. We must show that $(D^*, 0^*, '^*)$ is not isomorphic to $(\{0, 1, 2, \ldots\}, 0, ')$. Suppose there were an isomorphism of $(D^*, 0^*, '^*)$ to $(\{0, 1, 2, \ldots\}, 0, ')$. Say that in it (as a 1–1 correspondence of D^* to $\{0, 1, 2, \ldots\}$) $a_0^* \Leftrightarrow 0$, $a_1^* \Leftrightarrow 1$, $a_2^* \Leftrightarrow 2, \ldots$; so $D^* = \{a_0^*, a_1^*, a_2^*, \ldots\}$. Now what are the values in S^* of the numerals 0, 1, 2, . . . ? Since their values in S (computed from the usual zero and successor as the values of the symbols 0 and ′) are the natural numbers 0, 1, 2, . . . , their values in S^* (computed from 0^*, $'^*$ as the values of 0, ′) must be $a_0^*, a_1^*, a_2^*, \ldots$, by the properties of the isomorphism that $0^* \Leftrightarrow 0$, $x^* \Leftrightarrow x \rightarrow x^{*'*} \Leftrightarrow x'$. Now what is the value a^* in S^* of i? Because $0 \neq$ i, $1 \neq$ i, $2 \neq$ i, . . . are all t in S^*, $a_0^* \neq a^*, a_1^* \neq a^*, a_2^* \neq a^*, \ldots$. This is absurd, since $a^* \in D^*$ and $D^* = \{a_0^*, a_1^*, a_2^*, \ldots\}$.

A system like S^* which satisfies the axioms of number theory but is not isomorphic to the usual or "standard" number theory is called a *nonstandard model of number theory* (or *of arithmetic*), or a *Skolem model*. The theorem that any (finite or) countably infinite set of number-theoretic axioms in the first-order predicate calculus with equality has a nonstandard model was originally proved by Skolem in 1933 (finite case) and 1934, using a direct construction of the nonstandard model. Skolem's construction was employed by Ryll-Nardzewski 1952 in proving the non-(finite-axiomatizability) of number theory § 47.

The very compact proof above based on compactness is due to Henkin 1947 p. 70, 1950 p. 90.

It is surprising that the existence of nonstandard models of the usual axioms of elementary number theory was not widely recognized very early by juxtaposing Gödel's completeness theorem 1930 and his incompleteness theorem 1931, thus:[255]

{formal number theory} =
{predicate calculus} + {number-theoretic axiom system}.

By Gödel 1930, {predicate calculus} is complete.

By Gödel 1931, {formal number theory} is incomplete.

Therefore, {number-theoretic axiom system} is incomplete.

[255] Not however with quite the version in Gödel 1931, where the incompleteness is given for "Principia Mathematica [Whitehead and Russell 1910–13] and related systems". PM is not of the form {(first-order) predicate calculus} + {number-theoretic axiom system}. Number-theoretic systems of that form (like N in § 38) came in with Hilbert 1928, so they are sometimes called *Hilbert arithmetic*. In 1931–2 (reporting a colloquium held January 22, 1931), Gödel states his incompleteness theorem for such a formal system, with all the usual axioms of elementary number theory.

When the author pointed out the connection between Skolem's and Gödel's theorems in IM (1925b) p. 430 and 1956 § 18, he knew of no earlier mention of it in the literature. Recently, he has come across 2 lines in Gödel's review of Skolem 1933 (1934a p. 194 lines 10–11) and 3 in Henkin 1950 (p. 91 lines 8–10) alluding to such a connection.

To give this argument more explicitly for the **N** of § 38, consider the formula $\neg C_p$ of Theorem V § 43 (Gödel's incompleteness theorem), which is true under the standard interpretation of the number-theoretic symbolism, but unprovable in **N**. In Gödel's completeness theorem (say with equality, Theorem $37_=$ (b)) "If $E_0, E_1, E_2, \ldots (\leq \aleph_0)$-$\models$ F, then $E_0, E_1, E_2, \ldots \vdash$ F", take $E_0, E_1, E_2, \ldots$ to be the closed number-theoretic axioms, and F to be $\neg C_p$. Now $E_0, E_1, E_2, \ldots$ must be true under some interpretation (not the standard one) with a countable domain in which $\neg C_p$ is *not* true, or Gödel's completeness theorem would imply that $E_0, E_1, E_2, \ldots \vdash \neg C_p$ in the predicate calculus with equality,[251] and hence that $\vdash \neg C_p$ in **N**, contradicting Theorem V.[256]

From this standpoint, Gödel's incompleteness theorem is a phenomenon of the same kind as the unprovability of either Euclid's fifth postulate or its negation from the other postulates of Euclid (§ 36). Euclid's fifth postulate (or $\neg C_p$) is true under one interpretation of the axioms in question, false under another.

In any countable model $(D^*, 0^*, '^*, +^*, \cdot^*)$ of **N**, $\overline{\overline{D}} = \aleph_0$, since the axioms of **N** include Axioms 14 and 15. In any nonstandard model $(D^*, 0^*, '^*, +^*, \cdot^*)$ of **N**, the part $(D^*, 0^*, '^*)$ is nonstandard, i.e. not isomorphic to $(\{0, 1, 2, \ldots\}, 0, ')$, since the axioms of **N** include the recursive definitions of $+$ and $\cdot$ (Axs. 18–21). For, to make the closures of these four axioms all t when 0 and $'$ have their usual meanings with $D = \{0, 1, 2, \ldots\}$ as the domain, we are forced to give $+$ and $\cdot$ their usual meanings; and hence, if $(D^*, 0^*, '^*)$ were isomorphic to $(D, 0, ')$, then $(D^*, 0^*, '^*, +^*, \cdot^*)$ would be isomorphic to $(D, 0, ', +, \cdot)$. So we do get Skolem's theorem for **N**.

The proof of Skolem's theorem by juxtaposing Gödel's two theorems applies to lists $A_0, A_1, A_2, \ldots$ of formulas that can be the nonlogical axioms of a formal system of number theory. This entails an effectiveness requirement, used in proving Gödel's incompleteness theorem (cf. the discussion in § 43).[257] This effectiveness requirement does not detract from

[256] Still another pair of circumstances shows very simply that there must be nonstandard models. By the hierarchy theorem of Kleene 1943, the predicates definable in the symbolism of N fall into a hierarchy, with predicates of successively higher levels requiring more and more quantifiers in their definitions (§ 46 including Exercises 46.8, 46.9 (a)). This of course is proved by reasoning with the standard model of number theory. But by IM Theorem 40 pp. 401–402,[228] all the satisfying predicates in Gödel's completeness theorem are definable using just two quantifiers. These results would be incompatible, if the model for the latter were the standard one. This proof is given in detail in IM pp. 428–429.

[257] To any effective list $A_0, A_1, A_2, \ldots$ not sufficient to give us the formulas C_a of § 43 using in them only the parameters 0, $'$, $+$, $\cdot$, or not including Axs. 14, 15, 18–21, we can first add the axioms of N § 38.

There is a way of extending the argument to some noneffective lists. Cf. IM p. 431 Remark 1.

the significance of Skolem's theorem as showing that no list of axioms we could actually use can describe the number sequence completely. However, Skolem's proof and the Henkin proof given above apply to any list A_0, A_1, A_2, . . . , whether given effectively or not.

Peano's axioms for the natural numbers are commonly regarded as being categorical.[152] We formalized these axioms in setting up the system N § 38. This apparent contradiction between Skolem's theorem and the categoricity of Peano's axioms is explained away as follows. The fifth Peano axiom asserts that the principle of mathematical induction holds for all properties (i.e. 1-place predicates) of natural numbers. There are $2^{\aleph_0}$ such properties. The fifth Peano axiom is only incompletely formalized in **N**, since Axiom Schema 13 gives induction only for the $\aleph_0$ properties expressible by formulas A(x) of **N**.

Translating from Skolem's German, ". . . the number series is completely characterized, for example, by the Peano axioms, if one regards the notion 'set' or 'propositional function' as something given in advance with an absolute meaning independent of all principles of generation or axioms. But if one would make the axiomatics true to principle (konsequent), so that also the reasoning with the sets or propositional functions is axiomatized, then, as we have seen, the unique or complete characterization of the number series is impossible."[258]

Nonstandard models have recently become a standard branch of logical research; cf. Henkin 1950, Rabin 1958a, Kemeny 1958, Scott 1961, A. Robinson 1961, 1963.

For example, the type of order in the nonstandard models of arithmetic is known. Here we suppose the symbolism to include the predicate symbol $<$, and the formulas A_0, A_1, A_2, . . . to include at least the nonlogical axioms of N § 38 and the formula $a<b \sim \exists c\, c'+a=b$ ($<$ now being primitive). Then from their closures E_0, E_1, E_2, . . . , the formula $\boldsymbol{a}<\boldsymbol{b}$ is deducible in the predicate calculus for all pairs of numerals $\boldsymbol{a}$, $\boldsymbol{b}$ for natural numbers a, b such that $a < b$, and $\neg\boldsymbol{a}<\boldsymbol{b}$ for all pairs such that $a \geq b$ (IM pp. 196–197). What orderings can the nonstandard models S^* have under the order relation $a < b \equiv \{$a<b is t in S^* when a, b are the values in D^* of a, b$\}$? According to Kemeny 1958 (who says the result was found in 1947 by Henkin and him, but was known much earlier to Skolem), only one ordering is possible. Under this, D^* consists of the usual natural numbers, followed (in increasing order) by families of elements, the elements within each family having the same order as the integers, and the families having the same order as the rational numbers.

To describe the order type of the nonstandard models, of course we

[258] 1934 p. 160. Skolem like Peano used the positive integers instead of the natural numbers.

are thinking in terms of the standard natural numbers, whence the standard integers and rational numbers are constructed.

Theorem 38 denies the possibility of a *formal axiomatic* characterization of the natural number sequence *in the first-order predicate calculus.*

We argued in § 36 that formally axiomatized mathematics is not all of mathematics. An intuitive understanding of the natural number sequence is already presupposed in the statement of Theorem 38. Thus, in "$A_0, A_1, A_2, \ldots$", the reader is expected to understand what the three dots "..." mean. Also we wrote "$\overline{\overline{D}}{}^* = \aleph_0$". Also, we cannot contemplate, as we *ordinarily* do, languages such as the axioms are to be stated in, without using concepts basically equivalent to that of the natural numbers. Abstractly, the sequence of the natural numbers has the same structure as the sequence of expressions |, ||, |||, . . . , which we used to represent them on a Turing machine tape § 41. The possible expressions in a language such as the predicate calculus (unless we put a fixed upper bound on their lengths, so only finitely many of them exist) form a similar system, only with an alphabet of more than one symbol.

Robinson has now even a nonstandard analysis (theory of real numbers) 1961a, 1963, 1966. In 1966 (abstract 1964) Bernstein and Robinson found the answer to an open problem in functional analysis (Hilbert spaces) by use of nonstandard analysis.

THEOREM 39. (Henkin 1949.) *If a list of formulas* $A_0, A_1, A_2, \ldots$ *in the predicate calculus with equality admits arbitrarily large finite models, it also admits a countably infinite model.*

PROOF. Let $E_0, E_1, E_2, \ldots$ be the closures of $A_0, A_1, A_2, \ldots$. Let $Q_0, Q_1, Q_2, \ldots$ be the following formulas, which are respectively satisfied if and only if there are at least 2, 3, 4, . . . elements in the domain:

$$\exists x \exists y (x \neq y),\ \exists x \exists y \exists z (x \neq y \ \&\ x \neq z \ \&\ y \neq z),$$
$$\exists w \exists x \exists y \exists z (w \neq x \ \&\ w \neq y \ \&\ w \neq z \ \&\ x \neq y \ \&\ x \neq z \ \&\ y \neq z), \ldots.$$

Using the hypothesis, for each d, the first $d+1$ of

$$E_0, Q_0, E_1, Q_1, E_2, Q_2, \ldots$$

are simultaneously satisfiable in a respective domain D_d; we just use one of the given models having $\geq c+2$ elements, if $d = 2c$ or $d = 2c+1$. Hence by compactness (in Theorem $35_=$), all of the formulas last displayed are simultaneously satisfiable in a domain D with $0 < \overline{\overline{D}} \leq \aleph_0$. However, it is absurd that $\overline{\overline{D}} < \aleph_0$, as then Q_i for $i \geq \overline{\overline{D}}-1$ could not be satisfied.

EXERCISES. 53.1. Prove in detail for Theorem 38 that $D^* = \aleph_0$. (HINT: use Exercise 38.5.)

53.2. Show that the theorem in Footnote 253 cannot be improved by replacing "*ω-inconsistent*" by "(*simply*) *inconsistent*".

§ 54. Gentzen's theorem. In the course of proving Gödel's completeness theorem (§§ 48–51), we have been led to a new type of formalization of logic, represented by the Gentzen-type system *G*4. Indeed, combining Theorems 12_{Pd}, 34 and 36 Corollary (a), we have the equivalence of three notions, for any formula F not containing any variable both free and bound:[259]

$$\{\vdash_H \text{F}\} \overset{\text{C}^\circ}{\equiv} \{\models \text{F}\} \overset{\text{C}^\circ}{\equiv} \{\vdash_{G4} \rightarrow \text{F}\}.$$

The "C" indicates that classical (nonfinitary) model-theoretic reasoning is used.

The basic result that predicate logic can be formalized in a system like *G*4, or specifically the equivalence $\{\vdash_H \text{F}\} \equiv \{\vdash_{G4} \rightarrow \text{F}\}$ (or later, $\{\vdash_H \text{F}\} \equiv \{\vdash_{G4a} \rightarrow \text{F}\}$), we shall call GENTZEN'S THEOREM. From this equivalence, the principal results of Gentzen 1934–5 for classical logic, and some versions of Herbrand's theorem 1930 (§ 55), can easily be extracted.

Gentzen 1934–5 established the basic result entirely within finitary metamathematics, thus:

$$\{\vdash_H \text{F}\} \equiv \{\vdash_{G1} \rightarrow \text{F}\} \equiv \{\vdash_{G1 \text{ without cut}} \rightarrow \text{F}\}.$$

The system called "*LK*" by Gentzen and (with minor differences in the notion of formula) "*G*1" in IM is a sequent calculus generally similar to *G*4, except that it has a rule called "cut", as follows:

$$\frac{\Delta \rightarrow \Lambda, \text{C} \quad \text{C}, \Gamma \rightarrow \Theta}{\Delta, \Gamma \rightarrow \Lambda, \Theta} \text{Cut}$$

where C is any formula, and Δ, Λ, Γ, Θ are any lists of formulas. The presence of this rule makes it easy to prove $\{\vdash_H \text{F}\} \rightarrow \{\vdash_{G1} \rightarrow \text{F}\}$; the converse is essentially the same as our Theorem 36. In his "Hauptsatz" or "normal form theorem", Gentzen showed that the uses of this rule can be eliminated from any given proof in *G*1 to obtain a proof of the same sequent in "*G*1 without cut".

Herbrand and Gentzen used their theorems in giving consistency proofs from Hilbert's standpoint (like those cited in § 45 Footnote 194 and § 47 Footnote 216) and in other metamathematical applications. Such applications would lose their point if the theorems could not be proved metamathematically. Thus, for metamathematical purposes, our treatment at

[259] To deal with formulas containing variables both free and bound, we can use Theorem 25 as in the proof of Theorem 37.

this point leaves a gap to be filled, just as for Theorems I–VIII some gaps were left.[260]

Gentzen also gave a version "*LJ*" of his system for the intuitionistic logic, for which we gave a Hilbert-type system at the end of § 25.[261] Here our model-theoretic route (via "⊨ F") is not available, at least not without considerable effort to set up an intuitionistic model theory first (i.e. to give an intuitionistic analog of the classical notion "⊨ F").[262]

We formulated the 14 rules of our Gentzen-type system *G*4 so as to give ourselves as few choices as possible at each step *upward* in searching systematically for a counterexample to $\rightarrow$ F or to $E_1, \ldots, E_k \rightarrow F_1, \ldots, F_l$. For a different reason, we required the C for the axiom schema (×) to be prime.

When there is no counterexample, so $\vdash \rightarrow F$ or $\vdash E_1, \ldots, E_k \rightarrow F_1, \ldots, F_l$ in *G*4, the use of other Gentzen-type systems *G*4a and *G*4b (described next) may afford simpler proofs. Likewise, in

[260] As the chart shows, the gap can be filled by reading the proofs of IM Theorems 46 and 48 §§ 77, 78, and using for our applications the Gentzen-type systems *G*4a and *G*4b introduced below. For agreement with IM and Kleene 1952, we can optionally take "A ∼ B" in our systems to be an abbreviation for (A ⊃ B) & (B ⊃ A).

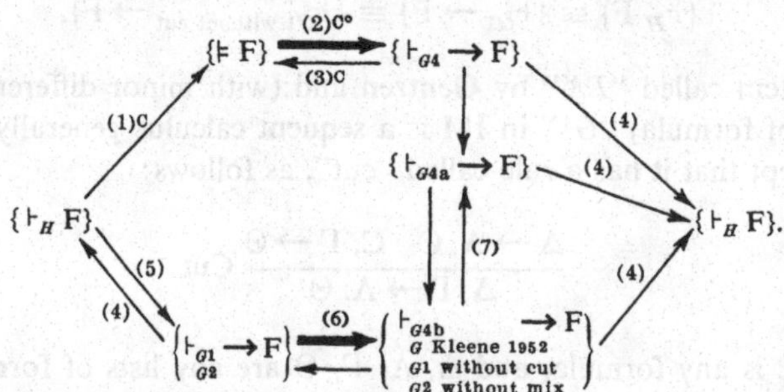

(1) Theorem 12_{Pd}. (2) Theorem 34 (Gödel's completeness theorem with a Gentzen-type formal system). (3) Theorem 33. (4) Theorem 36 Corollary (a) or IM Theorem 47. (5) IM Theorem 46. (6) IM Theorem 48 (Gentzen's Hauptsatz). (7) Cf. Exercise 49.1 (f) and (g), or Kleene 1952 Lemma 9. The systems grouped together are very similar, and easily proved equivalent. The unnumbered implications are trivial (one system is a subsystem of the next). The boldface arrows show the main part of the work on each route.

[261] In this chapter, the absence of a "°" on a theorem concerning provability in Gentzen-type systems means that there is a known intuitionistic version of the theorem (sometimes with qualifications), using the intuitionistic version of one of the systems *G*1, *G*2, *G*3, *G*3a, *G*. When we do not indicate otherwise, this intuitionistic version can be found in IM or Kleene 1952.

[262] For references to Beth's intuitionistic model theory or semantics, see Kleene and Vesley 1965 bottom p. 81.

attempting to construct proofs *downward*, and in performing manipulations on given proofs, the greater flexibility provided in *G*4a or *G*4b is often advantageous.

To get *G*4a from *G*4, we add four new rules of inference called "thinning (in antecedent or succedent)" and "contraction (in antecedent or succedent)":

$$\frac{\Gamma \to \Theta}{\Gamma \to \Theta, \mathrm{C}.}\to T \qquad\qquad \frac{\Gamma \to \Theta}{\mathrm{C}, \Gamma \to \Theta.}\ T\to$$

$$\frac{\Gamma \to \Theta, \mathrm{C}, \mathrm{C}}{\Gamma \to \Theta, \mathrm{C}.}\to C \qquad\qquad \frac{\mathrm{C}, \mathrm{C}, \Gamma \to \Theta}{\mathrm{C}, \Gamma \to \Theta.}\ C\to$$

Here C is any formula, and Γ and Θ any lists of formulas. As with the other rules, the order of formulas within antecedents and within succedents is immaterial. These rules (or inferences by them) we call *structural*; the former rules *logical*, or more specifically *proposition*(*al*) ($\to\supset, \ldots, \sim\to$) and *predicate* ($\to\forall, \ldots, \exists\to$). To save space that would be required in using the thinning rule separately, we shall usually write as one inference a series of applications of thinning followed by a logical inference. This has the effect of making the 14 logical rules applicable even with formulas in the premise(s) absent.

EXAMPLE 7. The following proof in *G*4 can be simplified to a proof of the same sequent in *G*4a by omitting the 13 boldface formulas. This gives the maximum simplification which *G*4a allows.

$$\dfrac{\dfrac{\dfrac{\dfrac{\dfrac{\dfrac{\mathrm{P(a)}, \mathbf{R}, \mathbf{\forall x(R\ \&\ P(x))}, \mathbf{Q(b)} \to \mathbf{\exists xP(x)}, \mathrm{P(a)}}{\mathrm{P(a)}, \mathbf{R}, \mathbf{\forall x(R\ \&\ P(x))}, \mathbf{Q(b)} \to \exists \mathrm{xP(x)}}\to\exists}{\mathrm{R\ \&\ P(a)}, \mathbf{\forall x(R\ \&\ P(x))}, \mathbf{Q(b)} \to \exists \mathrm{xP(x)}}\&\to}{\forall \mathrm{x(R\ \&\ P(x))}, \mathbf{Q(b)} \to \exists \mathrm{xP(x)}}\forall\to \qquad \dfrac{\mathrm{Q(b)}, \mathbf{\forall x(R\ \&\ P(x))} \to \mathbf{\exists xQ(x)}, \mathrm{Q(b)}}{\mathrm{Q(b)}, \mathbf{\forall x(R\ \&\ P(x))} \to \exists \mathrm{xQ(x)}}\to\exists}{\forall \mathrm{x(R\ \&\ P(x))}, \mathrm{Q(b)} \to \exists \mathrm{xP(x)} \vee \exists \mathrm{xQ(x)}}\to\vee}{\forall \mathrm{x(R\ \&\ P(x))} \to \mathrm{Q(b)} \supset \exists \mathrm{xP(x)} \vee \exists \mathrm{xQ(x)}}\to\supset}{\forall \mathrm{x(R\ \&\ P(x))} \to \forall \mathrm{x(Q(x)} \supset \exists \mathrm{xP(x)} \vee \exists \mathrm{xQ(x))}.}\to\forall$$

The $\to\vee$-inference as it appears after omitting the boldface formulas can be analyzed as consisting of two thinnings $T\to$ and one $\to\vee$, thus:

$$\dfrac{\dfrac{\forall \mathrm{x(R\ \&\ P(x))} \to \exists \mathrm{xP(x)}}{\forall \mathrm{x(R\ \&\ P(x))}, \mathbf{Q(b)} \to \exists \mathrm{xP(x)}}\ T\to \qquad \dfrac{\mathrm{Q(b)} \to \exists \mathrm{xQ(x)}}{\mathrm{Q(b)}, \mathbf{\forall x(R\ \&\ P(x))} \to \exists \mathrm{xQ(x)}}\ T\to}{\forall \mathrm{x(R\ \&\ P(x))}, \mathrm{Q(b)} \to \exists \mathrm{xP(x)} \vee \exists \mathrm{xQ(x)}.}\to\vee$$

In *G*4b, we further allow that the C for an axiom need not be prime.

THEOREMS 33 and 36 and COROLLARY (and LEMMAS 6 (a) and 7 with the new postulates included) extend immediately to *G*4a and *G*4b, since the treatment of the cases for →*T*, *T*→, →*C*, *C*→ and the liberalized axioms is obvious.

The *subformula*(*s*) *of* a formula F are F itself and all the formulas obtainable from it by repeated dissection (by removing one operator after another, all the way down to atoms); here, we allow the result of removing ∀x from ∀xA(x) or ∃x from ∃xA(x) to be A(r) for any term r free for x in A(x).

In other words: First, define the *immediate subformula*(*s*) *of* a composite formula F to be the formula(s) which can be side formula(s) to F as principal formula in any of the 14 logical rules →⊃, ... ∃→. Then the *subformula*(*s*) of any formula F are F itself, its immediate subformula(s) if it is composite, the immediate subformula(s) of each of the latter which is composite, etc.

For example, the subformulas of ∀a∀b(P(b) & Q(a)) are ∀a∀b(P(b) & Q(a)), all formulas ∀b(P(b) & Q(r)) where r is any term not containing b (free), all formulas P(u) & Q(r) where u is any term (with r as stated), and finally all formulas P(u) for such u and Q(r) for such r. In particular, Q(b) is not a subformula of ∀a∀b(P(b) & Q(a)); cf. Exercise 49.2.

Combining the definition of "subformula" with our observations early in § 49 of ancestral relationships in proofs in *G*4 (which exist likewise in *G*4a and *G*4b), we have:

LEMMA 13. (Ancestor and subformula property.) *In a proof in G*4, *G*4a *or G*4b, *in any given sequent*:

Each formula occurrence is identifiable as an ancestor of a specific formula occurrence in the endsequent; and the former formula is a subformula of the latter.

Each formula part (*or whole*) *is identifiable as an ancestral image of a specific part in the endsequent; the former formula part is identical with the latter, or comes from it by free substitutions of terms for free occurrences of variables.*

Each occurrence of an operator or predicate parameter is identifiable as an ancestor (*or ancestral image*) *of a specific occurrence of the same operator or predicate parameter in the endsequent.*

In contrast, Gentzen's system *LK* or *G*1 mentioned above does not have the subformula property, since the C of a cut need not be a subformula of any of the formulas in the conclusion. Likewise, the Hilbert-type system *H* does not have the corresponding subformula property, i.e. not every formula in a proof is a subformula of the formula proved. For, when we infer B from A and A ⊃ B by modus ponens, A ⊃ B is certainly not a subformula of B, and A is not necessarily one.

Thus proofs in *H* generally go outside the "parts" of the formula F being proved. We already mentioned this in § 40 as a reason why the decision problems for provability in formal systems like those we had considered (of Hilbert type) cannot be solved almost immediately from the definition of "provable formula". But for the *propositional* calculus formalized by *G*4, we can solve the decision problem this way, in view of the subformula property. In fact, in *G*4 our search procedure must simply terminate with outcome (I) or (II) after at most one step along each branch for each occurrence of a propositional connective in F. Using *G*4a, *G*4b or the systems in IM, there is slightly more to the proof of decidability, because of the contraction rules $\rightarrow C$, $C\rightarrow$.[263] Of course, for the classical propositional calculus, this only gives a new *proof* of the decidability, with a new *decision procedure*. For the intuitionistic propositional calculus, this is how the decidability was first discovered, by Gentzen 1934–5.[263]

This approach does not work on the predicate calculus, because the disection of ∀xA(x) or ∃xA(x) into A(r) gives rise to infinitely many subformulas.[264] No method can work (Theorem VII § 45).

A proof in a Gentzen-type subformula system is in a "certain, however by no means unique, normal form" (Gentzen 1934–5 p. 177). In it the only concepts introduced are ones which enter into its final result and which hence must be applied in winning that result.[265] Its result is built up progressively out of the components of the result (subformulas); nothing is first built up and then torn down. "It makes no detours (Umwege)."

We shall speak of such a proof as "direct".[266]

In informal mathematics and in mathematics as formalized in a Hilbert-type system, the proofs are not ordinarily "direct" in the present sense. As a hypothetical example, suppose we have proved in elementary number

[263] Cf. IM pp. 482–485.

[264] However, this approach is used in IM § 80 to show some classically provable formulas of the predicate calculus to be intuitionistically unprovable. The intuitionistic unprovability is shown by identifying an infinite branch in the sequent tree constructed upward using the Gentzen-type system *G*3. Also cf. Curry 1950.

[265] Exception: (individual and) function symbols may be introduced which are not in the endsequent, if they are allowed under the formation rules of the system in question. But, by our completeness argument (§§ 48–50), any provable (and hence valid) sequent can be proved in that system *G*4 which has only its own (individual,) function and predicate symbols. Or simply in proof theory, in a given proof we can replace the (individual,) function and predicate symbols not present in the endsequent (or in *H*, in the endformula), as in Theorem 31 § 29 and in the proof-theoretic treatment of substitution (Footnote 86 § 25).

[266] We are using "direct" here in a more general sense than at the ends of §§ 13, 25. There a proof or method of proof was considered "indirect" when, to prove F, we begin by deducing a contradiction from ¬F (thus proving ¬¬F, before using ¬-elimination) or use some variant of that device.

In §§ 13 (early), 21, 23 etc., "direct" has still another sense in "direct rule" vs. "subsidiary deduction rule".

The directness of a proof in *G*1 without cut compared to *G*1 using a cut is illustrated in IM pp. 448–449 Example 1.

theory that each prime number has a certain property *B*. We might put this result in a text book as Theorem 1066. Then, later we might apply Theorem 1066 to conclude that 7 has the property *B*. Would this be a direct proof (from first principles) of the latter proposition? Hardly. It could be very indirect indeed, if the proof of Theorem 1066 involved difficulties not having to be faced in the case of the prime 7, but perhaps affecting 61 and $2^{2281}-1$.

If E is the conjunction $E_1 \& \ldots \& E_k$ of the closures of the axioms $A_1, \ldots, A_k$ of number theory used in proving Theorem 1066, and also sufficing to establish that 7 is a prime, the foregoing proof can be formalized in the predicate calculus. First, $E \supset \forall x(Pr(x) \supset B(x))$ (the formula expressing Theorem 1066) is proved; then $E \supset Pr(7)$; and finally, using the axiom $\forall x(Pr(x) \supset B(x)) \supset (Pr(7) \supset B(7))$ by the $\forall$-schema, and steps in the propositional calculus, $E \supset B(7)$. This would not be a very direct proof of $E \supset B(7)$ in the predicate calculus.

Whether direct proofs are what we want in a given situation depends on our purposes.

If we have Theorem 1066 already proved anyway, it is much more efficient to *add* a short deduction of $E \supset B(7)$ than to construct a direct proof of $E \supset B(7)$ from first principles.

In developing mathematics, we usually strive to produce general theorems, which we store along the way for later use in particular applications or in establishing further general theorems. This is why the Hilbert-type systems lend themselves well to the formalization of mathematics as it is actually developed. When we are working in number theory, we would not carry the "$E \supset$" as in the above example, but would simply include $A_1, \ldots, A_k$ among the axioms of the system (like N of § 38).

Nevertheless, it was a major logical discovery by Gentzen 1934–5 that, when there is any (purely logical) proof of a proposition, there is a direct proof.[267] The immediate applications of this discovery are in theoretical logical investigations, rather than in building collections of proved formulas.[268] —

[267] Herbrand 1930 showed that any provable formula of the predicate calculus can be proved without using modus ponens in the part of the proof employing quantifiers (§ 55). Gentzen reached his "Hauptsatz" or "normal form theorem" by following his own direction of investigation, not using this result of Herbrand. Gentzen 1934–5 is crystal clear, while Herbrand 1930 has only recently been fully elucidated.[279]

[268] But Gentzen's work, and related work of others, has a practical side. We have seen this in connection with our format (B_2) end § 13. This can be obtained by taking a suitable Gentzen-type sequent system, and using the sequents without each time writing the formulas of the antecedent. Thus Gentzen 1934–5 obtained "natural deduction systems". Jaśkowski 1934 introduced such systems directly, at the suggestion of Łukasiewicz in seminars in 1926. Beth's semantic tableaux are closely related.[224] But we have preferred in our practice of logic to operate flexibly using derived rules, rather than to freeze our practice into a particular such system. Cf. § 25, Prawitz 1965.

Consider any (consecutive) part A of a given formula B not containing $\sim$ (or at least not containing $\sim$ with A in its scope), this part being a formula, an operator or a predicate parameter. As in Theorem 24 (for a formula part), we say A is a *positive* or *negative* part of B, according as A stands in the D of an even or odd number of parts of B of the forms $\neg D$ and $D \supset E$ (where D and E are formulas). For example, in $\exists x(P \supset Q(x)) \supset (P \supset \exists x Q(x))$, the first $\supset$ is negative, the other two positive; the first P is positive, the second negative; and the first Q(x) is negative, the second positive.

Now if we take B as one of the formulas Λ in a sequent $\Delta \rightarrow \Lambda$, the positive and negative parts of B are considered as *Positive* and *Negative* parts of the whole sequent $\Delta \rightarrow \Lambda$; but if B is one of the formulas Δ, we reverse these Signs. Equivalently, a part A of a sequent $\Delta \rightarrow \Lambda$ not containing $\sim$ is *Positive* or *Negative* according as it occurs in the dotted position relative to an even or odd number of the symbols $\neg \ldots$, $\ldots \supset$, $\ldots \rightarrow$. To emphasize the difference between these notions applied to A considered as a part of a formula B and considered as part of a sequent $\Delta \rightarrow \Lambda$ having B as a succedent or antecedent formula, we are capitalizing in the latter case. For example, in $\exists x(P \supset Q(x)) \rightarrow P \supset \exists x Q(x)$, the first P is negative (as a part of $\exists x(P \supset Q(x))$ but Positive (as a part of the whole sequent).

LEMMA 14. (Sign property.) *In a proof in* G4, G4a *or* G4b *of a sequent* $\Delta \rightarrow \Lambda$ *not containing* $\sim$, *each image of a Positive part of* $\Delta \rightarrow \Lambda$ *is Positive, and each image of a Negative part of* $\Delta \rightarrow \Lambda$ *is Negative.*

This may be observed in our examples, and verified as true in general by examining the 12 logical rules other than $\rightarrow\sim$ and $\sim\rightarrow$, and for *G*4a or *G*4b the 4 structural rules.

By Lemma 14, it is predetermined by the Sign of each formula part in a sequent $\Delta \rightarrow \Lambda$ not containing $\sim$ whether, as the rules are used upward from $\Delta \rightarrow \Lambda$, that part can have an image which is a succedent formula (the Sign is Positive) or an antecedent formula (the Sign is Negative).[269]

The two rules $\rightarrow\sim$ and $\sim\rightarrow$ are unlike the others in that, with $A \sim B$ as the principal formula on a given side of the arrow, we get two occurrences of A as side formula one on each side of the arrow, and likewise of B.

If we start with a logical problem stated using $\sim$, we can apply the foregoing notions of "Sign" etc. after replacing each part $A \sim B$ by

[269] The subformulas of a given formula without $\sim$ as a succedent or antecedent formula are accordingly classifiable as *succedent subformulas* (Positive) and *antecedent subformulas* (Negative). Cf. Kleene 1952 pp. 10–11, which the present discussion (including Lemma 14) follows in substance. The terms "ancestor", "descendant" and "image" are from that paper. "Subformula" should have been defined there as in IM or here (as noted by V. P. Orevkov); this is corrected in the 1967 edition.

(A ⊃ B) & (B ⊃ A), using *63a in Theorem 2. Equivalently, we can apply the foregoing discussion directly, by considering each formula part within the scope of exactly n occurrences of ∼ as having "multiplicity" 2^n; we imagine 2^n replicas of the same part superimposed, 2^{n-1} of them Positive and 2^{n-1} of them Negative.

EXERCISES. 54.1. Show that in *G*4a and *G*4b, without changing the class of provable sequents the axiom schema can be simplified to C → C, the two rules &→ and →∨ can be simplified by omitting either side formula (producing four rules), and the two rules ∀→ and →∃ can be simplified by omitting the principal formula in the premise. (From *G*4b, this gives a system *G*a differing from *G* of Kleene 1952 only in its having the two ∼-rules. In IM and Kleene 1952, ∼ is not a primitive symbol.)

54.2*. Show that, if ⊢ F in *H*, then there is a proof of F in which there occur only ⊃, ¬ and those of &, ∨, ∼, ∀, ∃ which occur in F, in the system *H* or (if ∀ but not & occurs in E) in the extension of *H* having the axiom schema ∀x(C ⊃ A(x)) ⊃ (C ⊃ ∀xA(x)) where C does not contain x free. (IM Theorem 49 p. 459.)

★§ 55. Permutability, Herbrand's theorem. Often logicians are concerned with the conditions under which certain kinds of formulas or sequents are provable.

Since proofs in the systems *G*4, *G*4a and *G*4b are "direct" (§ 54), it is easier to analyze their structure and thus to extract information from the supposition that a proof exists in these systems than in systems like *H*. So a logician may begin an investigation by using the Gentzen theorem to infer that, if a formula F is provable in *H* (or is valid), the sequent → F is provable in *G*4, *G*4a or *G*4b.

In a proof in *G*4, *G*4a or *G*4b, we say a given logical inference *belongs to* a given operator occurrence in the endsequent, if the principal operator of the inference is an ancestor of the operator occurrence. Using Lemma 13, a proof in *G*4 consists entirely of inferences belonging to the various operator occurrences in the endsequent, in some order. In *G*4a or *G*4b, there may also be structural inferences. By Lemma 14, without ∼ only a specific one of the 12 kinds of logical inferences other than →∼ and ∼→ can belong to a given operator occurrence in the endsequent, namely an inference by the succedent or antecedent rule for that operator according as the occurrence is Positive or Negative.[270]

For example, a proof in *G*4, *G*4a or *G*4b of the sequent

$$\rightarrow \exists_2 x(P \supset_1 Q(x)) \supset_5 (P \supset_4 \exists_3 x Q(x))$$

[270] The method of adapting the discussion to the presence of ∼ was indicated in the final paragraph of § 54.

(where we now number the operators for reference) consists of inferences

$$\exists_2\rightarrow,\ \supset_1\rightarrow,\ \rightarrow\supset_5,\ \rightarrow\supset_4,\ \rightarrow\exists_3$$

belonging to the respective operators (as indicated by the matching numbers), besides possibly structural inferences.

In general, there may be zero, one, or more than one inference belonging to a given operator occurrence in the endsequent.

To complete our discussion of Gentzen's "normal form" for proofs, it remains for us to consider the order in which the logical inferences belonging to the various operator occurrences are assembled (along with structural inferences).

We shall find that, in proofs in *G*4a or *G*4b with the extra freedom that the structural inferences give us, we can choose the order of the inferences to suit our purposes, within certain limitations (stated below). Thus by a rearrangement or "permutation" of the logical inferences in a given proof, we may be able to bring it to a form from which we can see more easily what is happening.

EXAMPLE 8. Suppose that the two inferences $\rightarrow\forall$ and $\rightarrow\vee$ shown below at the left occur as part of a proof in *G*4a or *G*4b. The principal formula of the upper inference is not a side formula of the lower one. This part can be replaced, without spoiling the whole as a proof in *G*4a or *G*4b, by the figure shown at the right with the $\rightarrow\vee$ over the $\rightarrow\forall$. Briefly, the $\rightarrow\vee$ can be lifted over the $\rightarrow\forall$, or "interchanged" with it.

$$\dfrac{\dfrac{\Gamma\rightarrow\Theta,\ P(b),\ Q}{\Gamma\rightarrow\Theta,\ \forall xP(x),\ Q}\rightarrow\forall \qquad \Gamma\rightarrow\Theta,\ \forall xP(x),\ R(a)}{\Gamma\rightarrow\Theta,\ \forall xP(x),\ Q\vee R(a).}\rightarrow\vee$$

$$\dfrac{\dfrac{\dfrac{\Gamma\rightarrow\Theta,\ P(b),\ Q \qquad \Gamma\rightarrow\Theta,\ \forall xP(x),\ R(a)}{\Gamma\rightarrow\Theta,\ P(b),\ \forall xP(x),\ Q\vee R(a)}\rightarrow\vee}{\Gamma\rightarrow\Theta,\ \forall xP(x),\ \forall xP(x),\ Q\vee R(a)}\rightarrow\forall}{\Gamma\rightarrow\Theta,\ \forall xP(x),\ Q\vee R(a).}\rightarrow C$$

In Example 8, we could not have simply interchanged the $\rightarrow\vee$ with the $\rightarrow\forall$, if the upper figure had had P(a) instead of P(b). For then the restriction on variables for the $\rightarrow\forall$ (§ 48) would not be met after the interchange.

Such difficulties will not occur in a *pure variable proof*, i.e. a proof in which no variable occurs both free and bound and in which the b of each $\rightarrow\forall$ or $\exists\rightarrow$ occurs only in sequents above its conclusion. (If the A(x) does not contain the x free, we can pick the b for the analysis toward meeting this condition.)

*In G*4, *G*4a *or G*4b, *a given proof of a sequent* $\Delta\rightarrow\Lambda$ *containing no variable both free and bound can be changed to a pure variable proof of* *the same sequent, simply by changing occurrences of variables in it to occurrences of other variables.* We establish this PURE VARIABLE LEMMA thus.

Consider a given proof of such a segment $\Delta \rightarrow \Lambda$. By Lemma 13 (the last part), every variable occurring bound anywhere in the proof occurs bound (and hence by the hypothesis, not free) in the endsequent $\Delta \rightarrow \Lambda$. Each of *these* variables that occurs free (elsewhere) in the proof can be changed in all its free occurrences to a different variable not previously occurring in the proof. Now, with each $\rightarrow\forall$- or $\exists\rightarrow$-inference in the *resulting* proof of $\Delta \rightarrow \Lambda$, we associate a family of sequents, as follows. If the A(x) of the inference does not contain the x free, the family is empty. Otherwise, the family consists of the premise of the inference and all other sequents containing the b free which can be reached from the premise by consecutive steps in the tree through such sequents. By the restriction on variables for the rules $\rightarrow\forall$ and $\exists\rightarrow$, such a family does not include the conclusion of the inference; and two such families with the same variable as the b do not overlap (or the restriction would be violated for the upper inference). Now to get a pure variable proof of $\Delta \rightarrow \Lambda$, we need only consider the families whose b occurs elsewhere in the proof, and change the b throughout each such family to a different variable not previously occurring in the proof.[271]

EXAMPLE 9. The following illustrates lifting a 1-premise inference over a 2-premise inference.

$$\dfrac{\dfrac{\Gamma \rightarrow \Theta, P(b), Q \qquad \Gamma \rightarrow \Theta, P(b), R(a)}{\Gamma \rightarrow \Theta, P(b), Q \vee R(a)}\rightarrow\vee}{\Gamma \rightarrow \Theta, \forall x P(x), Q \vee R(a)}\rightarrow\forall$$

$$\dfrac{\dfrac{\Gamma \rightarrow \Theta, P(b), Q}{\Gamma \rightarrow \Theta, \forall x P(x), Q}\rightarrow\forall \qquad \dfrac{\Gamma \rightarrow \Theta, P(b), R(a)}{\Gamma \rightarrow \Theta, \forall x P(x), R(a)}\rightarrow\forall}{\Gamma \rightarrow \Theta, \forall x P(x), Q \vee R(a).}\rightarrow\vee$$

If this interchange is performed in a pure variable proof, the result is not a pure variable proof. But it can be made one just by changing some variables, according to the pure variable lemma; indeed, we need only change all the b's over one of the two new $\rightarrow\forall$'s to a variable c not previously occurring in the proof.

EXAMPLE 10. Consider again the sequent

$$\rightarrow \exists_2 x(P \supset_1 Q(x)) \supset_5 (P \supset_4 \exists_3 x Q(x)).$$

[271] Our search procedure in §§ 49–50 will lead to a pure variable proof (when (II) holds), if it is modified in one respect. Whenever a branching (2-premise inference) occurs, we then split the part of the list $a_0, a_1, a_2, \ldots$ not yet activated into disjoint lists, one to be used with one premise and one with the other.

How much freedom do we have to choose the order of the inferences

$$\exists_2\!\rightarrow,\ \supset_1\!\rightarrow,\ \rightarrow\!\supset_5,\ \rightarrow\!\supset_4,\ \rightarrow\!\exists_3$$

in proving it? One of the permissible orders ($\rightarrow\exists_3$, $\supset_1\rightarrow$, $\exists_2\rightarrow$, $\rightarrow\supset_4$, $\rightarrow\supset_5$ reading down) is given in Example 2 § 48. Clearly, the inference(s) $\rightarrow\supset_5$ must come below all the others. For, the other operator occurrences in the endsequent are within the scope of $\supset_5$; so the inferences used to introduce those operators must come above in the process of building up the side formulas for $\rightarrow\supset_5$. For the same reason, each $\exists_2\rightarrow$ must come below each $\rightarrow\supset_1$; and $\rightarrow\supset_4$ must come below $\rightarrow\exists_3$. There is one other restriction. If in Example 2 we first lift $\supset_1\rightarrow$ over $\rightarrow\exists_3$ so that $\exists_2\rightarrow$ comes just under $\rightarrow\exists_3$, then we cannot next lift $\exists_2\rightarrow$ over $\rightarrow\exists_3$. For if we did, the restriction on variables for $\exists_2\rightarrow$ would be violated, because a_0 would be free in its conclusion.

Example 10 illustrates the only two obstacles to interchanging *adjacent* logical inferences, i.e. ones with at most structural inferences between. It can be verified that the following INTERCHANGE LEMMA holds: *When neither of these obstacles is present, two adjacent logical inferences can always be interchanged in a pure variable proof in* G4a *or* G4b, *preserving it as a pure variable proof of the same sequent.*[272]

Now we formulate what this allows us to do to a proof as a whole. It should be reasonably evident that we can get the following result, by a finite number of successive interchanges of adjacent logical inferences. (If persons of several families are in a line, the families can be brought together by a finite number of interchanges of adjacent people in the line.)

Let the operator occurrences in the endsequent $\Delta \rightarrow \Lambda$ *(without* $\sim$*) of a pure variable proof in* G4a *or* G4b *be put into q classes* $C_1, \ldots, C_q$ *from "highest" to "lowest" so that* (1) *each operator occurrence within the scope of another is in the same or a higher class than the latter, and* (2) *each operator occurrence of the kinds* $\forall\rightarrow$ *and* $\rightarrow\exists$ *is in the same or a higher class than any operator occurrence of the kinds* $\rightarrow\forall$ *or* $\exists\rightarrow$, *unless the latter occurrence is within the scope of the former. Then the proof can be rearranged (preserving it as a pure variable proof of* $\Delta \rightarrow \Lambda$*) so that along any path (zero or more) inferences (belonging to operator occurrences) of* C_1 *come highest, of* C_2 *next, ..., of* C_q *lowest.* This is the PERMUTABILITY THEOREM (Kleene 1952).

We apply this to a pure variable proof in *G*4a (obtained if necessary from a given proof by using the pure variable lemma) of a sequent $\Delta \rightarrow \Lambda$

[272] In Kleene 1952, the interchange lemma is established as Lemma 7 for a system *G* differing inessentially from *G*4b (cf. Exercise 54.1), using an exhaustive classification of the cases, and the permutability theorem (next) is proved thence as Theorem 2.

whose formulas are all prenex (§ 25). Let C_1 be all the propositional operators and C_2 all the predicate operators in $\Delta \rightarrow \Lambda$. Because the formulas of $\Delta \rightarrow \Lambda$ are prenex, (1) is satisfied. Because all predicate operators are put together, (2) is satisfied. So the proof can be rearranged to put all the propositional inferences above all the predicate inferences.[270] Reading up, no branching can occur until propositional inferences are reached. Now consider the sequent which is the premise of the highest predicate inference. It may contain formulas with quantifiers (necessarily at the front, since its formulas are subformulas of prenex formulas). But no ancestor of one of these formulas plays the role of a principal formula higher up (since all predicate inferences are below). Hence, since in *G*4a the C's for axioms are prime, no ancestor of one of these formulas plays the role of the C in an axiom.[273] So if all the quantified formulas are simply omitted from the sequent in question on up, and any resulting identical inferences (with premise and conclusion the same) are suppressed, we get a proof of the resulting sequent. From it by thinnings we can continue to the sequent in question. Thus:

*Given any proof in G*4a *of a sequent* $\Delta \rightarrow \Lambda$ *containing only prenex formulas, and containing no variable both free and bound, a pure variable proof in G*4a *of* $\Delta \rightarrow \Lambda$ *can be found in which there is a sequent* (called the *midsequent*) *containing no quantifier, with only propositional and structural inferences above it, and only predicate and structural inferences below it.*[274] This is GENTZEN'S SHARPENED (OR EXTENDED) HAUPTSATZ 1934–5.[275]

The significance of this result appears when we reflect on the rules by which the part of the proof from the midsequent down is conducted.

[273] For the approach via Gentzen's Hauptsatz,[260] it is necessary to proceed from *G*4b toward *G*4a only far enough to obtain quantifier-free axioms. (Cf. IM p. 461.)

[274] It is not hard to see that thinnings $\rightarrow T$ and $T \rightarrow$ can be avoided from the midsequent down (Exercise 55.3; solution in Kleene 1952 Lemma 4).

[275] Gentzen's "verschärfter Hauptsatz" (but proved entirely metamathematically) is a convenient starting point for consistency proofs for various systems weaker than N. Cf. § 54 Paragraph 4, and IM § 79.

Gentzen's sharpened Hauptsatz requires only the part of the permutability theorem by which propositional inferences can be moved up through predicate inferences whose principal formulas are not side formulas for the propositional inferences.

However, it has required only a little more space to formulate the more general permutability theorem (though of course more cases have to be checked in proving it). Gentzen 1934–5 p. 412 hinted at such a permutability theorem, without exploring the possibilities. Also cf. Curry 1952.

The classical and intuitionistic versions of the permutability theorem were used by Kleene 1952a in a rather complex consistency argument. (Lemma 9 is the proposition of Footnote 107 § 29, established metamathematically for both the classical and the intuitionistic systems.) This exercise convinced the author that Gentzen-type systems in essentially Gentzen's own form are a very effective tool for such purposes.

To give an illustration, suppose the formation rules in use provide (besides predicate parameters or ions, and variables) just one individual symbol $\varkappa$ and one 1-place function symbol λ.[276] Suppose A(w, x, y, z) is a formula containing no quantifiers. Suppose that $\vdash \exists w \forall x \exists y \forall z A(w, x, y, z)$ in *H*, and that after applying Gentzen's theorem and his sharpened Hauptsatz we have a pure variable proof in *G*4a of $\to \exists w \forall x \exists y \forall z A(w, x, y, z)$ in which the part from the midsequent down is as follows, with the variables a, b, c, d, e, f, w, x, y, z distinct from each other and from all other variables (if any) in A(w, x, y, z).

$$\begin{array}{cl}
\to A(\lambda(b), d, \varkappa, e), A(a, b, \lambda(c), f), A(a, b, b, c) & \\ \cline{1-1}
\to A(\lambda(b), d, \varkappa, e), \forall z A(a, b, \lambda(c), z), A(a, b, b, c) & \to\forall \\ \cline{1-1}
\to A(\lambda(b), d, \varkappa, e), \exists y \forall z A(a, b, y, z), A(a, b, b, c) & \to\exists \\ \cline{1-1}
\to \forall z A(\lambda(b), d, \varkappa, z), \exists y \forall z A(a, b, y, z), A(a, b, b, c) & \to\forall \\ \cline{1-1}
\to \exists y \forall z A(\lambda(b), d, y, z), \exists y \forall z A(a, b, y, z), A(a, b, b, c) & \to\exists \\ \cline{1-1}
\to \forall x \exists y \forall z A(\lambda(b), x, y, z), \exists y \forall z A(a, b, y, z), A(a, b, b, c) & \to\forall \\ \cline{1-1}
\to \forall x \exists y \forall z A(\lambda(b), x, y, z), \exists y \forall z A(a, b, y, z), \forall z A(a, b, b, z) & \to\forall \\ \cline{1-1}
\to \forall x \exists y \forall z A(\lambda(b), x, y, z), \exists y \forall z A(a, b, y, z) & \to\exists \\ \cline{1-1}
\to \exists w \forall x \exists y \forall z A(w, x, y, z), \exists y \forall z A(a, b, y, z) & \to\exists \\ \cline{1-1}
\to \exists w \forall x \exists y \forall z A(w, x, y, z), \forall x \exists y \forall z A(a, x, y, z) & \to\forall \\ \cline{1-1}
\to \exists w \forall x \exists y \forall z A(w, x, y, z). & \to\exists
\end{array}$$

Thus the midsequent is

(i) $\qquad \to A(\lambda(b), d, \varkappa, e), A(a, b, \lambda(c), f), A(a, b, b, c).$

Certain structural details which are significant here do not stand out clearly. We shall now see how (i) can be altered to a more revealing form (though it will no longer be a midsequent), by introducing symbols for functions called "Herbrand functions" or "Skolem functions".[277]

We begin by augmenting the symbolism of our Gentzen-type system *G*4a by two new function symbols $\beta(-)$ and $\delta(-, -)$. We consider $\beta(-)$ and $\delta(-, -)$ as corresponding to the universal quantifiers $\forall x$ and $\forall z$ in the formula of the endsequent, noting that it is a succedent formula. (For an antecedent formula, we would introduce function symbols corresponding to the existential quantifiers.)

[276] Since we are forced here to use such a large part of the Roman lower case alphabet "a", "b", "c", . . ., "x", "y", "z" as names for variables, we are now using lower case Greek letters "α", "β", "γ", . . . as names for function symbols (instead of "f", "g", "h", . . . as in § 28 etc.).

[277] Such functions were used by Herbrand 1930, and earlier by Skolem 1920, 1922–3, 1929 and implicitly by Löwenheim 1915. Cf. Footnote 252 Paragraph 3.

Now we take the fragment of a pure variable proof shown from endsequent up to midsequent, and perform on it a series of 5 substitutions, one to each $\rightarrow\forall$-inference. (More generally, we would do the like with all $\rightarrow\forall$- and $\exists\rightarrow$-inferences.) We begin with the lowest $\rightarrow\forall$, in which the variable b (not occurring below its premise) is *introduced* in the side formula $\exists y\forall z A(a, b, y, z)$ as we read the proof-fragment *upward*. Throughout the fragment, for all the 31 occurrences of b we substitute the term $\beta(a)$. Of course, this spoils the fragment as a part of a proof in *G*4a (actually, it spoils only the $\rightarrow\forall$ under consideration); but we don't care. We are using the fragment, and successively altering it, as a way of discovering a new sequent (ii), having a more revealing form than the midsequent (i). Now we take the next $\rightarrow\forall$ (reading up) in the already once-altered fragment; its altered side formula is $A(a, \beta(a), \beta(a), c)$. For all the 8 occurrences of c (which are only from there up) we substitute $\delta(a, \beta(a))$. We treat the remaining three $\rightarrow\forall$'s (in the now twice-altered fragment) in succession, similarly. That is, when the side formula of an $\rightarrow\forall$ has become $\exists y\forall z A(r, v, y, z)$ for some term r and variable v, we substitute for all v's the term $\beta(r)$; when it has become $A(r, s, t, v)$ for some terms r, s, t and variable v, we substitute for all v's the term $\delta(r, t)$. When the succession of 5 substitutions has been completed, the midsequent has become

$$\text{(ii)} \qquad \begin{aligned} \rightarrow\ & A(\lambda(\beta(a)), \beta(\lambda(\beta(a))), \varkappa, \delta(\lambda(\beta(a)), \varkappa)), \\ & A(a, \beta(a), \lambda(\delta(a, \beta(a))), \delta(a, \lambda(\delta(a, \beta(a))))), \\ & A(a, \beta(a), \beta(a), \delta(a, \beta(a))). \end{aligned}$$

The structure exhibited in (ii) provides a "history" of the steps in *G*4a by which we proceeded upward from the endsequent $\rightarrow \exists w\forall x\exists y\forall z A(w, x, y, z)$ to the midsequent (i). From the history exhibited in (ii), we can rediscover (i) and the steps from (i) down to $\rightarrow \exists w\forall x\exists y\forall z A(w, x, y, z)$, apart from inessential details (Exercise 55.1). From any similar history, we can do the like. This gives a preview of what we will do presently.

Meanwhile, let us examine the result we thus far have.

Imagine that at the beginning of the substitutions we had before us, not just the fragment from the midsequent (i) down, but a whole proof in *G*4a of $\rightarrow \exists w\forall x\exists y\forall z A(w, x, y, z)$ as described in Gentzen's sharpened Hauptsatz. The part from the midsequent (i) up consists of propositional and structural inferences and axioms. Suppose now that the substitutions of $\beta(a)$ for b, of $\delta(a, \beta(a))$ for c, . . . were carried out on the whole proof, including the part above the midsequent. Those substitutions do not invalidate the applications of the 10 propositional rules or the 4 structural rules of *G*4a, or of the axiom schema ($\times$). So, whereas originally the part above the midsequent was a proof of (i) in the system of *propositional calculus* $G4$a (i.e. the predicate calculus $G4$a minus the 4 predicate rules), after the substitutions it has become a proof of (ii) in the propositional calculus $G4$a (but with the symbolism augmented by β, δ). So by Theorem 33 the sequent (ii) is valid, and hence by Exercise 50.1 (b) the formula

$$A(\lambda(\beta(a)), \beta(\lambda(\beta(a))), \varkappa, \delta(\lambda(\beta(a)), \varkappa)) \vee$$
$$A(a, \beta(a), \lambda(\delta(a, \beta(a))), \delta(a, \lambda(\delta(a, \beta(a))))) \vee$$
$$A(a, \beta(a), \beta(a), \delta(a, \beta(a)))$$

is valid, i.e. is a tautology under the truth tables of the propositional calculus § 2 (or equivalently by §§ 11, 12 is provable in it § 9).[278]

The "only if" part of the following proposition is obtained by recognizing that for any proof of $\rightarrow \exists w\forall x\exists y\forall z A(w, x, y, z)$ as described in Gentzen's sharpened Hauptsatz, the result of our substitution method on the part from the midsequent down must lead to a valid disjunction of the form exhibited; i.e. our (i) and (ii) are "typical".

A prenex formula F *of the form* $\exists w\forall x\exists y\forall z A(w, x, y, z)$ (*with all its quantifiers shown*) *is provable in the predicate calculus H if and only if some disjunction of the form*

$$A(t_{11}, \beta(t_{11}), t_{12}, \delta(t_{11}, t_{12})) \vee \ldots \vee A(t_{l1}, \beta(t_{l1}), t_{l2}, \delta(t_{l1}, t_{l2}))$$

(*a Herbrand disjunction*) *is valid* (*or equivalently, is provable*) *in the propositional calculus H. Here* $t_{11}, \ldots, t_{l2}$ *are terms constructed from variables, the* (*individual,*) *function and predicate symbols of* F, *and the additional* 1-*place function symbol* β *and* 2-*place function symbol* δ. This is a version of HERBRAND'S THEOREM° 1930, illustrated by the case of $\exists w\forall x\exists y\forall z A(w, x, y, z)$. More precisely, it is one of the propositions which occur in the literature as partial versions of or as included in the "fundamental theorem" of Herbrand's thesis 1930. Another such version coincides with Gentzen's sharpened Hauptsatz (with Footnote 274) as applied to $\rightarrow$ F. There is still more to the theorem, however.[279]

[278] Alternatively, we can observe that Theorem 36 and Corollary hold for the propositional calculi $G4$a and H; i.e. their proofs require the postulates of the predicate calculus in H only when the given deduction in $G4$a uses the predicate rules $\rightarrow\forall, \ldots, \exists\rightarrow$. Applying Corollary (c) to (i), $A(\lambda(b), d, \varkappa, e) \vee A(a, b, \lambda(c), f) \vee A(a, b, b, c)$ is provable in H, and hence by Theorem 12 valid. Finally, we can observe that the substitutions of $\beta(a)$ for b, of $\delta(a, \beta(a))$ for c, ... simply effect a substitution for the atoms in the sense of Theorem 1 § 3.

[279] Another version, in Hilbert and Bernays 1939 pp. 163–178, follows from the present treatment without adding any major ideas. These several versions all apply to a prenex formula, while Herbrand stated and attempted to prove his theorem for an arbitrary formula.

Herbrand died in 1931, aged 23. Gentzen 1934–5 and Hilbert and Bernays 1939 gave clear treatments of the versions mentioned, proceeding from two other directions than

We still have to prove our "if" part.[280] So assume that some Herbrand disjunction as displayed is valid. We must show that then ∃w∀x∃y∀zA(w, x, y, z) is provable in *H*. There is no loss of generality in supposing the *l* disjunctands to be distinct, since if they were not we could suppress the duplications and still have a tautology. Likewise, we can assume all variables in the Herbrand disjunction to be distinct from w, x, y, z; for otherwise, using Theorem 1 § 3 we could make some changes in them to secure this.

Consider the distinct terms of the forms β(s) (s a term) and δ(s, t) (s, t terms) occurring in the Herbrand disjunction (or if w, x, y, z do not all occur in A(w, x, y, z), occurring in the terms shown in the expression for the Herbrand disjunction displayed above). List them as

$$t_0, \ldots, t_p$$

in such an order that *those terms occurring as parts of* s (*including possibly* s *itself*) *precede* β(s), *and* β(s) *and those terms occurring as parts of* t (*including possibly* t *itself*) *precede* δ(s, t). We can certainly list them in such an order. For, we could enumerate all the terms constructible using the variables, individual and function symbols of the Herbrand disjunction (including β, δ) by the method of digits with an alphabet in which β precedes δ ((A) § 32). The $p+1$ terms $t_0, \ldots, t_p$ under consideration would occur in that enumeration with the required kind of order.

Next, to the terms $t_0, \ldots, t_p$ we correlate respective variables

$$a_0, \ldots, a_p$$

Herbrand. These are the present writer's sources. Hilbert and Bernays (1939 p. 158) wrote, "Herbrand's argumentation (Beweisführung) is difficult to follow". In 1963 Dreben, Andrews and Aanderaa established that two of Herbrand's lemmas are false. It is only in the translation and recension by Dreben and van Heijenoort in van Heijenoort 1967 that Herbrand's text has become accessible without undue effort. Denton and Dreben 1968 work out the main ideas of Herbrand's thesis, obtaining among other things a connection between modus ponens (or cut) elimination and the consistency of number theory which is absent in Gentzen's consistency proofs 1936, 1938a.

[280] The proof of Gödel's completeness theorem in Kleene 1958 leads directly to (the present part of) Herbrand's theorem as the basic discovery (instead of the completeness of the Gentzen-type system *G4*). This it does by refining the easy axiom-of-choice proof of the Löwenheim-Skolem theorem (Footnote 252 Paragraph 3) to pick truth values of atoms toward building a satisfying assignment if there is one (Footnote 252 Paragraph 5). In this it is very similar to the proofs of the basic discovery in Skolem 1922–3 and 1929 (of which Kleene was unaware in 1958). The rest of the proof (in Kleene 1961) is the present proof of the "if" part of Herbrand's theorem.

The proof in IM pp. 389–393 is quite similar, though it is presented rather differently. It was written immediately after the appearance of Henkin 1949, adapting an idea from that to the proof of the first lemma.

distinct from each other, from all the variables in the Herbrand disjunction, and from w, x, y, z.

For any term r in the Herbrand disjunction, let $\bar{r}$ come from r by replacing each maximal (consecutive) part of r of the form $\beta(s)$ or $\delta(s, t)$ (s, t terms) by the correlated variable. A maximal such part is one not constituting a proper part of another such part. By the theory of proper pairing cited in § 38, it can be seen that any two distinct maximal such parts are nonoverlapping.

The Herbrand disjunction (assumed to be valid) will remain so when each of its atoms $P(r_1, \ldots, r_n)$ is replaced by $P(\bar{r}_1, \ldots, \bar{r}_n)$. For, like atoms will receive a like alteration, which makes the operation a substitution in the sense of Theorem 1 § 3. So $\models$ D in the propositional calculus *H*, where D is the resulting disjunction.

But β and δ are function symbols not in A(w, x, y, z). Hence the operation of replacing maximal parts $\beta(s)$ and $\delta(s, t)$ by the respective variables takes place entirely within the shown terms t_{i1}, $\beta(t_{i1})$, t_{i2}, $\delta(t_{i1}, t_{i2})$ ($i = 1, \ldots, l$) which are substituted for w, x, y, z; i.e. none of the shown terms is part of a larger term in a disjunctand $A(t_{i1}, \beta(t_{i1}), t_{i2}, \delta(t_{i1}, t_{i2}))$ which larger term gets replaced in toto. So the result of the replacement operation on the *i*th disjunctand can be written $A(\overline{t_{i1}}, \overline{\beta(t_{i1})}, \overline{t_{i2}}, \overline{\delta(t_{i1}, t_{i2})})$.

Now *either* by the completeness of the propositional calculus *H* (Theorem 14 § 12) and the Gentzen theorem (§ 54) and $\rightarrow\vee$ used upward $l-1$ times (as the permutation theorem allows), *or* directly by reasoning in §§ 48 and 49, the sequent shown at the top of the next figure (written with $l = 2$ to simplify the notation from this point on) is provable in *G*4a.

$$\begin{array}{cl}
\rightarrow A(\overline{t_{11}}, \overline{\beta(t_{11})}, \overline{t_{12}}, \overline{\delta(t_{11}, t_{12})}), A(\overline{t_{21}}, \overline{\beta(t_{21})}, \overline{t_{22}}, \overline{\delta(t_{21}, t_{22})}) & \\ \hline
\rightarrow \forall z A(\overline{t_{11}}, \overline{\beta(t_{11})}, \overline{t_{12}}, z), A(\overline{t_{21}}, \overline{\beta(t_{21})}, \overline{t_{22}}, \overline{\delta(t_{21}, t_{22})}) & \rightarrow\forall \\ \hline
\rightarrow \exists y \forall z A(\overline{t_{11}}, \overline{\beta(t_{11})}, y, z), A(\overline{t_{21}}, \overline{\beta(t_{21})}, \overline{t_{22}}, \overline{\delta(t_{21}, t_{22})}) & \rightarrow\exists \\ \hline
\rightarrow \exists y \forall z A(\overline{t_{11}}, \overline{\beta(t_{11})}, y, z), \forall z A(\overline{t_{21}}, \overline{\beta(t_{21})}, \overline{t_{22}}, z) & \rightarrow\forall \\ \hline
\rightarrow \exists y \forall z A(\overline{t_{11}}, \overline{\beta(t_{11})}, y, z) & \rightarrow\exists \\ \hline
\rightarrow \forall x \exists y \forall z A(\overline{t_{11}}, x, y, z) & \rightarrow\forall \\ \hline
\rightarrow \exists w \forall x \exists y \forall z A(w, x, y, z). & \rightarrow\exists
\end{array}$$

By our provision that the disjunctands in the Herbrand disjunction be distinct, the 2 formulas in the top sequent shown are distinct; so (t_{11}, t_{12}), (t_{21}, t_{22}) are different pairs of terms.

By the italicized property of the list of terms $t_0, \ldots, t_p$, the term $\delta(t_{i1}, t_{i2})$ comes later in the list than any part of t_{i1}, $\beta(t_{i1})$, t_{i2} ($i = 1, 2$).

So the "rightmost (free) variable" $\overline{\delta(t_{i1}, t_{i2})}$ in the *i*th formula of the top sequent (it is a variable, because $\delta(t_{i1}, t_{i2})$ is a maximal part $\delta(s, t)$) comes later in the list $a_0, \ldots, a_p$ than any other variable in that formula. So among the rightmost variables $\overline{\delta(t_{11}, t_{12})}$, $\overline{\delta(t_{21}, t_{22})}$ (which are distinct since (t_{11}, t_{12}), (t_{21}, t_{22}) are), one comes furthest out in the list $a_0, \ldots, a_p$ (indeed it must be a_p); to fix the notation, say this one is $\overline{\delta(t_{11}, t_{12})}$. All variables in $a_0, \ldots, a_p$ are distinct from the other variables in the sequent (i.e. the variables if any which were in the Herbrand disjunction and were not removed in the replacement of maximal parts $\beta(s)$ and $\delta(s, t)$), and from w, x, y, z (which do not occur in the sequent). Thus $\overline{\delta(t_{11}, t_{12})}$ occurs in the top sequent only as shown, and w, x, y, z do not occur. Hence the requirements are met for an application of →∀, with the sequent shown second (reading down) as conclusion. In this sequent, w, x, y, z occur only as shown, and thus no variables are bound in the parts $\overline{t_{11}}$, $\overline{\beta(t_{11})}$, $\overline{t_{12}}$, $\overline{t_{21}}$, $\overline{\beta(t_{21})}$, $\overline{t_{22}}$, $\overline{\delta(t_{21}, t_{22})}$. Using this fact (with respect to $\overline{t_{12}}$), we can apply →∃ in *G*4a (where the principal formula need not appear in the premise) to $\overline{t_{12}}$ (which we consider as "uncovered" by the conversion of $\overline{\delta(t_{11}, t_{12})}$ to z) to obtain the third sequent (reading down).

By reasoning as before, $\overline{\beta(t_{11})}$, $\overline{\delta(t_{21}, t_{22})}$ are variables further out in the list $a_0, \ldots, a_p$ than other variables in their respective formulas in the third sequent; and so one of them is furthest out of all (indeed it is a_{p-1}), and occurs only as shown (as do w, x, y, z). Say e.g. this one is $\overline{\delta(t_{21}, t_{22})}$. Then by →∀ we can infer the fourth sequent, with w, x, y, z occurring in it only as shown. While we had that (t_{11}, t_{12}), (t_{21}, t_{22}) are distinct pairs of terms, t_{11} and t_{21} are not necessarily distinct terms. Say e.g. they are not distinct. Then in the →∃ which we can now perform since $\overline{t_{22}}$ is uncovered, the first formula of the premise is the principal formula, so we get the fifth sequent as shown (or we could first use →∃ without regard to t_{11} and t_{21} being the same, and then →*C*).

Now an →∀ and →∃, each clearly legal, lead to the bottom sequent.

Thus this sequent is provable in *G*4a. By the Gentzen theorem (or specifically Theorem 36 Corollary (a)), ∃w∀x∃y∀zA(w, x, y, z) is provable in the predicate calculus *H*, as we aimed to show.

We used $l = 2$, and an illustrative series of assumptions. It should be clear that the process can always be carried out. Another illustration is provided by starting from (ii) with bars placed over the terms (Exercise 55.1).

We may summarize Herbrand's theorem by saying that it reduces the question of the provability of a particular formula with quantifiers (in the first instance, a prenex formula) to the question of the validity (or provability) in the propositional calculus of some one of a countably infinite class of quantifier-free formulas (the Herbrand disjunctions).

Among the applications of Herbrand's theorem are those in Hilbert and Bernays 1939 pp. 178 ff., in Kreisel 1958, and in Denton and Dreben 1967.

EXERCISES. 55.1. Apply the method of the "if" part of the proof of Herbrand's theorem to (ii), as follows. Take for $t_0, \ldots, t_p$ the list $\beta(a)$, $\delta(a, \beta(a))$, $\beta(\lambda(\beta(a)))$, $\delta(\lambda(\beta(a)), \varkappa)$, $\delta(a, \lambda(\delta(a, \beta(a))))$ (verify that it meets the requirements). Find steps by the method, leading downward in *G*4a to $\rightarrow \exists w \forall x \exists y \forall z A(w, x, y, z)$. This can be done without specifying the variables $a_0, \ldots, a_p$. Now take $a_0, \ldots, a_p$ to be b, c, d, e, f; and compare your result with the steps originally shown as leading up from $\rightarrow \exists w \forall x \exists y \forall z A(w, x, y, z)$ to (i).

55.2. Formulate a condition for each of the following, similar to that stated for $\vdash_H \exists w \forall x \exists y \forall z A(w, x, y, z)$ in our illustration of Herbrand's theorem.

(a) $\vdash_H \neg \exists w \forall x \exists y \forall z A(w, x, y, z)$ ($\equiv \vdash_{G4a} \exists w \forall x \exists y \forall z A(w, x, y, z) \rightarrow$).

(b) $\exists w \forall x \exists y \forall z A(w, x, y, z)$ is satisfiable.

(c) $\vdash_H \forall v B(v) \,\&\, \exists r \forall s \forall t \exists u C(r, s, t, u) \supset \exists w \forall x \exists y \forall z A(w, x, y, z)$.

55.3*. Prove the proposition of Footnote 274.

§ 56. Craig's interpolation theorem. In this section, we utilize the relationships which are present in proofs in *G*4 or *G*4a to establish (as Theorem 41) Craig's interpolation theorem or Craig's lemma 1957, 1957a, including a version derived from Lyndon 1959. Until Theorem 42, we shall deal with the predicate calculus without equality (without or with functions).[281] The main part of our work is put into proving Theorem 40.

The idea is this. Given a proof in *G*4 or *G*4a of E → F, we can split it vertically into two parts: the *E-part* we obtain by omitting in each sequent all the ancestors of the F in the endsequent E → F; the F-*part*, by omitting all the ancestors of the E. Of course, the E-part and the F-part won't in general be proofs. But now we ask whether we cannot repair each part, or at least one of them, to make it a proof, utilizing at least some of its essential anatomy. We aim to make the repair by restoring a minimum of what was cut away from the part in the splitting operation that cannot be spared for a proof, and by removing what is superfluous and in the way in the absence of the other part.

Clues to what we can do toward restoring the E-part or the F-part to become an E-*proof* or an F-*proof* will be provided by considering what happens to the axioms of the given proof in the splitting operation. We call an axiom C, Γ → Θ, C in the given proof of E → F an EF-*axiom* if

[281] The symbol = may be among our predicate symbols, but it is not to have a special status (as it did in § 29); i.e. it is to be counted as a predicate parameter, and no axioms are postulated for it (until Theorem 42).

one of its C's is an ancestor of E and the other of F; an E-*axiom* if both are ancestors of E; an F-*axiom* if both are ancestors of F. We have been assuming throughout our theory of the relationships within proofs in *G*4 or *G*4a that each proof is given with an analysis which determines the role of each formula occurrence in each step.[229]

First, consider an EF-axiom C, $\Gamma \rightarrow \Theta$, C, e.g. with the first C belonging to E and the second to F. Under the splitting operation, the E-part receives from this axiom C, $\Gamma_E \rightarrow \Theta_E$, where Γ_E are the ancestors of E among the formula occurrences Γ and Θ_E those among Θ. Similarly, the F-part receives $\Gamma_F \rightarrow \Theta_F$, C. Toward repairing the E-part to become a proof, we evidently should put the removed C back on the right (underlined) to obtain C, $\Gamma_E \rightarrow \Theta_E$, $\underline{C}$, and toward repairing the F-part on the left to obtain $\underline{C}$, $\Gamma_F \rightarrow \Theta_F$, C. This of course is just a first step toward repairing the two parts. But, when the original proof has only EF-axioms, the tops of all the branches in each part can be repaired in this manner. Then, as we shall see in proving Theorem 40, we can continue the restoration down through the two parts, performing a minimum of corresponding steps to combine the restored C's from the various EF-axioms into a single formula I (the *interpolated formula*). In this manner, we can make the E-part into an E-proof of E $\rightarrow$ I and the F-part into an F-proof of I $\rightarrow$ F. Here the I appears on opposite sides of the arrow in the endsequents of the E-proof and the F-proof, as did the restored C's at the tops of branches. It is convenient to construct the restored proofs in *G*4a (§ 54), even if the original proof is in *G*4.

Now consider an E-axiom C, $\Gamma \rightarrow \Theta$, C. In the splitting operation, the E-part receives from this axiom the sequent C, $\Gamma_E \rightarrow \Theta_E$, C and the F-part the sequent $\Gamma_F \rightarrow \Theta_F$. In the E-part, C, $\Gamma_E \rightarrow \Theta_E$, C is still an axiom. We don't need to make any restoration; both the C's, which were the vital parts at this position in the original proof, have been given to it. In the F-part, $\Gamma_F \rightarrow \Theta_F$ is not an axiom, at least not by the analysis used in the given proof of E $\rightarrow$ F. So (disregarding the possibility of a change in the analysis for the formulas Γ_F, Θ_F), we could make an axiom of $\Gamma_F \rightarrow \Theta_F$ only by bringing in two occurrences of some formula D to obtain D, $\Gamma_F \rightarrow \Theta_F$, D. In making an axiom this way, the sequent $\Gamma_F \rightarrow \Theta_F$ from the F-part is not contributing anything essential. We might as well have discarded it, and this is what we shall do. If there are only E-axioms in the given proof of E $\rightarrow$ F, then starting in the manner described at the tops of branches, and proceeding downward through the E-part removing some unnecessary sequents, we shall obtain an E-proof of E $\rightarrow$, while discarding the F-part altogether.

Similarly, if there are only F-axioms, we shall get an F-proof of $\rightarrow$ F by repairing the F-part, while abandoning the E-part.

If the axioms of the given proof of $E \rightarrow F$ include a mixture of two or three of the three kinds EF-axioms, E-axioms and F-axioms, then we shall get one of the results described for a pure set of axioms of a kind which is represented.[282]

In repairing the E-part and the F-part, one or both, we do the work step by step, corresponding to the sequents reading downward in the given proof in *G*4 or *G*4a of $E \rightarrow F$. So we generalize the conclusion we are aiming to establish to get a proposition applicable to each sequent $\Delta \rightarrow \Lambda$ in the given proof. Theorem 40 asserts that each sequent has this property.[283]

THEOREM 40.[284] *Suppose given a proof in G4 or G4a of* $E \rightarrow F$. *For each sequent* $\Delta \rightarrow \Lambda$ *in this proof, let* Δ_E, Λ_E *be those of* Δ, Λ *respectively which are ancestors of the E in the endsequent* $E \rightarrow F$; Δ_F, Λ_F *those which are ancestors of the* F. *For each sequent* $\Delta \rightarrow \Lambda$ *in the given proof*:

Either (CASE (EF)) *in the given proof there are* EF-*axioms over* $\Delta \rightarrow \Lambda$, *and there are a formula* J (*without* $\sim$) *and proofs in G*4a *of both* $\Delta_E \rightarrow \Lambda_E$, J *and* J, $\Delta_F \rightarrow \Lambda_F$ *such that*

(1) *the individual parameters of* J *are parameters of both* $\Delta_E \rightarrow \Lambda_E$ *and* $\Delta_F \rightarrow \Lambda_F$, *and*

(2) *in the proof of* $\Delta_E \rightarrow \Lambda_E$, J, *each atomic part of* J *is an image of one* C *of an axiom, the other* C *of which descends into an image in one of* Δ_E, Λ_E; *and similarly in the proof of* J, $\Delta_F \rightarrow \Lambda_F$.

Or (CASE (E)) *in the given proof there are* E-*axioms over* $\Delta \rightarrow \Lambda$, *and there is a proof in G*4a *of* $\Delta_E \rightarrow \Lambda_E$.

Or (CASE (F)) *in the given proof there are* F-*axioms over* $\Delta \rightarrow \Lambda$, *and there is a proof in G*4a *of* $\Delta_F \rightarrow \Lambda_F$.

PROOF. Starting at the tops of branches in the given proof of $E \rightarrow F$, and working down step by step in the tree, we can establish for each

[282] By making, for each E-axiom, an "unnecessary" restoration in the E-part and a restoration in the F-part to which its contribution is unnecessary, and similarly for each F-axiom, we can come out with an E-proof of $E \rightarrow I$ and an F-proof of $I \rightarrow F$ in all cases. (This amounts to using the treatment of Theorem 41 (a) Cases (E) and (F) in *G*4b at the stage of the axioms.) This procedure makes for unnecessarily complicated I's, so we prefer the procedure described in the text.

[283] By leaving out the part of the given proof from a certain sequent $\Delta \rightarrow \Lambda$ down to $E \rightarrow F$, we have a theorem about a proof in *G*4 or *G*4a of any sequent $\Delta \rightarrow \Lambda$ whose formula occurrences Δ, Λ are divided into two lists Δ_E, Λ_E and Δ_F, Λ_F.

[284] Schütte 1962 establishes the interpolation theorem (Theorem 41) for the intuitionistic predicate calculus, using a formal system without sequents resembling a Gentzen-type subformula system.

To adapt the present treatment to the intuitionistic predicate calculus, we can use Theorem 40 with the intuitionistic version of *G*3 or *G*, modifying Case (EF) to say that there are proofs *either* of both $\Delta_E \rightarrow \Lambda_E$, J and J, $\Delta_F \rightarrow \Lambda_F$, *or* of both J, $\Delta_E \rightarrow \Lambda_E$ and $\Delta_F \rightarrow \Lambda_F$, J.

sequent $\Delta \rightarrow \Lambda$ in its turn that it has the property ascribed to it by the theorem. We now describe what to do in each of the cases that can arise in the steps. The use of these instructions is illustrated in Example 11 below.

If the given proof is in *G*4a rather than *G*4, then for definiteness in our notation we shall suppose that the logical inferences in it are written as in *G*4, and all thinnings $\rightarrow T$, $T\rightarrow$ (as well as contractions $\rightarrow C$, $C\rightarrow$) are shown separately. But in writing the E-proof or the F-proof, thinnings may be assimilated into succeeding logical inferences.[285]

First, we show in the next table (under Cases 1a–2b) how to treat (what remains from) an axiom $C, \Gamma \rightarrow \Theta, C$ in the E-part (left column) and in the F-part (right column) after the splitting operation. Absence of an entry in one column indicates that in that part we discard the sequent resulting by the split (Cases 1a, 1b). Formulas added are underlined here for emphasis (but in later cases the J-notation will accomplish the same purpose, and in Example 11 the use of light- and boldface type). In Case 2a, the J for the theorem is C, and in Case 2b it is $\neg C$. Clearly (1) and (2) hold as required for Case (EF) of the theorem.

E-proof.	Axioms.	F-proof.
CASE 1a.	E-axiom.	
$C, \Gamma_E \rightarrow \Theta_E, C.$		
CASE 1b.	F-axiom.	
		$C, \Gamma_F \rightarrow \Theta_F, C.$
CASE 2a.	EF-axiom, first C in the E-part.	
$C, \Gamma_E \rightarrow \Theta_E, \underline{C}.$		$\underline{C}, \Gamma_F \rightarrow \Theta_F, C.$
CASE 2b.	EF-axiom, first C in the F-part.	
$\dfrac{\underline{C}, \Gamma_E \rightarrow \Theta_E, C}{\Gamma_E \rightarrow \Theta_E, C, \underline{\neg C}.} \rightarrow\neg$		$\dfrac{C, \Gamma_F \rightarrow \Theta_F, \underline{C}}{\underline{\neg C}, C, \Gamma_F \rightarrow \Theta_F.} \neg\rightarrow$

Now suppose we have treated all the sequents in the given proof down through the premise $\Delta_1 \rightarrow \Lambda_1$ or the two premises $\Delta_1 \rightarrow \Lambda_1$ and $\Delta_2 \rightarrow \Lambda_2$, inclusive, of an inference with conclusion $\Delta \rightarrow \Lambda$. We call a premise $\Delta_i \rightarrow \Lambda_i$ an EF-*premise* if this treatment has provided a proof of both $\Delta_{iE} \rightarrow \Lambda_{iE}$, J and J, $\Delta_{iF} \rightarrow \Lambda_{iF}$; an E-*premise* if it has provided a proof

[285] To deal directly with given proofs in *G*4a written with thinnings assimilated, the treatment in the cases below can be applied with formulas omitted from premises.

of $\Delta_{iE} \rightarrow \Delta_{iE}$; an F-*premise* if it has provided a proof of $\Delta_{iF} \rightarrow \Lambda_{iF}$. We classify the inference as an E-*inference* or an F-*inference*, according as its principal formula (and hence its side formula(s)) if it is logical, or its one or three C's if it is structural, belong to E or to F.

Consider the simple case (CASE 3a) of an E-inference with at least one F-premise. Since the principal and side formulas if it is logical (or the C's if it is structural) belong to E, in the F-part the premise(s) and conclusion each become simply $\Gamma_F \rightarrow \Theta_F$. By definition of "F-premise", our repair of the F-part above the inference in question gives a proof of $\Gamma_F \rightarrow \Theta_F$. So to extend the repair of the F-part to the conclusion of the inference in question, we only need to cut out the F-premise in question and connect the tree above it to the conclusion, while discarding the other premise if any with everything above it not already discarded. In the E-part, we discard the conclusion and everything above it not already discarded. This is summarized in the first row of the following table.

E-proof. Logical and structural inferences (simple cases). F-proof.

CASE 3a.	E-inference with at least one F-premise.	
		$\Gamma_F \rightarrow \Theta_F$.
CASE 3b.	F-inference with at least one E-premise.	
$\Gamma_E \rightarrow \Theta_E$.		

To combine the cases for the 10 propositional rules and 4 structural rules, when we can't just bypass the inference as above, we write the conclusion as $\Pi_1, \Gamma \rightarrow \Theta, \Pi_2$ where one of Π_1, Π_2 is the principal formula (or the C) and the other is empty. We write the premises similarly, where each of Σ_1, Σ_2 or $\Sigma_1, \Sigma_2, \Sigma_3, \Sigma_4$ is zero, one or two side formulas (or zero or two C's). For the rule $\supset\rightarrow$, then $\Pi_1, \Pi_2, \Sigma_1, \Sigma_2, \Sigma_3, \Sigma_4$ are $A \supset B$, $\varnothing$, $\varnothing$, A, B, $\varnothing$ respectively, where $\varnothing$ is the empty list. The result(s) of our previous treatment of the premise(s) are shown at the tops in each of the next figures (top p. 354), and the result of treating the conclusion at the bottom. As the figures show, we can get from the former to the latter by zero, one or two inferences in *G*4a. CASES 4b, 5b, 6b, 7b are symmetric (or "dual") to Cases 4a, 5a, 6a, 7a (as 3b, 8b are to 3a, 8a). In each of these cases with an EF-premise (and the inference propositional), (1) and (2) of Case (EF) in the theorem hold for the J in the conclusion in consequence of their holding in the premise(s).

An $\rightarrow\forall$-inference is treated under the appropriate one of Cases 3a–5b. By the restriction on variables, b does not occur free in the Γ, Θ. So if the inference is an E-inference, b does not occur free in $\Gamma_F \rightarrow \Theta_F$ (the

E-proof.	Propositional and structural inferences and certain predicate inferences (as explained below).	F-proof.

Case 4a. One premise E-inference with E-premise.

$$\frac{\Sigma_1, \Gamma_E \to \Theta_E, \Sigma_2}{\Pi_1, \Gamma_E \to \Theta_E, \Pi_2.}$$

Case 5a. One premise E-inference with EF-premise.

$$\frac{\Sigma_1, \Gamma_E \to \Theta_E, \Sigma_2, J}{\Pi_1, \Gamma_E \to \Theta_E, \Pi_2, J.} \qquad J, \Gamma_F \to \Theta_F.$$

Case 6a. E-inference with two E-premises.

$$\frac{\Sigma_1, \Gamma_E \to \Theta_E, \Sigma_2 \quad \Sigma_3, \Gamma_E \to \Theta_E, \Sigma_4}{\Pi_1, \Gamma_E \to \Theta_E, \Pi_2.}$$

Case 7a. E-inference with EF-premise (e.g. the first), E-premise.

$$\frac{\Sigma_1, \Gamma_E \to \Theta_E, \Sigma_2, J \quad \Sigma_3, \Gamma_E \to \Theta_E, \Sigma_4}{\Pi_1, \Gamma_E \to \Theta_E, \Pi_2, J.} \qquad J, \Gamma_F \to \Theta_F.$$

Case 8a. E-inference with two EF-premises.

$$\cfrac{\cfrac{\Sigma_1, \Gamma_E \to \Theta_E, \Sigma_2, J_1 \quad \Sigma_3, \Gamma_E \to \Theta_E, \Sigma_4, J_2}{\Pi_1, \Gamma_E \to \Theta_E, \Pi_2, J_1, J_2}}{\Pi_1, \Gamma_E \to \Theta_E, \Pi_2, J_1 \vee J_2.} \to\vee \qquad \frac{J_1, \Gamma_F \to \Theta_F \quad J_2, \Gamma_F \to \Theta_F}{J_1 \vee J_2, \Gamma_F \to \Theta_F.} \vee\to$$

Case 8b. F-inference with two EF-premises.

$$\frac{\Gamma_E \to \Theta_E, J_1 \quad \Gamma_E \to \Theta_E, J_2}{\Gamma_E \to \Theta_E, J_1 \& J_2.} \to\& \qquad \cfrac{\cfrac{J_1, \Sigma_1, \Gamma_F \to \Theta_F, \Sigma_2 \quad J_2, \Sigma_3, \Gamma_F \to \Theta_F, \Sigma_4}{J_1, J_2, \Pi_1, \Gamma_F \to \Theta_F, \Pi_2}}{J_1 \& J_2, \Pi_1, \Gamma_F \to \Theta_F, \Pi_2.} \&\to$$

F-part of the premise $\Delta_1 \to \Lambda_1$); if an F-inference, not in $\Gamma_E \to \Theta_E$. Hence with an EF-premise (Cases 5a, 5b), because (1) in Case (EF) of the theorem is satisfied for the premise, J does not contain b free. Hence the new $\to\forall$ satisfies the restriction on variables, and (1) is satisfied for the conclusion.[286]

[286] We have added to our preliminary plan (second paragraph of this section) to secure (1); i.e. individual parameters not common to the E-part and the F-part are removed from the J as soon as they cease to be common in treating $\forall\to$ and $\to\exists$. Otherwise, we would have to remove b (if present) in treating $\to\forall$ and $\exists\to$.

Now we treat $\forall\!\to$. Except in the circumstances described next, this can be handled under one of Cases 3a–5b, with the obvious modification of Cases 4a–5b that the Π_1, Π_2 (actually, Π_2 is empty now) appear in the premise also (Cases 4a′–5b′). Suppose the premise is an EF-premise, and the A(x) contains the x free. In the predicate calculus without functions, the r must be simply a variable. Say the inference is an E-inference, and that this variable r occurs free in $\Gamma_F \to \Theta_F$ but not in $\forall x A(x), \Gamma_E \to \Theta_E$. Then, under (1) for Case (EF) the J for the conclusion is not allowed to contain r free, but the J for the premise may. Supposing it does (Case 9a), let us write it "J(r)"; and let y be a variable (maybe x) free for r in J(r) and not occurring free in J(r) (unless y is r).

$\forall\!\to$ with r a variable free in
the J for the premise but not in both
the E-part and the F-part of the conclusion.

E-proof.	F-proof.
Case 9a. E-inference with EF-premise.	
$$\cfrac{\cfrac{A(r), \forall x A(x), \Gamma_E \to \Theta_E, J(r)}{\forall x A(x), \Gamma_E \to \Theta_E, J(r)}\ \forall\!\to}{\forall x A(x), \Gamma_E \to \Theta_E, \forall y J(y).}\ \to\!\forall$$	$$\cfrac{J(r), \Gamma_F \to \Theta_F}{\forall y J(y), \Gamma_F \to \Theta_E.}\ \forall\!\to$$
Case 9b. F-inference with EF-premise.	
$$\cfrac{\Gamma_E \to \Theta_E\, J(r)}{\Gamma_E \to \Theta_E, \exists y J(y).}\ \to\!\exists$$	$$\cfrac{\cfrac{J(r), A(r), \forall x A(x), \Gamma_F \to \Theta_F}{J(r), \forall x A(x), \Gamma_F \to \Theta_F}\ \forall\!\to}{\exists y J(y), \forall x A(x), \Gamma_F \to \Theta_F.}\ \exists\!\to$$

In Case 9a, because r does not occur free in $\forall x A(x), \Gamma_E \to \Theta_E$, the $\to\!\forall$ is legitimate. Similarly, in Case 9b (with r free in A(r) and $\Gamma_E \to \Theta_E$ and J(r) but not in $\forall x A(x), \Gamma_F \to \Theta_F$) the $\exists\!\to$ is legitimate. In the predicate calculus with functions (including individuals), more generally the r may contain free variables $c_1, \dots, c_m$ and individual symbols $e_{m+1}, \dots, e_n$ which are parameters of the J for the premise, but are not common to the E-part and the F-part of the conclusion (Cases 9a′, 9b′). Write r also as "$r(c_1, \dots, c_m, e_{m+1}, \dots, e_n)$", and the J for the premise as "$J(c_1, \dots, c_m, e_{m+1}, \dots, e_n)$". For Case 9a′ ($\forall x A(x), \Gamma_E \to \Theta_E$ not containing as parameters $c_1, \dots, c_m, e_{m+1}, \dots, e_n$), in the E-proof as constructed down through the premise of the inference in question, we can change $e_{m+1}, \dots, e_n$ throughout to respective distinct variables $c_{m+1}, \dots, c_n$ not previously occurring in that proof, to obtain instead a proof of $A(r(c_1, \dots, c_n)), \forall x A(x), \Gamma_E \to \Theta_E, J(c_1, \dots, c_n)$. Now, instead of inferring $\forall yJ(y)$ as principal formula from $J(r)$ as side formula by one inference in each proof (Case 9a), we can infer $\forall y_1 \ldots \forall y_n J(y_1, \ldots, y_n)$ from $J(c_1, \ldots, c_n)$ in the E-proof and from $J(c_1, \ldots, c_m, e_{m+1}, \ldots, e_n)$ in the F-proof by n inferences of the same respective kind.

The rules ∃→ and →∃ are treated similarly, partly under earlier cases, and partly as new CASES 10a, 10b, 10a′, 10b′.

Summarizing, we repair the E-part or the F-part or both to obtain an E-proof or an F-proof or both, by proceeding downward through the given proof step by step, applying the appropriate case at each step. This leads effectively to one of the three results described in Cases (EF), (E) and (F) of the theorem, determined by the given proof and the given analysis of it. This procedure may involve some unnecessary work in repairing the upper parts of branches that will later be discarded in treating two-premise inferences under Cases 3a, 3b. If we first classify the axioms as EF-, E- and F-, and the two-premise inferences as E- and F-, we can look ahead and anticipate which branches will be discarded.

EXAMPLE 11. First, we show a given proof of E → F in *G*4a, with the E-part in lightface, the F-part in boldface. To each numbered sequent $\Delta \rightarrow \Lambda$ in this proof, the like-numbered sequents in the E-proof and the F-proof (below it) are the ones reached by repairing the E-part and the F-part as far as the corresponding sequents $\Delta_E \rightarrow \Lambda_E$ and $\Delta_F \rightarrow \Lambda_F$.[287]

EF-axiom
1. **P(a), S, R → ∃xP(x),** P(a)
—— →∃
2. **P(a), S, R → ∃xP(x)**
—— →⊃
3. **P(a), S → R ⊃ ∃xP(x)**
—— &→
4. **S & P(a) → R ⊃ ∃xP(x)**
—— →*T*
5. **S & P(a) → R ⊃ ∃xP(x), S & Q(b)**

F-axiom: 6. **S,** Q(b), **P(a) → S** EF-axiom: 7. Q(b), S, **P(a) → Q(b)**
—— →&
8. **S,** Q(b), **P(a) → S & Q(b)**
—— &→
5. 9. Q(b), **S & P(a) → S & Q(b)**
—— ⊃→
10. (R ⊃ ∃xP(x)) ⊃ Q(b), **S & P(a) → S & Q(b)**
—— →⊃
11. (R ⊃ ∃xP(x)) ⊃ Q(b) **→ S & P(a) ⊃ S & Q(b)**
—— →∀
12. (R ⊃ ∃xP(x)) ⊃ Q(b) **→ ∀x(S & P(x) ⊃ S & Q(b)).**

[287] In this example, no repaired sequent is later discarded under Case 3a or 3b for a two-premise inference (at most duplications are suppressed), nor altered by changing $e_{m+1}, \ldots, e_n$ to $c_{m+1}, \ldots, c_n$ under Case 9a′, 9b′, 10a′ or 10b′.

The E-proof, with formulas replacing the F-part in boldface.

P(a), R → ∃xP(x), P(a)
——————————————— →¬
1. R → ∃xP(x), P(a), **¬P(a)**
——————————————— →∃
R → ∃xP(x), **¬P(a)**
——————————————— →∀
2. R → ∃xP(x), **∀x¬P(x)**
——————————————— →⊃
3, 4, 5. → R ⊃ ∃xP(x), **∀x¬P(x)**   7, 8, 9. Q(b) → **Q(b)**
——————————————————————————————— ⊃→
(R ⊃ ∃xP(x)) ⊃ Q(b) → **∀x¬P(x), Q(b)**
——————————————— →∨
10, 11, 12. (R ⊃ ∃xP(x)) ⊃ Q(b) → **∀x¬P(x) ∨ Q(b)**.

The F-proof, with formulas replacing the E-part in light face.

P(a), S → P(a)
——————————————— ¬→
1. ¬P(a), **P(a), S →**
——————————————— ∀→
2, 3. ∀x¬P(x), **P(a), S →**
——————————————— &→
4. ∀x¬P(x), **S & P(a) →**
——————————————— →*T*
5. ∀x¬P(x), **S & P(a) → S & Q(b)**

6. **S, P(a) → S**   7. Q(b), **S, P(a) → Q(b)**
——————————————————————————————— →&
8. Q(b), **S, P(a) → S & Q(b)**
——————————————— &→
5.   9. Q(b), **S & P(a) → S & Q(b)**
——————————————————————————————— ∨→
10. ∀x¬P(x) ∨ Q(b), **S & P(a) → S & Q(b)**
——————————————— →⊃
11. ∀x¬P(x) ∨ Q(b) → **S & P(a) ⊃ S & Q(b)**
——————————————— →∀
12. ∀x¬P(x) ∨ Q(b) → **∀x(S & P(x) ⊃ S & Q(b))**.

THEOREM 41.[288] (Craig's interpolation theorem 1957, 1957a.) *Suppose*

[288] Craig 1957, 1957a says that (b) (without the parenthetical version of the hypothesis) was first called to his attention by P. C. Gilmore. Using the Gentzen-type system *G*, (b) (with the parenthetical version, and essentially the present proof) is in Kleene 1952 in the form of Lemma 6 with Lemmas 3 and 4; and that combination of lemmas for both the classical and the intuitionistic cases is used repeatedly in Kleene 1952a (pp. 49, 50, 51, 53, 54).

We need the hypothesis "E *and* F *contain a common predicate parameter*" in (a) and "*neither* ⊢ ¬E *nor* ⊢ F" in (c), because our formation rules §§ 16, 28 give us no way to build a formula without its containing some predicate parameter. E.g. if E ⊃ F is (P ⊃ P) ⊃ (Q ⊃ Q), we can't find an I for (a) containing only common parameters; if E ⊃ F is ¬P ⊃ (Q ⊃ Q) ∨ P, we can satisfy (a) by using ¬P for I, but (c) would fail.

If we adopt formation rules which provide a proposition or predicate constant (like t, f or =) that does not count as a parameter, then we can dispense with the quoted hypotheses. Then e.g. t or ¬f or ∀x x = x ⊃ ∀x x = x could be the I for both
(P ⊃ P) ⊃ (Q ⊃ Q) and ¬P ⊃ (Q ⊃ Q) ∨ P.

Lyndon 1959 and Henkin 1963 use such rules (with propositional constants for truth and falsity). Lyndon 1959 (although he says he found his result using a Gentzen-type system) gives a model-theoretic treatment, in which his "main theorem" is incorrect (as shown by Taitslin 1960) but can be mended (according to Henkin 1963).

that in the predicate calculus without equality $\vdash \mathrm{E} \supset \mathrm{F}$. *Then:*

(a) *If* E *and* F *contain a common predicate parameter, then there is a formula* I *such that* $\vdash \mathrm{E} \supset \mathrm{I}$ *and* $\vdash \mathrm{I} \supset \mathrm{F}$, *and the individual and predicate parameters of* I *are common to* E *and* F.

(b) (Kleene 1952.) *If* E *and* F *contain no common predicate parameter (or for* $\mathrm{E} \supset \mathrm{F}$ *without* $\sim$, *even if no predicate parameter occurs positively in both or negatively in both),*[270] *then either* $\vdash \neg\mathrm{E}$ *or* $\vdash \mathrm{F}$.

(c) (Lyndon 1959.)[270] *For* $\mathrm{E} \supset \mathrm{F}$ *without* $\sim$, *if neither* $\vdash \neg\mathrm{E}$ *nor* $\vdash \mathrm{F}$, *then there is a formula* I *(without* $\sim$*) such that* $\vdash \mathrm{E} \supset \mathrm{I}$ *and* $\vdash \mathrm{I} \supset \mathrm{F}$, *and the individual parameters of* I *are common to* E *and* F, *and a predicate parameter occurs* $\left\{\begin{matrix}\textit{positively}\\ \textit{negatively}\end{matrix}\right\}$ *in* I *only if it occurs* $\left\{\begin{matrix}\textit{positively}\\ \textit{negatively}\end{matrix}\right\}$ *in both* E *and* F.

PROOF. We may suppose that $\mathrm{E} \supset \mathrm{F}$ contains no variable both free and bound.[259] Assume the hypothesis that $\vdash_H \mathrm{E} \supset \mathrm{F}$. Then by Gentzen's theorem, $\vdash_{G4a} \to \mathrm{E} \supset \mathrm{F}$, whence (by one $\to\supset$-step upward) $\vdash_{G4a} \mathrm{E} \to \mathrm{F}$. We apply Theorem 40 to a given proof of $\mathrm{E} \to \mathrm{F}$ with the endsequent $\mathrm{E} \to \mathrm{F}$ as the $\Delta \to \Lambda$.

(a) Now assume that E and F have a common predicate parameter. If Case (EF) of Theorem 40 holds, then $\vdash \mathrm{E} \to \mathrm{I}$ and $\vdash \mathrm{I} \to \mathrm{F}$ where I is the J for the bottom sequent as the $\Delta \to \Lambda$. By (1), I contains only individual parameters that are common to E and F. By (2), the predicate parameter of each atomic part of I is in each of E and F, by descent from the C of an axiom. If Case (E) or Case (F) holds instead, we begin by picking a predicate parameter K which is common to E and F, which we can do by hypothesis. Let D be $\forall x\mathrm{K}(x, \ldots, x)$ (or simply K if K takes 0 arguments). The following table indicates how in these cases we get $\vdash \mathrm{E} \supset \mathrm{I}$ and $\vdash \mathrm{I} \supset \mathrm{F}$, in $G4$b now.

When $\vdash \mathrm{E} \to$, take I to be $\neg(\mathrm{D} \supset \mathrm{D})$.

$$\dfrac{\mathrm{E} \to}{\mathrm{E} \to \neg(\mathrm{D} \supset \mathrm{D}).}\,{\to}T \qquad \dfrac{\dfrac{\mathrm{D} \to \mathrm{F}, \mathrm{D}}{\to \mathrm{F}, \mathrm{D} \supset \mathrm{D}}\,{\to}{\supset}}{\neg(\mathrm{D} \supset \mathrm{D}) \to \mathrm{F}.}\,{\neg}{\to}$$

When $\vdash \to \mathrm{F}$, take I to be $\mathrm{D} \supset \mathrm{D}$.

$$\dfrac{\mathrm{D}, \mathrm{E} \to \mathrm{D}}{\mathrm{E} \to \mathrm{D} \supset \mathrm{D}.}\,{\to}{\supset} \qquad \dfrac{\to \mathrm{F}}{\mathrm{D} \supset \mathrm{D} \to \mathrm{F}.}\,T{\to}$$

Again, I has only allowed parameters. Now in any case, using $\to\supset$ and Gentzen's theorem (or Theorem 36 Corollary (a)), $\vdash \mathrm{E} \supset \mathrm{I}$ and $\vdash \mathrm{I} \supset \mathrm{F}$.

(b) Suppose, for E ⊃ F without ∼, that no predicate parameter occurs positively in both E and F or negatively in both. We shall infer that there is no EF-axiom. It will follow that Case (EF) of Theorem 40 cannot apply; and Cases (E) and (F) lead respectively to $\vdash \neg$E and $\vdash$ F. So assume, for reductio ad absurdum, that there is an EF-axiom C, $\Gamma \rightarrow \Theta$, C in the given proof of E → F. Using Lemma 13 § 54, say that the Negative (antecedent) C in this axiom descends into an image in the E of E → F, and the Positive C into one in the F. Then by Lemma 14 the images are Negative and Positive, respectively; so as parts of E and F separately, they are both positive, giving rise to positive occurrences of the predicate parameter of C in both E and F. Similarly, if the Positive C descends into E and the Negative into F, we get negative occurrences of the parameter in both E and F.

(c) Suppose E ⊃ F does not contain ∼, and that neither $\vdash_H \neg$E nor $\vdash_H$ F. Then by Gentzen's theorem (and one use of →¬), neither $\vdash_{G4a}$ E → nor $\vdash_{G4a}$ → F. So it has to be Case (EF) of Theorem 40 which applies. Consider any predicate parameter occurrence in I; say C_I is the atomic part which contains it. Say C_I is positive as a part of I, and hence Positive as a part of E → I and Negative as a part of I → F. Applying (2) of Case (EF) with Lemma 14, in the proof of E → I the part C_I descends from the second (Positive) C of an axiom C, $\Gamma_E \rightarrow \Theta_E$, C, the first (Negative) C of which descends into an image C_E in the E of E → I. This image is Negative as a part of E → I but positive as a part of E. Similarly from the proof of I → F we are led to an image C_F of C which is positive as a part of F. Thus the parameter (assumed to occur positively in I) occurs positively in both E and F. Similarly, a predicate parameter occurring negatively in I occurs negatively in both E and F.—

In the predicate calculus with equality (with or without functions), the statement of Craig's interpolation theorem can be simplified, since the symbol = (which does not count there as a parameter) is available for building formulas.

THEOREM 42. (Craig's interpolation theorem with equality 1957a.) *In the predicate calculus with equality, if* $\vdash$ E ⊃ F, *then there is a formula* I *such that* $\vdash$ E ⊃ I *and* $\vdash$ I ⊃ F *and the parameters of* I *are common to* E *and* F.

PROOF. The method of the proof of Theorem 41 via Theorem 40 does not lend itself directly to excluding from I function parameters (with > 0 places) not common to E and F. However with equality available, we can first use the idea which we met in § 38 of paraphrasing statements employing functions to employ predicates instead.

Suppose that $\vdash$ E ⊃ F in the predicate calculus with equality. Let

$A_1, \ldots, A_m$ be the open equality axioms for $=$ and the other predicate and function symbols of $E \supset F$. By Theorem 31 § 29, then $\vdash E \supset F$ in the system consisting of the predicate calculus H with only these predicate and function symbols and with $A_1, \ldots, A_m$ as additional axioms. Say the (individual and) function symbols in $E \supset F$ are $f_k, \ldots, f_1$ ($k \geq 0$). Call the system just described S_k. As illustrated in § 45 (β), we can replace $f_k, \ldots, f_1$ as symbols for functions (whatever functions the symbols express in a given model) successively by $F_k, \ldots, F_1$ as symbols for the representing predicates of those functions. Thereby, we are led to a system S_0 in which the result $E' \supset F'$ of paraphrasing $E \supset F$ to use the representing-predicate symbols in place of the function symbols is provable. The formula E' contains the same free variables as E, no (individual and) function symbols, and (as predicate parameters) exactly the predicate parameters of E and the representing-predicate symbols replacing (individual and) function symbols of E. The parameters of F' are similarly related to those of F. The nonlogical axioms $B_1, \ldots, B_l$ of S_0 are the open equality axioms for $=$ and for the predicate symbols of $E' \supset F'$, and the formulas $\exists ! w F_i(x_1, \ldots, x_{n_i}, w)$ for each of the representing-predicate symbols $F_1, \ldots, F_k$. Thus each of $B_1, \ldots, B_l$ contains at most one predicate parameter, and that parameter occurs in $E' \supset F'$. Suppose the list $B_1, \ldots, B_l$ chosen so that in the first part of it $B_1, \ldots, B_k$ (possibly empty) only predicate parameters in E' occur, and in the second part $B_{k+1}, \ldots, B_l$ (possibly empty) only ones in F'. (Predicate parameters in both E' and F' can occur in either part.) Since $\vdash E' \supset F'$ in S_0, by $\forall$-eliminations $\forall B_1, \ldots, \forall B_l \vdash E' \supset F'$ in the predicate calculus H. Thence by the deduction theorem etc.,

$$\vdash \forall B_1 \mathbin{\&} \ldots \mathbin{\&} \forall B_k \mathbin{\&} E' \supset (\forall B_{k+1} \mathbin{\&} \ldots \mathbin{\&} \forall B_l \supset F') \text{ in } H.$$

Now we reason as for Theorem 41 (a), using the Gentzen-type system *G*4a with only the symbols of S_0 (and therefore no function symbols).[265] If Case (E) or (F) of Theorem 40 applies, we employ x=x in the role of the K(x, . . . , x) there. We obtain thus a formula I' such that in H

$$\vdash \forall B_1 \mathbin{\&} \ldots \mathbin{\&} \forall B_k \mathbin{\&} E' \supset I' \quad \text{and} \quad \vdash I' \supset (\forall B_{k+1} \mathbin{\&} \ldots \mathbin{\&} \forall B_l \supset F')$$

and I' contains only individual and predicate parameters common to $\forall B_1 \mathbin{\&} \ldots \forall B_k \mathbin{\&} E'$ and $\forall B_{k+1} \mathbin{\&} \ldots \mathbin{\&} \forall B_l \supset F'$, and no function parameters. These common parameters are free variables, which clearly must be common to E' and F', and predicate parameters, also common to E' and F' by the way we split the list $B_1, \ldots, B_l$.

Now $\forall B_1, \ldots, \forall B_k \vdash E' \supset I'$ and $\forall B_{k+1}, \ldots, \forall B_l \vdash I' \supset F'$ in H. So both $E' \supset I'$ and $I' \supset F'$ are deducible in H from $\forall B_1, \ldots, \forall B_l$, and hence are provable in S_0. By unparaphrasing, we obtain a formula I such that

E ⊃ I and I ⊃ F are both provable in S_k, and hence in the predicate calculus with equality. That we can thus get back to the original $A_1, \ldots, A_m$, E and F is given by the theorem on paraphrasing.[289] The free variables of I are the same as of I′ and hence are common to E′, F′, E, F; and the other parameters of I are those predicate parameters of I′ which are common predicate parameters of E′, F′, E, F, and the (individual and) function symbols of E and F which in the passage from S_k to S_0 were replaced by representing-predicate symbols common to E′, F′. Thus all the parameters of I are common to E and F.

EXERCISES. 56.1. Treat in the manner of Example 11:
(a) The proof in *G*4 in Example 7 § 54.
(b) The proof of (P(b) ⊃ S)· P(b)& → ∀xQ(x) ⊃ R ∨ Q(b) obtained by using in order from the bottom up &→, →⊃, ⊃→ and (in one branch) ∀→, →∨.

56.2. Show that in the predicate calculus (for E not containing ∼): If ⊢ F, then some predicate parameter occurs in F both positively and negatively.

§ 57. Beth's theorem on definability, Robinson's consistency theorem. The fifth postulate of Euclid is *independent* of the other postulates, i.e. it cannot be deduced by logic from them. This was demonstrated, after more than two millenia of speculation, by giving an interpretation which makes the other postulates all true and the fifth postulate false. The Cayley-Kline model for non-Euclidean geometry 1871 (and for the geometry of a bounded part of the plane, Beltrami's model 1868) is such an interpretation (§ 36). Since then, this has become the standard method in formal axiomatics of showing the independence of one of a list of axioms from the others, or more generally of a given statement from a given list of axioms.

To put this method in our terms, take the case of an axiomatic theory formalizable in the (first-order) predicate calculus without or with equality, say with $\aleph_0$ axioms. Say the axioms are expressed by formulas A_0, A_1, $A_2, \ldots$ (with their free variables if any having the generality interpretation) and the statement to be proved independent of them by F. Let E_0, E_1, $E_2, \ldots$ be the closures of $A_0, A_1, A_2, \ldots$. The traditional method then simply applies the consistency theorem (cf. end § 52),

$$\{E_0, E_1, E_2, \ldots \vdash F\} \rightarrow \{E_0, E_1, E_2, \ldots \models F\},$$

with informal use of *12a (contraposition), *82b and *55c (§§ 3, 25).

[289] IM Theorem 43 p. 417. The present E′ and F′ are the results of applying to our E and F the ′ of IM p. 411 successively with respect to each of $f_k, \ldots, f_1$, and I is the result of applying to our I′ the ° of IM p. 417 successively with respect to each of $F_1, \ldots, F_k$.

Similarly, the completeness theorem,

$$\{E_0, E_1, E_2, \ldots \models F\} \rightarrow \{E_0, E_1, E_2, \ldots \vdash F\},$$

when contraposed, says that in principle the traditional method must always work; i.e., whenever the statement in question does not follow by logic from the axioms, it has to be false under an interpretation which makes all the axioms true (cf. § 53).

Just as we usually wish to have the axioms of a formal theory independent, we would also like to have the primitive concepts independent, i.e. such that no one of them can be defined from the others. How can it be shown that, in the theory based on certain axioms $A_0, A_1, A_2, \ldots$, a certain concept q cannot be defined from the other concepts $p_0, p_1, p_2, \ldots$ of the theory. Padoa 1900 used a counterpart of the more familiar method of showing axioms to be independent. Namely, he gave two interpretations (with the same domain) such that the axioms $A_0, A_1, A_2, \ldots$ are all true under both, $p_0, p_1, p_2, \ldots$ have the same values under both, but q has different values under the two. For, if there were a definition of q from $p_0, p_1, p_2, \ldots$ in the theory with $A_0, A_1, A_2, \ldots$ as the axioms, that definition should determine the value of q from the values of $p_0, p_1, p_2, \ldots$ for any values of the latter which are compatible with $A_0, A_1, A_2, \ldots$ all being true. Thus it should not be possible to make q have different values when $p_0, p_1, p_2, \ldots$ have the same values and $A_0, A_1, A_2, \ldots$ are all true.

We now study this method in modern terms. We generalize it to discuss the definability of q from $p_0, p_1, p_2, \ldots$ when the closures $E_0, E_1, E_2, \ldots$ of the axioms $A_0, A_1, A_2, \ldots$ may obtain still other parameters $r_0, r_1, r_2, \ldots$. Our notation is for the most general case, of $\aleph_0$ axioms and $\aleph_0$ parameters in each of the two lists. The reader can supply the modifications when one or more of these lists are finite or even empty. We take q and all the listed parameters to be distinct. For the present, each parameter in the two lists (which together with q must include all occurring in the closed axioms $E_0, E_1, E_2, \ldots$) shall be a (proposition or) predicate symbol or an individual symbol, q shall be an n-place predicate symbol Q (with $n \geq 0$), and the logic shall be the predicate calculus without equality.

Padoa's method is equivalent (by contraposition etc.) to the implication: {Q is definable from $p_0, p_1, p_2, \ldots$ in the theory based on $E_0, E_1, E_2, \ldots$ as axioms} → {each two interpretations of the parameters Q, $p_0, p_1, p_2, \ldots$, $r_0, r_1, r_2, \ldots$ which make $E_0, E_1, E_2, \ldots$ all true and give $p_0, p_1, p_2, \ldots$ the same values give Q the same value}, or briefly $\text{Dfb}_E \rightarrow \text{Dfd}_I$.

First we analyze Dfb_E. We take it to mean that a suitable defining expression can be given for $Q(x_1, \ldots, x_n)$ (as *definiendum*). Such an expression shall be in the language of the theory, i.e. the predicate calculus with the named parameters. It shall contain as free variables only $x_1, \ldots, x_n$, and as other parameters only ones from the list $p_0, p_1, p_2, \ldots$, of which it can contain only finitely many, say only $p_0, \ldots, p_s$. Say the defining expression or *definiens* is $R(x_1, \ldots, x_n, p_0, \ldots, p_s)$. While the logic is the predicate calculus, we must have $s \geq 0$ with at least one predicate parameter among $p_0, \ldots, p_s$, as a definiens cannot be constructed without using a predicate parameter. Finally, we must express that $R(x_1, \ldots, x_n, p_0, \ldots, p_s)$ is a definiens for $Q(x_1, \ldots, x_n)$ in the theory based on $E_0, E_1, E_2, \ldots$. This means that, for each domain D and assignment which satisfy all of $E_0, E_1, E_2, \ldots$, the formula $R(x_1, \ldots, x_n, p_0, \ldots, p_s)$ shall have the same truth value as $Q(x_1, \ldots, x_n)$, for all values in D of $x_1, \ldots, x_n$. Thus we are led to the following rendering of Dfb_E.[290]

Dfb_E^M: For some formula $R(x_1, \ldots, x_n, p_0, \ldots, p_s)$ containing only the parameters shown,

$$E_0, E_1, E_2, \ldots \models Q(x_1, \ldots, x_n) \sim R(x_1, \ldots, x_n, p_0, \ldots, p_s).$$

We say in this case that (*model-theoretically*) Q is *definable explicitly* from $p_0, p_1, p_2, \ldots$ *in* the theory based on $E_0, E_1, E_2, \ldots$ as the closed axioms, or that $E_0, E_1, E_2, \ldots$ *make* Q so *definable*. Further, for such an $R(x_1, \ldots, x_n, p_0, \ldots, p_s)$, we say that (*model-theoretically*) $E_0, E_1, E_2, \ldots$ *define* Q *explicitly from* $p_0, \ldots, p_s$ *as* $R(x_1, \ldots, x_n, p_0, \ldots, p_s)$.

We can replace "$E_0, E_1, E_2, \ldots \models$" equivalently by "for some d, $E_0, \ldots, E_d \vdash$" [by the completeness and consistency of the predicate calculus], and further by "for some d, $\vdash E_0 \,\&\, \ldots \,\&\, E_d \supset$" [by $\supset$-introd. etc.]. Only finitely many parameters can occur in $E_0 \,\&\, \ldots \,\&\, E_d$; we can increase the s in $R(x_1, \ldots, x_n, p_0, \ldots, p_s)$ if necessary so only $p_0, \ldots, p_s, r_0, \ldots, r_t$ occur in $E_0 \,\&\, \ldots \,\&\, E_d$. By these steps, Dfb_E^M is equivalent to the following.

Dfb_E^P: For some finite conjunction $E(Q, p_0, \ldots, p_s, r_0, \ldots, r_t)$ of $E_0, E_1, E_2, \ldots$ and some formula $R(x_1, \ldots, x_n, p_0, \ldots, p_s)$, each containing only the parameters shown,

$$\vdash E(Q, p_0, \ldots, p_s, r_0, \ldots, r_t) \supset [Q(x_1, \ldots, x_n) \sim R(x_1, \ldots, x_n, p_0, \ldots, p_s)].$$

We say in this case that (*proof-theoretically*) Q is *definable explicitly from* $\left\{\begin{matrix} p_0, p_1, p_2, \ldots \\ p_0, \ldots, p_s \end{matrix}\right\}$ *in* the theory based on $\left\{\begin{matrix} E_0, E_1, E_2, \ldots \\ E(Q, p_0, \ldots, p_s, r_0, \ldots, r_t) \end{matrix}\right\}$, or that $\left\{\begin{matrix} E_0, E_1, E_2, \ldots \textit{ make} \\ E(Q, p_0, \ldots, p_s, r_0, \ldots, r_t) \textit{ makes} \end{matrix}\right\}$ Q so *definable*. Further, (*proof-theoretically*) $\left\{\begin{matrix} E_0, E_1, E_2, \ldots \textit{ define} \\ E(Q, p_0, \ldots, p_s, r_0, \ldots, r_t) \textit{ defines} \end{matrix}\right\}$ Q *explicitly from*

[290] Since $E_0, E_1, E_2, \ldots$ are closed, "$E_0, E_1, E_2, \ldots \models Q(x_1, \ldots, x_n) \sim R(x_1, \ldots, x_n, p_0, \ldots, p_s)$" is equivalent to "$E_0, E_1, E_2, \ldots \models \forall x_1 \ldots \forall x_n[Q(x_1, \ldots, x_n) \sim R(x_1, \ldots, x_n, p_0, \ldots, p_s)]$", and similarly with "$\vdash$" in place of "$\models$". It is convenient to use the shorter expressions with the free $x_1, \ldots, x_n$.

$p_0, \ldots, p_s$ *as* $R(x_1, \ldots, x_n, p_0, \ldots, p_s)$. The lower version here can be used with any formula $E(Q, p_0, \ldots, p_s, r_0, \ldots, r_t)$ with only the parameters shown, whether or not it is a finite conjunction of $E_0, E_1, E_2, \ldots$.[291]

Next we analyze Dfd_I. Let $Q', r'_0, r'_1, r'_2, \ldots$ be new parameters of the same respective kinds (predicate or individual) and numbers of argument places as $Q, r_0, r_1, r_2, \ldots$; and let $E'_0, E'_1, E'_2, \ldots$ come from $E_0, E_1, E_2, \ldots$ by substituting $Q', r'_0, r'_1, r'_2, \ldots$ for $Q, r_0, r_1, r_2, \ldots$, respectively. Instead of having two interpretations of $Q, p_0, p_1, p_2, \ldots, r_0, r_1, r_2, \ldots$ which make $E_0, E_1, E_2, \ldots$ all true and give $p_0, p_1, p_2, \ldots$ the same values, it is equivalent to have one interpretation of $Q, Q', p_0, p_1, p_2, \ldots, r_0, r'_0, r_1, r'_1, r_2, r'_2, \ldots$ which make $E_0, E'_0, E_1, E'_1, E_2, E'_2, \ldots$ all true. This device leads to the following rendering of Dfd_I.

Dfd_I^M: $\quad E_0, E'_0, E_1, E'_1, E_2, E'_2, \ldots \models Q(x_1, \ldots, x_n) \sim Q'(x_1, \ldots, x_n)$.

We say in this case that (*model-theoretically*) Q is *defined implicitly from* $p_0, p_1, p_2, \ldots$ (*with* $r_0, r_1, r_2, \ldots$ *as auxiliary parameters*) *by*, or *in* the theory based on, $E_0, E_1, E_2, \ldots$.

By the same transformations as we used on Dfb_E^M, this condition Dfd_I^M is equivalent to the following, where $E(Q', p_0, \ldots, p_s, r'_0, \ldots, r'_t)$ is the result of substituting $Q', r'_0, \ldots, r'_t$ for $Q, r_0, \ldots, r_t$ in $E(Q, p_0, \ldots, p_s, r_0, \ldots, r_t)$.

Dfd_I^P: For some finite conjunction $E(Q, p_0, \ldots, p_s, r_0, \ldots, r_t)$ of $E_0, E_1, E_2, \ldots$ containing only the parameters shown,

$$\vdash E(Q, p_0, \ldots, p_s, r_0, \ldots, t_t) \,\&\, E(Q', p_0, \ldots, p_s, r'_0, \ldots, r'_t) \supset [Q(x_1, \ldots, x_n) \sim Q'(x_1, \ldots, x_n)].$$

We say in this case that (*proof-theoretically*) Q is *defined implicitly from* $\left\{\begin{matrix} p_0, p_1, p_2, \ldots \\ p_0, \ldots, p_s \end{matrix}\right\}$ (*with* $\left\{\begin{matrix} r_0, r_1, r_2, \ldots \\ r_0, \ldots, r_t \end{matrix}\right\}$ *as auxiliary parameters*) *by*, or *in* the theory based on, $\left\{\begin{matrix} E_0, E_1, E_2, \ldots \\ E(Q, p_0, \ldots, p_s, r_0, \ldots, r_t) \end{matrix}\right\}$.

Summarizing, Padoa's method rests on an implication which (after contraposition) we have rendered in model theory by $\text{Dfb}_E^M \to \text{Dfd}_I^M$, or equivalently (using $\text{Dfb}_E^M \equiv \text{Dfb}_E^P$ and $\text{Dfd}_I^M \equiv \text{Dfd}_I^P$) in proof theory by $\text{Dfb}_E^P \to \text{Dfd}_I^P$.

It is a simple exercise to corroborate Padoa's method proof-theoretically by establishing $\text{Dfb}_E^P \to \text{Dfd}_I^P$. In fact, any formula $E(Q, p_0, \ldots, p_s, r_0, \ldots, r_t)$ which makes Q definable explicitly from $p_0, \ldots, p_s$ also defines Q implicitly from $p_0, \ldots, p_s$ (Exercise 57.1).

[291] In like manner we could introduce $E(Q, p_0, \ldots, p_s, r_0, \ldots, r_t)$ into the model-theoretic formulation Dfb_E^M (also into Dfd_I^M below). If there are only finitely many axioms to begin with, $E(Q, p_0, \ldots, p_s, r_0, \ldots, r_t)$ can be simply their conjunction.

Our analysis thus far has been a straightforward application of results which have been available since 1930 (consistency with respect to validity e.g. Hilbert and Ackermann 1928 bottom p. 73, the deduction theorem 1930, Gödel's completeness theorem 1930).

Now we inquire whether Padoa's method is complete. That is, will it always work, whenever one primitive concept in an axiomatic theory is in fact independent of the others? An affirmative answer would be equivalent via contraposition to saying that, whenever it is impossible to satisfy the closed axioms $E_0, E_1, E_2, \ldots$ by two values of Q compatibly with given values of $p_0, p_1, p_2, \ldots$, the resulting implicit model-theoretic dependence of Q on $p_0, p_1, p_2, \ldots$ must be expressible by an explicit definition within the syntactical limitations of the language. This is certainly not obvious.

However, the completeness of Padoa's method was established with a trivial exception by Beth in 1953.[292] The exception is when $p_0, \ldots, p_s$ include no predicate parameter; without a predicate parameter, we can't build a definiens $R(x_1, \ldots, x_n, p_0, \ldots, p_s)$ in the predicate calculus without equality. The provision for auxiliary parameters $r_0, \ldots, r_t$ in the implicit definition not appearing in the definiens, and some other generalizations of Beth's result, were given by Craig in 1957a.

By the foregoing preliminary analysis, what we need is $\mathrm{Dfd}_{\mathrm{I}}^{\mathrm{M}} \rightarrow \mathrm{Dfb}_{\mathrm{E}}^{\mathrm{M}}$, or equivalently $\mathrm{Dfd}_{\mathrm{I}}^{\mathrm{P}} \rightarrow \mathrm{Dfb}_{\mathrm{E}}^{\mathrm{P}}$. For the latter, it will suffice to show that any formula $E(Q, p_0, \ldots, p_s, r_0, \ldots, r_t)$ which (proof-theoretically) defines Q implicitly from $p_0, \ldots, p_s$ also makes Q definable explicitly from $p_0, \ldots, p_s$, as we state in:

THEOREM 43. (Beth's theorem on definability 1953.) *Let the symbolism and logic be those of the predicate calculus without equality but allowing individuals. Let the notation be as explained above (in particular, each formula shall contain only the distinct parameters shown). Suppose that*

$$\vdash E(Q, p_0, \ldots, p_s, r_0, \ldots, r_t) \,\&\, E(Q', p_0, \ldots, p_s, r'_0, \ldots, r'_t) \supset [Q(x_1, \ldots, x_n) \sim Q'(x_1, \ldots, x_n)],$$

i.e. $E(Q, p_0, \ldots, p_s, r_0, \ldots, r_t)$ *defines* Q *implicitly from* $p_0, \ldots, p_s$ *with* $r_0, \ldots, r_t$ *as auxiliary parameters. Then:*

(a) *If one of* $p_0, \ldots, p_s$ *is a predicate parameter, there is a formula* $R(x_1, \ldots, x_n, p_0, \ldots, p_s)$ *such that*

$$\vdash E(Q, p_0, \ldots, p_s, r_0, \ldots, r_t) \supset [Q(x_1, \ldots, x_n) \sim R(x_1, \ldots, x_n, p_0, \ldots, p_s)],$$

i.e. $E(Q, p_0, \ldots, p_s, r_0, \ldots, r_t)$ *makes* Q *definable explicitly from* $p_0, \ldots, p_s$.

[292] Tarski 1934 dealt with the essentially simpler case of theories based on higher-order predicate calculus (type theory).

The present proof (of Theorem 43), using Craig's interpolation theorem, is essentially the same as in Craig 1957a.

(b) *If none of* $p_0, \ldots, p_s$ *is a predicate parameter occurring in* $E(Q, p_0, \ldots, p_s, r_0, \ldots, r_t)$, *then*

$$\textit{either} \vdash E(Q, p_0, \ldots, p_s, r_0, \ldots, r_t) \supset Q(x_1, \ldots, x_n)$$
$$\textit{or} \vdash E(Q, p_0, \ldots, p_s, r_0, \ldots, r_t) \supset \neg Q(x_1, \ldots, x_n),$$

i.e. $E(Q, p_0, \ldots, p_s, r_0, \ldots, r_t)$ *defines* Q *either as the constant predicate "truth" or as the constant predicate "falsity".*

PROOF. Using the main hypothesis with the propositional calculus,

(i) $\vdash E(Q, p_0, \ldots, p_s, r_0, \ldots, r_t) \;\&\; Q(x_1, \ldots, x_n) \supset$
$$[E(Q', p_0, \ldots, p_s, r'_0, \ldots, r'_t) \supset Q'(x_1, \ldots, x_n)].$$

This we take as the "$\vdash E \supset F$" for Craig's interpolation theorem (Theorem 41).

(a_1) Suppose one of $p_0, \ldots, p_s$ is a predicate parameter occurring in $E(Q, p_0, \ldots, p_s, r_0, \ldots, r_t)$. Then by (a) of Theorem 41, there is a formula I such that $\vdash E \supset I$ and $\vdash I \supset F$ and I contains only parameters common to E and F. But at most $x_1, \ldots, x_n, p_0, \ldots, p_s$ are common to E and F. So letting $R(x_1, \ldots, x_n, p_0, \ldots, p_s)$ be this I, the structural requirement is met, and

(ii) $\vdash E(Q, p_0, \ldots, p_s, r_0, \ldots, r_t) \;\&\; Q(x_1, \ldots, x_n) \supset$
$$R(x_1, \ldots, x_n, p_0, \ldots, p_s),$$

(iii) $\vdash R(x_1, \ldots, x_n, p_0, \ldots, p_s) \supset$
$$[E(Q', p_0, \ldots, p_s, r'_0, \ldots, r'_t) \supset Q'(x_1, \ldots, x_n)].$$

The proof of the formula in (iii) remains a proof on substituting in it $Q, r_0, \ldots, r_t$ for $Q', r'_0, \ldots, r'_t$.[86] So

(iv) $\vdash R(x_1, \ldots, x_n, p_0, \ldots, p_s) \supset$
$$[E(Q, p_0, \ldots, p_s, r_0, \ldots, r_t) \supset Q(x_1, \ldots, x_n)].$$

From (ii) and (iv) by propositional calculus,

(v) $\vdash E(Q, p_0, \ldots, p_s, r_0, \ldots, r_t) \supset$
$$[Q(x_1, \ldots, x_n) \sim R(x_1, \ldots, x_n, p_0, \ldots, p_s)].$$

(a_2) If one of $p_0, \ldots, p_s$ is a predicate parameter, but no predicate parameter among $p_0, \ldots, p_s$ occurs in $E(Q, p_0, \ldots, p_s, r_0, \ldots, r_t)$, then (b) (to be established next) applies, after which we can trivially construct an $R(x_1, \ldots, x_n, p_0, \ldots, p_s)$.

(b) Suppose no predicate parameter among $p_0, \ldots, p_s$ occurs in $E(Q, p_0, \ldots, p_s, r_0, \ldots, r_t)$. Then by Theorem 41 (b),

(vi-1) either $\vdash \neg[E(Q, p_0, \ldots, p_s, r_0, \ldots, r_t) \;\&\; Q(x_1, \ldots, x_n)]$

(vi-2) or $\vdash E(Q', p_0, \ldots, p_s, r'_0, \ldots, r'_t) \supset Q'(x_1, \ldots, x_n)$.

Using propositional calculus (*60, *49; or IM *58b) on (vi-1), and substituting into (vi-2),

(vii-1) either $\vdash E(Q, p_0, \ldots, p_s, r_0, \ldots, r_t) \supset \neg Q(x_1, \ldots, x_n)$

(vii-2) or $\vdash E(Q, p_0, \ldots, p_s, r_0, \ldots, r_t) \supset Q(x_1, \ldots, x_n)$. —

Now we take the case of the predicate calculus with equality and functions. The parameter q for Padoa's method can now be either an n-place predicate symbol Q or an n-place function symbol g (with $n \geq 0$).

THEOREM 44. (Beth's theorem on definability, in the predicate calculus with equality and functions.) *In the predicate calculus with equality and functions, and with the notation as explained:*

(A) *If*

$$\vdash E(Q, p_0, \ldots, p_s, r_0, \ldots, r_t) \ \&\ E(Q', p_0, \ldots, p_s, r_0', \ldots, r_t') \supset [Q(x_1, \ldots, x_n) \sim Q'(x_1, \ldots, x_n)],$$

then there is a formula $R(x_1, \ldots, x_n, p_0, \ldots, p_s)$ *such that*

$$\vdash E(Q, p_0, \ldots, p_s, r_0, \ldots, r_t) \supset [Q(x_1, \ldots, x_n) \sim R(x_1, \ldots, x_n, p_0, \ldots, p_s)].$$

(B) *If*

$$\vdash E(g, p_0, \ldots, p_s, r_0, \ldots, r_t) \ \&\ E(g', p_0, \ldots, p_s, r_0', \ldots, r_t') \supset g(x_1, \ldots, x_{n-1}) = g'(x_1, \ldots, x_{n-1}),$$

then there is a formula $R(x_1, \ldots, x_n, p_0, \ldots, p_s)$ *such that*

$$\vdash E(g, p_0, \ldots, p_s, r_0, \ldots, r_t) \supset [g(x_1, \ldots, x_{n-1}) = x_n \sim R(x_1, \ldots, x_n, p_0, \ldots, p_s)].$$

PROOF. (a) As before, except using Theorem 42, which gives us an $R(x_1, \ldots, x_n, p_0, \ldots, p_s)$ in all cases.

(b) In the predicate calculus with equality, $g(x_1, \ldots, x_{n-1}) = g'(x_1, \ldots, x_{n-1})$ is equivalent to $g(x_1, \ldots, x_{n-1}) = x_n \sim g'(x_1, \ldots, x_{n-1}) = x_n$. Now the above reasoning applies with $g(x_1, \ldots, x_{n-1}) = x_n$ in place of $Q(x_1, \ldots, x_n)$. —

As we remarked in § 38,[157] in **N** with the symbol · and its axioms 20 and 21 omitted, the representing predicate $a{\cdot}b = c$ of $a{\cdot}b$ cannot be expressed. Hence by Theorem 44 (B) (contraposed),[251] Padoa's method applies; i.e. there are two models of **N**, with the same domain D and the same assignment to 0, ′, + (with = meaning equality), in which · has different functions as values. —

As a second application of Craig's interpolation theorem, we establish the consistency theorem of A. Robinson 1956, 1963 p. 114. Robinson proved this independently of Craig's theorem, and used it to prove the theorem of Beth.

The problem is this. Suppose we have two consistent formal systems S_1 and S_2, with axioms $A_0, A_1, A_2, \ldots$ and $B_0, B_1, B_2, \ldots$ and nonvariable parameters $p_0, p_1, p_2, \ldots$ and $q_0, q_1, q_2, \ldots$, respectively. (The reader may substitute finite lists.) Will we obtain a consistent system $S_1 \cup S_2$ by combining S_1 and S_2 so the axioms of $S_1 \cup S_2$ are $A_0, B_0, A_1, B_1, A_2, B_2, \ldots$ and the nonvariable parameters are $p_0, q_0, p_1, q_1, p_2, q_2, \ldots$ (each list with repetitions allowed)?

Not necessarily. For, let S_1 be number theory **N** with the addition of Gödel's formula $\neg C_p$ (which is true in the standard model, but unprovable in **N**), and S_2 be **N** with C_p added. The systems S_1 and S_2 are each consistent, using the consistency of **N** (cf. § 47, where S_2 is called "**M**"). But the union $S_1 \cup S_2$ is not. Clearly the difficulty here is that the common portion **N** of S_1 and S_2 is incomplete, so that in S_1 and in S_2 we could extend **N** in two different ways, each consistent with **N**, but inconsistent with each other.

This example suggests an additional hypothesis under which we can hope to answer the question affirmatively: the two formal systems should be in "complete agreement" on topics of mutual concern; i.e., for each closed formula I which contains only parameters common to S_1 and S_2, the same one of the two formulas I and $\neg$I should be provable in each of S_1 and S_2.

The reasoning here, as in Gödel's completeness theorem and most other passages in this chapter dealing with $\aleph_0$ formulas, does not depend on the formulas being given effectively, as the axioms of a formal system are required to be (cf. §§ 37, 43). So in stating the theorem, we allow S_1 and S_2 alternatively to be sets of formulas based on the predicate calculus, analogously to (the set of provable formulas of) a formal system based on the predicate calculus, but without the effectiveness requirement for the axioms.

THEOREM 45. (Robinson's consistency theorem, 1956.) *Let* S_1 *and* S_2 *be formal systems (or sets of formulas) based on the predicate calculus without or with functions and equality. Suppose that* S_1 *and* S_2 *are each simply consistent. Suppose they are in complete agreement, i.e. for each closed formula* I *in the common portion (or intersection) of their symbolisms, one of* I *and* $\neg$I *is provable in both* S_1 *and* S_2. *Then the combined system (or union)* $S_1 \cup S_2$ *is simply consistent.*

PROOF. We begin with the case of the predicate calculus without functions and equality but admitting individuals. Let $E_0, E_1, E_2, \ldots$ and $F_0, F_1, F_2, \ldots$ be the closures of the axioms of S_1 and S_2, respectively. Suppose $S_1 \cup S_2$ is not consistent, so a contradiction K & $\neg$K is deducible from $E_0, F_0, E_1, F_1, E_2, F_2, \ldots$ in the predicate calculus. Let $E_0, \ldots, E_c$, $F_0, \ldots, F_d$ be those of $E_0, F_0, E_1, F_1, E_2, F_2, \ldots$ which are used in a given deduction of $K \& \neg K$. Thus $E_0, \ldots, E_c, F_0, \ldots, F_d \vdash K \& \neg K$ in the predicate calculus. Thence by propositional calculus,

$$\vdash E_0 \& \ldots \& E_c \supset \neg(F_0 \& \ldots \& F_d).$$

We take this as the "$\vdash E \supset F$" for Theorem 41. If no predicate parameter is common to $E_0 \& \ldots \& E_c$ and $\neg(F_0 \& \ldots \& F_d)$, then by Theorem 41 (b) $\vdash \neg(E_0 \& \ldots \& E_c)$ or $\vdash \neg(F_0 \& \ldots \& F_d)$. But then S_1 or S_2, respectively, is inconsistent, contrary to hypothesis. If some predicate parameter is common, then by Theorem 41 (a) there is a formula I such that

$$\text{(i)}\ \vdash E_0 \& \ldots \& E_c \supset I \text{ and (ii)}\ \vdash I \supset \neg(F_0 \& \ldots \& F_d)$$

and I contains only parameters common to $E_0 \& \ldots \& E_c$ and $\neg(F_0 \& \ldots \& F_d)$. So I is closed, since $E_0, E_1, E_2, \ldots$ and $F_0, F_1, F_2, \ldots$ are closed. By the hypothesis of complete agreement, in the predicate calculus

either (iii-1) $E_0, E_1, E_2, \ldots \vdash I$ and (iv-1) $F_0, F_1, F_2, \ldots \vdash I$

or (iii-2) $E_0, E_1, E_2, \ldots \vdash \neg I$ and (iv-2) $F_0, F_1, F_2, \ldots \vdash \neg I$.

But in the first case, (ii) and (iv-1) are incompatible with the consistency of S_2; for, contraposing (ii) (*13 in § 24), $\vdash F_0 \& \ldots \& F_d \supset \neg I$. In the second case, (i) and (iii-2) are incompatible with the consistency of S_1.

For the predicate calculus with functions and equality, we use Theorem 42 instead.

EXERCISE 57.1. Show that, if $E(Q_0, p_0, \ldots, p_s, r_0, \ldots, r_t)$ defines Q explicitly from $p_0, \ldots, p_s$, it does so implicitly.

BIBLIOGRAPHY

ACKERMANN, WILHELM (also see Hilbert and —)

1924–5. *Begründung des "tertium non datur" mittels der Hilbertschen Theorie der Widerspruchsfreiheit.* **Mathematische Annalen,** vol. 93, pp. 1–36.

1940. *Zur Widerspruchsfreiheit der Zahlentheorie.* Ibid., vol. 117, pp. 162–194.

ADDISON, J. W.

1960. *The theory of hierarchies.* **Logic, methodology and philosophy of science: Proceedings of the 1960 International Congress** (Stanford, Aug. 24–Sept. 2), ed. by Nagel, Suppes and Tarski, Stanford, Calif. (Stanford Univ. Press) 1962, pp. 26–37.

ADDISON, J. W., HENKIN, LEON and TARSKI, ALFRED

1965. (editors) **The theory of models, Proceedings of 1963 the International Symposium at Berkeley.** Amsterdam (North-Holland Pub. Co.), XV+494 pp.

AMBROSE, ALICE and LAZEROWITZ, MORRIS

1948. **Fundamentals of symbolic logic.** New York (Rinehart), ix+310 pp.

BACHMANN, HEINZ

1955. **Transfinite Zahlen.** Ergebnisse der Mathematik und ihrer Grenzgebiete, n.s., no. 1, Berlin, Göttingen and Heidelberg (Springer-Verlag), VII+204 pp.

BALL, W. W. ROUSE

1892. **Mathematical recreations and essays.** 1st. ed. 1892; 11th. ed., revised by H. S. M. Coxeter, New York (Macmillan) 1939, xvi+418 pp.

BAR-HILLEL, Y., PERLES, M. and SHAMIR, E.

1961. *On formal properties of simple phase structure grammars.* **Zeitschrift für Phonetik, Sprachwissenschaft und Kommunikationsforschung,** vol. 14, pp. 143–172.

BENACERRAF, PAUL and PUTNAM, HILARY

1964. (editors) **Philosophy of mathematics, selected readings.** Englewood Cliffs, N.J. (Prentice-Hall), vii+536 pp.

BERNAYS, PAUL (also — and Fraenkel; Hilbert and —)

1937–1954. *A system of axiomatic set theory—Parts I–VII.* **The journal of symbolic logic,** vol. 2 (1937), pp. 65–77, vol. 6 (1941), pp. 1–17, vol. 7 (1942), pp. 65–89, 133–145, vol. 8 (1943), pp. 89–106, vol. 13 (1948), pp. 65–79, vol. 19 (1954), pp. 81–96.

BERNAYS, PAUL (and FRAENKEL, ABRAHAM A.)

1958. **Axiomatic set theory,** by Bernays, with a historical introduction by Fraenkel (pp. 3–35). Amsterdam (North-Holland Pub. Co.), VIII+226 pp.

BERNSTEIN, ALLEN R. and ROBINSON, ABRAHAM

1966 (abstract 1964). *Solution of an invariant subspace problem of K. T. Smith and P. R. Halmos.* **Pacific journal of mathematics,** vol. 16, pp. 421–431. Abstract in **Notices of the American Mathematical Society,** vol. 11, p. 586.

BETH, EVERT W.

1951. *A topological proof of the theorem of Löwenheim-Skolem-Gödel.* **Koninklijke Nederlandse Akademie van Wetenschappen** (Amsterdam), **Proceedings, ser. A,** vol. 54 (or **Indagationes mathematicae,** vol. 13), pp. 436–444.

1953. *On Padoa's method in the theory of definition.* Ibid., vol. 56 (or vol. 15), pp. 330–339.

1955. *Semantic entailment and formal derivability.* **Mededelingen der Koninklijke Nederlandse Akademie van Wetenschappen** (Amsterdam), **Afd. letterkunde,** n.s., vol. 18, no. 13, pp. 309–342.

1959. **The foundations of mathematics.** Amsterdam (North-Holland Pub. Co.), XXVI+741 pp.

BOCHEŃSKI, I. M.

1956. **Formale Logik.** Freiburg and Munich (Verlag Karl Alber), XV+640 pp. Eng. tr. by Ivo Thomas, **A history of formal logic,** Notre Dame, Ind. (Univ. of Notre Dame Press) 1961, xxii+567 pp.

BOLZANO, BERNARD

1851. **Paradoxien des Unendlichen.** Berlin, 12+134 pp. Eng. tr. **Paradoxes of the infinite,** with hist. introd. (pp. 1–55), by Donald A. Steele, London (Routledge and Paul) 1950, ix+189 pp.

BOOLE, GEORGE

1847. **The mathematical analysis of logic, being an essay toward a calculus of deductive reasoning.** Cambridge and London, 82 pp. Reprinted Oxford (Basil Blackwell) and New York (Philos. Libr.) 1948.

BOONE, WILLIAM W. (also —, Haken and Poénaru)

1954–7. *Certain simple, unsolvable problems of group theory, I–VI.* **Kon. Ned. Akad. Wet.** (Amsterdam), **Proc., ser. A.,** vol. 57 (1954), pp. 231–237, 492–497, vol. 58 (1955), pp. 252–256, 571–577, vol. 60 (1957), pp. 22–27, 227–232 (or **Indag. math.,** vols. 16, 17, 19, resp.).

1958. *An analysis of Turing's "The word problem in semi-groups with cancellation".* **Annals of mathematics,** 2 s., vol. 67, pp. 195–202.

1959. *The word problem.* Ibid., vol. 70, pp. 207–265.

1966. *Word problems and recursively enumerable degrees of unsolvability. A first paper on Thue systems.* Ibid., vol. 83, pp. 520–571.

1966a. *Word problems and recursively enumerable degrees of unsolvability. A sequel on finitely presented groups.* Ibid., vol. 84, pp. 49–84.

1967. *Decision problems about algebraic and logical systems as a whole and recursively enumerable degrees of unsolvability.* To appear in the proceedings of the 1966 "Kolloquium über Logik und Grundlagen der Mathematik" in Hannover.

BOONE, W. W., HAKEN, WOLFGANG and POÉNARU, VALENTIN

1967. *On recursively unsolvable decision problems in topology and their classification.* Ibid.

BOREL, ÉMILE

1898. **Leçons sur la théorie des fonctions.** Paris, 8+136 pp. 4th ed. **(Leçons sur la théorie des fonctions; principes de la théorie des ensembles en vue des applications à la théorie des fonctions),** Paris (Gauthier-Villars) 1950, xii+295 pp.

BRITTON, J. L.

1956–8. *Solution of the word problem for certain types of groups II.* **Proceedings of the Glasgow Mathematical Association,** vol. 3, pp. 68–90.

1958. *The word problem for groups.* **Proceedings of the London Mathematical Society,** 3 s., vol. 8 (1958), pp. 493–506.

1963. *The word problem.* **Ann. of math.,** 2 s., vol. 77, pp. 16–32.

BROUWER, L. E. J.

1908. *De onbetrouwbaarheid der logische principes* (The untrustworthiness of the principles of logic). **Tijdschrift voor wijsbegeerte,** vol. 2, pp. 152–158. Reprinted in **Wiskunde, waarheid, werkelijkheid,** by L. E. J. Brouwer, Groningen (P. Noordhoff) 1919, 12 pp.

1923. *Über die Bedeutung des Satzes vom ausgeschlossenen Dritten in der Mathematik, insbesondere in der Funktionentheorie.* **Journal für die reine und angewandte Mathematik,** vol. 154 (1925), pp. 1–7. Original in Dutch 1923. Eng. tr. in van Heijenoort 1967.

1923a. *Begründung der Funktionenlehre unabhängig vom logischen Satz vom ausgeschlossenen Dritten.* **Verhandelingen der Koninklijke Nederlandsche Akademie van Wetenschappen te Amsterdam (Eerste sectie),** vol. 13, no. 2, 24 pp. (Errata in 1924 Footnote 1.)

1924. *Beweis, dass jede volle Funktion gleichmässig stetig ist.* **Kon. Ned. Akad. Wet. Amsterdam, Proc. Sect. Sci.,** vol. 27, pp. 189–193.

1928. *Intuitionistische Betrachtungen über den Formalismus,* **Sitzungsberichte der Preussischen Akademie der Wissenschaften, Physikalisch-mathematische Klasse,** 1928, pp. 48–52. Also **Kon. Ned. Akad. Wet. Amsterdam, Proc. Sect. Sci.,** vol. 31, pp. 374–379. Eng. tr. in van Heijenoort 1967.

BÜCHI, J. RICHARD

1962. *Turing-machines and the Entscheidungsproblem.* **Math. Ann.,** vol. 148, pp. 201–213.

BURALI-FORTI, CESARE

1897. *Una questione sui numeri transfiniti.* **Rendiconti del Circolo Matematico di Palermo,** vol. 11, pp. 154–164 (cf. p. 260). Eng. tr. in van Heijenoort 1967.

CANTOR, GEORG

1874. *Über eine Eigenschaft des Inbegriffes aller reellen algebraischen Zahlen.* **Jour. reine angew. Math.,** vol. 77, pp. 258–262. Reprinted in **Georg Cantor Gesammelte Abhandlungen,** 1932, pp. 115–118.

1895–7. *Beiträge zur Begründung der transfiniten Mengenlehre.* **Math. Ann.,** vol. 46 (1895), pp. 481–512, and vol. 49 (1897), pp. 207–246. Reprinted in **Gesam. Abh.,** pp. 282–356. Eng. tr. by Ph. E. B. Jourdain, **Contributions to the founding of the theory of transfinite numbers,** Chicago and London (Open Court), 1915, ix+211 pp.

CARNAP, RUDOLPH

1934. **The logical syntax of language.** New York (Harcourt, Brace) and London (Kegan Paul, Trench, Trubner) 1937, xvi+352 pp. Tr. by Amethe Smeaton from the German original 1934, with additions.

1935. *Ein Gültigkeitskriterium für die Sätze der klassischen Mathematik.* **Monatshefte für Mathematik und Physik,** vol. 42, pp. 163–190.

CHURCH, ALONZO (also — and Kleene)

1936. *An unsolvable problem of elementary number theory.* **American journal of mathematics,** vol. 58, pp. 345–363. Reprinted in Davis 1965, pp. 88–107.

1936a. *A note on the Entscheidungsproblem.* **Jour. symbolic logic,** vol. 1, pp. 40–41. *Correction,* ibid., pp. 101–102. Reprinted in Davis 1965, pp. 108–115.

1938. *The constructive second number class.* **Bulletin of the American Mathematical Society,** vol. 44, pp. 224–232.

1956. **Introduction to mathematical logic,** vol. I. Princeton, N.J. (Princeton Univ. Press), x+376 pp.

Church, Alonzo and Kleene, S. C.

1936. *Formal definitions in the theory of ordinal numbers*. **Fund. Math.**, vol. 28, pp. 11–21.

Clapham, C. R. J.

1964. *Finitely presented groups with word problems of arbitrary degrees of insolubility*. **Proc. London Math. Soc.**, 3 s., vol. 14, pp. 633–676.

Clark, Romane and Welsh, Paul

1962. **Introduction to logic.** Princeton, Toronto, New York and London (Van Nostrand), xii+268 pp.

Cohen, Paul J.

1963–4. *The independence of the continuum hypothesis*, and ibid., *II*. **Proceedings of the National Academy of Sciences,** vol. 50, pp. 1143–1148 (1963) and vol. 51, pp. 105–110 (1964).

1966. **Set theory and the continuum hypothesis.** New York and Amsterdam (W. A. Benjamin, Inc.), vi+154 pp.

Craig, William

1957. *Linear reasoning. A new form of the Herbrand-Gentzen theorem*. **Jour. symbolic logic,** vol. 22, pp. 250–268.

1957a. *Three uses of the Herbrand-Gentzen theorem in relating model theory and proof theory*. Ibid., pp. 269–285.

Curry, Haskell B.

1950. *A theory of formal deducibility*. Notre Dame mathematical lectures, no. 6, Univ. of Notre Dame, Notre Dame, Ind., ix+126 pp.

1952. *The permutability of rules in the classical inferential calculus*. **Jour. symbolic logic,** vol. 17, pp. 245–248.

Davis, Martin (also — and Putnam)

1950. *On the theory of recursive unsolvability*. Ph.D. thesis, Princeton Univ., 96 pp.

1965. (editor) **The undecidable. Basic papers on undecidable propositions, unsolvable problems and computable functions.** Hewlitt, N.Y. (Raven Press), v+440 pp. Cf. Stefan Bauer-Mengelberg in **Jour. symbolic logic,** vol. 31 (1966), pp. 484–494.

Davis, M. and Putnam, H.

1960. *A computing procedure for quantification theory*. **Journal of the Association for Computing Machinery,** vol. 7, pp. 201–215.

Dedekind, Richard

1888. **Was sind und was sollen die Zahlen?** Braunschweig (6th ed. 1930). Also in **Dedekind Gesammelte mathematische Werke,** vol. III (1932), pp. 335–391. Eng. tr. in **Essays on the theory of numbers,** Chicago (Open Court) 1901, pp. 29–115.

Dehn, Max (also Pasch and —)

1912. *Über unendliche diskontinuierliche Gruppen*. **Math. Ann.,** vol. 71, pp. 116–144.

De Morgan, Augustus

1847. **Formal logic: or, the calculus of inference, necessary and probable.** London, xvi+336 pp. Reprinted (ed. A. E. Taylor) Chicago and London (Open Court) 1926, 392 pp.

Denton, John and Dreben, Burton

1968. **The Herbrand theorem and the consistency of arithmetic.** Forthcoming 1968(?).

Dodgson, Charles Lutwidge (= Lewis Carroll)

1887, 1897. **The game of logic,** London (Macmillan) 1887, 96 pp. **Symbolic logic, Part I, Elementary,** 4th ed. London, New York (Macmillan) 1897, xxxi+201 pp.

Both reprinted in **Mathematical recreations,** by Lewis Carroll, 2 vols. New York (Dover) 1958.

DREBEN, BURTON, ANDREWS, PETER and AANDERAA, STÅL (also Denton and Dreben)

1963. *False lemmas in Herbrand.* **Bull. Amer. Math. Soc.,** vol. 69, pp. 699–706.

FEFERMAN, SOLOMON

1958 abstracts. *Ordinal logics re-examined* and *On the strength of ordinal logics.* **Jour. symbolic logic,** vol. 23, pp. 105–106.

1960. *Arithmetization of metamathematics in a general setting.* **Fund. math.,** vol. 49, pp. 35–92.

1962. *Transfinite recursive progressions of axiomatic theories.* **Jour. symbolic logic,** vol. 27, pp. 259–316.

FEYS, ROBERT

1965. **Modal logics.** Ed. with some complements by Joseph Dopp, Louvain (E. Nauwelaerts) and Paris (Gauthier-Villars), XIV+219 pp.

FRAENKEL, ABRAHAM ADOLF (also Bernays and —; — and Bar-Hillel)

1922. *Der Begriff "definit" und die Unabhängigkeit des Auswahlaxioms.* **Sitz. Preuss. Akad. Wiss., Phys.-math. Kl.,** 1922, pp. 253–257. Eng. tr. in van Heijenoort 1967.

1961. **Abstract set theory,** 2nd ed. Amsterdam (North-Holland Pub. Co.), VIII+295 pp. (1st ed. 1953, XII+479 pp.)

FRAENKEL, ABRAHAM A. and BAR-HILLEL, YEHOSHUA

1958. **Foundations of set theory.** Amsterdam (North-Holland Pub. Co.), X+415 pp.

FREGE, GOTTLOB

1879. *Begriffsschrift, eine der arithmetischen nachgebildete Formelsprache des reinen Denkens.* Halle, viii+88 pp. Eng. tr. in van Heijenoort 1967.

1884. **Die Grundlagen der Arithmetik, eine logisch-mathematische Untersuchung über den Begriff der Zahl.** Breslau, xix+119 pp. Reprinted Breslau (M. & H. Marcus) 1934. Eng. tr. by J. L. Austin (with German original): **The foundations of arithmetic. A logico-mathematical enquiry into the concept of number.** Oxford (Basil Blackwell) and New York (Philosophical Library) 1950, (xii+XI+119) × 2 pp.

FRIEDBERG, RICHARD M.

1956 article concerning. **Time,** vol. 67, no. 12 (March 19, 1956), p. 83.

1957 (abstract 1956). *Two recursively enumerable sets of incomparable degrees of unsolvability (solution of Post's problem,* 1944). **Proc. Nat. Acad. Sci.,** vol. 43, pp. 236–238. Abstract in **Bull. Amer. Math. Soc.,** vol. 62, p. 260.

GENTZEN, GERHARDT

1932. *Über die Existenz unabhängiger Axiomensysteme zu unendlichen Satzsystemen.* **Math. Ann.,** vol. 107, pp. 329–350.

1934–5. *Untersuchungen über das logische Schliessen.* **Mathematische Zeitschrift,** vol. 39, pp. 176–210, 405–431. Fr. tr. by Jean Ladrière (with notes by R. Feys and Ladrière), **Recherches sur la déduction logique,** Paris (Presses Universitaires de France) 1955, XI+170 pp.

1936. *Die Widerspruchsfreiheit der reinen Zahlentheorie.* **Math. Ann.,** vol. 112, pp. 493–565.

1938a. *Neue Fassung des Widerspruchsfreiheitsbeweises für die reine Zahlentheorie.* **Forschungen zur Logik und zur Grundlegung der exakten Wissenschaften,** n.s., no. 4, Leipzig (Hirzel), pp. 19–44.

GÖDEL, KURT

1930. *Die Vollständigkeit der Axiome des logischen Funktionenkalküls.* **Monatsh. Math. Phys.,** vol. 37, pp. 349–360. Eng. tr. in van Heijenoort 1967.

1931. *Über formal unentscheidbare Sätze der Principia Mathematica und verwandter Systeme I.* Ibid., vol. 38, pp. 173–198. Eng. tr. in Davis 1965 pp. 4–38, and in van Heijenoort 1967.

1931–2. *Über Vollständigkeit und Widerspruchsfreiheit.* **Ergebnisse eines mathematischen Kolloquiums,** Heft 3 (for 1930–1, pub. 1932), pp. 12–13.

1931–2b. *Eine Eigenschaft der Realisierung des Aussagenkalküls.* Ibid., pp. 20–21.

1932–3. *Zur intuitionistischen Arithmetik und Zahlentheorie.* Ibid., Heft 4 (for 1931–2, pub. 1933), pp. 34–38. Eng. tr. in Davis 1965, pp. 75–81.

1934. **On undecidable propositions of formal mathematical systems.** Notes by S. C. Kleene and Barkley Rosser on lectures at the Institute for Advanced Study 1934, mimeographed, Princeton, N.J., 30 pp. Reprinted (with a postscript) in Davis 1965, pp. 39–74.

1934a. Review of Skolem 1933. **Zentralblatt für Mathematik und ihre Grenzgebiete,** vol. 7, pp. 193–194.

1938. *The consistency of the axiom of choice and of the generalized continuum-hypothesis.* **Proc. Nat. Acad. Sci.,** vol. 24, pp. 556–557.

1939. *Consistency-proof for the generalized continuum-hypothesis.* Ibid., vol. 25, pp. 220–224.

1940. **The consistency of the axiom of choice and of the generalized continuum-hypothesis with the axioms of set theory.** Notes by George W. Brown on lectures at the Institute for Advanced Study 1938–9. Annals of Mathematics studies, no. 3, Princeton, N.J. (Princeton Univ. Press), 66 pp. Second printing 1951, v+69 pp.

1947. *What is Cantor's continuum problem?* **American mathematical monthly,** vol. 54, pp. 515-525. Reprinted with additions in Benacerraf and Putnam 1964, pp. 258-273.

HALL, MARSHALL, JR.

1959. **The theory of groups.** New York (Macmillan), xiii+434 pp.

HENKIN, LEON

1947. **The completeness of formal systems.** Ph.D. thesis, Princeton, ii+75 pp.

1949. *The completeness of the first-order functional calculus.* **Jour. symbolic logic,** vol. 14, pp. 159–166.

1950. *Completeness in the theory of types.* Ibid., vol. 15, pp. 81–91.

1963. *An extension of the Craig-Lyndon interpolation* theorem. Ibid., vol. 28, pp. 201–216.

HERBRAND, JACQUES

1928. *Sur la théorie de la démonstration.* **Comptes rendus hebdomadaires des séances de l'Académie des Sciences** (Paris), vol. 186, pp. 1274–1276.

1930. **Recherches sur la théorie de la démonstration.** Travaux de la Société des Sciences et des Lettres de Varsovie, Classe III sciences mathématiques et physiques, no. 33, 128 pp. Eng. tr. of Chapter 5 (extensively annotated) by Burton Dreben and J. van Heijenoort in van Heijenoort 1967.

1931–2. *Sur la non-contradiction de l'arithmétique.* **Jour. reine angew. Math.,** vol. 166, pp. 1–8. Eng. tr. in van Heijenoort 1967.

HERTZ, PAUL

1929. *Über Axiomensysteme für beliebige Satzsysteme.* **Math. Ann.,** vol. 101, pp. 457–514.

HEYTING, AREND

1930. *Die formalen Regeln der intuitionistischen Logik.* **Sitz. Preuss. Akad. Wiss., Phys.-math. Kl.,** 1930, pp. 42–56.

1930a. *Die formalen Regeln der intuitionistischen Mathematik.* Ibid., pp. 57–71, 158–169.

1934. **Mathematische Grundlagenforschung. Intuitionismus. Beweistheorie.** Ergeb. Math. Grenzgeb., vol. 3, no. 4, Berlin (Springer), iv+73 pp.

1955. **Les fondements des mathématiques. Intuitionnisme. Théorie de la démonstration.** Paris (Gauthier-Villars) and Louvain (E. Nauwelaerts), 91 pp. The second ed. of 1934.

1956. *Intuitionism. An introduction.* Amsterdam (North-Holland Pub. Co.), VIII+133 pp.

HIGMAN, G.

1961. *Subgroups of finitely presented groups.* **Proceedings of the Royal Society of London, ser. A,** vol. 262, pp. 455–475.

HILBERT, DAVID (also — and Ackermann; — and Bernays)

1899. **Grundlagen der Geometrie.** 7th ed. (1930), Leipzig and Berlin (Teubner), vii+326 pp. Eng. tr. by E. J. Townsend, **The foundations of geometry,** Chicago (Open Court) 1902, vii+143 pp.

1900a. *Mathematical problems. Lecture delivered before the International Congress of Mathematicians at Paris in* 1900. Eng. tr. from the original Ger., **Bull. Amer. Math. Soc.,** vol. 8 (1901–2), pp. 437–479. *Sur les problèmes futurs des mathématiques,* Fr. tr. with some modifications and additions, **Compte rendus du Deuxième Congrès International des Mathématiciens tenu à Paris du 6 au 12 août 1900,** Paris 1902, pp. 58–114.

1904. *Über die Grundlagen der Logik und der Arithmetik.* **Verhandlungen des Dritten Internationalen Mathematiker-Kongresses in Heidelberg vom 8. bis 13. August 1904,** Leipzig 1905, pp. 174–185. Reprinted in 7th ed. of 1899, pp. 247–261. Eng. tr. in van Heijenoort 1967.

1918. *Axiomatisches Denken.* **Math. Ann.,** vol. 78, pp. 405–415. Reprinted in **David Hilbert Gesammelte Abhandlungen,** vol. 3, Berlin (Springer) 1935, pp. 146–156.

1926. *Über das Unendliche.* **Math. Ann.,** vol. 95, pp. 161–190. Reprinted with some revisions in 7th ed. of 1899, pp. 262–288. Eng. tr. in Benacerraf and Putnam 1964 pp. 134–151 (abridged) and in van Heijenoort 1967.

1928. *Die Grundlagen der Mathematik.* **Abhandlungen aus dem Mathematischen Seminar der Hamburgischen Universität,** vol. 6, pp. 65–85. Reprinted with abridgements in 7th ed. of 1899, pp. 289–312. Eng. tr. in van Heijenoort 1967.

HILBERT, DAVID and ACKERMANN, WILHELM

1928, 1938, 1949. **Grundzüge der theoretischen Logik.** Berlin (Springer), viii+120 pp. 2nd. ed. 1938, viii+133 pp. Reprinted New York (Dover Pubs.) 1946. 3rd. ed. Berlin, Göttingen, Heidelberg (Springer) 1949, viii+155 pp. Eng. tr. of 2nd. ed., **Principles of mathematical logic,** New York (Chelsea Pub. Co.), 1950, xii+172 pp.

HILBERT, DAVID and BERNAYS, PAUL

1934, 1939. **Grundlagen der Mathematik.** Vol. 1 1934, xii+471 pp. Vol. 2, 1939, xii+498 pp. Berlin (Springer). Reprinted Ann Arbor, Mich. (J. W. Edwards) 1944.

HINTIKKA, K. JAAKKO J.

1955. *Form and content in quantification theory.* **Two papers on symbolic logic,** Acta philosophica Fennica no. 8, Helsinki 1955, pp. 7–55.

1955a. *Notes on quantification theory.* **Societas Scientiarum Fennica, Commentationes physico-mathematicae,** vol. 17, no. 12, 13 pp.

JAŚKOWSKI, STANISŁAW

1934. **On the rules of suppositions in formal logic.** Studia logica, no. 1, Warsaw, 32 pp.

KALMÁR, LASLÓ

1928–9. *Eine Bemerkung zur Entscheidungstheorie.* **Acta litterarum ac scientiarum Regiae Universitatis Hungaricae Francisco-Josephinae, Sectio scientiarum mathematicarum** (Szeged), vol. 4, pp. 248–252.

1934–5. *Über die Axiomatisierbarkeit des Aussagenkalküls.* Ibid., vol. 7, pp. 222–243.

1959. *An argument against the plausibility of Church's thesis.* **Constructivity in mathematics, Proceedings of the Colloquium held at Amsterdam,** [Aug. 26–31] **1957** (A. Heyting, ed.), Amsterdam (North-Holland Pub. Co.), pp. 72–80.

KANGER, STIG

1957. **Provability in logic.** Acta Universitatis Stockholmiensis, Stockholm studies in philosophy 1, Stockholm (Almqvist and Wiksell), 47 pp.

KEMENY, JOHN G.

1958. *Undecidable problems of elementary number theory.* **Math. Ann.,** vol. 135, pp. 160–169.

KETONEN, OIVA

1944. *Untersuchungen zum Prädikatenkalkäl.* **Annales Academia Scientiarum Fennicae,** ser. A, **I. Mathematica-physica** 23, Helsinki, 71 pp.

KLEENE, STEPHEN COLE (also Church and —; — and Post; — and Vesley)

1934. *Proof by cases in formal logic.* **Ann. of Math.,** 2 s., vol. 35, pp. 529–544.

1935. *A theory of positive integers in formal logic.* **Amer. jour. math.,** vol. 57, pp. 153–173, 219–244.

1936. *General recursive functions of natural numbers.* **Math. Ann.,** vol. 112, pp. 727–742. Reprinted in Davis 1965, pp. 236–253.

1936a. *λ-definability and recursiveness.* **Duke mathematical journal,** vol. 2, pp. 340–353.

1938. *On notation for ordinal numbers.* **Jour. symbolic logic,** vol. 3, pp. 150–155.

1943 (abstract 1940). *Recursive predicates and quantifiers.* **Transactions of the American Mathematical Society,** vol. 53, pp. 41–73. Reprinted in Davis 1965, pp. 254–287. Abstract in **Bull. Amer. Math. Soc.,** vol. 46, p. 885.

1950. *A symmetric form of Gödel's theorem.* **Kon. Ned. Akad. Wet.** (Amsterdam), **Proc. Sect. Sci.,** vol. 53, pp. 800–802 (or **Indag. math.,** vol. 12, pp. 244–246).

1952. *Permutability of inferences in Gentzen's calculi LK and LJ.* **Two papers on the predicate calculus,** by S. C. Kleene (Memoirs Amer. Math. Soc., no. 10), pp. 1–26. In Footnote 15 line 2 "subformula" should be "side formula", and in line 4 the two ¬→'s (or ⊃→'s) should have the same principal formula, as noted by G. E. Minc.[269] These and other corrections of **Two papers** are made in the 1957 edition.

1952a. *Finite axiomatizability of theories in the predicate calculus using additional predicate symbols,* Ibid., pp. 27–68. Amplifications are required on p. 53 (to treat →∃'s and →&'s belonging to operators in N(x)) and helpful on p. 58 middle, as given by G. E. Minc in the Russian tr. of **Two papers** in **Matematičeskaja teorija logičeskogo vyvoda** (The mathematical theory of logical inference), Moscow (Nauka) 1967(?).

1952b. **Introduction to metamathematics.** Amsterdam (North-Holland Pub. Co.), Groningen (Noordhoff), New York and Toronto (Van Nostrand), X+550 pp. Some errata in early printings, and emendations, are listed in Kleene and Vesley 1965 pp. 1, 192 and here in Footnote 86.

1955b. *Hierarchies of number-theoretic predicates.* **Bull. Amer. Math. Soc.,** vol. 61, pp. 193–213.

1956. *A note on computable functionals.* **Kon. Ned. Akad. Wet.** (Amsterdam), **Proc., Ser. A,** vol. 59 (or **Indag. math.,** vol. 18), pp. 275–280.

1956a. **Sets, logic, and mathematical foundations.** Notes by H. William Oliver on lectures at a N.S.F. Summer Institute for Teachers of Secondary and College

Mathematics, Williams College, Williamstown, Mass., mimeographed, v+169 pp.

1957b. *Mathematics, Foundations of.* **Encyclopaedia Britannica,** 1957 and subsequent printings.

1958. *Mathematical logic: constructive and non-constructive operations.* **Proceedings of the International Congress of Mathematicians,** Edinburgh, **14–21 August 1958,** Cambridge (Cambridge Univ. Press) 1960, pp. 137–153.

1961. **Mathematical logic.** Notes by Edward Pols on lectures at a N.S.F. Summer Institute, Bowdoin College, Brunswick, Maine, mimeographed, ix+217 pp.

1964. *Computability.* **The Voice of America Forum Lectures, Philosophy of Science Series,** no. 6, 8 pp. (broadcast 1963, pub. 1964). The series reprinted New York (The Editors of Basic Books) 1967(?), Sidney Morgenbesser ed.

KLEENE, S. C. and POST, EMIL.

1954. *The upper semi-lattice of degrees of recursive unsolvability.* **Ann. of Math.,** 2 s., vol. 59, pp. 379–407.

KLEENE, STEPHEN COLE and VESLEY, RICHARD EUGENE

1965. **The foundations of intuitionistic mathematics, especially in relation to recursive functions.** Amsterdam (North-Holland Pub. Co.), VIII+206 pp. Erratum: p. 113 line 11, for "no" read "each".

KOLMOGOROV, A. N.

1924–5. *O principe tertium non datur* (*Sur le principe de tertium non datur*). **Matematičeskiĭ sbornik (Recueil mathématique de la Société Mathématique de Moscou),** vol. 32, pp. 646–667 (with brief French abstract). Eng. tr. in van Heijenoort 1967.

KÖNIG, DÉNES

1926. *Sur les correspondences multivoques des ensembles.* **Fund. math.,** vol. 8, pp. 114–134.

KREISEL, GEORG

1951–2. *On the interpretation of non-finitist proofs.* **Jour. symbolic logic,** vol. 16 (1951), pp. 241–267, vol. 17 (1952), pp. 43–58.

1953. *A variant to Hilbert's theory of the foundations of arithmetic.* **The British journal for the philosophy of science,** vol. 4 (1953–4), pp. 107–129, 357.

1958. *Mathematical significance of consistency proofs.* **Jour. symbolic logic,** vol. 23, pp. 155–182.

1958c. *Ordinal logics and the characterization of informal concepts of proof.* **Proc. Internat. Congress Math. Edinburgh 1958,** pp. 289–299.

1965. *Mathematical logic.* **Lectures on modern mathematics,** vol. III (ed. by T. L. Saaty), New York (Wiley), pp. 95–195.

LEWIS, CLARENCE IRVING (also — and Langford)

1912. *Implication and the algebra of logic.* **Mind,** n.s., vol. 21, pp. 522–531.

1917. *The issues concerning material implication.* **The journal of philosophy, psychology and scientific method,** vol. 14, pp. 350–356.

1918. **A survey of symbolic logic.** Berkeley, Calif. (Univ. of Calif. Press), vi+406 pp.

LEWIS, CLARENCE IRVING and LANGFORD, COOPER HAROLD

1932. **Symbolic logic.** New York and London (The Century Co.) xi+506 pp. Reprinted New York (Dover Pubs.) 1951.

LÖWENHEIM, LEOPOLD

1915. *Über Möglichkeiten im Relativkalkül.* **Math. Ann.,** vol. 76, pp. 447–470. Eng. tr. (with commentary) by Stefan Bauer-Mengelberg in van Heijenoort 1967.

ŁUKASIEWICZ, JAN

1920. *O logice trójwartościowej* (On three-valued logic). **Ruch filozoficzny** (Lwów), vol. 5, pp. 169–171.

1921. *Logika dwuwartościowa* (Two-valued logic). **Przegląd filozoficzny,** vol. 23, pp. 189–205.

1934. *Zur Geschichte der Aussagenlogik.* **Erkenntnis,** vol. 5 (1935–6), pp. 111–131. Original in Polish 1934.

LUSIN, NICHOLAS N.

1930. **Leçons sur les ensembles analytiques et leurs applications.** Paris (Gauthier-Villars), xv+328 pp.

LYNDON, ROGER C.

1959. *An interpolation theorem in the predicate calculus.* **Pacific jour. math.,** vol. 9, pp. 129–142.

MACCOLL, HUGH

1896–7. *The calculus of equivalent statements.* (*Fifth paper.*) **Proc. London Math. Soc.,** vol. 28, pp. 156–183.

MACDUFFEE, CYRUS COLTON

1954. **Theory of equations.** New York (Wiley) and London (Chapman and Hall), vii+120 pp.

MALCEV, A.

1936. *Untersuchungen aus dem Gebiete der mathematischen Logik.* **Recueil mathématique (Matématičeskiĭ sbornik),** n.s., vol. 1, pp. 323–336.

MARKOV, A. A.

1947. *On the impossibility of certain algorithms in the theory of associative systems.* **Comptes rendus (Doklady) de l'Académie des Sciences de l'URSS,** n.s., vol. 55, pp. 583–586 (tr. from the Russian).

1951c. *Theory of algorithms.* **American Mathematical Society translations,** ser. 2, vol. 15 (1960), pp. 1–14. Russian orig. 1951.

1958. *Nerazrešimost' problemy gomeomorfii* (Unsolvability of the problem of homeomorphy). **Proc. Internat. Congress Math.,** Edinburgh **1958,** pp. 300–306. Eng. tr. available, Math. Dept., Princeton Univ.

MENDELSON, ELLIOTT

1963. *On some recent criticism of Church's thesis.* **Notre Dame journal of formal logic,** vol. 4, pp. 201–205.

MOSTOWSKI, ANDRZEJ (also — and Tarski; Tarski, — and Robinson)

1947. *On definable sets of positive integers.* **Fund. math.,** vol. 34, pp. 81–112.

1948b. **Logika matematyczna. Kurs uniwersytecki.** Monografie matematyczne t. 18, Varsovie et Wrocław, VIII+388 pp.

1951a. *On the rules of proof in the pure functional calculus of the first order.* **Jour. symbolic logic,** vol. 16, pp. 107–111.

1954. *Development and applications of the "projective" classification of sets of integers.* **Proceedings of the International Congress of Mathematicians, Amsterdam Sept. 2–9, 1954,** vol. III (1956), pp. 280–288.

MOSTOWSKI, ANDRZEJ and TARSKI, ALFRED

1949 abstract. *Undecidability in the arithmetic of integers and in the theory of rings.* **Jour. symbolic logic,** vol. 14, p. 76.

MUČNIK, A. A.

1956. *Nerazrešimost' problemy svodimosti teorii algoritmov* (Negative answer to the problem of reducibility of the theory of algorithms). **Doklady Akademii Nauk S.S.S.R.,** n.s., vol. 108, pp. 194–197.

1958. *Solution of Post's reduction problem and some other problems of the theory of algorithms. I.* **Amer. Math. Soc. translations,** s. 2, vol. 29 (1963), pp. 197–215. Russian orig. 1958.

Nagel, Ernest and Newman, James R.

1956, 1958. *Gödel's proof.* **Scientific american,** vol. 194, no. 6 (June 1956), pp. 71–86. Expanded into a book, New York (New York Univ. Press) 1958, ix+118 pp.

Neumann, John von: see von Neumann, John

Novikov, P. S.

1955. *On the algorithmic unsolvability of the word problem in group theory.* **Amer. Math. Soc. translations,** ser. 2, vol. 9 (1958), pp. 1–122. Russian original 1955.

Padoa, A.

1900. *Essai d'une théorie algébrique des nombres entiers, précédé d'une introduction logique à une théorie déductive quelconque.* **Bibliothèque du Congrès Internationale de Philosophie, Paris 1900,** Paris (1903), vol. 3, pp. 309–365. The introduction is tr. into Eng. in van Heijenoort 1967.

Pasch, Moritz (and Dehn, Max)

1882. **Vorlesungen über neuere Geometrie.** Leipzig (Teubner), iv+202 pp. Reprinted in **Vorlesungen über neuere Geometrie** by Moritz Pasch and Max Dehn, Berlin (Springer) 1926, x+275 pp.

Peano, Giuseppe

1889. **Arithmetices principia, nova methodo exposita.** Turin (Bocca), xvi+20 pp. Eng. tr. in van Heijenoort 1967.

Peirce, Charles Sanders

1885. *On the algebra of logic: A contribution to the philosophy of notation.* **Amer. jour. math.,** vol. 7, pp. 180–202. Reprinted with an added note in **Collected papers of Charles Sanders Peirce** (ed. by Charles Hartschorne and Paul Weiss), vol. 3, Cambridge, Mass, 1933, pp. 210–249.

Popper, Karl R.

1954. *Self-reference and meaning in ordinary language.* **Mind,** n.s., vol. 43, pp. 162–169. Cf. J. F. Thomson in **Jour. Symbolic logic,** vol. 21 (1956), p. 381.

Post, Emil (also Kleene and —)

1921. *Introduction to a general theory of elementary propositions.* **Amer. jour. math.,** vol. 43, pp. 163–185. Reprinted in van Heijenoort 1967.

1936. *Finite combinatory processes—formulation* 1. **Jour. symbolic logic,** vol. 1, pp. 103–105. Reprinted in Davis 1965, pp. 288–291.

1943. *Formal reductions of the general combinatorial decision problem.* **Amer. jour math.,** vol. 65, pp. 197–215.

1944. *Recursively enumerable sets of positive integers and their decision problems.* **Bull. Amer. Math. Soc.,** vol. 50, pp. 284–316. Reprinted in Davis 1965, pp. 304–337.

1947. *Recursive unsolvability of a problem of Thue.* **Jour. symbolic logic,** vol. 12, pp. 1–11. Reprinted in Davis 1965, pp. 292–303.

1948 abstract. *Degrees of recursive unsolvability.* Preliminary report. **Bull. Amer. Math. Soc.,** vol. 54, pp. 641–642.

Prawitz, Dag

1965. **Natural deduction. A proof-theoretical study.** Acta Universitatis Stockholmiensis, Stockholm studies in philosophy 3, Stockholm, Göteborg, Uppsala (Almqvist & Wicksell), 113 pp.

Presburger, M.

1930. *Über die Vollständigkeit eines gewissen Systems der Arithmetik ganzer Zahlen, in welchem die Addition als einzige Operation hervortritt.* **Sprawozdanie z I Kongresu**

Matematyków Krajów Słowiańskich (Comptes-rendus du I Congrès des Mathématiciens des Pays Slaves), **Warszawa 1929,** Warsaw 1930, pp. 92–101, 395.

QUINE, WILLARD VAN ORMAN

1940. **Mathematical logic.** New York (Norton), xiii+348 pp. Rev. ed., Harvard Univ. Press, 1951, xii+346 pp.

1950. **Methods of logic.** New York (Henry Holt and Co.), xx+264 pp. Rev. ed. 1959, xx+272 pp.

RABIN, MICHAEL O. (also — and Scott)

1958. *Recursive unsolvability of group theoretic problems.* **Ann. of Math.,** 2 s., vol. 67, pp. 172–194.

1958a. *On recursively enumerable and arithmetic models of set theory.* **Jour. symbolic logic,** vol. 23, pp. 408–416.

RABIN, M. O. and SCOTT, D.

1959. *Finite automata and their decision problems.* **IBM journal,** vol. 3, pp. 114–125.

RAMSEY, F. P.

1926. *The foundations of mathematics.* **Proc. London Math. Soc.,** ser. 2, vol. 25, pp. 338–384. Reprinted as pp. 1–61 in **The foundations of mathematics and other logical essays** by F. P. Ramsey (ed. by R. B. Braithwaite) 1931, reprinted London (Routledge and Kegan Paul) and New York (Humanities Press) 1950.

RASIOWA, H. and SIKORSKI, R.

1950. *A proof of the completeness theorem of Gödel.* **Fund. math.,** vol. 37, pp. 193–200.

RICHARD, JULES

1905. *Les principes des mathématiques et le problème des ensembles.* **Revue générale des sciences pures et appliquées,** vol. 16, pp. 541–543. Also in **Acta mathematica,** vol. 30 (1906), pp. 295–296. Eng. tr. in van Heijenoort 1967.

RICHARDSON, DANIEL

1966 abstract. *Some unsolvable problems involving functions of a real variable.* **Notices Amer. Math. Soc.,** vol. 13, p. 135.

RIEGER, LADISLAV

1951. *On free $\aleph_\xi$-complete Boolean algebras (with an application to logic).* **Fund. math.,** vol. 38, pp. 35–52.

ROBINSON, ABRAHAM (also Bernstein and —)

1951. **On the metamathematics of algebra.** Amsterdam (North-Holland Pub. Co.), IX+195 pp.

1956. *A result on consistency and its application to the theory of definition.* **Kon. Ned. Akad.** (Amsterdam), **Proc., Ser. A.,** vol. 59 (or **Indag. math.,** vol. 18), pp. 47–58.

1961. *Model theory and non-standard arithmetic.* **Infinitistic methods, Proceedings of the Symposium on Foundations of Mathematics, Warsaw 2–9 September 1959,** Oxford, London, New York, Paris (Pergamon Press), Warszawa (Państwowe Wydawnictwo Naukowe) 1961, pp. 265–302.

1961a. *Non-standard analysis.* **Kon. Ned. Akad. Wet.** (Amsterdam), **Proc., Ser. A.,** vol. 64 (or **Indag. math.,** vol. 23), pp. 432–440.

1963. **Introduction to model theory and to the metamathematics of algebra.** Amsterdam (North-Holland Pub. Co.), ix+284 pp.

1966. **Non-standard analysis.** Amsterdam (North-Holland Pub. Co.), XI+293 pp.

ROBINSON, J. A.

1963. *Theorem-proving on the computer.* **Jour. Assoc. Comput. Mach.,** vol. 10, pp. 163–174.

1965. *A machine-oriented logic based on the resolution principle.* Ibid., vol. 12, pp. 23–41.

Robinson, Raphael M. (also Tarski, Mostowski and —)

1950 abstract. *An essentially undecidable axiom system.* **Proceedings of the International Congress of Mathematicians (Cambridge, Mass. U.S.A., Aug. 30–Sept. 6, 1950)**, Providence, R. I. (Amer. Math. Soc.) 1952, vol. 1, pp. 729–730.

Rosser, J. Barkley (also — and Turquette)

1935. *A mathematical logic without variables.* **Ann. Math.**, 2 s., vol. 36, pp. 127–150 and **Duke math. jour.**, vol. 1, pp. 328–355.

1936. *Extensions of some theorems of Gödel and Church.* **Jour. symbolic logic,** vol. 1, pp. 87–91.

Rosser, J. Barkley and Turquette, Atwell R.

1952. **Many-valued logics.** Amsterdam (North-Holland Pub. Co.), vii+124 pp.

Rotman, Joseph J.

1965. **Theory of groups: an introduction.** Boston (Allyn and Bacon), xiii+305 pp.

Russell, Bertrand (also Whitehead and —)

1902. *On finite and infinite cardinal numbers* (Section III of A. N. Whitehead's *On cardinal numbers*). **Amer. jour. math.**, vol. 24, pp. 378–383.

1902a. Letter to Frege, pub. in van Heijenoort 1967.

1906. *Les paradoxes de la logique.* **Revue de métaphysique et de morale,** vol. 14, pp. 627–650.

1906a. *On some difficulties in the theory of transfinite numbers and order types.* **Proc. London. Math. Soc.**, 2 s., vol. 4, pp. 29–53.

1919. **Introduction to mathematical philosophy.** London (G. Allen and Unwin) and New York (Macmillan), viii+208 pp. 2nd. ed. 1920.

Ryll-Nardzewski, Czesław

1952. *The role of the axiom of induction in the elementary arithmetic.* **Fund. math.**, vol. 39, pp. 239–263.

Sacks, Gerald E.

1963. **Degrees of unsolvability.** Annals of Mathematics studies, no. 55, Princeton, N.J. (Princeton Univ. Press), xi+174 pp.

Scarpellini, Bruno

1963. *Zwei unentscheidbare probleme der Analysis.* **Zeitschrift für mathematische Logik und Grundlagen der Mathematik,** vol. 9, pp. 265–289.

Schröder, Ernst

1877. **Der Operationskreis des Logikkalkuls.** Leipzig, v+37 pp.

1895. **Vorlesungen über die Algebra der Logik (exakte Logik).** Vol. 3 **Algebra und Logik der Relativ** part 1, Leipzig, viii+649 pp.

Schütte, Kurt

1950. *Schlussweisen-Kalküle der Prädikatenlogik*, **Math. Ann.**, vol. 122, pp. 47–65.

1956. *Ein System des verknüpfenden Schliessens.* **Archiv für mathematische Logik und Grundlagenforschung,** vol. 2, nos. 2–4, pp. 34–67. Also **Archiv für philosophie,** vol. 5, no. 4, pp. 375–387.

1960. **Beweistheorie.** Berlin, Göttingen and Heidelberg (Springer-Verlag), X+355 pp.

1962. *Der Interpolationssatz der intuitionistischen Prädikatenlogik.* **Math. Ann.,** vol. 148, pp. 192–200.

Scott, Dana (also Rabin and —)

1961. *On constructing models for arithmetic.* **Infinitistic methods** (cf. A. Robinson 1961), pp. 235–255.

1966. **A proof of the independence of the continuum hypothesis.** Math. Dept., Stanford Univ., Stanford, Calif., mimeographed, 43 pp.

1967. *Existence and description in formal logic.* To appear in **Bertrand Russell: philosopher of the century** (Ralph Schoenman, ed.), London (Allen & Unwin) 1967(?).

SHEPHERDSON, J. C.

1965. *Machine configuration and word problems of given degree of unsolvability.* **Zeitsch. math. Logik Grundlagen Math.**, vol. 11, pp. 149–175.

SKOLEM, THORALF

1920. *Logisch-kombinatorische Untersuchungen über die Erfüllbarkeit oder Beweisbarkeit mathematische Sätze nebst einem Theoreme über dichte Mengen.* **Skrifter utgit av Videnskapsselskapet i Kristiania, I. Matematisk-naturvidenskabelig klasse 1920**, no. 4, 36 pp. Eng. tr. in van Heijenoort 1967.

1922–3. *Einige Bemerkungen zur axiomatischen Begründung der Mengenlehre.* **Wissenschaftliche Vorträge gehalten auf dem Fünften Kongress der Skandinavischen Mathematiker in Helsingfors vom 4. bis 7. Juli 1922**, Helsingfors 1923, pp. 217–232. Eng. tr. in van Heijenoort 1967.

1928. *Über die mathematische Logik.* **Norsk matematisk tidsskrift**, vol. 10, pp. 125–142. Eng. tr. in van Heijenoort 1967.

1929. *Über einige Grundlagenfragen der Mathematik.* **Skr. Oslo** (= Kristiania), **Mat.-natur. kl. 1929**, no. 4, 49 pp.

1933. *Über die Unmöglichkeit einer vollständigen Charakterisierung der Zahlenreihe mittels eines endlichen Axiomensystems.* **Norsk matematisk forenings skrifter**, ser. 2, no. 10, pp. 73–82.

1934. *Über die Nicht-charakterisierbarkeit der Zahlenreihe mittels endlich oder abzählbar unendlich vieler Aussagen mit ausschliesslich Zahlenvariablen.* **Fund. math.**, vol. 23, pp. 150–161.

SMULLYAN, RAYMOND M.

1961. **Theory of formal systems.** Rev. ed., Annals of Mathematics studies, no. 47, Princeton, N.J. (Princeton Univ. Press), xi+147 pp. Orig. ed., ibid., xi+142 pp.

SPECKER, ERNST

1949. *Nicht konstruktiv beweisbare Sätze der Analysis*, **Jour. symbolic logic**, vol. 14, pp. 145–158.

STRAWSON, P. F.

1952. **Introduction to logical theory.** London (Methuen) and New York (Wiley), x+266 pp.

SUPPES, PATRICK

1957. **Introduction to logic.** Princeton, Toronto, London and New York (Van Nostrand), xviii+312 pp.

SZMIELEW, WANDA

1948. *Decision problem in group theory.* **Proceedings of the Xth International Congress of Philosophy (Amsterdam, Aug. 11–18, 1948)**, Amsterdam (North-Holland Pub. Co.) 1949, fasc. 2, pp. 763–766.

1955. *Elementary properties of Abelian groups.* **Fund. math.**, vol. 41, pp. 203–271.

TAITSLIN, M. A.

1960. Review of Lyndon 1959. **Jour. symbolic logic**, vol. 25, pp. 273–274.

TARSKI, ALFRED (also Mostowski and —; —, Mostowski and Robinson)

1930. *Über einige fundamentalen Begriffe der Metamathematik.* **Comptes rendus des séances de la Société des Sciences et des Lettres de Varsovie**, Classe III, vol. 23, pp. 22–29. Eng. tr. in 1956, pp. 30–37.

1933. *Der Wahrheitsbegriff in den formalisierten Sprachen.* **Studia philosophica**, vol. 1 (1936), pp. 261–405 (offprints dated 1935). Ger. tr. by L. Blaustein from the Polish original 1933, with a postscript added. Eng. tr. in 1956, pp. 152–278.

1934. *Einige methodologische Untersuchungen über die Definierbarkeit der Begriffe*. **Erkenntnis,** vol. 5, pp. 80–100. Orig. in Polish 1934. Eng. tr. in 1956, pp. 296–319.

1935. *Über den Begriff der logischen Folgerung*. **Actes du Congrès International de Philosophie Scientifique, VII Logique,** Actualités scientifique et industrielles, 394, Paris (Hermann and Cie) 1936, pp. 1–11. Eng. tr. in 1956, pp. 409–420.

1949 abstract. *On essential undecidability*. **Jour. symbolic logic,** vol. 14, pp. 75–76.

1956. **Logic, semantics, metamathematics,** papers 1923–1938 tr. into Eng. by J. H. Woodger, Oxford (Clarendon), xiv+471 pp.

TARSKI, ALFRED, (MOSTOWSKI, ANDRZEJ and ROBINSON, RAPHAEL M.)

1953. **Undecidable theories.** Amsterdam (North-Holland Pub. Co.), XI+98 pp.

THOMAS, IVO

1958. *A 12th century paradox of the infinite*. **Jour. symbolic logic,** vol. 23, pp. 133–134.

TURING, ALAN MATHISON

1936–7. *On computable numbers, with an application to the Entscheidungsproblem*. **Proc. London Math. Soc.,** ser. 2, vol. 42 (1936–7), pp. 230–265. *A correction*, ibid., vol. 43 (1937), pp. 544–546. Reprinted in Davis 1965, pp. 115–154.

1937. *Computability and λ-definability*. **Jour. symbolic logic,** vol. 2, pp. 153–163.

1939. *Systems of logic based on ordinals*. **Proc. London Math. Soc.,** 2 s., vol. 45, pp. 161–228. Reprinted in Davis 1965, pp. 155–222.

1950. *The word problem in semi-groups with cancellation*. **Ann. of math.,** 2 s., vol. 52, pp. 491–505. Clarifications in Boone 1958.

VAN HEIJENOORT, JEAN

1967. (editor) **From Frege to Gödel, A source book in mathematical logic, 1879–1931.** Cambridge, Mass. (Harvard Univ. Press), xii+600 pp.

VENN, JOHN

1881. **Symbolic logic.** London, xxxix+446 pp. 2nd. ed. London 1894, xxxviii+527 pp.

VON NEUMANN, JOHN

1925. *Eine Axiomatisierung der Mengenlehre*. **Jour. reine angew. Math.,** vol. 154, pp. 219–240. *Berichtigung*, ibid., vol. 155 (1926), p. 128. Eng. tr. in van Heijenoort 1967.

1927. *Zur Hilbertschen Beweistheorie*. **Math. Zeit.,** vol. 26, pp. 1–46.

VON WRIGHT, GEORG H.

1951. **An essay in modal logic,** Amsterdam (North-Holland Pub. Co.), VII+90 pp.

WANG, HAO

1960. *Toward mechanical mathematics*. **IBM journal,** vol. 4, pp. 2–22.

WEYL, HERMANN

1926. **Die heutige Erkenntnislage in der Mathematik.** Sonderdrucke des Symposion, Erlangen (im Weldkreis-Verlag), Heft 3 (1926), 32 pp. Also in **Symposion** (Berlin), vol. 1 (1925–7), pp. 1–32.

1946. *Mathematics and logic. A brief survey serving as a preface to a review of "The Philosophy of Bertrand Russell"*. **Amer. math. monthly,** vol. 53, pp. 2–13.

1949. **Philosophy of mathematics and natural science.** Princeton, N.J. (Princeton Univ. Press), x+311 pp. Revised and augmented Eng. ed., based on a tr. by Olaf Helmer from the German original 1926.

WHITEHEAD, ALFRED NORTH and RUSSELL, BERTRAND

1910–13. **Principia mathematica.** Vol. 1 1910, xv+666 pp. (2nd. ed. 1925). Vol. 2 1912, xxiv+772 pp. (2nd. ed. 1927). Vol. 3 1913, x+491 pp. (2nd. ed. 1927). Cambridge, Eng. (Cambridge Univ. Press.)

WITTGENSTEIN, LUDWIG

1921. *Logisch-philosophische Abhandlung*. **Annalen der Naturphilosophie** (Leipzig), vol. 14, pp. 185–262. Reprinted with Eng. tr. as **Tractatus logico-philosophicus,** New York and London, 1922, 189 pp.

WRIGHT, GEORG H. VON: see von Wright, Georg H.

YOUNG, JOHN WESLEY

1911. **Lectures on fundamental concepts of algebra and geometry.** New York (Macmillan), vii+247 pp.

ZERMELO, ERNST

1904. *Beweis, dass jede Menge wohlgeordnet werden kann.* **Math. Ann.,** vol. 59, pp. 514–516. Eng. tr. in van Heijenoort 1967.

1908a. *Neuer Beweis für die Möglichkeit einer Wohlordnung*, Ibid., vol. 65, pp. 107–128. Eng. tr. in van Heijenoort 1967.

1908. *Untersuchung über die Grundlagen der Mengenlehre I.* Ibid., pp. 261–281. Eng. tr. in van Heijenoort 1967.

THEOREM AND LEMMA NUMBERS: PAGES

LIST OF POSTULATES

Propositional calculus, pp. 33–34, 15–16, 18.

1a. $A \supset (B \supset A)$.

1b. $(A \supset B) \supset ((A \supset (B \supset C)) \supset (A \supset C))$.

3. $A \supset (B \supset A \& B)$.

5a. $A \supset A \vee B$.

5b. $B \supset A \vee B$.

7. $(A \supset B) \supset ((A \supset \neg B) \supset \neg A)$.

9a. $(A \supset B) \supset ((B \supset A) \supset (A \sim B))$.

MODUS PONENS $\dfrac{A,\ A \supset B}{B.}$ or $\supset$-RULE.

4a. $A \& B \supset A$.

4b. $A \& B \supset B$.

6. $(A \supset C) \supset ((B \supset C) \supset (A \vee B \supset C))$.

8. $\neg\neg A \supset A$.

10a. $(A \sim B) \supset (A \supset B)$.

10b. $(A \sim B) \supset (B \supset A)$.

Predicate calculus (additional), pp. 107, 94–96, 150.

$\forall$-SCHEMA. $\forall x A(x) \supset A(r)$.

$\exists$-SCHEMA. $A(r) \supset \exists x A(x)$.

The term r must be free for the variable x in the formula A(x).

$\forall$-RULE. $\dfrac{C \supset A(x)}{C \supset \forall x A(x).}$

$\exists$-RULE. $\dfrac{A(x) \supset C}{\exists x A(x) \supset C.}$

The variable x must not occur free in the formula C.

Equality (additional), p. 154.

(a) $x=x$.

(b) $x=y \supset (x=z \supset y=z)$.

(c_i^n) $x=y \supset (P(a_1, \ldots, a_{i-1}, x, a_{i+1}, \ldots, a_n) \supset P(a_1, \ldots, a_{i-1}, y, a_{i+1}, \ldots, a_n))$.

(d_i^n) $x=y \supset f(a_1, \ldots, a_{i-1}, x, a_{i+1}, \ldots, a_n)=f(a_1, \ldots, a_{i-1}, y, a_{i+1}, \ldots, a_n)$.

Formal number theory (additional to predicate calculus), p. 206.

13. $A(0) \& \forall x(A(x) \supset A(x')) \supset A(x)$.

14. $a'=b' \supset a=b$.

15. $\neg a'=0$.

16. $a=b \supset (a=c \supset b=c)$.

17. $a=b \supset a'=b'$.

18. $a+0=a$.

19. $a+b'=(a+b)'$.

20. $a \cdot 0=0$.

21. $a \cdot b'=a \cdot b+a$.

The derived introduction and elimination rules for the propositional and predicate calculi are on pp. 44–45, 118.

The postulates for the Gentzen-type system *G*4 are on pp. 289–290, 306, the additional postulates for *G*4a, *G*4b on p. 333.

The nonlogical axioms for groups are on pp. 215, 218, 219.

For intuitionistic systems, replace Ax. Sch. 8 by: 8^I. $\neg A \supset (A \supset B)$.

SYMBOLS AND NOTATIONS

USED IN (NAMES FOR) EXPRESSIONS OF AN OBJECT LANGUAGE[6,293]

Symbol	Pages	Symbol	Pages	Symbol	Pages
$\sim$, $\supset$	5, 9, 63	$\vert$	213	P, Q, R, . . .	5, 78
&, ∨, ¬	5, 9, 63–64	$\vert$ (short)	202, 220–221	P(—),Q(—), . . .	77–78
≁, ⊅	24	0, ′, +	202, 207	a, b, . . . , x, . . .	78
□, ◊	49	·	202, 207, 215; 262	$\mathscr{a}, \mathscr{b}, \dots, \mathscr{x}, \dots$	202
→	286	$^{-1}$	215	***a*, *b*, . . . , *x*, . . .**	248
∀	74, 85, 141; 105	1	205, 215	r, r_1, r_2, . . .	78
∀′	105	2, 3, . . .	205	C_a	248
∃	74, 86, 141	A, B, C, . . .	6, 7, 80	Pr	213
∃!	154	A(—),B(—),. .	80,94,96	f(—), g(—), . . .	148
=	151, 165	$\mathscr{A}, \mathscr{B}, \mathscr{C}, \dots$	220	$\mathscr{f}$(—), $\mathscr{g}$(—), . . .	221
≠, <	205	$\mathscr{A}$(—),$\mathscr{B}$(—), . . .	220	ι	169

USED OTHERWISE[293]

Symbol	Pages	Symbol	Pages	Symbol	Pages
$\vert$	228; 235; 298	=	135, 157–165, 207	**AG, AGp**	219
≡, →	24, 229	<	185, 207, 254, 268	$\mathfrak{f}$	9, 85–86
&, ∨	229	$\sim$	184	$f_1(x), f_2(x), \dots$	149
—	229; 140	≃	159	**G, Gp**	215, 218
(*x*), (*Ex*)	229	∈, ∉	135, 165, 183	*G*4, *H*	306, 289–290
⊨	14, 26, 88, 101, 314	⊆, ⊂	135, 165, 183	*G*4a, *G*4b, *G*	333, 338
$\overline{\overline{D}}$-⊨	89, 314	{*a*, *b*, . . .}	135, 183	$\mathfrak{l}(x, y), \mathfrak{l}_1(x), \dots$	85–92
$(\leq\aleph_0)$-⊨	318	∅	135, 183	max, min	230
$\models^{x_1 \dots x_q}$	106	∪, ∩	140, 185, 368	$\mathfrak{M}_i$	243
⊢	34–36, 108, 314	$\hat{x}C(x)$	137, 190	**N**	201–206, 215
$\vdash^{x_1 \dots x_q}$	108	$\overline{\overline{M}}$	184	**Pd, Pdf**	220–221
∴	59, 68	2^M	184	**Pp**	220
$\therefore^{x_1 \dots x_q}$	136–137, 146	0, 1, 2, . . .	176, 207	$\mathfrak{t}$	9, 85–86
→⊃, . . .	287, 289, 306	$\aleph_0$	185	*T*(*i*, *a*, *x*)	243
→*T*, . . . , *C*→	333	**0, 0′, d, d′**	268–269	Γ, Δ, . . .	44, 286, 289
√, ×	287, 290, 306	°, ^{C}	49, 331, 332	$\varphi_i(a)$	243–244

Notations relating to Turing machines 233–238, 242–244

[293] Exceptions occur to this classification-as-to-use. Thus on p. 196 "∨" and "∃" were used as though they were in the second list (before we had Footnote 165); for *G*4 etc. as the object language, "Γ", "Δ", . . . belong in the first list; and other exceptions are noted in Footnotes 64 and 149.

INDEX

A CATALOG OF SELECTED

DOVER BOOKS

IN SCIENCE AND MATHEMATICS

Astronomy

CHARIOTS FOR APOLLO: The NASA History of Manned Lunar Spacecraft to 1969, Courtney G. Brooks, James M. Grimwood, and Loyd S. Swenson, Jr. This illustrated history by a trio of experts is the definitive reference on the Apollo spacecraft and lunar modules. It traces the vehicles' design, development, and operation in space. More than 100 photographs and illustrations. 576pp. 6 3/4 x 9 1/4. 0-486-46756-2

EXPLORING THE MOON THROUGH BINOCULARS AND SMALL TELESCOPES, Ernest H. Cherrington, Jr. Informative, profusely illustrated guide to locating and identifying craters, rills, seas, mountains, other lunar features. Newly revised and updated with special section of new photos. Over 100 photos and diagrams. 240pp. 8 1/4 x 11. 0-486-24491-1

WHERE NO MAN HAS GONE BEFORE: A History of NASA's Apollo Lunar Expeditions, William David Compton. Introduction by Paul Dickson. This official NASA history traces behind-the-scenes conflicts and cooperation between scientists and engineers. The first half concerns preparations for the Moon landings, and the second half documents the flights that followed Apollo 11. 1989 edition. 432pp. 7 x 10. 0-486-47888-2

APOLLO EXPEDITIONS TO THE MOON: The NASA History, Edited by Edgar M. Cortright. Official NASA publication marks the 40th anniversary of the first lunar landing and features essays by project participants recalling engineering and administrative challenges. Accessible, jargon-free accounts, highlighted by numerous illustrations. 336pp. 8 3/8 x 10 7/8. 0-486-47175-6

ON MARS: Exploration of the Red Planet, 1958-1978--The NASA History, Edward Clinton Ezell and Linda Neuman Ezell. NASA's official history chronicles the start of our explorations of our planetary neighbor. It recounts cooperation among government, industry, and academia, and it features dozens of photos from Viking cameras. 560pp. 6 3/4 x 9 1/4. 0-486-46757-0

ARISTARCHUS OF SAMOS: The Ancient Copernicus, Sir Thomas Heath. Heath's history of astronomy ranges from Homer and Hesiod to Aristarchus and includes quotes from numerous thinkers, compilers, and scholasticists from Thales and Anaximander through Pythagoras, Plato, Aristotle, and Heraclides. 34 figures. 448pp. 5 3/8 x 8 1/2. 0-486-43886-4

AN INTRODUCTION TO CELESTIAL MECHANICS, Forest Ray Moulton. Classic text still unsurpassed in presentation of fundamental principles. Covers rectilinear motion, central forces, problems of two and three bodies, much more. Includes over 200 problems, some with answers. 437pp. 5 3/8 x 8 1/2. 0-486-64687-4

BEYOND THE ATMOSPHERE: Early Years of Space Science, Homer E. Newell. This exciting survey is the work of a top NASA administrator who chronicles technological advances, the relationship of space science to general science, and the space program's social, political, and economic contexts. 528pp. 6 3/4 x 9 1/4. 0-486-47464-X

STAR LORE: Myths, Legends, and Facts, William Tyler Olcott. Captivating retellings of the origins and histories of ancient star groups include Pegasus, Ursa Major, Pleiades, signs of the zodiac, and other constellations. "Classic." – *Sky & Telescope.* 58 illustrations. 544pp. 5 3/8 x 8 1/2. 0-486-43581-4

A COMPLETE MANUAL OF AMATEUR ASTRONOMY: Tools and Techniques for Astronomical Observations, P. Clay Sherrod with Thomas L. Koed. Concise, highly readable book discusses the selection, set-up, and maintenance of a telescope; amateur studies of the sun; lunar topography and occultations; and more. 124 figures. 26 halftones. 37 tables. 335pp. 6 1/2 x 9 1/4. 0-486-42820-6

Chemistry

MOLECULAR COLLISION THEORY, M. S. Child. This high-level monograph offers an analytical treatment of classical scattering by a central force, quantum scattering by a central force, elastic scattering phase shifts, and semi-classical elastic scattering. 1974 edition. 310pp. 5 3/8 x 8 1/2. 0-486-69437-2

HANDBOOK OF COMPUTATIONAL QUANTUM CHEMISTRY, David B. Cook. This comprehensive text provides upper-level undergraduates and graduate students with an accessible introduction to the implementation of quantum ideas in molecular modeling, exploring practical applications alongside theoretical explanations. 1998 edition. 832pp. 5 3/8 x 8 1/2. 0-486-44307-8

RADIOACTIVE SUBSTANCES, Marie Curie. The celebrated scientist's thesis, which directly preceded her 1903 Nobel Prize, discusses establishing atomic character of radioactivity; extraction from pitchblende of polonium and radium; isolation of pure radium chloride; more. 96pp. 5 3/8 x 8 1/2. 0-486-42550-9

CHEMICAL MAGIC, Leonard A. Ford. Classic guide provides intriguing entertainment while elucidating sound scientific principles, with more than 100 unusual stunts: cold fire, dust explosions, a nylon rope trick, a disappearing beaker, much more. 128pp. 5 3/8 x 8 1/2. 0-486-67628-5

ALCHEMY, E. J. Holmyard. Classic study by noted authority covers 2,000 years of alchemical history: religious, mystical overtones; apparatus; signs, symbols, and secret terms; advent of scientific method, much more. Illustrated. 320pp. 5 3/8 x 8 1/2. 0-486-26298-7

CHEMICAL KINETICS AND REACTION DYNAMICS, Paul L. Houston. This text teaches the principles underlying modern chemical kinetics in a clear, direct fashion, using several examples to enhance basic understanding. Solutions to selected problems. 2001 edition. 352pp. 8 3/8 x 11. 0-486-45334-0

PROBLEMS AND SOLUTIONS IN QUANTUM CHEMISTRY AND PHYSICS, Charles S. Johnson and Lee G. Pedersen. Unusually varied problems, with detailed solutions, cover of quantum mechanics, wave mechanics, angular momentum, molecular spectroscopy, scattering theory, more. 280 problems, plus 139 supplementary exercises. 430pp. 6 1/2 x 9 1/4. 0-486-65236-X

ELEMENTS OF CHEMISTRY, Antoine Lavoisier. Monumental classic by the founder of modern chemistry features first explicit statement of law of conservation of matter in chemical change, and more. Facsimile reprint of original (1790) Kerr translation. 539pp. 5 3/8 x 8 1/2. 0-486-64624-6

MAGNETISM AND TRANSITION METAL COMPLEXES, F. E. Mabbs and D. J. Machin. A detailed view of the calculation methods involved in the magnetic properties of transition metal complexes, this volume offers sufficient background for original work in the field. 1973 edition. 240pp. 5 3/8 x 8 1/2. 0-486-46284-6

GENERAL CHEMISTRY, Linus Pauling. Revised third edition of classic first-year text by Nobel laureate. Atomic and molecular structure, quantum mechanics, statistical mechanics, thermodynamics correlated with descriptive chemistry. Problems. 992pp. 5 3/8 x 8 1/2. 0-486-65622-5

ELECTROLYTE SOLUTIONS: Second Revised Edition, R. A. Robinson and R. H. Stokes. Classic text deals primarily with measurement, interpretation of conductance, chemical potential, and diffusion in electrolyte solutions. Detailed theoretical interpretations, plus extensive tables of thermodynamic and transport properties. 1970 edition. 590pp. 5 3/8 x 8 1/2. 0-486-42225-9

Engineering

FUNDAMENTALS OF ASTRODYNAMICS, Roger R. Bate, Donald D. Mueller, and Jerry E. White. Teaching text developed by U.S. Air Force Academy develops the basic two-body and n-body equations of motion; orbit determination; classical orbital elements, coordinate transformations; differential correction; more. 1971 edition. 455pp. 5 3/8 x 8 1/2. 0-486-60061-0

INTRODUCTION TO CONTINUUM MECHANICS FOR ENGINEERS: Revised Edition, Ray M. Bowen. This self-contained text introduces classical continuum models within a modern framework. Its numerous exercises illustrate the governing principles, linearizations, and other approximations that constitute classical continuum models. 2007 edition. 320pp. 6 1/8 x 9 1/4. 0-486-47460-7

ENGINEERING MECHANICS FOR STRUCTURES, Louis L. Bucciarelli. This text explores the mechanics of solids and statics as well as the strength of materials and elasticity theory. Its many design exercises encourage creative initiative and systems thinking. 2009 edition. 320pp. 6 1/8 x 9 1/4. 0-486-46855-0

FEEDBACK CONTROL THEORY, John C. Doyle, Bruce A. Francis and Allen R. Tannenbaum. This excellent introduction to feedback control system design offers a theoretical approach that captures the essential issues and can be applied to a wide range of practical problems. 1992 edition. 224pp. 6 1/2 x 9 1/4. 0-486-46933-6

THE FORCES OF MATTER, Michael Faraday. These lectures by a famous inventor offer an easy-to-understand introduction to the interactions of the universe's physical forces. Six essays explore gravitation, cohesion, chemical affinity, heat, magnetism, and electricity. 1993 edition. 96pp. 5 3/8 x 8 1/2. 0-486-47482-8

DYNAMICS, Lawrence E. Goodman and William H. Warner. Beginning engineering text introduces calculus of vectors, particle motion, dynamics of particle systems and plane rigid bodies, technical applications in plane motions, and more. Exercises and answers in every chapter. 619pp. 5 3/8 x 8 1/2. 0-486-42006-X

ADAPTIVE FILTERING PREDICTION AND CONTROL, Graham C. Goodwin and Kwai Sang Sin. This unified survey focuses on linear discrete-time systems and explores natural extensions to nonlinear systems. It emphasizes discrete-time systems, summarizing theoretical and practical aspects of a large class of adaptive algorithms. 1984 edition. 560pp. 6 1/2 x 9 1/4. 0-486-46932-8

INDUCTANCE CALCULATIONS, Frederick W. Grover. This authoritative reference enables the design of virtually every type of inductor. It features a single simple formula for each type of inductor, together with tables containing essential numerical factors. 1946 edition. 304pp. 5 3/8 x 8 1/2. 0-486-47440-2

THERMODYNAMICS: Foundations and Applications, Elias P. Gyftopoulos and Gian Paolo Beretta. Designed by two MIT professors, this authoritative text discusses basic concepts and applications in detail, emphasizing generality, definitions, and logical consistency. More than 300 solved problems cover realistic energy systems and processes. 800pp. 6 1/8 x 9 1/4. 0-486-43932-1

THE FINITE ELEMENT METHOD: Linear Static and Dynamic Finite Element Analysis, Thomas J. R. Hughes. Text for students without in-depth mathematical training, this text includes a comprehensive presentation and analysis of algorithms of time-dependent phenomena plus beam, plate, and shell theories. Solution guide available upon request. 672pp. 6 1/2 x 9 1/4. 0-486-41181-8

HELICOPTER THEORY, Wayne Johnson. Monumental engineering text covers vertical flight, forward flight, performance, mathematics of rotating systems, rotary wing dynamics and aerodynamics, aeroelasticity, stability and control, stall, noise, and more. 189 illustrations. 1980 edition. 1089pp. 5 5/8 x 8 1/4. 0-486-68230-7

MATHEMATICAL HANDBOOK FOR SCIENTISTS AND ENGINEERS: Definitions, Theorems, and Formulas for Reference and Review, Granino A. Korn and Theresa M. Korn. Convenient access to information from every area of mathematics: Fourier transforms, Z transforms, linear and nonlinear programming, calculus of variations, random-process theory, special functions, combinatorial analysis, game theory, much more. 1152pp. 5 3/8 x 8 1/2. 0-486-41147-8

A HEAT TRANSFER TEXTBOOK: Fourth Edition, John H. Lienhard V and John H. Lienhard IV. This introduction to heat and mass transfer for engineering students features worked examples and end-of-chapter exercises. Worked examples and end-of-chapter exercises appear throughout the book, along with well-drawn, illuminating figures. 768pp. 7 x 9 1/4. 0-486-47931-5

BASIC ELECTRICITY, U.S. Bureau of Naval Personnel. Originally a training course; best nontechnical coverage. Topics include batteries, circuits, conductors, AC and DC, inductance and capacitance, generators, motors, transformers, amplifiers, etc. Many questions with answers. 349 illustrations. 1969 edition. 448pp. 6 1/2 x 9 1/4. 0-486-20973-3

BASIC ELECTRONICS, U.S. Bureau of Naval Personnel. Clear, well-illustrated introduction to electronic equipment covers numerous essential topics: electron tubes, semiconductors, electronic power supplies, tuned circuits, amplifiers, receivers, ranging and navigation systems, computers, antennas, more. 560 illustrations. 567pp. 6 1/2 x 9 1/4. 0-486-21076-6

BASIC WING AND AIRFOIL THEORY, Alan Pope. This self-contained treatment by a pioneer in the study of wind effects covers flow functions, airfoil construction and pressure distribution, finite and monoplane wings, and many other subjects. 1951 edition. 320pp. 5 3/8 x 8 1/2. 0-486-47188-8

SYNTHETIC FUELS, Ronald F. Probstein and R. Edwin Hicks. This unified presentation examines the methods and processes for converting coal, oil, shale, tar sands, and various forms of biomass into liquid, gaseous, and clean solid fuels. 1982 edition. 512pp. 6 1/8 x 9 1/4. 0-486-44977-7

THEORY OF ELASTIC STABILITY, Stephen P. Timoshenko and James M. Gere. Written by world-renowned authorities on mechanics, this classic ranges from theoretical explanations of 2- and 3-D stress and strain to practical applications such as torsion, bending, and thermal stress. 1961 edition. 560pp. 5 3/8 x 8 1/2. 0-486-47207-8

PRINCIPLES OF DIGITAL COMMUNICATION AND CODING, Andrew J. Viterbi and Jim K. Omura. This classic by two digital communications experts is geared toward students of communications theory and to designers of channels, links, terminals, modems, or networks used to transmit and receive digital messages. 1979 edition. 576pp. 6 1/8 x 9 1/4. 0-486-46901-8

LINEAR SYSTEM THEORY: The State Space Approach, Lotfi A. Zadeh and Charles A. Desoer. Written by two pioneers in the field, this exploration of the state space approach focuses on problems of stability and control, plus connections between this approach and classical techniques. 1963 edition. 656pp. 6 1/8 x 9 1/4. 0-486-46663-9

Mathematics–Bestsellers

HANDBOOK OF MATHEMATICAL FUNCTIONS: with Formulas, Graphs, and Mathematical Tables, Edited by Milton Abramowitz and Irene A. Stegun. A classic resource for working with special functions, standard trig, and exponential logarithmic definitions and extensions, it features 29 sets of tables, some to as high as 20 places. 1046pp. 8 x 10 1/2. 0-486-61272-4

ABSTRACT AND CONCRETE CATEGORIES: The Joy of Cats, Jiri Adamek, Horst Herrlich, and George E. Strecker. This up-to-date introductory treatment employs category theory to explore the theory of structures. Its unique approach stresses concrete categories and presents a systematic view of factorization structures. Numerous examples. 1990 edition, updated 2004. 528pp. 6 1/8 x 9 1/4. 0-486-46934-4

MATHEMATICS: Its Content, Methods and Meaning, A. D. Aleksandrov, A. N. Kolmogorov, and M. A. Lavrent'ev. Major survey offers comprehensive, coherent discussions of analytic geometry, algebra, differential equations, calculus of variations, functions of a complex variable, prime numbers, linear and non-Euclidean geometry, topology, functional analysis, more. 1963 edition. 1120pp. 5 3/8 x 8 1/2. 0-486-40916-3

INTRODUCTION TO VECTORS AND TENSORS: Second Edition--Two Volumes Bound as One, Ray M. Bowen and C.-C. Wang. Convenient single-volume compilation of two texts offers both introduction and in-depth survey. Geared toward engineering and science students rather than mathematicians, it focuses on physics and engineering applications. 1976 edition. 560pp. 6 1/2 x 9 1/4. 0-486-46914-X

AN INTRODUCTION TO ORTHOGONAL POLYNOMIALS, Theodore S. Chihara. Concise introduction covers general elementary theory, including the representation theorem and distribution functions, continued fractions and chain sequences, the recurrence formula, special functions, and some specific systems. 1978 edition. 272pp. 5 3/8 x 8 1/2. 0-486-47929-3

ADVANCED MATHEMATICS FOR ENGINEERS AND SCIENTISTS, Paul DuChateau. This primary text and supplemental reference focuses on linear algebra, calculus, and ordinary differential equations. Additional topics include partial differential equations and approximation methods. Includes solved problems. 1992 edition. 400pp. 7 1/2 x 9 1/4. 0-486-47930-7

PARTIAL DIFFERENTIAL EQUATIONS FOR SCIENTISTS AND ENGINEERS, Stanley J. Farlow. Practical text shows how to formulate and solve partial differential equations. Coverage of diffusion-type problems, hyperbolic-type problems, elliptic-type problems, numerical and approximate methods. Solution guide available upon request. 1982 edition. 414pp. 6 1/8 x 9 1/4. 0-486-67620-X

VARIATIONAL PRINCIPLES AND FREE-BOUNDARY PROBLEMS, Avner Friedman. Advanced graduate-level text examines variational methods in partial differential equations and illustrates their applications to free-boundary problems. Features detailed statements of standard theory of elliptic and parabolic operators. 1982 edition. 720pp. 6 1/8 x 9 1/4. 0-486-47853-X

LINEAR ANALYSIS AND REPRESENTATION THEORY, Steven A. Gaal. Unified treatment covers topics from the theory of operators and operator algebras on Hilbert spaces; integration and representation theory for topological groups; and the theory of Lie algebras, Lie groups, and transform groups. 1973 edition. 704pp. 6 1/8 x 9 1/4. 0-486-47851-3

A SURVEY OF INDUSTRIAL MATHEMATICS, Charles R. MacCluer. Students learn how to solve problems they'll encounter in their professional lives with this concise single-volume treatment. It employs MATLAB and other strategies to explore typical industrial problems. 2000 edition. 384pp. 5 3/8 x 8 1/2. 0-486-47702-9

NUMBER SYSTEMS AND THE FOUNDATIONS OF ANALYSIS, Elliott Mendelson. Geared toward undergraduate and beginning graduate students, this study explores natural numbers, integers, rational numbers, real numbers, and complex numbers. Numerous exercises and appendixes supplement the text. 1973 edition. 368pp. 5 3/8 x 8 1/2. 0-486-45792-3

A FIRST LOOK AT NUMERICAL FUNCTIONAL ANALYSIS, W. W. Sawyer. Text by renowned educator shows how problems in numerical analysis lead to concepts of functional analysis. Topics include Banach and Hilbert spaces, contraction mappings, convergence, differentiation and integration, and Euclidean space. 1978 edition. 208pp. 5 3/8 x 8 1/2. 0-486-47882-3

FRACTALS, CHAOS, POWER LAWS: Minutes from an Infinite Paradise, Manfred Schroeder. A fascinating exploration of the connections between chaos theory, physics, biology, and mathematics, this book abounds in award-winning computer graphics, optical illusions, and games that clarify memorable insights into self-similarity. 1992 edition. 448pp. 6 1/8 x 9 1/4. 0-486-47204-3

SET THEORY AND THE CONTINUUM PROBLEM, Raymond M. Smullyan and Melvin Fitting. A lucid, elegant, and complete survey of set theory, this three-part treatment explores axiomatic set theory, the consistency of the continuum hypothesis, and forcing and independence results. 1996 edition. 336pp. 6 x 9. 0-486-47484-4

DYNAMICAL SYSTEMS, Shlomo Sternberg. A pioneer in the field of dynamical systems discusses one-dimensional dynamics, differential equations, random walks, iterated function systems, symbolic dynamics, and Markov chains. Supplementary materials include PowerPoint slides and MATLAB exercises. 2010 edition. 272pp. 6 1/8 x 9 1/4. 0-486-47705-3

ORDINARY DIFFERENTIAL EQUATIONS, Morris Tenenbaum and Harry Pollard. Skillfully organized introductory text examines origin of differential equations, then defines basic terms and outlines general solution of a differential equation. Explores integrating factors; dilution and accretion problems; Laplace Transforms; Newton's Interpolation Formulas, more. 818pp. 5 3/8 x 8 1/2. 0-486-64940-7

MATROID THEORY, D. J. A. Welsh. Text by a noted expert describes standard examples and investigation results, using elementary proofs to develop basic matroid properties before advancing to a more sophisticated treatment. Includes numerous exercises. 1976 edition. 448pp. 5 3/8 x 8 1/2. 0-486-47439-9

THE CONCEPT OF A RIEMANN SURFACE, Hermann Weyl. This classic on the general history of functions combines function theory and geometry, forming the basis of the modern approach to analysis, geometry, and topology. 1955 edition. 208pp. 5 3/8 x 8 1/2. 0-486-47004-0

THE LAPLACE TRANSFORM, David Vernon Widder. This volume focuses on the Laplace and Stieltjes transforms, offering a highly theoretical treatment. Topics include fundamental formulas, the moment problem, monotonic functions, and Tauberian theorems. 1941 edition. 416pp. 5 3/8 x 8 1/2. 0-486-47755-X

Mathematics–Logic and Problem Solving

PERPLEXING PUZZLES AND TANTALIZING TEASERS, Martin Gardner. Ninety-three riddles, mazes, illusions, tricky questions, word and picture puzzles, and other challenges offer hours of entertainment for youngsters. Filled with rib-tickling drawings. Solutions. 224pp. 5 3/8 x 8 1/2. 0-486-25637-5

MY BEST MATHEMATICAL AND LOGIC PUZZLES, Martin Gardner. The noted expert selects 70 of his favorite "short" puzzles. Includes The Returning Explorer, The Mutilated Chessboard, Scrambled Box Tops, and dozens more. Complete solutions included. 96pp. 5 3/8 x 8 1/2. 0-486-28152-3

THE LADY OR THE TIGER?: and Other Logic Puzzles, Raymond M. Smullyan. Created by a renowned puzzle master, these whimsically themed challenges involve paradoxes about probability, time, and change; metapuzzles; and self-referentiality. Nineteen chapters advance in difficulty from relatively simple to highly complex. 1982 edition. 240pp. 5 3/8 x 8 1/2. 0-486-47027-X

SATAN, CANTOR AND INFINITY: Mind-Boggling Puzzles, Raymond M. Smullyan. A renowned mathematician tells stories of knights and knaves in an entertaining look at the logical precepts behind infinity, probability, time, and change. Requires a strong background in mathematics. Complete solutions. 288pp. 5 3/8 x 8 1/2. 0-486-47036-9

THE RED BOOK OF MATHEMATICAL PROBLEMS, Kenneth S. Williams and Kenneth Hardy. Handy compilation of 100 practice problems, hints and solutions indispensable for students preparing for the William Lowell Putnam and other mathematical competitions. Preface to the First Edition. Sources. 1988 edition. 192pp. 5 3/8 x 8 1/2. 0-486-69415-1

KING ARTHUR IN SEARCH OF HIS DOG AND OTHER CURIOUS PUZZLES, Raymond M. Smullyan. This fanciful, original collection for readers of all ages features arithmetic puzzles, logic problems related to crime detection, and logic and arithmetic puzzles involving King Arthur and his Dogs of the Round Table. 160pp. 5 3/8 x 8 1/2. 0-486-47435-6

UNDECIDABLE THEORIES: Studies in Logic and the Foundation of Mathematics, Alfred Tarski in collaboration with Andrzej Mostowski and Raphael M. Robinson. This well-known book by the famed logician consists of three treatises: "A General Method in Proofs of Undecidability," "Undecidability and Essential Undecidability in Mathematics," and "Undecidability of the Elementary Theory of Groups." 1953 edition. 112pp. 5 3/8 x 8 1/2. 0-486-47703-7

LOGIC FOR MATHEMATICIANS, J. Barkley Rosser. Examination of essential topics and theorems assumes no background in logic. "Undoubtedly a major addition to the literature of mathematical logic." – *Bulletin of the American Mathematical Society.* 1978 edition. 592pp. 6 1/8 x 9 1/4. 0-486-46898-4

INTRODUCTION TO PROOF IN ABSTRACT MATHEMATICS, Andrew Wohlgemuth. This undergraduate text teaches students what constitutes an acceptable proof, and it develops their ability to do proofs of routine problems as well as those requiring creative insights. 1990 edition. 384pp. 6 1/2 x 9 1/4. 0-486-47854-8

FIRST COURSE IN MATHEMATICAL LOGIC, Patrick Suppes and Shirley Hill. Rigorous introduction is simple enough in presentation and context for wide range of students. Symbolizing sentences; logical inference; truth and validity; truth tables; terms, predicates, universal quantifiers; universal specification and laws of identity; more. 288pp. 5 3/8 x 8 1/2. 0-486-42259-3

Mathematics–Algebra and Calculus

VECTOR CALCULUS, Peter Baxandall and Hans Liebeck. This introductory text offers a rigorous, comprehensive treatment. Classical theorems of vector calculus are amply illustrated with figures, worked examples, physical applications, and exercises with hints and answers. 1986 edition. 560pp. 5 3/8 x 8 1/2. 0-486-46620-5

ADVANCED CALCULUS: An Introduction to Classical Analysis, Louis Brand. A course in analysis that focuses on the functions of a real variable, this text introduces the basic concepts in their simplest setting and illustrates its teachings with numerous examples, theorems, and proofs. 1955 edition. 592pp. 5 3/8 x 8 1/2. 0-486-44548-8

ADVANCED CALCULUS, Avner Friedman. Intended for students who have already completed a one-year course in elementary calculus, this two-part treatment advances from functions of one variable to those of several variables. Solutions. 1971 edition. 432pp. 5 3/8 x 8 1/2. 0-486-45795-8

METHODS OF MATHEMATICS APPLIED TO CALCULUS, PROBABILITY, AND STATISTICS, Richard W. Hamming. This 4-part treatment begins with algebra and analytic geometry and proceeds to an exploration of the calculus of algebraic functions and transcendental functions and applications. 1985 edition. Includes 310 figures and 18 tables. 880pp. 6 1/2 x 9 1/4. 0-486-43945-3

BASIC ALGEBRA I: Second Edition, Nathan Jacobson. A classic text and standard reference for a generation, this volume covers all undergraduate algebra topics, including groups, rings, modules, Galois theory, polynomials, linear algebra, and associative algebra. 1985 edition. 528pp. 6 1/8 x 9 1/4. 0-486-47189-6

BASIC ALGEBRA II: Second Edition, Nathan Jacobson. This classic text and standard reference comprises all subjects of a first-year graduate-level course, including in-depth coverage of groups and polynomials and extensive use of categories and functors. 1989 edition. 704pp. 6 1/8 x 9 1/4. 0-486-47187-X

CALCULUS: An Intuitive and Physical Approach (Second Edition), Morris Kline. Application-oriented introduction relates the subject as closely as possible to science with explorations of the derivative; differentiation and integration of the powers of x; theorems on differentiation, antidifferentiation; the chain rule; trigonometric functions; more. Examples. 1967 edition. 960pp. 6 1/2 x 9 1/4. 0-486-40453-6

ABSTRACT ALGEBRA AND SOLUTION BY RADICALS, John E. Maxfield and Margaret W. Maxfield. Accessible advanced undergraduate-level text starts with groups, rings, fields, and polynomials and advances to Galois theory, radicals and roots of unity, and solution by radicals. Numerous examples, illustrations, exercises, appendixes. 1971 edition. 224pp. 6 1/8 x 9 1/4. 0-486-47723-1

AN INTRODUCTION TO THE THEORY OF LINEAR SPACES, Georgi E. Shilov. Translated by Richard A. Silverman. Introductory treatment offers a clear exposition of algebra, geometry, and analysis as parts of an integrated whole rather than separate subjects. Numerous examples illustrate many different fields, and problems include hints or answers. 1961 edition. 320pp. 5 3/8 x 8 1/2. 0-486-63070-6

LINEAR ALGEBRA, Georgi E. Shilov. Covers determinants, linear spaces, systems of linear equations, linear functions of a vector argument, coordinate transformations, the canonical form of the matrix of a linear operator, bilinear and quadratic forms, and more. 387pp. 5 3/8 x 8 1/2. 0-486-63518-X

Mathematics–Probability and Statistics

BASIC PROBABILITY THEORY, Robert B. Ash. This text emphasizes the probabilistic way of thinking, rather than measure-theoretic concepts. Geared toward advanced undergraduates and graduate students, it features solutions to some of the problems. 1970 edition. 352pp. 5 3/8 x 8 1/2. 0-486-46628-0

PRINCIPLES OF STATISTICS, M. G. Bulmer. Concise description of classical statistics, from basic dice probabilities to modern regression analysis. Equal stress on theory and applications. Moderate difficulty; only basic calculus required. Includes problems with answers. 252pp. 5 5/8 x 8 1/4. 0-486-63760-3

OUTLINE OF BASIC STATISTICS: Dictionary and Formulas, John E. Freund and Frank J. Williams. Handy guide includes a 70-page outline of essential statistical formulas covering grouped and ungrouped data, finite populations, probability, and more, plus over 1,000 clear, concise definitions of statistical terms. 1966 edition. 208pp. 5 3/8 x 8 1/2. 0-486-47769-X

GOOD THINKING: The Foundations of Probability and Its Applications, Irving J. Good. This in-depth treatment of probability theory by a famous British statistician explores Keynesian principles and surveys such topics as Bayesian rationality, corroboration, hypothesis testing, and mathematical tools for induction and simplicity. 1983 edition. 352pp. 5 3/8 x 8 1/2. 0-486-47438-0

INTRODUCTION TO PROBABILITY THEORY WITH CONTEMPORARY APPLICATIONS, Lester L. Helms. Extensive discussions and clear examples, written in plain language, expose students to the rules and methods of probability. Exercises foster problem-solving skills, and all problems feature step-by-step solutions. 1997 edition. 368pp. 6 1/2 x 9 1/4. 0-486-47418-6

CHANCE, LUCK, AND STATISTICS, Horace C. Levinson. In simple, non-technical language, this volume explores the fundamentals governing chance and applies them to sports, government, and business. "Clear and lively ... remarkably accurate." – *Scientific Monthly.* 384pp. 5 3/8 x 8 1/2. 0-486-41997-5

FIFTY CHALLENGING PROBLEMS IN PROBABILITY WITH SOLUTIONS, Frederick Mosteller. Remarkable puzzlers, graded in difficulty, illustrate elementary and advanced aspects of probability. These problems were selected for originality, general interest, or because they demonstrate valuable techniques. Also includes detailed solutions. 88pp. 5 3/8 x 8 1/2. 0-486-65355-2

EXPERIMENTAL STATISTICS, Mary Gibbons Natrella. A handbook for those seeking engineering information and quantitative data for designing, developing, constructing, and testing equipment. Covers the planning of experiments, the analyzing of extreme-value data; and more. 1966 edition. Index. Includes 52 figures and 76 tables. 560pp. 8 3/8 x 11. 0-486-43937-2

STOCHASTIC MODELING: Analysis and Simulation, Barry L. Nelson. Coherent introduction to techniques also offers a guide to the mathematical, numerical, and simulation tools of systems analysis. Includes formulation of models, analysis, and interpretation of results. 1995 edition. 336pp. 6 1/8 x 9 1/4. 0-486-47770-3

INTRODUCTION TO BIOSTATISTICS: Second Edition, Robert R. Sokal and F. James Rohlf. Suitable for undergraduates with a minimal background in mathematics, this introduction ranges from descriptive statistics to fundamental distributions and the testing of hypotheses. Includes numerous worked-out problems and examples. 1987 edition. 384pp. 6 1/8 x 9 1/4. 0-486-46961-1

Mathematics–Geometry and Topology

PROBLEMS AND SOLUTIONS IN EUCLIDEAN GEOMETRY, M. N. Aref and William Wernick. Based on classical principles, this book is intended for a second course in Euclidean geometry and can be used as a refresher. More than 200 problems include hints and solutions. 1968 edition. 272pp. 5 3/8 x 8 1/2. 0-486-47720-7

TOPOLOGY OF 3-MANIFOLDS AND RELATED TOPICS, Edited by M. K. Fort, Jr. With a New Introduction by Daniel Silver. Summaries and full reports from a 1961 conference discuss decompositions and subsets of 3-space; n-manifolds; knot theory; the Poincaré conjecture; and periodic maps and isotopies. Familiarity with algebraic topology required. 1962 edition. 272pp. 6 1/8 x 9 1/4. 0-486-47753-3

POINT SET TOPOLOGY, Steven A. Gaal. Suitable for a complete course in topology, this text also functions as a self-contained treatment for independent study. Additional enrichment materials make it equally valuable as a reference. 1964 edition. 336pp. 5 3/8 x 8 1/2. 0-486-47222-1

INVITATION TO GEOMETRY, Z. A. Melzak. Intended for students of many different backgrounds with only a modest knowledge of mathematics, this text features self-contained chapters that can be adapted to several types of geometry courses. 1983 edition. 240pp. 5 3/8 x 8 1/2. 0-486-46626-4

TOPOLOGY AND GEOMETRY FOR PHYSICISTS, Charles Nash and Siddhartha Sen. Written by physicists for physics students, this text assumes no detailed background in topology or geometry. Topics include differential forms, homotopy, homology, cohomology, fiber bundles, connection and covariant derivatives, and Morse theory. 1983 edition. 320pp. 5 3/8 x 8 1/2. 0-486-47852-1

BEYOND GEOMETRY: Classic Papers from Riemann to Einstein, Edited with an Introduction and Notes by Peter Pesic. This is the only English-language collection of these 8 accessible essays. They trace seminal ideas about the foundations of geometry that led to Einstein's general theory of relativity. 224pp. 6 1/8 x 9 1/4. 0-486-45350-2

GEOMETRY FROM EUCLID TO KNOTS, Saul Stahl. This text provides a historical perspective on plane geometry and covers non-neutral Euclidean geometry, circles and regular polygons, projective geometry, symmetries, inversions, informal topology, and more. Includes 1,000 practice problems. Solutions available. 2003 edition. 480pp. 6 1/8 x 9 1/4. 0-486-47459-3

TOPOLOGICAL VECTOR SPACES, DISTRIBUTIONS AND KERNELS, François Trèves. Extending beyond the boundaries of Hilbert and Banach space theory, this text focuses on key aspects of functional analysis, particularly in regard to solving partial differential equations. 1967 edition. 592pp. 5 3/8 x 8 1/2. 0-486-45352-9

INTRODUCTION TO PROJECTIVE GEOMETRY, C. R. Wylie, Jr. This introductory volume offers strong reinforcement for its teachings, with detailed examples and numerous theorems, proofs, and exercises, plus complete answers to all odd-numbered end-of-chapter problems. 1970 edition. 576pp. 6 1/8 x 9 1/4. 0-486-46895-X

FOUNDATIONS OF GEOMETRY, C. R. Wylie, Jr. Geared toward students preparing to teach high school mathematics, this text explores the principles of Euclidean and non-Euclidean geometry and covers both generalities and specifics of the axiomatic method. 1964 edition. 352pp. 6 x 9. 0-486-47214-0

Mathematics-History

THE WORKS OF ARCHIMEDES, Archimedes. Translated by Sir Thomas Heath. Complete works of ancient geometer feature such topics as the famous problems of the ratio of the areas of a cylinder and an inscribed sphere; the properties of conoids, spheroids, and spirals; more. 326pp. 5 3/8 x 8 1/2. 0-486-42084-1

THE HISTORICAL ROOTS OF ELEMENTARY MATHEMATICS, Lucas N. H. Bunt, Phillip S. Jones, and Jack D. Bedient. Exciting, hands-on approach to understanding fundamental underpinnings of modern arithmetic, algebra, geometry and number systems examines their origins in early Egyptian, Babylonian, and Greek sources. 336pp. 5 3/8 x 8 1/2. 0-486-25563-8

THE THIRTEEN BOOKS OF EUCLID'S ELEMENTS, Euclid. Contains complete English text of all 13 books of the Elements plus critical apparatus analyzing each definition, postulate, and proposition in great detail. Covers textual and linguistic matters; mathematical analyses of Euclid's ideas; classical, medieval, Renaissance and modern commentators; refutations, supports, extrapolations, reinterpretations and historical notes. 995 figures. Total of 1,425pp. All books 5 3/8 x 8 1/2.
Vol. I: 443pp. 0-486-60088-2
Vol. II: 464pp. 0-486-60089-0
Vol. III: 546pp. 0-486-60090-4

A HISTORY OF GREEK MATHEMATICS, Sir Thomas Heath. This authoritative two-volume set that covers the essentials of mathematics and features every landmark innovation and every important figure, including Euclid, Apollonius, and others. 5 3/8 x 8 1/2.
Vol. I: 461pp. 0-486-24073-8
Vol. II: 597pp. 0-486-24074-6

A MANUAL OF GREEK MATHEMATICS, Sir Thomas L. Heath. This concise but thorough history encompasses the enduring contributions of the ancient Greek mathematicians whose works form the basis of most modern mathematics. Discusses Pythagorean arithmetic, Plato, Euclid, more. 1931 edition. 576pp. 5 3/8 x 8 1/2. 0-486-43231-9

CHINESE MATHEMATICS IN THE THIRTEENTH CENTURY, Ulrich Libbrecht. An exploration of the 13th-century mathematician Ch'in, this fascinating book combines what is known of the mathematician's life with a history of his only extant work, the Shu-shu chiu-chang. 1973 edition. 592pp. 5 3/8 x 8 1/2. 0-486-44619-0

PHILOSOPHY OF MATHEMATICS AND DEDUCTIVE STRUCTURE IN EUCLID'S ELEMENTS, Ian Mueller. This text provides an understanding of the classical Greek conception of mathematics as expressed in Euclid's Elements. It focuses on philosophical, foundational, and logical questions and features helpful appendixes. 400pp. 6 1/2 x 9 1/4. 0-486-45300-6

BEYOND GEOMETRY: Classic Papers from Riemann to Einstein, Edited with an Introduction and Notes by Peter Pesic. This is the only English-language collection of these 8 accessible essays. They trace seminal ideas about the foundations of geometry that led to Einstein's general theory of relativity. 224pp. 6 1/8 x 9 1/4. 0-486-45350-2

HISTORY OF MATHEMATICS, David E. Smith. Two-volume history – from Egyptian papyri and medieval maps to modern graphs and diagrams. Non-technical chronological survey with thousands of biographical notes, critical evaluations, and contemporary opinions on over 1,100 mathematicians. 5 3/8 x 8 1/2.
Vol. I: 618pp. 0-486-20429-4
Vol. II: 736pp. 0-486-20430-8

Physics

THEORETICAL NUCLEAR PHYSICS, John M. Blatt and Victor F. Weisskopf. An uncommonly clear and cogent investigation and correlation of key aspects of theoretical nuclear physics by leading experts: the nucleus, nuclear forces, nuclear spectroscopy, two-, three- and four-body problems, nuclear reactions, beta-decay and nuclear shell structure. 896pp. 5 3/8 x 8 1/2. 0-486-66827-4

QUANTUM THEORY, David Bohm. This advanced undergraduate-level text presents the quantum theory in terms of qualitative and imaginative concepts, followed by specific applications worked out in mathematical detail. 655pp. 5 3/8 x 8 1/2. 0-486-65969-0

ATOMIC PHYSICS AND HUMAN KNOWLEDGE, Niels Bohr. Articles and speeches by the Nobel Prize–winning physicist, dating from 1934 to 1958, offer philosophical explorations of the relevance of atomic physics to many areas of human endeavor. 1961 edition. 112pp. 5 3/8 x 8 1/2. 0-486-47928-5

COSMOLOGY, Hermann Bondi. A co-developer of the steady-state theory explores his conception of the expanding universe. This historic book was among the first to present cosmology as a separate branch of physics. 1961 edition. 192pp. 5 3/8 x 8 1/2. 0-486-47483-6

LECTURES ON QUANTUM MECHANICS, Paul A. M. Dirac. Four concise, brilliant lectures on mathematical methods in quantum mechanics from Nobel Prize-winning quantum pioneer build on idea of visualizing quantum theory through the use of classical mechanics. 96pp. 5 3/8 x 8 1/2. 0-486-41713-1

THE PRINCIPLE OF RELATIVITY, Albert Einstein and Frances A. Davis. Eleven papers that forged the general and special theories of relativity include seven papers by Einstein, two by Lorentz, and one each by Minkowski and Weyl. 1923 edition. 240pp. 5 3/8 x 8 1/2. 0-486-60081-5

PHYSICS OF WAVES, William C. Elmore and Mark A. Heald. Ideal as a classroom text or for individual study, this unique one-volume overview of classical wave theory covers wave phenomena of acoustics, optics, electromagnetic radiations, and more. 477pp. 5 3/8 x 8 1/2. 0-486-64926-1

THERMODYNAMICS, Enrico Fermi. In this classic of modern science, the Nobel Laureate presents a clear treatment of systems, the First and Second Laws of Thermodynamics, entropy, thermodynamic potentials, and much more. Calculus required. 160pp. 5 3/8 x 8 1/2. 0-486-60361-X

QUANTUM THEORY OF MANY-PARTICLE SYSTEMS, Alexander L. Fetter and John Dirk Walecka. Self-contained treatment of nonrelativistic many-particle systems discusses both formalism and applications in terms of ground-state (zero-temperature) formalism, finite-temperature formalism, canonical transformations, and applications to physical systems. 1971 edition. 640pp. 5 3/8 x 8 1/2. 0-486-42827-3

QUANTUM MECHANICS AND PATH INTEGRALS: Emended Edition, Richard P. Feynman and Albert R. Hibbs. Emended by Daniel F. Styer. The Nobel Prize–winning physicist presents unique insights into his theory and its applications. Feynman starts with fundamentals and advances to the perturbation method, quantum electrodynamics, and statistical mechanics. 1965 edition, emended in 2005. 384pp. 6 1/8 x 9 1/4. 0-486-47722-3

Physics

INTRODUCTION TO MODERN OPTICS, Grant R. Fowles. A complete basic undergraduate course in modern optics for students in physics, technology, and engineering. The first half deals with classical physical optics; the second, quantum nature of light. Solutions. 336pp. 5 3/8 x 8 1/2. 0-486-65957-7

THE QUANTUM THEORY OF RADIATION: Third Edition, W. Heitler. The first comprehensive treatment of quantum physics in any language, this classic introduction to basic theory remains highly recommended and widely used, both as a text and as a reference. 1954 edition. 464pp. 5 3/8 x 8 1/2. 0-486-64558-4

QUANTUM FIELD THEORY, Claude Itzykson and Jean-Bernard Zuber. This comprehensive text begins with the standard quantization of electrodynamics and perturbative renormalization, advancing to functional methods, relativistic bound states, broken symmetries, nonabelian gauge fields, and asymptotic behavior. 1980 edition. 752pp. 6 1/2 x 9 1/4. 0-486-44568-2

FOUNDATIONS OF POTENTIAL THERY, Oliver D. Kellogg. Introduction to fundamentals of potential functions covers the force of gravity, fields of force, potentials, harmonic functions, electric images and Green's function, sequences of harmonic functions, fundamental existence theorems, and much more. 400pp. 5 3/8 x 8 1/2. 0-486-60144-7

FUNDAMENTALS OF MATHEMATICAL PHYSICS, Edgar A. Kraut. Indispensable for students of modern physics, this text provides the necessary background in mathematics to study the concepts of electromagnetic theory and quantum mechanics. 1967 edition. 480pp. 6 1/2 x 9 1/4. 0-486-45809-1

GEOMETRY AND LIGHT: The Science of Invisibility, Ulf Leonhardt and Thomas Philbin. Suitable for advanced undergraduate and graduate students of engineering, physics, and mathematics and scientific researchers of all types, this is the first authoritative text on invisibility and the science behind it. More than 100 full-color illustrations, plus exercises with solutions. 2010 edition. 288pp. 7 x 9 1/4. 0-486-47693-6

QUANTUM MECHANICS: New Approaches to Selected Topics, Harry J. Lipkin. Acclaimed as "excellent" (*Nature*) and "very original and refreshing" (*Physics Today*), these studies examine the Mössbauer effect, many-body quantum mechanics, scattering theory, Feynman diagrams, and relativistic quantum mechanics. 1973 edition. 480pp. 5 3/8 x 8 1/2. 0-486-45893-8

THEORY OF HEAT, James Clerk Maxwell. This classic sets forth the fundamentals of thermodynamics and kinetic theory simply enough to be understood by beginners, yet with enough subtlety to appeal to more advanced readers, too. 352pp. 5 3/8 x 8 1/2. 0-486-41735-2

QUANTUM MECHANICS, Albert Messiah. Subjects include formalism and its interpretation, analysis of simple systems, symmetries and invariance, methods of approximation, elements of relativistic quantum mechanics, much more. "Strongly recommended." – *American Journal of Physics.* 1152pp. 5 3/8 x 8 1/2. 0-486-40924-4

RELATIVISTIC QUANTUM FIELDS, Charles Nash. This graduate-level text contains techniques for performing calculations in quantum field theory. It focuses chiefly on the dimensional method and the renormalization group methods. Additional topics include functional integration and differentiation. 1978 edition. 240pp. 5 3/8 x 8 1/2. 0-486-47752-5

Physics

MATHEMATICAL TOOLS FOR PHYSICS, James Nearing. Encouraging students' development of intuition, this original work begins with a review of basic mathematics and advances to infinite series, complex algebra, differential equations, Fourier series, and more. 2010 edition. 496pp. 6 1/8 x 9 1/4. 0-486-48212-X

TREATISE ON THERMODYNAMICS, Max Planck. Great classic, still one of the best introductions to thermodynamics. Fundamentals, first and second principles of thermodynamics, applications to special states of equilibrium, more. Numerous worked examples. 1917 edition. 297pp. 5 3/8 x 8. 0-486-66371-X

AN INTRODUCTION TO RELATIVISTIC QUANTUM FIELD THEORY, Silvan S. Schweber. Complete, systematic, and self-contained, this text introduces modern quantum field theory. "Combines thorough knowledge with a high degree of didactic ability and a delightful style." – *Mathematical Reviews.* 1961 edition. 928pp. 5 3/8 x 8 1/2. 0-486-44228-4

THE ELECTROMAGNETIC FIELD, Albert Shadowitz. Comprehensive undergraduate text covers basics of electric and magnetic fields, building up to electromagnetic theory. Related topics include relativity theory. Over 900 problems, some with solutions. 1975 edition. 768pp. 5 5/8 x 8 1/4. 0-486-65660-8

THE PRINCIPLES OF STATISTICAL MECHANICS, Richard C. Tolman. Definitive treatise offers a concise exposition of classical statistical mechanics and a thorough elucidation of quantum statistical mechanics, plus applications of statistical mechanics to thermodynamic behavior. 1930 edition. 704pp. 5 5/8 x 8 1/4. 0-486-63896-0

INTRODUCTION TO THE PHYSICS OF FLUIDS AND SOLIDS, James S. Trefil. This interesting, informative survey by a well-known science author ranges from classical physics and geophysical topics, from the rings of Saturn and the rotation of the galaxy to underground nuclear tests. 1975 edition. 320pp. 5 3/8 x 8 1/2. 0-486-47437-2

STATISTICAL PHYSICS, Gregory H. Wannier. Classic text combines thermodynamics, statistical mechanics, and kinetic theory in one unified presentation. Topics include equilibrium statistics of special systems, kinetic theory, transport coefficients, and fluctuations. Problems with solutions. 1966 edition. 532pp. 5 3/8 x 8 1/2. 0-486-65401-X

SPACE, TIME, MATTER, Hermann Weyl. Excellent introduction probes deeply into Euclidean space, Riemann's space, Einstein's general relativity, gravitational waves and energy, and laws of conservation. "A classic of physics." – *British Journal for Philosophy and Science.* 330pp. 5 3/8 x 8 1/2. 0-486-60267-2

RANDOM VIBRATIONS: Theory and Practice, Paul H. Wirsching, Thomas L. Paez and Keith Ortiz. Comprehensive text and reference covers topics in probability, statistics, and random processes, plus methods for analyzing and controlling random vibrations. Suitable for graduate students and mechanical, structural, and aerospace engineers. 1995 edition. 464pp. 5 3/8 x 8 1/2. 0-486-45015-5

PHYSICS OF SHOCK WAVES AND HIGH-TEMPERATURE HYDRO DYNAMIC PHENOMENA, Ya B. Zel'dovich and Yu P. Raizer. Physical, chemical processes in gases at high temperatures are focus of outstanding text, which combines material from gas dynamics, shock-wave theory, thermodynamics and statistical physics, other fields. 284 illustrations. 1966–1967 edition. 944pp. 6 1/8 x 9 1/4. 0-486-42002-7